ENVIRONMENTAL SOIL AND WATER CHEMISTRY

ENVIRONMENTAL SOIL AND WATER CHEMISTRY

PRINCIPLES AND APPLICATIONS

V. P. EVANGELOU
University of Kentucky
Lexington, Kentucky

A Wiley-Interscience Publication

JOHN WILEY & SONS, INC.
New York · Chichester · Weinheim · Brisbane · Singapore · Toronto

Library of Congress Cataloging-in-Publication Data
Evangelou, V. P.
    Environmental soil and water chemistry : principles and applications / Bill Evangelou.
        p. cm.
    Includes bibliographical references and index.
    ISBN 0-471-16515-8 (cloth : alk. paper)
        1. Soil pollution. 2. Soil chemistry. 3. Water—Pollution. 4. Water chemistry. I. Title.
TD878.E93 1998
628.5—dc21   98-13433
                                                                                CIP

To my late brother, P. Evangelou, M.D., who taught me how to read and write.
His gift is passed on!
Thank you to my wife Shelly, daughter Julia, and son Peter, with love

Do you feel the need to read because you understand
or do you feel the need to understand and therefore you read?

Contents

Preface

For the past 18 years I have been involved in educating undergraduate and graduate students in the field of soil–water chemistry. Early in my teaching/research career, students in the college of agriculture in the field of soils had primarily a farming background. With the passing of time, however, the number of such students declined dramatically and most universities and colleges across the country established environmental science units in some form or another. Some of these units represented the reorganization of soil science departments, forestry departments, and so on; others represented independent environmental or natural resources departments. Similar reorganization took place or is currently taking place in geology and engineering schools. This field reorganization created a need for new textbooks with an emphasis on examining soil and water as natural resources. In my view, we have not succeeded in introducing an appropriate textbook on the subject of soil and water chemistry to cover the needs of this new type of student.

This book is designed to serve as a beginning textbook for college seniors and beginning graduate students in environmental sciences, and is tailored specifically to the disciplines of soil science, environmental science, agricultural engineering, environmental engineering, and environmental geology.

The textbook contains reviews of all the necessary fundamental principles of chemistry required for understanding soil–water chemistry and quality and soil–water treatments of chemically polluted soils and waters, for example, heavy-metal contaminated soil–water, acid drainage, and restoration of sodic soils and brackish waters. The purpose of the book is to educate college seniors and beginning graduate students about the toxicity, chemistry, and control of pollutants in the soil–water environment and about the application of such knowledge to environmental restoration. Special emphasis is placed on the educational level at which the book is written so that it can be understood by seniors and beginning graduate students majoring in environmental science.

The book consists of two major sections—Principles and Application. Each section covers several major subject areas. The Principles section is divided into the following parts: I. Water Chemistry and Mineral Solubility; II. Soil Minerals and Surface Chemical Properties; and III. Electrochemistry and Kinetics. The Application section also covers several subject areas: IV. Soil Dynamics and Agricultural–Organic Chemicals; V. Colloids and Transport Processes in Soils; VI. Land-Disturbance Pollution and Its Control; VII. Soil and Water: Quality and Treatment Technologies. Each subject area contains one to three chapters.

Some of the parts in the Principles section are written at a level that would be challenging to a beginning graduate student. After going through these parts, the student may find it helpful to follow up with the following books, which are also listed in the reference section: M. B. McBride, *Environmental Chemistry of Soils*; F. M. M. Morel and J. G. Hering, *Principles and Applications of Aquatic Chemistry*; G. Sposito, *The Thermodynamics of Soil Solutions* and *The Surface Chemistry of Soils*; and W. Stumm and J. J. Morgan, *Aquatic Chemistry*.

For the upperclass student or beginning graduate student whose environmental field does not require detailed knowledge of chemistry, the easiest subsections in the Principles section (at the instructor's discretion) should be read so that the student obtains a good conceptual knowledge of soil–water chemistry.

The Application section should be read by all students to familiarize themselves with (1) current outstanding environmental soil–water problems, (2) concepts of soil–water chemistry in solving environmental soil–water problems, and (3) current technologies for soil–water environmental problems.

The Application section alone contains adequate material to be taught as an undergraduate level course. The Principles section may also be taught as a separate course.

I hope that this book gives the reader a quantitative understanding of the principles involved in environmental soil–water chemistry dealing with modeling soil nutrient availability to plants, soil transport processes, fertilizer management, and soil physical stability. It should also justify the need for knowledge about the physical chemistry and natural behavior of potential soil–water contaminants. This requires a background in water chemistry, soil mineralogy, mineral surface chemistry, chemistry of natural and/or anthropogenic contaminants, and knowledge of soil–water remediation technologies and the scientific principles on which they are based.

I wish to thank several people who helped with various aspects of producing this book. My secretary, Marsha Short, helped with the endless typing and corrections. My graduate students, postdoctoral candidates, and technicians, Dr. Louis McDonald, Dr. Ananto Seta, and Mr. Martin Vandiviere, reviewed the material and contributed data. Dr. Chris Amrhein provided a review of portions of the manuscript and made many important points and suggestions concerning the technical aspects of the book.

I am also grateful to the administration of the University of Kentucky for its support over the years of my soil–water chemistry research, which made it possible for me to write this book.

V. P. EVANGELOU
Lexington, Kentucky

About the Author

Bill Evangelou was born and raised in Olympias, Greece and obtained his B.S. in 1972 and M.S. in 1974 in Agriculture and Plant Science, respectively, from California State University, Chico, California. In 1981 he received his Ph.D in Soil Science, specializing in mineralogy and soil–water physical chemistry, from the University of California at Davis.

Dr. Evangelou is currently Professor of Soil–Water Physical Chemistry at the University of Kentucky. He has served as major professor to numerous graduate students and supervisor of a number of postdoctoral fellows. He teaches courses in soil chemistry, soil physical chemistry, and environmental soil–water chemistry.

Dr. Evangelou's research is focused on cation-exchange equilibria and kinetics of soils and clay minerals, the surface chemistry of soils, the physical behavior of soil colloids, plant root cell wall–metal ion interactions and acquisition by plants, kinetics of pyrite oxidation and surface processes controlling rates of oxidation reactions, and recently, organometallic complexes and herbicide colloid suspension interactions and behavior. He has published more than 100 scientific papers on these subjects and has conducted more than 30 short courses on the subjects of environmental soil–water chemistry, pyrite chemistry, and acid mine drainage (AMD) for government and private industry professionals from the United States, Canada, Europe, and South Africa. More than 2000 professionals have attended Dr. Evangelou's short courses. He has been recognized for his scientific contributions with a number of awards, including the Marion L. & Chrystie M. Jackson Soil Science Award, Soil Science Society of America, for outstanding contributions in the areas of soil chemistry and mineralogy and graduate student education; Fellow, American Society of Agronomy; Fellow, Soil Science Society of America; U.S. Patent on "Peroxide Induced Oxidation Proof Phosphate Surface Coating on Iron Sulfides"; U.S. and Canadian Patent on "Oxidation Proof Silicate Surface Coating on Iron Sulfides"; Senior Fulbright Scholar Award; and Thomas Poe Cooper Award, University of Kentucky, College of Agriculture, 1994, for distinguished achievement in research.

ENVIRONMENTAL SOIL AND WATER CHEMISTRY

PRINCIPLES

PART I
Water Chemistry and Mineral Solubility

1 Physical Chemistry of Water and Some of Its Constituents

1.1 ELEMENTS OF NATURE

It is necessary to understand the behavior of soil–water and its mineral components (e.g., nutrients, contaminants) for the purpose of developing conceptual and/or mechanistic process models. Such models can be used to predict nutrient fate in soil–water or contamination–decontamination of soil–water and to develop soil–water remediation–decontamination technologies. To gain an understanding of the soil–water mineral components, their physical and chemical properties need to be known.

Nature is made out of various elements and scientists have agreed on a classification scheme based on atomic mass and electron orbital configuration, which are related to some of the important physicochemical properties of the elements. Classification of elements is given by the periodic table (Table 1.1), which is separated into groups, and for the purpose of this book they are represented by three major classes. The first class represents the *light metals* composed of groups 1, 2, and aluminum (Al). They are located on the left-hand side of the periodic table, except for Al. The second class represents the *heavy* or *transition metals*, located in the middle of the periodic table. Also included in this class are the elements Ga, In, Ti, Sn, Pb, and Bi, which are referred to as *post-transition metals*. The third class represents the *nonmetals* or *metalloids* (right-hand side of the periodic table), which includes groups 3–7. Finally, a subclass represents those elements found in the atmosphere. It includes the *noble gases* (group 8) (furthest right side of the periodic table) as well as nitrogen (N) and oxygen (O_2) gases (Table 1.2).

Table 1.1   Periodic Table of the Elements and Their ATomic Weights

Transition metals

Group 1	Group 2	Group 3	Group 4	Group 5	Group 6	Group 7	Group 8	Group		Group	Group	Group 3	Group 4	Group 5	Group 6	Group 7	Group 8
1 H 1.0079																	2 He 4.002602
3 Li 6.941	4 Be 9.01218											5 B 10.811	6 C 12.011	7 N 14.00674	8 O 15.9994	9 F 18.9984032	10 Ne 20.1797
11 Na 22.98977	12 Mg 24.305											13 Al 26.981539	14 Si 28.0855	15 P 30.973762	16 S 32.066	17 Cl 35.4527	18 Ar 39.948
19 K 39.0983	20 Ca 40.078	21 Sc 44.95591	22 Ti 47.88	23 V 50.9415	24 Cr 51.9961	25 Mn 54.93805	26 Fe 55.847	27 Co 58.9332	28 Ni 58.6934	29 Cu 63.546	30 Zn 65.39	31 Ga 69.723	32 Ge 72.61	33 As 74.92159	34 Se 78.96	35 Br 79.904	36 Kr 83.80
37 Rb 85.4678	38 Sr 87.62	39 Y 88.9059	40 Zr 91.224	41 Nb 92.9064	42 Mo 95.94	43 Tc (98)	44 Ru 101.07	45 Rh 102.90550	46 Pd 106.42	47 Ag 107.8682	48 Cd 112.411	49 In 114.82	50 Sn 118.710	51 Sb 121.757	52 Te 127.60	53 I 126.90447	54 Xe 131.29
55 Cs 132.9054	56 Ba 137.327	57 La 138.9055	72 Hf 178.49	73 Ta 180.9479	74 W 183.85	75 Re 186.207	76 Os 190.2	77 Ir 192.22	78 Pt 195.08	79 Au 196.9665	80 Hg 200.59	81 Tl 204.3833	82 Pb 207.2	83 Bi 208.98037	84 Po (209)	85 At (210)	86 Rn (222)
87 Fr (223)	88 Ra (226)	89 Ac (227)	104 Unq (261)	105 Unp (262)	106 Unh (263)	107 Uns (262)	108 Uno (265)	109 Une (266)									

Lanthanide Series

58 Ce 140.115	59 Pr 140.90765	60 Nd 144.24	61 Pm (145)	62 Sm 150.36	63 Eu 151.965	64 Gd 157.25	65 Tb 158.92534	66 Dy 162.50	67 Ho 164.93032	68 Er 167.26	69 Tm 168.93421	70 Yb 173.04	71 Lu 174.967

Actinide Series

90 Th 232.0381	91 Pa 231.03588	92 U 238.0289	93 Np (237)	94 Pu (244)	95 Am (243)	96 Cm (247)	97 Bk (247)	98 Cf (251)	99 Es (252)	100 Fm (257)	101 Md (258)	102 No (259)	103 Lr (262)

Molar Volume of Ideal gas at STP = 22.414 liter	Ideal Gas Constant:
Faraday Constant, $F = 9.6486 \times 10^4$ C/mol electrons	$R = 8.3145$ J·K^{-1}·mol^{-1}
Avogadro's Number, $N = 6.0221 \times 10^{23}$ mol^{-1}	$R = 1.987$ cal·K^{-1}·mol^{-1}
Planck's Constant, $h = 6.6261 \times 10^{-34}$ J·s	$R = 8.206 \times 10^{-2}$ liter·atm·K^{-1}·mol^{-1}

Velocity of light, $c = 2.9979 \times 10^8$ m·s^{-1}
Rydberg Constant, $R_H = 2.1799 \times 10^{-18}$ J
Electronic Charge, $e = 1.6021 \times 10^{-19}$ C
Atomic mass unit, $u = 1.6606 \times 10^{-24}$ g

TABLE 1.2. Elements of the Atmosphere

	Group	Volume in Air (%)	Uses
N_2	5	78.08	**Synthesis of NH_3.** Packaging of foods such as instant coffee to preserve flavor; liquid N_2 used as coolant (safer than liquid air)
O_2	6	20.95	**Making steel.** Life support systems, wastewater treatment, high-temperature flames
He	8	0.000524	Balloons, dirgibles
Ne	8	0.00182	Neon signs
Ar	8	0.934	Provides inert atmosphere in light bulbs, arc welding of Al, Mg
Kr	8	0.000114	High-speed flash bulbs
Xe	8	0.000009	Experimental anesthetic; ^{133}Xe used as radioactive tracer in medical diagnostic studies

Source: Masterson et al., 1981.

1.1.1 Light Metals [Groups 1, 2, and Aluminum (Al)]

Light metals have low densities (< 3 g cm^{-3}) and occur in nature mainly as ionic compounds (e.g., Na^+ and Ca^{2+}) associated with Cl^-, SO_4^{2-}, CO_3^{2-}, PO_4^{3-}, NO_3^- and so on. Aluminum is commonly associated with the oxide ion O_2^{2-} (e.g., soil minerals). Light metals are used in industrial applications and some serve as nutrients to various organisms and higher plants. Additional information on light metals is given in Table 1.3.

TABLE 1.3. Properties and Sources of the Light Metals

Metal	Group	Density (g cm^{-3})	Principle Source
Li	1	0.534	Complex silicates (natural and primary minerals)
Na	1	0.971	NaCl
K	1	0.862	KCl
Rb	1	1.53	RbCl (with K^+)
Cs	1	1.87	CsCl (with K^+)
Be	2	1.85	Complex silicates (natural and primary minerals)
Mg	2	1.74	Mg^{2+} in seawater; $MgCO_3$
Ca	2	1.55	$CaCO_3$ (limestone) $CaSO_4 \cdot 2 H_2O$ (gypsum)
Sr	2	2.60	$SrCO_3$, $SrSO_4$
Ba	2	3.51	$BaCO_3$, $BaSO_4$
Al	3	2.70	Al_2O_3 (bauxite)

Source: Masterson et al., 1981.

1.1.2 Heavy Metals (Transition Metals)

Heavy metals have a density greater than 3 g cm^{-3}. They are found in nature as elements such as gold or as metal sulfides (e.g., CuS_2, PbS_2, and FeS) or as metal oxides (e.g., MnO_2, Cr_2O_3, and Fe_2O_3). Heavy metals are widely used in various industries and also serve as micronutrients to microorganisms and higher plants.

1.1.3 Nonmetals or Metalloids

Metalloids are extracted from water and the earth's solid surface. Some metalloids are environmentally important because they react with oxygen to form *oxyanions*. Some oxyanions are toxic to organisms (e.g., arsenite, AsO_3; arsenate, AsO_4; chromate, CrO_4), others may serve as nutrients (e.g., phosphate, PO_4; nitrate, NO_3), while still others may serve as nutrients at low concentrations but become quite toxic at high concentrations (e.g., selenite, SeO_3; selenate, SeO_4). Oxyanions are commonly associated with light or heavy metals. Additional information on metalloids is given in Table 1.4 (see also Chapter 13).

1.2 CHEMICAL BONDING

Chemical substances are made out of molecules. For example, water is made out of molecules composed of one oxygen atom and two hydrogen atoms (H_2O). An *atom* is

TABLE 1.4. Nonmetals and Metalloids Found in the Earth's Crust

	Group	Principal Source
B	3	$Na_2B_4O_7 \cdot 10\ H_2O$ (borax)
C	4	Coal, petroleum, natural gas
Si	4	SiO_2 (sand, quartz)
Ge	4	Sulfides
P	5	$Ca_3(PO_4)_2$ (phosphate rock)
As	5	As_2S_3, other sulfides
Sb	5	Sb_2S_3
S	6	Free element
Se	6	PbSe, other selenides
Te	6	PbTe, other tellurides
H_2		Natural gas, petroleum, H_2O
F_2	7	CaF_2 (fluorite), Na_3AlF_6 (cryolite)
Cl_2	7	NaCl, Cl$^-$ in ocean
Br_2	7	Br$^-$ in salt brines
I_2	7	I$^-$ in salt brines

Source: Masterson et al., 1981.

the smallest particle of an element that can exist either alone or in combination with similar particles of the same or a different element, or the smallest particle that enters into the composition of molecules. Any atom possesses an *atomic number* which is characteristic of an element and represents the positive charge of the nucleus. The atomic number of an atom equals the number of protons in the nucleus or the number of electrons outside the nucleus when the atom is neutral.

An atom is also characterized by an *atomic weight* which represents the relative weight of an element in nature in reference to the hydrogen taken as standard. An atom is made up of neutrons, protons, and electrons. The positive charge of the nucleus is balanced by electrons (e–) which swarm about the nucleus in orbitals. Only two electrons may occupy a particular orbital. The potential of an atom of any given element to react depends on the affinity of its nucleus for electrons and the strong tendency of the atom to gain maximum stability by filling its outer electron shells.

Generally, when the outer shell of an atom contains a complete set of paired electrons and the total number of electrons of all orbiting shells exceeds the number of the positively charged protons in the nucleus, the atom is referred to as a negatively charged ion (*anion*). The magnitude of the difference between electrons and protons is commonly referred to as anion charge (e.g., 1–, 2–, 3–) (Table 1.5). On the other hand, when the number of protons exceeds the sum of all the orbiting electrons and the latter are complete sets of pairs, the atom is referred to as a positively charged ion (*cation*). The magnitude of the difference between protons and electrons is commonly referred to as cation charge (e.g., 1+, 2+, 3+, or K^+, Na^+, Ca^{2+}, Mg^{2+}, Al^{3+}) (Table 1.5). The attraction between two oppositely charged ions forms what is known as an *ionic bond*, which is a characteristic of salts such as NaCl, KCl, and $NaNO_3$ (Fig. 1.1). It is generally known to be a weak bond, which explains the high solubility of most such salts. Generally, ionic bonding is a characteristic of light metals and exhibits different degrees of strength, depending on the charges of the ions involved and the type of anions (nonmetals) they associate with. The data in Table 1.6 show relative solubilities of compounds commonly encountered in nature.

When atoms possess an incomplete outer shell (e.g., nonpaired electrons), yet their net charge is zero, attraction between such atoms takes place because of their strong tendency to complete their outer electron orbital shell by sharing their unpaired electrons. This gives rise to a *covalent bond*. One example of a covalent bond is the bimolecular chlorine gas (Cl_2) (Fig. 1.1). Covalent bonding is a characteristic of some nonmetals or metalloids (bimolecular molecules), but may also arise between any two atoms when one of the atoms shares its outer-shell electron pair (Lewis base) with a second atom that has an empty outer shell (Lewis acid). Such bonds are known as *coordinated covalent bonds* or *polar covalent bonds*. They are commonly weaker than the covalent bond of two atoms which share each other's unpaired outer-shell electrons (e.g., F_2 and O_2). Coordinated covalent bonds often involve organometallic complexes.

Bonding strength between ions forming various solids or minerals implies degree of solubility. A way to qualitatively assess bonding strength is through electronegativity. *Electronegativity* is defined as the ability of an atom to attract to itself the electrons in a covalent bond. In a covalent bond of any biomolecular species (e.g., Cl_2, F_2, and O_2), the complex formed is nonpolar because the electrons are equally shared.

TABLE 1.5. International Atomic Weights for the Most Environmentally Important Elements

Element	Symbol	Atomic Number	Oxidation State	Atomic Weight[a]
Aluminum	Al	13	3	26.9815
Arsenic	As	33	−3, 0	74.9216
Barium	Ba	56	3	137.34
Beryllium	Be	4	2	9.0122
Boron	B	5	3	10.811
Bromine	Br	35	1, 3, 5, 7	79.909
Cadmium	Cd	48	2	112.40
Calcium	Ca	20	2	40.08
Carbon	C	6	2, 3, 4	12.01115
Cesium	Cs	55	1	132.905
Chlorine	Cl	17	−1, 1, 3, 5, 7	35.453
Chromium	Cr	24	2, 3, 6	51.996
Cobalt	Co	27	2, 3	58.9332
Copper	Cu	29	1, 2	63.54
Fluorine	F	9	−1	18.9984
Gold	Au	79	1, 3	196.967
Helium	He	2	0	4.0026
Hydrogen	H	1	1	1.00797
Iodine	I	53	−1, 1, 3, 5, 7	126.9044
Iron	Fe	26	2, 3	55.847
Lead	Pb	82	4, 2	207.19
Lithium	Li	3	1	6.939
Magnesium	Mg	12	2	24.312
Manganese	Mn	25	7, 6, 4, 3, 2	54.9380
Mercury	Hg	80	1, 2	200.59
Molybdenum	Mo	42	2, 3, 4, 5, 6	95.94
Nickel	Ni	28	2, 3	58.71
Nitrogen	N	7	−3, 3, 5, 4, 2	14.0067
Oxygen	O	8	−2	15.9994
Phosphorus	P	15	−3, 5, 4	30.9738
Platinum	Pt	78	2, 4	195.09
Plutonium	Pu	94	6, 5, 4, 3	(244)
Potassium	K	19	1	39.102
Radon	Rn	86	0	(222)
Rubidium	Rb	37	1	85.47
Selenium	Se	34	−2, 4, 6	78.96
Silicon	Si	14	4	28.086
Silver	Ag	47	1	107.870
Sodium	Na	11	1	22.9898
Strontium	Sr	38	2	87.62
Sulfur	S	16	−2, 2, 4, 6	32.064
Tin	Sn	50	4, 2	118.69
Tungsten	W	74	6, 5, 4, 3, 2	183.85
Uranium	U	92	6, 5, 4, 3	238.03
Zinc	Zn	30	2	65.37

[a]Numbers in parentheses indicate the mass number of the most stable known isotope.

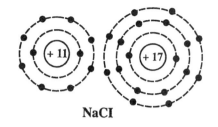

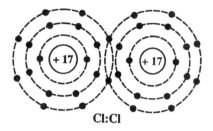

Figure 1.1. Schematic of an ionic bond between Na⁺ and Cl⁻ and a covalent bond between two chlorine atoms.

However, many covalent bonds do not equally share electrons; such covalent bonds, as pointed out above, are referred to as polar covalent bonds or bonds of partial ionic character. Electronegativity is rated on a relative scale ranging from 4 (most electronegative, fluorine) to 0.7 (least electronegative, cesium) (Table 1.7). In general, the greater the difference in electronegativity between two elements, the more ionic will be the bond between them (Fig. 1.2).

TABLE 1.6. Solubilities of Compounds of the Group 1 and Group 2 Metals[a]

	Li^+	Na^+	K^+	Rb^+	Cs^+	Be^{2+}	Mg^{2+}	Ca^{2+}	Sr^{2+}	Ba^{2+}
F^-	ss	S	S	S	S	S	I	I	I	I
Cl^-	S	S	S	S	S	S	S	S	S	S
Br^-	S	S	S	S	S	S	S	S	S	S
I^-	S	S	S	S	S	S	S	S	S	S
NO_3^-	S	S	S	S	S	S	S	S	S	S
SO_4^{2-}	S	S	S	S	S	S	S	ss	I	I
OH^-	S	S	S	S	S	I	I	ss	ss	S
CO_3^{2-}	ss	S	S	S	S	ss	I	I	I	I
PO_4^{3-}	ss	S	S	S	S	S	I	I	I	I

Source: Masterson et al., 1981.
[a]S = soluble (> 0.1 M); ss = slightly soluble (0.1–0.01 M); I = insoluble (< 0.01 M).

TABLE 1.7. Electronegativity Values

H						
2.1						
Li	Be	B	C	N	O	F
1.0	1.5	2.0	2.5	3.0	3.5	4.0
Na	Mg	Al	Si	P	S	Cl
0.9	1.2	1.5	1.8	2.1	2.5	3.0
K	Ca	Sc	Ge	As	Se	Br
0.8	1.0	1.3	1.8	2.0	2.4	2.8
Rb	Sr	Y	Sn	Sb	Te	I
0.8	10	1.2	1.8	1.9	2.1	2.5
Cs	Ba	La–Lu	Pb	Bi	Po	At
0.7	0.9	1.0–1.2	1.9	1.9	2.0	2.2

In nature, elements exist in various oxidation states (Table 1.5) which determine many of the properties of the molecule or compound formed. The *oxidation state* of an element is characterized by its *valence*, denoted by the so-called *valency*, which is the oxidation number or the number of electrons lost or gained. An example of the role of element valence on the properties of molecules can be demonstrated through the behavior of manganese (Mn) in nature. Manganese can be found in nature under three oxidation states, Mn^{2+}, Mn^{3+}, and Mn^{4+}. Each one of these manganese species reacts differently in nature. For example, Mn^{4+} reacts strongly with oxygen to form manganese oxide (MnO_2), which is extremely insoluble under oxidizing conditions. Similarly, Mn^{3+} reacts strongly with oxygen and hydroxyl (OH) to form manganese oxyhydroxide (MnOOH), also an insoluble mineral. On the other hand, Mn^{2+} reacts with OH^- to form manganese dihydroxide [$Mn(OH)_2$], which is very soluble at circumneutral pH (see also Chapter 2).

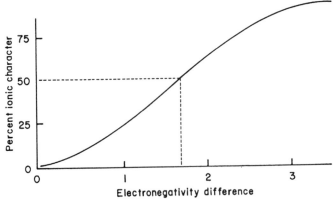

Figure 1.2. Relationship between the ionic character of a bond and the difference in electronegativity of the bonded atoms (see also Table 1.7) (after Masterson et al., 1981, with permission).

TABLE 1.8. Classification of Metals and Metalloids

	Acids	Bases
Hard	H^+, Li^+, Na^+, K^+, Mg^{2+}, Ca^{2+}, Sr^{2+}, Al^{3+}, Cr^{3+}, Mn^{3+}, Fe^{3+}	H_2O, OH^-, F^-, $CH_3CO_2^-$, PO_4^{3-}, SO_4^{2-}, Cl^-, CO_3^{2-}, ClO_4^-, NO_3^-, $NH3$
Borderline	Cr^{2+}, Mn^{2+}, Fe^{2+}, Ni^{2+}, Cu^{2+}, Zn^{2+}	$C_6H_5NH_2$, C_5N_5N, N_3^-, Br^-, NO_2^-, SO_3^{2-}
Soft	Cu^+, Ag^+, Cd^{2+}, Hg^{2+}, Pb^{2+}	R_2S, RSH, I^-, SCN^-, CN^-

Source: Sposito, 1981.

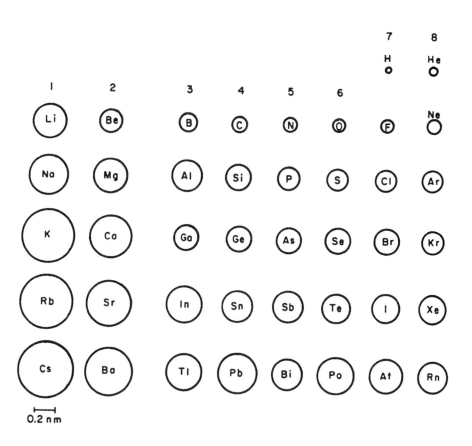

Figure 1.3. Atomic radii of the main-group elements. Atomic radii increase as one goes down a group and in general decrease going across a row in the Periodic Table. Hydrogen has the smallest atom and cesium the largest.

One method to predict bond formation between charged species is the hard and soft acid and base theory (HSAB). It separates the metals (Lewis acids) and ligands (Lewis bases) into hard, soft, and borderline groups (Table 1.8). This characterization is based on ion electronegativity, ion polarizability, and oxidation potential (Pearson, 1963, 1966). Polarizability denotes electronic orbital deformation potential by an electric field. A hard Lewis acid is a molecule of relatively small size, high oxidation state, high electronegativity, and low polarizability. A soft Lewis acid is a molecule of relatively small size, low electronegativity, and high polarizability. A hard Lewis base is a molecule of high electronegativity and low polarizability; it is difficult to oxidize and does not possess empty low-energy electronic orbitals. A soft Lewis base is a molecule of low electronegativity, high polarizability, and relatively strong tendency to oxidize. In general, the HSAB theory states that hard acids prefer hard bases and soft acids prefer soft bases. The only exception is metals of the top transition group (Tables 1.1 and 1.8). They can bind with either soft or hard bases.

Organic ligands in soil with oxygen as the ligating atom (e.g., simple organic acids of carboxylic groups or phenolic groups, see Chapter 3) behave as hard bases and prefer hard metals. However, ligands with sulfur or nitrogen as the ligating atom behave as soft bases and prefer soft acids (Buffle, 1984; Buffle and Stumm, 1984). Finally, inorganic ligands with oxygen as the ligating atom also behave as hard bases and prefer hard metals. Relative hardness within a group of elements can be determined by the term z^2/r, where z denotes charge and r denotes ionic radius. At any given z, the larger r is, the lower the hardness. For any given r, the larger z is, the greater the hardness. The data in Figure 1.3 shows that atomic radii increase as one goes down a group, and in general decrease going across a row in the periodic table. For this reason, the preference for hard metals by ligand atoms decreases in the order $F > O > N$ (hard ligating atoms) $> Br > I >$ (soft ligating atoms).

1.3   REVIEW OF CHEMICAL UNITS

The most common units describing elements, chemical constituents, or contaminants in the environment are:

a. moles per liter (mol L^{-1} or M)
b. millimoles per liter (mmol L^{-1} or mM), estimated by multiplying mol L^{-1} times 1000
c. micromoles per liter (μmol L^{-1} or μM), estimated by multiplying mol L^{-1} times 1,000,000 or mmol L^{-1} times 1000

These units are preferred by chemists because they denote quantities that describe chemical reactions. In other words, chemical reactions occur because 1 mol of a given reactant always reacts with a given number of moles of a second reactant, depending on the nature of the reactants. For example,

$$AgNO_{3aq} + NaCl_{aq} \rightarrow AgCl_s + NaNO_{3aq} \qquad (1.1)$$

where the subscript aq denotes dissolved and the subscript s denotes solid. Reaction 1.1 shows that 1 mol of silver nitrate ($AgNO_3$) in solution reacts with 1 mol of sodium chloride in solution to produce 1 mol of silver chloride solid ($AgCl_s$) and 1 mol of sodium nitrate in solution. For the reaction

$$CaCO_{3s} + 2HCl_{aq} \rightarrow CaCl_{2aq} + CO_{2\,gas} + H_2O \qquad (1.2)$$

1 mol of limestone ($CaCO_{3s}$) reacts with 2 mol of hydrochloric acid in solution to produce 1 mol of $CaCl_2$ in solution, 1 mol of carbon dioxide gas, and 1 mol of water.

The number of moles of a given reactant that would react with a given number of moles of a second reactant is dependent on mass–balance, which denotes that reactions always occur on a mole-charge basis. Thus, 1 mol of positive charge always reacts with 1 mol of negative charge. This is necessary because solution electroneutrality is present at the beginning of the reaction and must be maintained at the end of the reaction. For example, in Reaction 1.1, 1 mol of silver (Ag^+) (which equals 1 mol of positive charge) reacts with 1 mol of chloride (Cl^-) since it represents 1 mol of negative charge. In Reaction 1.2, 1 mol of calcium (Ca^{2+}), which represents 2 mol of charge, reacts with 2 mol of Cl^- because the latter is a monovalent anion. Furthermore, 1 mol of CO_3^{2-}, which is 2 mol of charge, reacts with 2 mol of H^+. Based on the above, environmental chemists often give concentration units in equivalents per liter (eq L^{-1} or mol_c L^{-1}), milliequivalents per liter (meq L^{-1} or $mmol_c$ L^{-1}), or microequivalents per liter (μeq L^{-1} or μmol_c L^{-1}). The relationship between moles and equivalents is

$$[\text{mol } L^{-1}] \times [\text{valence } (z)] = \text{eq } L^{-1} \qquad (1.3)$$

Examples are given below on how one might use this information to prepare a solution of a given concentration. Assume that one needs to prepare 1 mol L^{-1} NaCl solution. The first thing that is needed is the molecular weight (MW) of NaCl, which is the sum of the atomic weights of Na^+ (22.99 g mol^{-1}) (g = grams) and Cl^- (35.45 g mol^{-1}) (MW NaCl = 58.44 g). Therefore, to make 1 mol L^{-1} NaCl solution, one needs to dissolve 58.44 g of NaCl in sufficient solute (e.g., distilled water) to make a total volume of 1 L. Hence, 1 mol L^{-1} NaCl is also 1 eq L^{-1} NaCl or one mol L^{-1} Na and 1 mol L^{-1} Cl. In the preceding statement, the symbols for the elements do not include valence numbers.

If one prepares a solution of 1 mol L^{-1} NaCl, it is not clear whether there will be 1 mol L^{-1} Na^+ or 1 mol L^{-1} Cl^-, because the two ions may react with each other or with other chemical species in solution to form additional solution species with different valencies. For example, assuming that the NaCl solution contains also lead (Pb^{2+}), Pb^{2+} and Cl^- would react with each other to form the dissolved chemical species $PbCl^+$, $PbCl_2^0$, and so on. Chemists distinguish the two situations (free vs. paired solution species) by referring to the total dissolved concentration of an element as *formality* (F), and to the concentration of certain known dissolved chemical species (e.g., Na^+ and Pb^{2+}) as *molarity* (M) (Table 1.9). Field practitioners of environmental chemistry almost always refer to concentrations of elements because it is total dissolved concen-

TABLE 1.9. Review of Concentration Units

Unit	Definition	Comment
Mole (M)	An Avogadro's number of "things" = 6.02×10^{23}	Fundamental to chemical reactions is the fact that a given number of "things" (atoms, molecules, electrons, ions, etc.) react with a given number of other reactant "things" to yield an exact number of product "things"
Atomic weight (AW)	Weight in grams of a mole of selected atoms (e.g., AW of Zn = 65.37 g)	
Formula weight	Weight in grams of a mole of the selected compound (e.g., FW of NaCl = 58.4428)	
Molarity (M)	1 mole of the solute dissolved in sufficient solvent to give a total volume of 1 L	A 1 M solution of $CaCl_2$ is also 1 M in Ca^{2+} but is 2 M in Cl^-. This term is often given the additional restriction that is represents only the species indicated. Thus, 10^{-5} moles of $AlCl_3$ dissolved and made to a volume of 1 L with water would be almost exactly 3×10^{-5} M in Cl^- because the chloride ion does not complex or ion pair significantly with aluminum ions in solution. On the other hand, Al^{3+} is considerably less than 10^{-5} M because the hydrated aluminum ion hydrolyzes significantly to form the $AlOH^{2+}$ ion. The solution could properly be described as 10^{-5} F in $AlCl_3$ (see definition of F)
Molality (m)	1 mole of solute plus 1 kg of solvent	
Formality (F)	1 formula weight (mole) of solute dissolved in sufficient solvent to make a total volume of 1 L	
Equivalent	The quantity of reactant which will give 1 mole of reaction defined by a specific chemical equation	If the reaction is a redox reaction, an equivalent of reactant either gives up or accepts 1 mole of electrons

tration that the government regulates. However, often it is the concentration of certain chemical species and not elemental concentrations that control toxicities.

When a solution of 1 mol L^{-1} $CaCl_2$ is needed, dissolve 1 mol $CaCl_2$ (MW $CaCl_2$ = 110.98 g) in sufficient solvent (e.g., distilled water) to make a total volume of 1 L. However, because $CaCl_2$ is a nonsymmetrical electrolyte (Ca and Cl possess different valencies), 1 mol L^{-1} $CaCl_2$ would give 1 mol L^{-1} Ca, 2 eq L^{-1} Ca and 2 mol L^{-1} Cl, or 2 eq L^{-1} Cl. It follows that if one needs to convert moles per liter to grams per liter, one needs to multiply the moles per liter with the molecular weight of the salt or the atomic weight (AW) of the particular element. Therefore,

1 mol L^{-1} CaCl$_2$ contains 110.98 g L^{-1} CaCl$_2$ or
110.98 × 10^3 mg L^{-1} CaCl$_2$ (parts per million, ppm) or
110.98 × 10^6 µg L^{-1} CaCl$_2$ (parts per billion, ppb)

40.08 g L^{-1} Ca or
40.08 × 10^3 mg L^{-1} Ca (ppm) or
40.08 × 10^6 µg L^{-1} Ca (ppb)

70.90 g L^{-1} Cl or
70.90 × 10^3 mg L^{-1} Cl (ppm) or
70.90 × 10^6 µg L^{-1} Cl (ppb)

It is often necessary to convert quantities of a given chemical reagent to equivalent quantities of another chemical reagent. For example, assume that one has a choice of two chemical reagents to neutralize a given amount of acid water and these chemical reagents are potassium hydroxide (KOH), (MW = 56.09 g) and CaCO$_3$ (MW = 100.06 g). Assume further that an analytical laboratory determined that 100 kg of CaCO$_3$ are needed to carry out this acid-water neutralization task. How many kilograms of KOH will be needed to carry out the same neutralization process? To answer this question, one needs to know the reactions and their stoichiometry. Consider

$$H_2SO_4 + CaCO_3 \rightarrow CaSO_4 + CO_{2\,gas} + H_2O \qquad (1.4a)$$

and

$$H_2SO_4 + 2KOH \rightarrow K_2SO_4 + 2H_2O \qquad (1.4b)$$

Reactions 1.4a and 1.4b reveal that for every mole of CaCO$_3$ needed to neutralize 1 mol of acid (H$_2$SO$_4$), 2 mol KOH are needed to neutralize the same amount of H$_2$SO$_4$. Based on this finding, to convert kilograms of CaCO$_3$ to kilograms of KOH, one needs to use the *gravimetric formula*:

$$Q_{unknown} = (Q_{known})\,[(MW_{unknown}/MW_{known})\,(mol_{unknown}/mol_{known}] \qquad (1.5)$$

where

$Q_{unknown}$ is quantity (grams, kilograms, tons, etc.) of unknown (KOH)
Q_{known} is quantity (grams, kilograms, tons, etc.) of known (CaCO$_3$)
$MW_{unknown}$ is the molecular weight of unknown (KOH)
MW_{known} is the molecular weight of known (CaCO$_3$)
$mol_{unknown}$ is number of moles of unknown (KOH) required to complete the reaction
mol_{known} is the number of moles of the known (CaCO$_3$) needed to complete the reaction

Thus

$$[100] \, [56.09/100.06] \, [2/1] = 112.11 \text{ kg KOH} \qquad (1.6)$$

1.4 BASIC INFORMATION ABOUT WATER CHEMISTRY

Water is made up of two hydrogens and one oxygen. Oxygen has six frontier electrons. Four of these electrons come in pairs of two; the other two electrons are unpaired. A chemical bond between two elements takes place when the elements donate electrons to each other so that all frontier electrons are paired. In the case of water, the oxygen's two unpaired electrons are paired by bonding with two hydrogens, each donating an electron. After the covalent bonds of the oxygen with the two hydrogens are formed, the oxygen has four sets of paired electrons and each hydrogen has one set of paired electrons. This makes the water molecule stable.

Paired electrons exert repulsive forces against each other. Bond-forming electron pairs exert less repulsive force than unshared pairs of electrons. It follows that electron-pair distribution in the oxygen becomes skewed and the water molecule gains a positive and a negative pole (Fig. 1.4). This arrangement makes the water molecule "the universal solvent." The two unshared pairs of electrons attract hydrogens of other water molecules, forming weak hydrogen bonds. When many H_2O molecules are

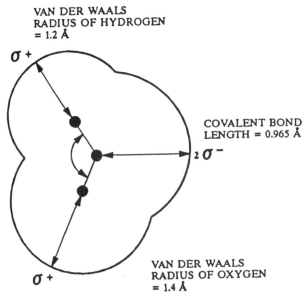

Figure 1.4. Model of a water molecule. The curved lines represent borders at which van der Waals attractions are counterbalanced by repulsive forces (after Hillel, 1980, with permission).

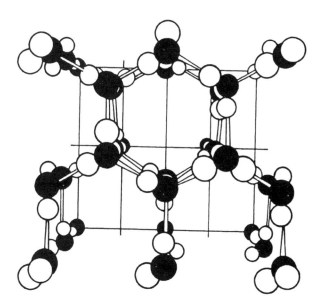

Figure 1.5. Schematic of an ice crystal. The oxygen atoms are shown in black and the hydrogen atoms in white (after Hillel, 1980, with permission).

present they create a three-dimensional "scaffolding" of molecules held together by the weak hydrogen bonds (Fig. 1.5). The force created by these weak hydrogen bonds is known as *cohesion*. Hydrogen bonds are also created between water and solid substances such as soil minerals (inorganic and/or organic). The force that binds water to other solid substances (e.g., soil minerals) is called *adhesion*. Generally, substances exhibiting adhesion are known as *hydrophillic*, while substances not capable of adhesion are known as *hydrophobic*. Cohesion and adhesion as well as hydrophobicity are part of many important natural occurrences, such as water retention and movement in soil, as well as solubility and mobility of pollutants in the groundwater.

1.4.1   Physical States and Properties of Water

Water is encountered in nature in three states: (1) the vapor state [H_2O, $(H_2O)_2$ or $(H_2O)_3$] at or above 100°C, (2) the solid state (ice sheets of puckered hexagonal rings, Fig. 1.5, at or below 0°C), and (3) the liquid state (between 0° and 100°C) which is described by the flickering cluster model [monomers and up to $(H_2O)_{40}$ molecules] with an average life of 10^{-10} to 10^{-11} sec (Fig. 1.6).

   The forces holding water molecules together and the ideal molecular structure of water, as shown in Figure 1.5, give rise to some of the most important properties of water contributing to supporting life, as we know it, on earth. For example, Table 1.10 shows that water exhibits a rather large surface tension relative to other liquids, which helps explain the potential of water molecules to attract each other or stay together

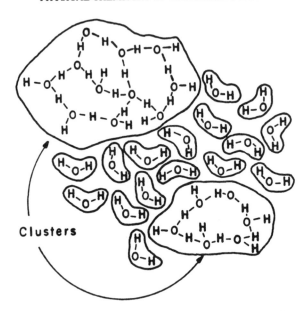

Figure 1.6. Polymers of water molecules demonstrating the "flickering clusters" model (after Hillel, 1980, with permission).

under tension and thus its ability to reach the highest leaves on a tall tree (e.g., redwoods). The data in Table 1.11 show that water possesses the highest specific heat capacity in comparison to the other substances listed, which may help explain freezing of lakes and oceans only on the surface, thus protecting aquatic life. Similarly, the viscosity of water is not being affected dramatically by temperature until it reaches the boiling or freezing point (Table 1.12). Finally, the data in Table 1.13 reveal the large transformation heat that water possesses relative to some other liquids. Thus, even under extremely droughty conditions, one may find water in its liquid phase. Also, because of water's high heat of transformation, it is used to heat buildings and to protect crops from freezing.

The potential of water to dissolve other polar substances can be explained on the basis of its *dielectric constant*. A dielectric constant is a measure of the amount of

TABLE 1.10. Surface Tension of Water Relative to Other Liquids

Substance	Surface Tension $(dyne \cdot cm^{-1})^a$
Water	72.7
Ethanol	22
Mercury	430

^{a}Dyne = $g \cdot cm \cdot sec^{-2}$.

TABLE 1.11. Specific Heat Capacity of Water Relative to Other Substances

Substance	Specific Heat Capacity (cal·deg^{-1}·gm^{-1})[a]
Water	1.0
Ice	0.50
Iron	0.11
Dry soil	0.20
Air	0.17

[a]Calorie = amount of heat required to raise the temperature of 1 g of H_2O 1°C. Hydrogen bonds require 4.5 kcal mol^{-1} in order to break. H–O bonds (covalent character) require 110 kcal mol^{-1} in order to break.

electrical charge a given substance can withstand at a given electric field strength. For the purpose of this book, a dielectric constant regulates the force of attraction between two oppositely charged particles (e.g., Ca^{2+} and SO_4^{2-}) in a liquid medium (e.g., water). This force of attraction can be predicted by Coulomb's Law:

TABLE 1.12. Viscosity of Water Under Various Temperatures

Temperature (°C)	g·cm^{-1}·sec^{-1}
10	1.30
15	1.14
20	1.00
25	0.89
35	0.80

TABLE 1.13. Heat of Transformation Relative to Other Liquids

Substance	Liquid to Gaseous State (cal·gm^{-1})
Water[a]	540
Methanol	263
Ethanol	204
Acetone	125

[a]Heat of transformation from solid to liquid for H_2O = 80 cal·gm^{-1}. (In other words, to thaw 1 g of ice, 80 cal must be supplied.)

TABLE 1.14. Dielectric Constant of Water Relative to Other Liquids

Substance	Dielectric Constant[a]
Water	80
Methanol	33
Ethanol	24
Acetone	21.4
Benzene	2.3

[a]Dielectric constant = capacitance H_2O/capacitance vacuum. Capacitance = ability of a nonelectrical conductor to store electrical energy.

$$F = e_1 e_2 / Dr^2 \tag{1.7}$$

where

F = force of attraction

e_1, e_2 = charges of the ions

r = distance between ions

D = dielectric constant

Equation 1.7 demonstrates that the force of attraction between oppositely charged particles is inversely related to the dielectric constant. The data in Table 1.14 show that water possesses the highest dielectric constant in comparison to the other liquids reported in the table. This explains why, for example, gypsum ($CaSO_4 \cdot 2H_2O$) dissolves in water at 2.2 g L^{-1} while its solubility in alcohol is negligible.

1.4.2 Effects of Temperature, Pressure, and Dissolved Salts

The physical properties of water are subject to change as temperature and/or pressure changes. The major physical changes, commonly observed under changing temperature, pressure, and salt content include:

1. Molecular clusters decrease as temperature and pressure decrease
2. Boiling point increases as pressure increases
3. Freezing point decreases as salt content increases
4. Volume increases as temperature increases
5. Boiling point increases as salt content increases
6. Surface tension increases as salt content increases
7. Viscosity increases as salt content increases
8. Osmotic pressure increases as salt content increases

Even though water is affected by temperature and pressure, such effects are minimized until the boiling or freezing point is reached. Furthermore, some of these effects are not as obvious as one might expect. For example, water reaches a minimum volume at 4°C, and below 4°C its volume starts to increase again, explaining the potential of ice to float in water, helping to protect aquatic life.

The solubility of inert gases in water (e.g., oxygen, O_2) also depends on pressure and temperature. This can be explained by the ideal gas law:

$$n = PV/RT \qquad (1.8)$$

where

n = amount of gas

P = pressure

V = volume

T = temperature

R = universal gas constant

Considering that water possesses a certain "free" space because of its molecular arrangement (Fig. 1.5), and assuming that this "free" space is negligibly affected by temperature, Equation 1.8 demonstrates that under a constant atmospheric pressure (P), as temperature increases, the expansion potential of the gas causes its apparent solubility to decrease. This explains large fish kills in shallow waters during extremely hot weather, a condition that suppresses the solubility of atmospheric air.

1.4.3 Hydration

Because of its polarity, water tends to hydrate ions. The phenomenon of hydration is demonstrated in Figure 1.7, which shows three types of water surrounding the sodium ion (Na^+). The first water layer, nearest the ion, is very rigid owing to its strong attraction to the cation's electronic sphere. Some researchers equate this water's structural arrangement to that of ice. The dielectric constant of this water is reported to be as low as 6, as opposed to 80 for pure liquid water (Table 1.14). The next water layer is somewhat rigid with slightly higher dielectric constant (e.g., 20), and finally, the third water layer is made of "free" water. One may envision the same triple-layer water arrangement on hydrophillic solid surfaces (e.g., wet soil minerals). Generally speaking, the greater the charge density of an ion, the more heavily hydrated it will be. Anions are hydrated less than cations because of lesser charge density. Cations are heavily hydrated because of their higher charge density, and the process can be demonstrated as follows:

$$Na^+ + (n + 4H_2O) \rightarrow Na(H_2O)_4 (H_2O)_n^+ \qquad (1.9)$$

Commonly, two processes take place when a metal salt is added to water:

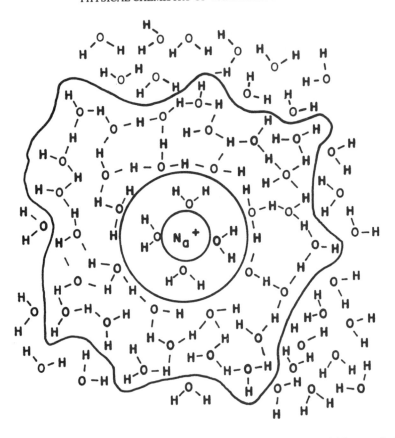

Figure 1.7. A model showing the hydration sphere of sodium—an inner rigid water shell; an outer, somewhat rigid water shell, with the whole assembly floating in a sea of "free" water (after Hillel, 1980, with permission).

1. Hydration (H_2O molecules adsorb onto the ions)
2. Hydrolysis (degree to which adsorbed H_2O dissociates to satisfy ion electronegativity).

$$Ca^{2+} + H_2O \Leftrightarrow CaOH^+ + H^+ \tag{1.10}$$

1.5   CHEMICAL PROPERTIES OF WATER

Water is an amphoteric substance (acts as acid or base) depending on the substance that the water reacts with. Water molecules may dissociate as shown below:

$$H_2O \Leftrightarrow H^+ + OH^- \tag{1.11}$$

TABLE 1.15. Ion Activity Product of H_2O at Various Temperatures[a]

°C	K_W	pK_W
0	0.12×10^{-14}	14.93
15	0.45×10^{-14}	14.35
20	0.68×10^{-14}	14.17
25	1.01×10^{-14}	14.0
30	1.47×10^{-14}	13.83
50	5.48×10^{-14}	13.26

[a]The pH of neutrality shifts with respect to standard pressure and temperature.

and

$$K_w = 10^{-14} = (H^+)(OH^-) \tag{1.12}$$

where K_w is the water dissociation constant and is somewhat dependent on temperature (Table 1.15).

1.6 BRONSTED–LOWRY AND LEWIS DEFINITIONS OF ACIDS AND BASES

A *Bronsted–Lowry acid* is any substance that is capable of donating a proton, whereas a *Bronsted–Lowry base* is any substance that is capable of accepting a proton. The loss of a proton by an acid gives rise to an entity that is a potential proton acceptor and thus a base; it is called the conjugate base of the parent acid. Examples of acids reacting with bases are given in Table 1.16. The reactions listed in Table 1.16 are spontaneous in the direction that favors production of the weaker acid and base. Compounds that may act as bases and acids are referred to as amphoteric.

Bronsted–Lowry acids and bases are also classified according to the extent that they react with solvents (H_2O). Commonly, they are classified into *strong acids and bases* and *weak acids and bases*. Strong acids are 100% dissociated in water. For example, hydrochloric acid (HCl), a strong acid, dissociates as follows:

$$0.01 \ M \ HCl \rightarrow 0.01 \ M \ H^+ + 0.01M \ Cl^- \tag{1.13}$$

Reaction 1.13 reveals that 0.01 mol L^{-1} HCl dissociates to give 0.01 mol L^{-1} H^+ and 0.01 mol L^{-1} Cl^-. Examples of strong Bronsted–Lowry acids of some interest to environmental scientists include nitric acid (HNO_3), hydrochloric acid (HCl), and sulfuric acid (H_2SO_4). Examples of strong Bronsted–Lowry bases of interest to environmental scientists include potassium hydroxide (KOH), sodium hydroxide (NaOH), and ammonium hydroxide (NH_4OH).

TABLE 1.16. Examples of Acid–Base Reactions

H_2O + Acid	$\rightarrow$	Conjugate Base	+	H_3O^+
Base + acid	$\rightarrow$	Base	+	Acid
H_2O + HCl	$\rightarrow$	Cl^-	+	H_3O^+
$H_2O + Al(H_2O)_6^{3+}$	$\rightarrow$	$(Al(OH)(H_2O)_5^{2+}$	+	H_3O^+
$H_2O + H_2PO_4^-$	$\rightarrow$	HPO_4^{2-}	+	H_3O^+
$H_2O + NH_4^+$	$\rightarrow$	NH_3	+	H_3O^+
$NH_3 + H_2O$	$\rightarrow$	OH^-	+	NH_4^+
$CO_3^{2-} + H_2O$	$\rightarrow$	OH^-	+	HCO_3^-

Weak acids are less than 100% dissociative. Generally, they are 1% dissociative or less. For example, a weak acid (HA) is 1% dissociated in water when only 1% would give H^+ and A^- (A^- is the conjugate base):

$$0.01 \, M \, HA \Leftrightarrow 0.0001 \, M \, H^+ + 0.0001 \, M \, A^- \tag{1.14}$$

Examples of weak Bronsted–Lowry acids of considerable interest to environmental scientists include carbonic acid (H_2CO_3), phosphoric acid (H_3PO_4), silicic acid (H_4SiO_4), boric acid (H_3BO_3), hydrogen sulfide (H_2S), and bisulfate (HSO_4). Examples of weak Bronsted–Lowry bases of considerable interest to environmental scientists include carbonate (CO_3^{2-}), acetate (OAc^-), and sulfide (S^{2-}).

A *Lewis acid* is any substance that is capable of accepting an electron pair while a *Lewis base* is any substance that is capable of donating an electron pair. Examples of relatively strong Lewis acids include cations with valence higher than 2, such as iron III (Fe^{3+}) and aluminum (Al^{3+}). Examples of relatively strong Lewis bases include anions with valence higher than 2, such as phosphate (PO_4^{3-}) and arsenate (AsO_4^{3-}). Examples of relatively weak Lewis acids of significant importance to environmental scientists include cations with valence lower than or equal to 2, for example, potassium (K^+), sodium (Na^+), calcium (Ca^{2+}), magnesium (Mg^{2+}), barium (Ba^{2+}), and strontium (Sr^{2+}). On the other hand, examples of weak Lewis bases of significant importance to environmental scientists include anions with valence lower than or equal to 2, such as chloride (Cl^-), nitrate (NO_3^-), sulfate (SO_4^{2-}), and bicarbonate (HCO_3^{2-}).

1.6.1 Weak Monoprotic Acids

A weak monoprotic acid is any Bronsted–Lowry acid that possesses a single dissociable H^+. The dissociation of a weak acid can be expressed in terms of the acid dissociation constant, K_a:

$$HA \Leftrightarrow H^+ + A^- \tag{1.15}$$

and

TABLE 1.17. Selected Dissociation Constants for Weak Acids

Element	Formula of Acid	pK_a
C	H_2CO_3	6.3
	HCO_3^-	10.33
P	H_3PO_4	2.23
	$H_2PO_4^-$	7.2
	HPO_4^{2-}	12.3
S	HSO_4^-	2.0
	H_2S	7.0
	HS^-	12.9
N	NH_4^+	9.2
B	H_2BO_3	9.2
Si	H_4SiO_4	9.5
O	H_2O	14.0

$$K_a = [(H^+)(A^-)]/(HA) \tag{1.16}$$

where A^- denotes any conjugate base. The K_a is constant for any given weak acid; and the larger K_a is, the greater the dissociation of the acid. The K_a values generally range from 10^{-1} to 10^{-14} and measure the relative tendency of an acid to dissociate. For convenience, K_a is commonly expressed in the negative log form or pK_a, for example,

$$pK_a = -\log K_a \tag{1.17}$$

The pK_a values of weak Bronstead–Lowry acids generally range from 1 to 14 (Table 1.17).

1.6.2   Weak Polyprotic Acids

Weak polyprotic Bronsted–Lowry acids are those that possess more than one dissociable H^+. Examples of polyprotic acids include carbonic acid (H_2CO_3, a diprotic acid) and phosphoric acid (H_3PO_4), a triprotic acid (Table 1.17). In the case of phosphoric acid, the following dissociation steps can be demonstrated:

$$H_3PO_4 \Leftrightarrow H^+ + H_2PO_4^- \tag{1.18}$$

where H_3PO_4 is the acid form, $H_2PO_4^-$ is the base form, and the pK_a is 2.23. The second dissociation step is

$$H_2PO_4^- \Leftrightarrow H^+ + HPO_4^{2-} \qquad (1.19)$$

where $H_2PO_4^-$ is the acid form, HPO_4^{2-} is the base form, and the pK_a is 7.2. The third dissociation step is:

$$HPO_4^{2-} \Leftrightarrow H^+ + PO_4^{3-} \qquad (1.20)$$

where HPO_4^{2-} is the acid form, PO_4^{3-} is the base form, and the pK_a is 12.3. A given weak acid has a fixed K_a value and is expressed as previously shown by Equation 1.16, which can be manipulated to obtain

TABLE 1.18. Dissociation Constants for Acids

		Dissociation Constant (25°C)		
Name	Formula	K_1	K_2	K_3
Acetic	CH_3COOH	1.75×10^{-5}		
Arsenic	H_3AsO_4	6.0×10^{-3}	1.05×10^{-7}	3.0×10^{-12}
Arsenious	H_3AsO_3	6.0×10^{-10}	3.0×10^{-14}	
Benzoic	C_6H_5COOH	6.14×10^{-5}		
1–Butanoic	$CH_3CH_2CH_2COOH$	1.51×10^{-5}		
Chloroacetic	$ClCH_2COOH$	1.36×10^{-3}		
Citric	$HOOC(OH)C(CH_2COOH)_2$	7.45×10^{-4}	1.73×10^{-5}	4.02×10^{-7}
Ethylenediamine- tetraacetic	HY_4	1.0×10^{-2}	2.1×10^{-2} $\quad K_4 = 5.5 \times 10^{-11}$	6.9×10^{-7}
Formic	$HCOOH$	1.77×10^{-4}		
Fumaric	$trans$-$HOOCCH{:}CHCOOH$	9.6×10^{-4}	4.1×10^{-5}	
Glycolic	$HOCH_2COOH$	1.48×10^{-4}		
Hydrazoic	HN_3	1.9×10^{-5}		
Hydrogen cyanide	HCN	2.1×10^{-9}		
Hydrogen fluoride	H_2F_2	7.2×10^{-4}		
Hydrogen peroxide	H_2O_2	2.7×10^{-12}		
Hypochlorous	$HOCl$	3.0×10^{-8}		
Iodic	HIO_3	1.7×10^{-1}		
Lactic	$CH_3CHOHCOOH$	1.37×10^{-4}		
Maelic	cis-$HOOCCH{:}CHCOOH$	1.20×10^{-2}	5.96×10^{-7}	
Malic	$HOOCCHOHCH_2COOH$	4.0×10^{-4}	8.9×10^{-6}	
Malonic	$HOOCCH_2COOH$	1.40×10^{-3}	2.01×10^{-6}	
Mandelic	$C_6H_5CHOHCOOH$	3.88×10^{-4}		
Nitrous	HNO_2	5.1×10^{-4}		
Oxalic	$HOOCCOOH$	5.36×10^{-2}	5.42×10^{-5}	
Periodic	H_5IO_6	2.4×10^{-2}	5.0×10^{-9}	
Phenol	C_6H_5OH	1.00×10^{-10}		

$$\log K_a = \log[(H^+)\,(A^-)/HA] \tag{1.21}$$

or

$$\log K_a = \log (H^+) + \log[A^-/HA] \tag{1.22}$$

or

$$-\log K_a = -\log (H^+) - \log[A^-/HA] \tag{1.23}$$

Considering that $-\log K_a = pK_a$, and $-\log (H^+) = pH$, Equation 1.23 can be rewritten as

$$pK_a = pH - \log[A^-/HA] \tag{1.24}$$

or

$$pH = pK_a + \log[A^-/HA] \tag{1.25}$$

or

$$pH = pK_a + \log[(\text{base form})/(\text{acid form})] \tag{1.26}$$

The last equation is the Henderson–Hasselbalch equation. Its use will be demonstrated in the sections that follow. The ratio of A^- to HA can be determined if the K_a and H^+ values are fixed. It is the H^+ concentration, therefore, that determines the net charge (ratio of charged to uncharged species). Conversely, the H^+ can be determined if the A^- to HA ratio is fixed. Table 1.18 lists various organic and inorganic acids of the monoprotic or polyprotic form along with their pK_a values. Note that in the equations above and throughout this chapter, it is assumed that concentration equals activity. These two terms are discussed in detail in Chapter 2.

1.6.3 Titration Curve

One of the techniques used to characterize an unknown base or acid is *titration*. Titration is when a strong base (e.g., sodium hydroxide, $NaOH \rightarrow Na^+ + OH^-$) is added to a weak acid (e.g., HA in water). For example,

$$HA + OH^- \rightarrow A^- + H_2O \tag{1.27}$$

The OH^- in Equation 1.27 reacts completely with HA to convert it to A^-. The data in Table 1.19 represent titration of 0.1 M HA with a pK_a of 5. Nearly every point in between the starting and end points have HA and A^- present so the Henderson–Hasselbalch equation can be used:

$$pH = 5 + \log[(A^-)/(HA)] \tag{1.28}$$

TABLE 1.19. Potentiometric Titration Data of a Monoprotic Acid with a pK_a of Approximately 5

	(OH⁻) Added	(HA) (M)	(A⁻) (M)	pH
Starting point	0	0.10	0	3
	0.02	0.08	0.02	4.4
	0.04	0.06	0.04	4.8
Midpoint	0.05	0.05	0.05	5
	0.06	0.04	0.06	5.2
	0.08	0.02	0.08	5.6
Endpoint	0.10	0	0.10	9.0

Note that at the midpoint, HA is equal to A⁻ and the pH is equal to the pK_a. Note also that at this pH (5), as much A⁻ was obtained as OH⁻ was added. At the point where 0.08 M OH⁻ was added (Table 1.19),

$$pH = 5 + \log[(0.08)/(0.02)] = 5.6 \qquad (1.29)$$

A titration curve can be plotted with such information. The line that passes through the points is characteristic of every weak acid. An example of an actual experimental

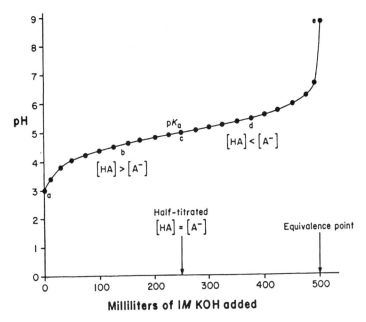

Figure 1.8. Potentiometric titration plot of a weak monoprotic acid with a pK_a of 4.8 (after Segel, 1976, with permission).

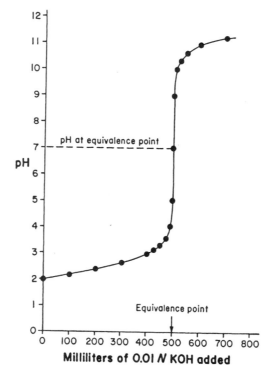

Figure 1.9. Potentiometric titration plot of a strong acid (after Segel, 1976, with permission).

titration plot of a weak acid with a pK_a of 4.8 is shown in Figure 1.8. The titration plot reveals three important points. The first point labeled *a* is the beginning of the titration and the compound is in the acid form (HA). When strong base is added, the acid form is consumed and the base form (A⁻) is formed. At point *c*, HA = A⁻ and pH = pK_a. Above point *c*, the base form (A⁻) predominates, and at point *e* all the acid form (HA) has converted to the base form (A⁻). Points *a* and *e* are also known as *equivalence points*.

Classical potentiometric titrations have a major weakness—their inability to characterize weak acids with a pK_a less than 3 and greater than 10. This is because water in those pH regions is capable of acting as weak acid or weak base, respectively (Fig. 1.9). In such cases colorimetric techniques are more suitable.

1.6.4 Environmental Water Buffers

Environmental water buffers are compounds that exhibit pK_a values near pH 7 (physiological pH). This is because a buffer with a pK_a of 7 has a strong tendency to maintain the physiological pH, hence, an environmental buffer. In addition to its unique pK_a (approximately 7), an environmental buffer must be nontoxic to the biological world. There are two such buffers in natural water systems. One is the phosphate buffer and the other is the bicarbonate buffer.

Phosphoric acid, as pointed out previously, exhibits three pK_a values, 2.23, 7.2, and 12.3, and its titration plot is shown in Figure 1.10. As expected, it shows three pK_a values and four equivalence points. The only pK_a that is of environmental importance is that at slightly above 7.2 (marked with an X). However, phosphate is not a desirable environmental buffer because of its eutrophication potential and its strong tendency to precipitate in natural water systems as metal–phosphate (where metal denotes any divalent or trivalent cations) (Stumm and Morgan, 1981). In most cases, its concentration in natural waters is less than 1 ppm.

Bicarbonate is much more suitable as a major environmental water pH buffer. Its parent acid is H_2CO_3, which is diprotic (Fig. 1.11). The dissociation of carbonic acid is described as follows:

$$H_2CO_{3aq} \Leftrightarrow H^+ + HCO_3^- \tag{1.30}$$

where the subscript aq denotes aqueous or dissolved. Reaction 1.30 exhibits a pK_a of 3.8. The second dissociation reaction of H_2CO_3 is

$$HCO_3^- \Leftrightarrow H^+ + CO_3^{2-} \tag{1.31}$$

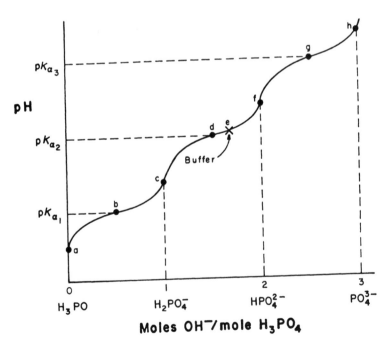

Figure 1.10. Potentiometric titration plot of a weak triprotic acid (after Segel, 1976, with permission).

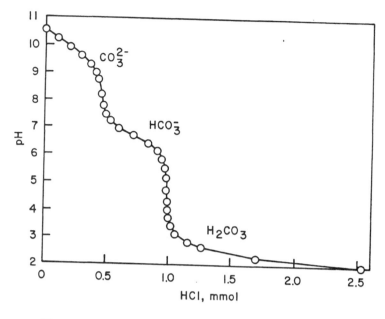

Figure 1.11. Potentiometric titration plot of a Na_2CO_3 solution.

with a pK_a of 10.3. These two pK_a values do not appear to be consistent with a good environmental water pH buffer system (pK_a values significantly different than physiological pH). However, two additional factors need to be considered. Note that

$$CO_{2aq} + H_2O \Leftrightarrow H_2CO_{3aq} \tag{1.32}$$

with a K_{eq} of 5×10^{-3}. Based on the magnitude of this K_{eq}, one would conclude that the equilibrium strongly favors the presence of CO_{2aq}. Since most of the compound is in the $CO_{2\,aq}$ form, the effective concentration of carbonic acid ($H_2CO_{3\,aq}$) is low and the pH is higher. By summing Equations 1.30 and 1.32, the overall reaction is

$$CO_{2aq} + H_2O \Leftrightarrow H^+ + HCO_3^- \tag{1.33}$$

The pK_a of Reaction 1.33 is 6.3, which is the sum of the pK_a values of Reactions 1.30 and 1.32. This sum of the two reactions produces a better environmental buffer, but is still not ideal if one considers ideal a pH buffer with a pK_a of 7.

Nature at times employs a clever mechanism in controlling the pH in natural water systems. It does so by controlling the partial pressure of CO_2 gas. The CO_{2aq}, a product of microbiological respiration, has a tendency to move toward equilibrium with the CO_2 gas in the atmosphere:

$$(CO_2)_{aq} \Leftrightarrow (CO_2)_{gas} \tag{1.34}$$

Thus, the CO_{2aq} in a natural water system is directly proportional to the pressure of the CO_2 gas in the atmosphere, that is,

$$(CO_2)_{aq} = k \cdot pCO_2 \tag{1.35}$$

where pCO_2 is the partial pressure of CO_2 in the atmosphere (0.0003 or 0.03%). The k value for Reaction 1.35, also known as Henry's constant, is 3×10^{-5}. Through the tendency of gases to move from a region of high partial pressure or concentration to a region of low partial pressure or concentration, the pCO_2 in water could be maintained above that of the atmosphere at some constant value by its rate of production in relationship to its rate of diffusion and/or mass flow, that is, it could escape to the atmosphere (perhaps because of a steady state sustained under certain conditions, such as temperature, pressure, mass flow, etc.). The HCO_3^- species, in biologically active natural water systems and in the absence of limestone, could be the direct product of a number of biotic processes. For example, during NO_3^- uptake by an organism, the latter (organism) would release HCO_3^- in order to sustain electroneutrality.

1.6.5 Open and Closed Systems

Consider a biologically active natural water system where pCO_2 and HCO_3^- are in control of pH. Upon adding a strong acid,

$$HCO_3^- + H^+ \Leftrightarrow (H_2CO_{3aq}) \Leftrightarrow CO_{2aq} + H_2O \tag{1.36}$$

bicarbonate is converted to carbon dioxide by the H^+. Assuming that the CO_2 is kept constant by letting off the extra CO_2 into the atmosphere, the system is referred to as an *open system*. In contrast, a *closed system* is one in which the CO_2 produced is not allowed to escape to the atmosphere, thus pCO_2 increases proportionally. The question that one needs to address is the pH outcome in an open or closed system. As an example, let us consider a system under the following conditions: A natural water containing $0.02\,M\,HCO_3^-$ with a pH 7.0 (pK_a of $H_2CO_3 = 6.3$). Upon adding $0.01\,M\,H^+$ (e.g., acid rain), what will the pH be under a closed system? What will it be under an open system? Because the acid will react with the HCO_3^- (base form), one first needs to know the concentration of the species involved. This can be determined using the Henderson–Hasselbalch equation:

$$pH = pK_{a1} + \log[(HCO_3^-)/(CO_{2aq})] \tag{1.37}$$

$$7.0 = 6.3 + \log[(HCO_3^-)/(CO_{2aq})] \tag{1.38}$$

$$0.7 = \log(HCO_3^-)/(CO_{2aq}) \tag{1.39}$$

Taking the antilog, 5:1 is the ratio of $(HCO_3^-):(CO_{2aq})$. This is reasonable because there should be more base form than acid form when the pH is considerably above the pK_a (pH 7.0 versus pK_a 6.3). The initial concentration of HCO_3^- and CO_{2aq} can now be calculated as follows:

$$HCO_3^- = [5/6] \times 0.02\ M = 0.017\ M \tag{1.40}$$

$$CO_{2aq} = [1/6] \times 0.02\ M = 0.0033\ M \tag{1.41}$$

The added H^+, in the form of acid rain, reacts with an equal amount of HCO_3^-:

$$0.01\ M\ H^+ + 0.01\ M\ HCO_3^- \rightarrow 0.01\ M\ CO_{2aq} \tag{1.42}$$

In the case of a closed system, the new HCO_3^- concentration could be calculated by subtracting from the estimated HCO_3^- concentration (Eq. 1.40) the concentration of HCO_3^- converted to CO_2 by reacting with the 0.01 M H^+ added (Eq. 1.42)

$$(HCO_3^-) = 0.017 - 0.01 = 0.007\ M \tag{1.43}$$

and the new CO_{2aq} concentration can be estimated by adding to the estimated CO_{2aq} (Eq. 1.41) the amount of HCO_3^- converted to CO_{2aq} by reacting with the 0.01 M H^+ added (Eq. 1.42)

$$CO_{2aq} = 0.0033\ M + 0.01\ M = 0.0133\ M \tag{1.44}$$

and

$$pH = 6.3 + \log[0.007/0.0133] = 6.02 \tag{1.45}$$

It is shown that in the case of a closed system, the pH decreased significantly (from 7.0 to 6.02), nearly one pH unit.

In the case of an open system, the new HCO_3^- concentration could also be calculated by subtracting from the estimated HCO_3^- concentration (Eq. 1.41) the concentration of HCO_3^- converted to CO_2 by reacting with the 0.01 M H^+ added (Eq. 1.42):

$$HCO_3^- = 0.017 - 0.01 = 0.007\ M \tag{1.46}$$

but since the system is treated as open, it is assumed that the CO_2 formed escapes to the atmosphere. Therefore,

$$pH = 6.3 + \log[0.007/0.0033] = 6.63 \tag{1.47}$$

This example shows that under an open system, even though approximately 60% of the HCO_3^- was consumed, it had a negligible effect on water pH (initial pH of 7.0 versus final pH of 6.63).

ACID–BASE CHEMISTRY PROBLEMS

All acid–base problems involving aqueous solutions of weak acids (HA) and/or their corresponding base forms (A^-) fall into three distinct types: type 1, Solutions of acid form only (e.g., HA); type 2, solutions of base form only (e.g., A^-); type 3, solutions of both acid and base forms (e.g., HA and A^-). For all three types of problems, two equilibrium conditions must be satisfied:

$$HA \Leftrightarrow H^+ + A^- \tag{A}$$

with

$$K_a = (H^+)(A^-)/(HA) \tag{B}$$

and

$$H_2O \Leftrightarrow H^+ + OH^- \tag{C}$$

with

$$K_w = (H^+)(OH^-) \tag{D}$$

In solving any equilibria problems involving acids and bases, two terms must always be considered, *charge–balance* and *mass–balance*. To understand the use of these two terms, two examples are given below.

Consider the addition of a small amount of a strong acid (e.g., 10^{-8} mol L^{-1} HCl) to deionized water. Since hydrochloric acid is a strong acid, it completely dissociates to equamolar concentrations of H^+ and Cl^-. The charge–balance term is

$$H^+ = Cl^- + OH^- \tag{E}$$

and the mass–balance term is

$$Cl^- = HCl \tag{F}$$

Substituting the mass–balance term (Eq. F) into the charge–balance term (Eq. E) gives

$$H^+ = HCl + OH^- \tag{G}$$

Equation G reveals that the concentration of H^+ equals the concentration of HCl plus OH^- contributed by H_2O. Therefore,

$$H^+ = HCl + K_w/H^+ \tag{H}$$

Multiplying both sides by H^+ and rearranging gives

$$(H^+)^2 - HCl(H^+) - K_w \tag{I}$$

Equation I is quadratic and can be solved by the quadratic formula

$$H^+ = [-a^2 \pm (b^2 - 4ac)^{1/2}]/2a \tag{J}$$

where $a = 1$, $b = 10^{-8}$, and $c = 10^{-14}$. Introducing these values into Equation J and solving for H^+ gives $10^{-6.9}$. Since pH is the negative logarithm of H^+, pH = 6.9. Note that equation G is applicable when HCl is less than 100 times greater than OH^-. If HCl is more than 100 times greater than OH^-,

$$H^+ = HCl \tag{K}$$

and pH is the negative logarithm of HCl.

Equilibria problems involving weak acids or weak bases also include charge–balance and mass–balance terms. For example, consider Reaction A. The charge–balance term is

$$H^+ = A^- + OH^- \tag{L}$$

and the mass–balance term is

$$A = HA \tag{M}$$

Note that A is not equal A^- because HA is a weak acid and only a fraction of its initial quantity dissociates. The charge–balance equation (Eq. L) reveals that when OH^- (contributed by water) is less than 10% of A^-, H^+ equals A^-. However, when OH^- is greater than 10% of A^-, the solution is as follows:
  Considering that

$$K_{eq} = (H^+)(A^-)/(HA) \tag{N}$$

and

$$A^- = K_{eq}(HA)/(H^+) \tag{O}$$

where HA is the concentration at equilibrium (this is the original quantity of HA added per liter minus the amount per liter that dissociated). Substituting Equation O and K_w/H^+ into the charge–balance equation (Eq. L) gives

$$H^+ = K_{eq}(HA)/(H^+) + K_w/H^+ \tag{P}$$

Multiplying both sides of Equation P by H^+ and rearranging gives

$$(H^+) = [K_{eq}(HA) + K_w]^{1/2} \tag{Q_a}$$

When $K_{eq}(HA)$ is at least 100 times greater than K_w, Equation Q_a becomes

$$H^+ = [K_{eq}(HA)]^{1/2} \tag{Q_b}$$

Note that when HA initial (HA_{init}) is at least 100 times greater than equilibrium HA, (HA_e), then HA in Equation Q is approximately HA_{init}. Some examples of how to approximate HA at equilibrium are given below.

Examples of Monoprotic Acid–Base Chemistry

Type 1. What is the pH of a 0.01 M solution of a weak acid (HA) with a pK_a of 5.0? This is a type 1 problem because initially only the acid form (HA) is present. Hydrogen ions (H^+) arise from the dissociation of both HA and H_2O. Water is a very weak acid, and most of the H^+ will come from the dissociation of HA; thus, the H_2O dissociation can be ignored (see Eq. Q_b). Each mole of HA that dissociates gives 1 mol of H^+ and 1 mol of A^-. The HA dissociates until equilibrium is met.

	HA	H^+	A^-
Initial conditions	0.01 M	~0	~0
Change due to dissociation	$-X$	$+X$	$+X$
Concentration at equilibrium	$0.01-X$	X	X

Since $H^+ = A^- = X$

$$K_a = \frac{(H^+)(A^-)}{(HA)} = \frac{(H^+)^2}{(0.01 - H^+)} = 10^{-5} \tag{R}$$

Equation R is quadratic and can be solved for H^+ using the quadratic equation (Eq. J). If, however, H^+ is much less than (HA) initial, H^+ in the denominator can be neglected and the equation is simplified. For most situations, H^+ in the denominator can be ignored whenever it is less than 10% of HA; H^+ in the denominator will turn out to be less than 10% of HA whenever $(HA)_{init}$ is at least 100 times greater than K_a. For the problem given above $(HA) = (1000 \times K_a)$; thus, H^+ in the denominator can be ignored. Solving for X,

$$K_a = (H^+)^2/0.01 = 10^{-5} \tag{S}$$

$$H^+ = 10^{-3.5} \tag{T}$$

and

$$pH = 3.5 \tag{U}$$

The degree of dissociation (a) is the fraction of HA that dissociates:

$$a = \frac{A^-}{(HA)_{init}} = \frac{H^+}{(HA)_{init}} = \frac{10^{-3.5}}{10^{-2}} = 0.032 \qquad (V)$$

Thus, HA is 3.2% dissociated. The degree of dissociation depends on the concentration of HA_{init}. As the concentration of HA_{init} is decreased, a larger fraction of it must dissociate to satisfy the equilibrium conditions.

Type 2. What is the pH of a 0.01 M solution of NaA (pK_a of HA = 5.0)? The salt NaA completely dissociates in H_2O to give Na^+ and A^-. This is a type 2 problem because only the base form (A^-) is initially present. HA is a weak acid, the A^- will tend to combine with any available H^+ to form HA. The only H^+ available, however, comes from H_2O dissociation, and H_2O is such a weak acid that only a limited amount of H^+ will become available. The two reactions below will proceed simultaneously until equilibrium is obtained.

$$H_2O \Leftrightarrow H^+ + OH^- \qquad (W)$$

with $K_{eq} = K_w = 10^{-14}$, and

$$H^+ + A^- \Leftrightarrow HA \qquad (X)$$

with $K_{eq} = 1/K_a = 10^5$. The overall reaction is the sum of the two (W and X):

$$A^- + H_2O \Leftrightarrow HA + OH^- \qquad (Y)$$

and

$$K_{eq} = \frac{(HA)(OH^-)}{(A^-)} = \frac{1}{K_a} \cdot K_w \qquad (Z)$$

or

$$K_{eq} = (K_w)(1/K_a) = 10^{-9} \qquad (A')$$

Note that the overall K_{eq} is the product of the K_{eq} values for each reaction. This problem can now be solved in the same manner as a type 1 problem.

	A^-	HA	OH–
Initial concentration	0.01 M	~0	~0
Change due to dissociation	–X	+X	+X
Concentration at equilibrium	0.01–X	X	X

Since HA = OH^- = X,

$$K_{eq} = \frac{(HA)(OH^-)}{(A^-)} = \frac{(OH^-)^2}{0.01 - OH^-} = 10^{-9} \qquad (B')$$

Ignoring OH^- in the denominator because 0.01 is at least 100 times greater than 10^{-9} (see justification for ignoring H^+ in the denominator of Eq. R),

$$(OH^-)^2 = (0.01)(10^{-9}) = 10^{-11} = 10 \times 10^{-12} \qquad (C')$$

$$OH^- = 3.2 \times 10^{-6} \qquad (D')$$

since

$$H^+ = K_w/OH^- = \frac{10^{-14}}{3.2 \times 10^{-6}} = 3.2 \times 10^{-9} \qquad (E')$$

$$pH = -\log(3.2 \times 10^{-9}) = 8.5 \qquad (F')$$

This net reaction (Reaction Y) is often referred to as hydrolysis and A^- in this case is acting as a base because it "dissociates" to give OH^-. The equilibrium constant (K_{eq} for Equation B') is then referred to as K_b. Thus,

$$K_{eq} = K_w/K_a = K_b \qquad (G')$$

Therefore, for a weak acid or its corresponding base, you only need to know K_a in order to solve either an acid or a base problem. There is never any need to know K_b.

Type 3. What is the pH of a solution containing 0.02 M HA and 0.01 M A^-? The pK_a value for HA is 5.0. Since both the acid form (HA) and base form of (A^-) are present, this is a type 3 problem. The easiest way to solve these problems is to treat them formally as a type 1 problem in which the initial concentration of A^- is no longer zero.

	HA	H^+	A^-
Initial concentration	0.02 M	~0	~0
Change due to dissociation	$-X$	$+X$	$+X$
Concentration at equilibrium	$0.02 - X$	X	$0.01 + X$

Since $H^+ = A^- = X$,

$$K_a = \frac{(H^+)\,(A^-)}{(HA)} = \frac{(H^+)\,(0.01 + H^+)}{(0.02 - H^+)} = 10^{-5} \qquad \text{(H')}$$

It is usually possible to ignore H^+ in both the denominator and in the (A^-) term of the numerator since 0.02 is at least 100 times greater than 10^{-5}. Thus,

$$(H^+)(0.01)/(0.02) = 10^{-5} \qquad \text{(I')}$$

$$H^+ = (10^{-5})0.02/0.01 = 2 \times 10^{-5} \qquad \text{(J')}$$

and

$$pH = -\log(2 \times 10^{-5}) = 4.7 \qquad \text{(K')}$$

Type 3 problems reduce to a very simple form because the value of H^+ depends only on K_a and the initial ratio of A^-/HA. Thus, unlike type 1 and type 2 problems, the value of H^+ does not depend on the actual concentrations of A^- or HA (provided both A^- and HA are large enough so that the H^+ can be ignored). Since this type of problem is the one most frequently encountered, the Henderson–Hasselbalch equation is commonly employed:

$$pH = pK_a + \log[A^-/HA] \qquad \text{(L')}$$

or

$$pH = pK_a + \log[(\text{base form})/(\text{acid form})] \qquad \text{(M')}$$

Examples of Diprotic Acid–Base Chemistry

Diprotic acids can be treated as type 1, 2, or 3 problems. One additional feature of diprotic acids (the midpoint) requires special treatment (type 4). Consider a diprotic acid H_2A,

$$H_2A \Leftrightarrow H^+ + HA^- \qquad K_{a1};\ pK_{a1} \qquad \text{(N')}$$

$$HA^- \Leftrightarrow H^+ + A^{2-} \qquad K_{a2};\ pK_{a2} \qquad \text{(O')}$$

In the reactions above, the following types of problems are possible: H_2A only—type 1; $H_2A + HA^-$—type 3 (use pK_{a1}); HA^- only—type 4; $HA^- + A^{2-}$—type 3 (use pK_{a2}); A^{2-} only—type 2. HA^- only is a special case because it is a weak acid and can undergo dissociation type 1, and it is also the base form of H_2A and can therefore undergo dissociation type 2.

One way to look at this problem is to consider that A^{2-} and H_2A will be equal during addition of HA^- to water at concentration levels where H_2O is an insignificant contributor of H^+ or OH^-. The pH can then be calculated as follows: Considering the Henderson–Hasselbalch equations,

$$pH = pK_{a1} + \log \left\{ \frac{(HA^-)}{H_2A} \right\} \qquad (P')$$

and

$$pH = pK_{a2} + \log \left\{ \frac{(A^{2-})}{(HA^-)} \right\} \qquad (Q')$$

Summing Reactions P' and Q' produces

$$2\,pH = pK_{a1} + pK_{a2} + \log \left\{ \frac{(HA^-)\,(A^{2-})}{(H_2A)\,(HA^-)} \right\} \qquad (R')$$

and cancelling HA^- gives

$$2\ pH = pK_{a1} + pK_{a2} + \log \left\{ \frac{(A^{2-})}{(H_2A)} \right\} \qquad (S')$$

However, since (A^{2-}) is very small and approximately equal to H_2A, the last term is zero and Equation S' becomes

$$pH \simeq [pK_{a1} + pK_{a2}]/2 \qquad (T')$$

In the case of a diaprotic acid (e.g., H_2A), where the concentration of H_2A is not equal to A^{2-}, pH can be calculated using mass- and charge-balance equations (Skoog and West, 1976). For example, upon introducing NaHA to water, the mass–balance equation is

$$NaHA = (HA^-) + (H_2A) + (A^{2-}) \qquad (U')$$

and the charge-balance equation is

$$(Na^+) + (H^+) = (HA^-) + 2(A^{2-}) + (OH^-) \qquad (V')$$

Setting Equations U' and V' equal and rearranging gives

$$(H_2A) = (A^{2-}) + (OH^-)-(H^+) \qquad (W')$$

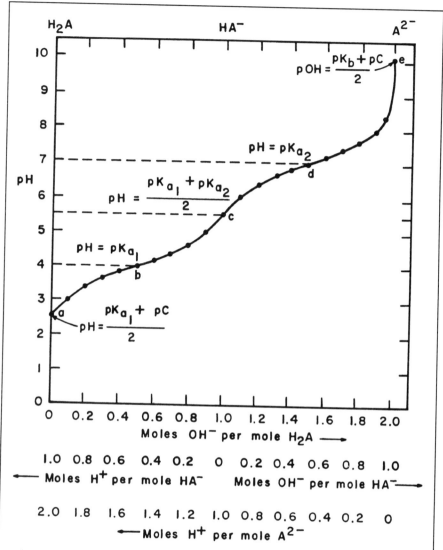

Figure 1A. Potentiometric titration plot of a weak diprotic acid with pK_a values of 4.0 and 7 (after Segel, 1976, with permission).

Substituting the appropriate equilibria constants into Equation W' and rearranging gives

$$(H^+) = [K_{a1}K_{a2}(NaHA) + K_{a1}K_w]/[(NaHA) + K_{a1}] \qquad (X')$$

where $K_{a1} = (HA^-)(H^+)/(H_2A)$, $K_{a2} = (A^{2-})(H^+)/(HA^-)$, and $K_w = (H^+)(OH^-)$. In deriving Equation X', it is assumed that $HA^- \cong NaHA$. Also, when $K_{a1} <<< NaHA$, Equation X' reduces to T' and

$$H^+ = [K_{a1}K_{a2}]^{1/2} \qquad\qquad\qquad (Y')$$

or

$$pH = [pK_{a1} + pK_{a2}]/2 \qquad\qquad\qquad (Z')$$

Note that Equation Z' is similar to Equation T'.

The discussion above, dealing with a diprotic acid (H_2A), is summarized in Figure 1A. The figure reveals three equivalence points (noted as a, c, and e) and includes the equations for estimating pH at these points. Furthermore, the figure shows two pK_a values (identified as b and d) and provides the equations for estimating pH at these points.

PROBLEMS AND QUESTIONS

1. Convert the following chemical units to milliequivalents per liter (meq L^{-1}) and milligrams per liter (mg L^{-1}) or parts per million (ppm).

   a. A water sample containing 10 mmol L^{-1} sodium (Na):
      i. meq L^{-1}
      ii. mg L^{-1}

   b. A water sample containing 10 mmol L^{-1} calcium (Ca):
      i. meq L^{-1}
      ii. mg L^{-1}

   c. A water sample containing 10 mmol L^{-1} chloride (Cl):
      i. meq L^{-1}
      ii. mg L^{-1}

   d. A water sample containing 10 mmol L^{-1} sulfate (SO_4):
      i. meq L^{-1}
      ii. mg L^{-1}

2. Calculate grams (or milligrams) of NaCl that you would need to dissolve in 1 L of water to make a solution of 200 meq L^{-1} Na. Convert the dissolved Na to mg L^{-1}.

   a. mg L^{-1} NaCl
   b. mg L^{-1} Na

3. Calculate grams (or milligrams) of $CaCl_2 \cdot 2H_2O$ that you would need to dissolve in 1 L of water to make a solution of 200 meq L^{-1} Ca. Convert the dissolved Ca to mg L^{-1}.

   a. mg L^{-1} $CaCl_2 \cdot 2H_2O$
   b. mg L^{-1} Ca

4. Calculate grams (or milligrams) of $CaCl_2 2H_2O$ that you would need to dissolve in 1 L of water to make a solution of 200 meq L^{-1} Cl. Convert the dissolved Cl to mg L^{-1}.

   a. mg L^{-1} $CaCl_2 2H_2O$

   b. mg L^{-1} Cl

5. Use the gravimetric formula to answer the following question. NO_3–N (note, NO_3–N means N in the NO_3 form) in drinking water is regulated by the federal government at 10 ppm N. A given water sample is known to contain 50 ppm NO_3. Calculate its nitrogen content in units of ppm N.

6. Calculate the total dissolved solids in water assuming all the following metals are associated with chloride:

   Ca = 75 ppm

   Mg = 120 ppm

   Na = 50 ppm

   K = 39 ppm

   Fe = 80 ppm

7. Would you rather have a high covalent character bond between a pollutant and a soil mineral or a high ionic character bond? Why?

8. Consider a water sample with 10 mmol L^{-1} hydrochloric acid (HCl). Calculate the following:

   Cl = ppm

   H^+ = mol L^{-1}

   pH =

   H^+ = ppm

9. Calculate the pH of a water sample containing 10^{-8} mol L^{-1} nitric acid (HNO_3). Nitric acid is one of the acid forms in rain.

10. A water sample containing 10 mmol L^{-1} hydrochloric acid (HCl) requires neutralization. Calculate the amount of $Ca(OH)_2$ (in grams) needed to neutralize 1 L of this water.

11. A water sample containing 10 mmol L^{-1} hydrochloric acid (HCl) requires neutralization. Calculate the amount of $CaCO_3$ (in grams) needed to neutralize 1 L of this water.

12. Consider

$$H_2CO_3 \Leftrightarrow HCO_3^- + H^+ \qquad K_a = 10^{-6.4}$$

and

$$HCO_3^- \Leftrightarrow CO_3^{2-} + H^+ \qquad K_a = 10^{-10.2}$$

Consider a water sample with 2 mmol L^{-1} carbonic acid (H_2CO_3). Calculate its pH.

13. Calculate the pH of a water sample containing 10^{-3} mol L^{-1} sodium bicarbonate ($NaHCO_3$) (baking soda).

14. Consider a water sample containing 10 mmol L^{-1} $NaHCO_3$ and 5 mmol L^{-1} H_2CO_3; calculate its pH.

15. Calculate the pH of a water sample made by mixing equal volumes of 10 mmol L^{-1} disodium carbonate (Na_2CO_3) and 10 mmol L^{-1} $NaHCO_3$ (baking soda).

2 Solution/Mineral–Salt Chemistry

2.1 INTRODUCTION

Environmental soil–water scientists are interested in acquiring the knowledge necessary for predicting soil–water system processes. All such processes are products of
biophysical–chemical reactions. This chapter covers mineral solubility, a process
related to ion availability to soil microbes and higher plants, ion release to ground or
surface water, mineral precipitation, soil mineral weathering, and/or soil formation.
The chapter deals specifically with chemical equilibria and its purpose is to provide
students with the tools necessary for quantifying such reactions.

   Two parameters in soil–water chemistry must be understood in order to make
reaction predictions. One parameter is *concentration*, which refers to the total dissolved quantity of a given element; the other is *single-ion activity*, which describes the
thermodynamic behavior of a particular chemical species in solution. Single-ion
activity (α_j) is estimated by

$$\alpha_j = \gamma_j m_j \tag{2.1}$$

where γ_j is the single-ion activity coefficient of ionic species j and m_j is the molar
concentration of species j. The single-ion activity coefficient is estimated by the
equations shown below. Equation 2.2 describes ionic strength (I):

$$I = 1/2 \sum_{j=1}^{n} Z_j^2 \, m_j \tag{2.2}$$

where Z_j and m_j are the charge and molar concentration of the ionic species (Sposito,
1981c; 1984b). The single-ion activity coefficient γ_j is then calculated by the Debye–
Huckle equation:

$$\log \gamma_j = -A Z_j^2 \, [I^{1/2}/(1 + B \text{å} I^{1/2})] \tag{2.3a}$$

where A is a constant which at 25°C equals 0.512; $B = \simeq 0.33$ in water at 25°C; å =
adjustable parameter corresponding to the size of the ion. There are several equations
for estimating single-ion activity coefficients. These equations and some of their
limitations are given in Table 2.1. Tabulated single-ion activity coefficient values using
the extended Debye–Huckle equation with the appropriate constants (Tables 2.1 and
2.2) are given in Table 2.3.

TABLE 2.1. Equations for Estimating Single-Ion Activity Coefficients

Type of Equation	Equation[a]	Approximate Ionic Strength (I) Range
Debye–Huckle	$\log \gamma = - AZ^2(I)^{1/2}$	$<10^{-2.3}$
Extended Debye–Huckle	$\log \gamma = - AZ^2/[(I)^{1/2}/1 + B\mathring{a}(I)^{1/2}]$	10^{-1}
Guntelberg	$\log \gamma = - AZ^2/[(I)^{1/2}/1 + (I)^{1/2}]$	10^{-1} in mixed electrolyte systems
Davies	$\log \gamma = - AZ^2 [I^{1/2} / (1 + I^{1/2}) - 0.3I]$	<0.5

Source: Stumm and Morgan, 1970.

[a] $A = 0.512$ for water at 25°C; $B = = 0.33$ in water at 25°C; $\mathring{a}$ = adjustable parameter corresponding to the size of the ion.

TABLE 2.2. Values of the Parameter $\mathring{a}$ Used in the Extended Debye–Huckel Equation (Eq. 2.3a) for Selected Ions

$\mathring{a}$	Ions and Charge
	Charge 1
9	H^+
6	Li^+
4	Na^+, $CdCl^+$, ClO_2^-, IO_3^-, HCO_3^-, $H_2PO_4^-$, HSO_3^-, $H_2AsO_4^-$, $Co(NH_3)_4(NO_2)_2^+$
3	OH^-, F^-, CNS^-, CNO^-, HS^-, ClO_3^-, ClO_4^-, BrO_3^-, IO_4^-, MnO_4^-, K^+, Cl^-, Br^-, I^-, CN^-, NO_2^-, NO_3^-, Rb^+, Cs^+, NH_4^+, Tl^+, Ag^+
	Charge 2
8	Mg^{2+}, Be^{2+}
6	Ca^{2+}, Cu^{2+}, Zn^{2+}, Sn^{2+}, Mn^{2+}, Fe^{2+}, Ni^{2+}, Co^{2+}
5	Sr^{2+}, Ba^{2+}, Ra^{2+}, Cd^{2+}, Hg^{2+}, S^{2-}, $S_2O_4^{2-}$, WO_4^{2-}, Pb^{2+}, CO_3^{2-}, SO_3^{2-}, MoO_4^{2-}, $Co(NH_3)_5Cl^{2+}$, $Fe(CN)_5NO^{2-}$
4	Hg_2^{2+}, SO_4^{2-}, $S_2O_3^{2-}$, $S_2O_8^{2-}$, SeO_4^{2-}, CrO_4^{2-}, HPO_4^{2-}, $S_2O_6^{2-}$
	Charge 3
9	Al^{3+}, Fe^{3+}, Cr^{3+}, Se^{3+}, Y^{3+}, La^{3+}, In^{3+}, Ce^{3+}, Pr^{3+}, Nd^{3+}, Sm^{3+}
4	PO_4^{3-}, $Fe(CN)_6^{3-}$, $Cr(NH_3)_6^{3+}$, $Co(NH_3)_5H_2O^{3+}$
	Charge 4
11	Th^{4+}, Zr^{4+}, Ce^{4+}, Sn^{4+}
6	$Co(S_2O_3)(CN)_5^{4-}$

Source: Novozamsky et al., 1976.

TABLE 2.3. Single-Ion Activity Coefficients Calculated from the Extended Debye–Huckel Equation at 25°C

å	0.001	0.0025	0.005	0.01	0.025	0.05	0.1
				Ionic Strength			
				Charge 1			
9	0.967	0.950	0.933	0.914	0.88	0.86	0.83
8	0.966	0.949	0.931	0.912	0.88	0.85	0.82
7	0.965	0.948	0.930	0.909	0.875	0.845	0.81
6	0.965	0.948	0.929	0.907	0.87	0.835	0.80
5	0.964	0.947	0.928	0.904	0.865	0.83	0.79
4	0.964	0.947	0.927	0.901	0.855	0.815	0.77
3	0.964	0.945	0.925	0.899	0.85	0.805	0.755
				Charge 2			
8	0.872	0.813	0.755	0.69	0.595	0.52	0.45
7	0.872	0.812	0.753	0.685	0.58	0.50	0.425
6	0.870	0.809	0.749	0.675	0.57	0.485	0.405
5	0.868	0.805	0.744	0.67	0.555	0.465	0.38
4	0.867	0.803	0.740	0.660	0.545	0.445	0.355
				Charge 3			
9	0.738	0.632	0.54	0.445	0.325	0.245	0.18
6	0.731	0.620	0.52	0.415	0.28	0.195	0.13
5	0.728	0.616	0.51	0.405	0.27	0.18	0.115
4	0.725	0.612	0.505	0.395	0.25	0.16	0.095
				Charge 4			
11	0.588	0.455	0.35	0.255	0.155	0.10	0.065
6	0.575	0.43	0.315	0.21	0.105	0.055	0.027
5	0.57	0.425	0.31	0.20	0.10	0.048	0.021

Source: Novozamsky et al., 1976.

Generally, for simple but strong (dissociable) electrolyte solutions (e.g., NaCl, KCl, $CaCl_2$, and $MgCl_2$) single-ion activity coefficients can be approximated by the equations given in Table 2.1. Choose the correct one by following the restrictions given in the table and using as m_j in Equation 2.2, the total molar concentration of the particular elements in solution, assuming they exist as fully dissociated species. In the case of complex mixed electrolyte solutions, however, the most appropriate way to estimate single-ion activity coefficients is to employ the Davies equation (Davies, 1962; Sposito, 1981c; 1984b):

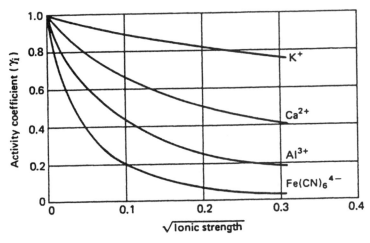

Figure 2.1. Relationship between solution ionic strength and single-ion activity coefficients of ions with different valencies. Calculated utilizing the extended Debye–Huckle equation) (from Skoog and West, 1976, with permission).

$$\log \gamma_j = -A Z_j^2 \left[I^{1/2}/(1 + I^{1/2}) - 0.3I \right] \tag{2.3b}$$

where I is also calculated using Equation 2.2, but m_j denotes *effective molar concentration*. The latter necessitates estimation of ion pairs and complexes (see Section 2.1.3). This requires the simultaneous solution of Equations 2.1, 2.2, and 2.3b, along with the equations predicting the molar concentration of ionic species (see the section entitled Iteration Example in this chapter) by ion-association models (e.g., GEO-CHEM, Sposito and Mattigod, 1979; Parker et al., 1995; MINTEQ, Allison et al., 1991; SOILCHEM, Sposito and Coves, 1988). Note that direct measurement of single-ion activity in mixed electrolyte systems cannot be made. Therefore, the validity of ion-association models can only be demonstrated indirectly, for example, by experimentally quantifying the precipitation of a given metal under a given set of conditions (i.e., pH, ionic strength, and solution ionic composition) and then comparing these values to those predicted by ion-association models.

The importance of single-ion activity in predicting the chemical behavior of a particular ionic species is demonstrated in Figure 2.1. For example, the γ value of any divalent or monovalent ions at the highest possible ionic strength (I) causes 60% suppression in the activity of the divalent ion and 25% reduction in the activity of the monovalent ion. This implies that as γ_j decreases, the apparent solubility of any given mineral increases, as demonstrated later in this chapter.

2.1.1 Mineral Solubility

Mineral solubility is dependent on the type of ions composing a particular mineral. The parameter used to predict mineral solubility is the solubility product constant or

K_{sp}. The use of K_{sp} in estimating mineral solubility in units of single-ion activity (mol L^{-1}) is demonstrated below. Consider the solid silver chloride (AgCls) which, when introduced to water, undergoes dissociation as shown below:

$$AgCls \Leftrightarrow Ag^+ + Cl^- \qquad (2.4)$$

Equation 2.4 shows that AgCls undergoes dissociation to produce an equamolar concentration of Ag^+ and Cl^-. The K_{sp} of AgCls can be expressed as

$$(Ag^+)(Cl^-) = 1.82 \times 10^{-10} = K_{sp} \qquad (2.5)$$

where the parenthesis denote single-ion activity. Setting

$$Ag^+ = X \qquad (2.6)$$

since

$$Ag^+ = Cl^- \qquad (2.7)$$

then

$$Cl^- = X \qquad (2.8)$$

Substituting Equations 2.6 and 2.8 into Equation 2.5:

$$(X)(X) = 1.82 \times 10^{-10} = K_{sp} \qquad (2.9)$$

or

$$(X)^2 = 1.82 \times 10^{-10} = K_{sp} \qquad (2.10)$$

and

$$X = (1.82 \times 10^{-10})^{1/2} = 1.35 \times 10^{-5} \text{ mol } L^{-1} \qquad (2.11)$$

According to Equations 2.6 and 2.8, $Ag^+ = 1.35 \times 10^{-5}$ mol L^{-1} and $Cl^- = 1.35 \times 10^{-5}$ mol L^{-1}. Note, the parenthesis around the units for Ag^+ or Cl^- denote single-ion activity. To convert single-ion activity to concentration, divide single-ion activity by the corresponding single-ion activity coefficient (see Eqs. 2.1 and 2.3).

In the case of AgCls, the two ions involved (Ag^+ and Cl^-) exhibit the same valence. A mineral solubility example is presented below where the cation is monovalent but the anion is divalent. Consider the solid silver sulfide Ag_2Ss,

$$Ag_2Ss \Leftrightarrow 2Ag^+ + S^{2-} \qquad (2.12)$$

Equation 2.12 shows that when 1 mol of Ag_2Ss dissociates, it gives 2 mol of Ag^+ and 1 mol of S^{2-}. The K_{sp} of Ag_2S is given by

$$(Ag^+)^2 (S^{2-}) = K_{sp} \qquad (2.13)$$

Setting

$$(S^{2-}) = X \qquad\qquad (2.14)$$

since 1 mol of Ag_2Ss produces 2 mol Ag^+ and 1 mol of S^{2-}, then

$$(Ag^+) = 2X \qquad\qquad (2.15)$$

Substituting Equations 2.14 and 2.15 into Equation 2.13 gives

$$(2X)^2 (X) = K_{sp} \qquad\qquad (2.16)$$

or

$$4(X)^3 = K_{sp} \qquad\qquad (2.17)$$

and

$$X = (K_{sp}/4)^{1/3} = (6.0 \times 10^{-50}/4)^{1/3} = 2.47 \times 10^{-7} \text{ mol L}^{-1} \qquad (2.18)$$

Solving for X (by substituting the K_{sp} of Ag_2Ss) and considering Equations 2.14 and 2.15 gives $S^{2-} = 2.47 \times 10^{-7}$ and $Ag^+ = 4.94 \times 10^{-7}$ mol L^{-1} activity. To obtain concentration, single-ion activity needs to be divided by the corresponding single-ion activity coefficient (see Eqs. 2.1 and 2.3).

When one of the ions of a particular mineral is trivalent and the other is monovalent, such as ferric hydroxide solid $[Fe(OH)_3s]$, its solubility can be expressed as follows:

$$Fe(OH)s \Leftrightarrow Fe^{3+} + 3OH^- \qquad\qquad (2.19)$$

$$(Fe^{3+})(OH^-)^3 = K_{sp} \qquad\qquad (2.20)$$

Setting

$$(Fe^{3+}) = X \qquad\qquad (2.21)$$

since 1 mol $Fe(OH)_3s$ produces 3 mol OH^- and 1 mol Fe^{3+}, then

$$(OH^-) = 3X \qquad\qquad (2.22)$$

Substituting Equations 2.21 and 2.22 into Equation 2.20,

$$(X) (3X)^3 = K_{sp} \qquad\qquad (2.23)$$

or

$$27(X)^4 = K_{sp} \qquad\qquad (2.24)$$

and

$$X = (K_{sp}/27)^{1/4} = (4 \times 10^{-38}/27)^{1/4} = 5.92 \times 10^{-39} \tag{2.25}$$

Solving for X and considering Equations 2.21 and 2.22 gives $OH^- = 1.78 \times 10^{-38}$ and $Fe^{3+} = 5.93 \times 10^{-39}$ in units of moles per liter of single-ion activity. Estimating dissolved concentrations requires the use of single-ion activity coefficients (see Eqs. 2.1 and 2.3). If the answers obtained by Equation 2.25 represent a good approximation of the mineral's solubility, the charge-balance equation, shown below, should be satisfied:

$$3(Fe^{3+}) + H^+ = OH^- \tag{2.26}$$

By substituting the estimated single-ion activity values into Equation 2.26,

$$1.78 \times 10^{-38} + (10^{-14}/1.78 \times 10^{-38}) \neq 1.78 \times 10^{-38} \tag{2.27}$$

or

$$1.78 \times 10^{-38} + 5.62 \times 10^{23} \neq 1.78 \times 10^{-38} \tag{2.28}$$

where $10^{-14} = K_w$ and $K_w/(OH^-) = (H^+)$. Therefore, the answer obtained by Equation 2.25 does not satisfy Equation 2.26 [e.g., H^+ is too large (Eq. 2.28)], and the solution to the problem is therefore incorrect. A correct approach to solve this problem is to assume that $H^+ = OH^-$ and the activity of Fe^{3+} can be calculated by rearranging Equation 2.20 and substituting values for K_{sp} and OH^- activity at pH 7. Therefore,

$$(Fe^{3+}) = K_{sp}/(10^{-7})^3 = 4 \times 10^{-38}/10^{-21} = 4 \times 10^{-17} \tag{2.29}$$

and by substituting the newly estimated concentration values into Equation 2.26, we obtain

$$3(4 \times 10^{-17}) + 10^{-7} \cong 10^{-7} \tag{2.30}$$

This equation shows that the solution of Equation 2.29 appears to satisfy Equation 2.26. Hence, in a natural system, $Fe(OH)_3s$ is not soluble enough to have any adverse effects on the pH or concentration of Fe^{3+}. Note that environmental concerns are raised when total iron concentration is greater than 10^{-5} mol L^{-1}.

2.1.2 Single-Ion Activity Coefficient

Mineral solubility in soil–water systems varies and depends on conditions controlling single-ion activity coefficients. Gypsum ($CaSO_4 \cdot 2H_2Os$), a common natural mineral, is used here as an example to demonstrate the influence of single-ion activity coefficients on mineral solubility. The solubility of $CaSO_4 \cdot 2H_2Os$ in water is expressed as

$$CaSO_4 \cdot 2H_2Os \Leftrightarrow Ca^{2+} + SO_4^{2-} + 2H_2O \tag{2.31}$$

Reaction 2.31 at equilibrium can be described by

$$K_{sp} = (Ca^{2+})(SO_4^{2-})(H_2O)^2/(CaSO_4 \cdot 2H_2Os) \tag{2.32}$$

Considering that at the standard state (25°C and 1 atmo pressure) the activity of water (H_2O) and the activity of gypsum ($CaSO_4 \cdot 2H_2Os$) are set by convention to 1, the solubility of $CaSO_4 \cdot 2H_2Os$ is expressed by

$$K_{sp} = [Ca^{2+}]\gamma_{Ca^{2+}}[SO_4^{2-}]\gamma_{SO_4^{2-}} = 2.45 \times 10^{-5} \tag{2.33}$$

where the brackets denote dissociated concentration. Substituting the K_{sp} of gypsum and rearranging

$$[Ca^{2+}]\gamma_{Ca^{2+}} = [SO_4^{2-}]\gamma_{SO_4^{2-}} = (2.45 \times 10^{-5})^{1/2} = 4.95 \times 10^{-3} \text{ mol L}^{-1} \tag{2.34}$$

The single-ion activity coefficient (γ) for Ca^{2+} or SO_4^{2-} for a system in equilibrium with gypsum is approximately 0.5 (see the section entitled Iteration Example in this chapter). Therefore, dissociated Ca^{2+} or SO_4^{2-} concentration equals

$$4.95 \times 10^{-3} \text{ mol L}^{-1}/0.5 = 9.9 \times 10^{-3} \text{ mol L}^{-1} \tag{2.35}$$

Note that when water is added to $CaSO_4 \cdot 2H_2Os$, the latter dissolves until its rate of dissolution is equal to its rate of precipitation. This is, by definition, the chemical equilibrium point. If, at this point, a certain amount of NaCl (a very water-soluble salt) is added, it suppresses the single-ion activity coefficient of Ca^{2+} and SO_4^{2-}. Hence, the

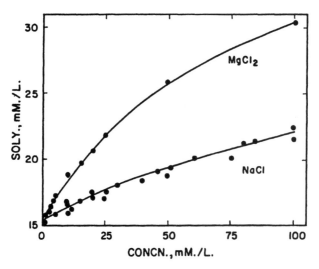

Figure 2.2. Gypsum ($CaSO_4 \cdot 2H_2Os$) solubility data demonstrating the ionic strength effect (salt effect, NaCl) and complexation effect ($MgCl_2$) (from Tanji, 1969b, with permission).

activity of the two ions is also suppressed, and the rate of $CaSO_4 \cdot 2H_2Os$ dissolution exceeds its rate of precipitation. Under these conditions, the apparent solubility of the mineral increases to meet a new equilibrium state. Experimental evidence shows that $CaSO_4 \cdot 2H_2Os$ solubility in the presence of 50 mmol L^{-1} NaCl increases by 33% in relationship to its solubility in distilled water (Fig. 2.2). This solubility enhancement phenomenon is known as the *salt effect* or *ionic strength effect*.

2.1.3 Ion Pair or Complex Effects

When water is added to any mineral (e.g., $CaSO_4 \cdot 2H_2Os$), a portion of the mineral ionizes, forming charged species (e.g., Ca^{2+}, SO_4^{2-}) which tend to associate with each other, forming pairs. A pair is an association of two oppositely charged ions, with each of the ions retaining their hydration sphere, hence, it is a weak complex (Fig. 2.3). The ability of ions to associate with each other is shown in Table 2.4. Tables 2.5 and 2.6 list some of the most commonly encountered chemical species (ion pairs) in fresh water. The importance of ion pairing on mineral solubility is demonstrated below. From Table 2.4,

$$CaSO_4^0 \Leftrightarrow Ca^{2+} + SO_4^{2-} \qquad (2.36)$$

with $K_{eq} = 5.25 \times 10^{-3}$. Substituting the K_{eq} value into Equation 2.36 and rearranging,

$$(Ca^{2+})\,(SO_4^{2-})/(5.25 \times 10^{-3}) = CaSO_4^0 \qquad (2.37)$$

where the parentheses again denote single-ion activity. Substituting single-ion activity values for Ca^{2+} and SO_4^{-1} from Equation 2.34 into Equation 2.37,

$$(CaSO_4^0) = (4.95 \times 10^{-3})\,(4.95 \times 10^{-3})/(5.25 \times 10^{-3}) = 4.67 \times 10^{-3}\,mol\,L^{-1} \quad (2.38)$$

Therefore, the total calcium concentration $[Ca_T]$ resulting from gypsum's solubility is the sum of all the calcium species in solution:

$$[Ca_T] = [(Ca^{2+})/\gamma_{Ca^{2+}}] + [(CaSO_4^0)/\gamma_{CaSO_4^0}] + \ldots \qquad (2.39)$$

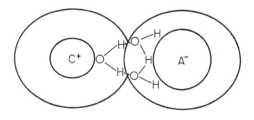

Figure 2.3. Schematic of an ion pair.

TABLE 2.4.　Ion Pair Equilibrium Constants
in Waters

Reaction	K_{eq}[a]
$NaSO_4^- = Na^+ + SO_4^{2-}$	2.3×10^{-1}
$HSO_4^- = H^+ + SO_4^{2-}$	1.2×10^{-2}
$CuSO_4^0 = Cu^{2+} + SO_4^{2-}$	4.36×10^{-3}
$ZnSO_4^0 = Zn^{2+} + SO_4^{2-}$	5.0×10^{-3}
$FeSO_4^0 = Fe^{2+} + SO_4^{2-}$	5.0×10^{-3}
$CaSO_4^0 = Ca^{2+} + SO_4^{2-}$	5.25×10^{-3}
$MnSO_4^0 = Mn^{2+} + SO_4^{2-}$	5.25×10^{-3}
$MgSO_4^0 = Mg^{2+} + SO_4^{2-}$	5.88×10^{-3}
$AlSO_4^+ = Al^{3+} + SO_4^{2-}$	6.30×10^{-4}
$FeSO_4^- = Fe^{3+} + SO_4^{2-}$	7.1×10^{-5}
$MgCl^+ = Mg^{2+} + Cl^-$	3.2×10^{-1}
$KSO_4^- = K^+ + SO_4^{2-}$	1.1×10^{-1}
$KCl^0 = K^+ + Cl^-$	Extremely high

[a]These values were selected from Adams, 1971.

Similarly, total sulfate concentration $[SO_{4T}]$ is the sum of all the sulfate species in solution:

$$[SO_{4T}] = [(SO_4^{2-})/\gamma_{SO_4^{2-}}] + [(CaSO_4^0)/\gamma_{CaSO_4^0}] + \cdots \qquad (2.40)$$

where the parentheses denote molar activity for the species shown and γ denotes the single-ion activity coefficient. Although Equations 2.39 and 2.40 contain only one ion pair, there are many other ion pairs in such systems (see Tables 2.5 and 2.6).

TABLE 2.5.　Chemical Species in Fresh Waters

	Quantitatively Important Species	Known to Exist in Minor Quantities
Ca	Ca^{2+}, $CaSO_4^0$	$CaCO_3^0$, $CaHCO_3^+$, $CaHPO_4^0$, $CaH_2PO_4^+$
Mg	Mg^{2+}, $MgSO_4^0$	$MgCO_3^0$, $MgHCO_3^+$, $MgHPO_4^0$
Na	Na^+	$NaHCO_3^0$, $NaCO_3^-$, $NaSO_4^-$
K	K^+	KSO_4^-
N(H)	NH_4^+, NH_3	
Mn	Mn^{2+}	$MnSO_4^-$, $MnCO_3^0$
Fe	Fe^{2+}, Fe^{3+}, $Fe(OH)^{2+}$, $Fe(OH)_2^+$, $Fe_2(OH)_2^{4+}$	$FeSO_4^0$, SOC
Zn	Zn^{2+}, $Zn(OH)^+$	$ZnSO_4^0$
Cu	SOC[a]	Cu^{2+}, $CuOH^+$, $CuSO_4^0$
Al	Al^{3+}, $AlOH^+$, $Al(OH)_4^-$	

[a]SOC = soluble organic matter complex or chelate.
Source: Selected and adapted from L. G. Sillen and A. E. Martell, *Stability Constants of Metal–Ion Complexes*, Special Publication No. 17, The Chemical Society, London, 1964.

TABLE 2.6. Chemical Species in Fresh Waters

Quantitatively Important Species		Known to Exist in Minor Quantities
S	SO_4^{2-}, $CaSO_4^0$, $MgSO_4^0$	HSO_4^-, $NaSO_4^-$, KSO_4^-, $MnSO_4^0$, $FeSO_4^0$, $ZnSO_4^0$, $CuSO_4^0$, $H_2S(aq)$, HS^-
N	NO_3, N_2 (aq)	NO_2^-, N_2O (aq)
Cl	Cl^-	
P	$H_2PO_4^-$, HPO_4^{2-}, $MgHPO_4^0$, $CaHPO_4^0$, $CaH_2PO_4^+$, $SOAP^a$	$AlH_2PO_4^{2+}$+, $FeH_2PO_4^{2+}$, $FeHPO_4^+$
Si	H_4SiO_4, $H_3SiO_4^-$	
B	H_3BO_3, $B(OH)_4^-$	SOC^b
C	CO_2 (aq), H_2CO_3, HCO_3^-, CO_3^{2-}	$SOAP$, $CaCO_3^0$, $CaHCO_3^+$, $MgCO_3^0$, $MgHCO_3^+$, $NaHCO_3^0$, $NaCO_3^-$

^{a}SOAP = soluble organic anions and polyanions (e.g., low FW carboxylates such as acetate, uronides, and phenolates as well as high FW fulvates and humates).
^{b}SOC = Soluble organic complex (especially sugars and organic ligands exposing hydroxyls).

The parameters K_{sp}, K_{eq}, and γ play a major role in the solubility of all minerals. By comparing experimental values of Ca_T (obtained by equilibrating $CaSO_4 \cdot 2H_2Os$ in solutions with varying electrolytes) with computer-generated data [by considering γ, K_{sp} of $CaSO_4 \cdot 2H_2Os$, and K_{eq} of solution pairs (Table 2.4)], excellent agreement between experimental and predicted values was observed (Fig. 2.4). The computer data in Figure 2.4 was produced by a mass balance–iteration procedure which is demonstrated later in this chapter.

The potential influence of ionic strength and ion pairing on single-ion activity could be demonstrated through computer simulations. For these situations, a constant concentration of the cation in question under variable ionic strength is assumed, and variability in ionic strength is attained by increasing the concentration of NaCl or Na_2SO_4. Ion pairs considered along with their stability constants are shown in Table 2.4. The data in Figure 2.5A show that when dissolved potassium is kept constant under increasing ionic strength, K^+ activity decreases. This decrease is greater in the SO_4^{2-} system than in the Cl^- system. The difference is due to the greater stability of KSO_4^- (Table 2.4; Fig. 2.5C) as opposed to KCl^0. Additionally, the decrease in K^+ activity with respect to ionic strength can be considered linear.

The data in Figure 2.5A also show that there is a significant decrease in Mg^{2+} activity as ionic strength increases; furthermore, this decrease is biphasic. Furthermore, the decrease in Mg^{2+} activity is much greater in the SO_4^{2-} system than in the Cl^- system. The data in Figure 2.5B demonstrate that at 20 mmol$_c$ L^{-1} background electrolyte (NaCl or Na_2SO_4), nearly 30% of the total dissolved Mg is in the $MgSO_4^0$ form and only about 3% is in the $MgCl^+$ form. With respect to Mg^{2+} activity, the data in Figure 2.5A reveal that at a background electrolyte (NaCl or Na_2SO_4) of 20 mmol$_c$ L^{-1}, Mg^{2+} activity represents 52% of the total dissolved Mg in the Cl^- system as opposed to only 34% in the SO_4^{2-} system.

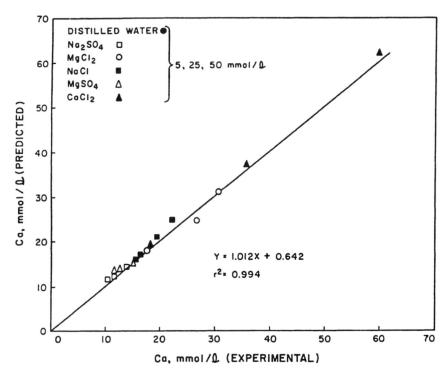

Figure 2.4. Relationship between experimental and computer-simulated CaSO₄·2H₂Os solubility at different concentrations of various salt solutions employing an ion association model (from Evangelou et al., 1987, with permission).

Thus, if a certain amount of $MgCl_2$ is added to a solution in equilibrium with gypsum, the former ($MgCl_2$) ionizes and interacts with Ca^{2+} and SO_4^{2-} forming magnesium sulfate pairs ($MgSO_4^0$) and calcium chloride pairs ($CaCl^+$). When this occurs, the rate of $CaSO_4·2H_2Os$ dissolution exceeds its rate of precipitation and a new equilibrium point is established by dissolving more $CaSO_4·2H_2Os$. Experimental evidence shows that $CaSO_4·2H_2Os$ solubility in the presence of 50 mmol L^{-1} $MgCl_2$ increases by 69% in relationship to its solubility in distilled water; in the presence of an equivalent concentration of NaCl, solubility increases by only 33% (Fig. 2.2). This difference in gypsum solubility by the two salts ($MgCl_2$ vs. NaCl) is due to the relatively high pairing potential of divalent ions (Mg^{2+} and SO_4^{2-}) versus the weak pairing potential of divalent–monovalent ions (Ca^{2+} and Cl^- or Na^+ and SO_4^{2-}) or monovalent–monovalent ions (Na^+ and Cl^-). The solubility enhancement phenomenon is known as the *ion-pairing effect*.

If a certain amount of $MgSO_4$ or $CaCl_2$, or Na_2SO_4 (all three salts are highly soluble) is added to a solution in equilibrium with gypsum, the added SO_4^{2-} interacts with the Ca^{2+} released from gypsum, or the added Ca^{2+} (added as $CaCl_2$) interacts with the SO_4^{2-} released from gypsum, and the rate of $CaSO_4·2H_2Os$ precipitation exceeds its

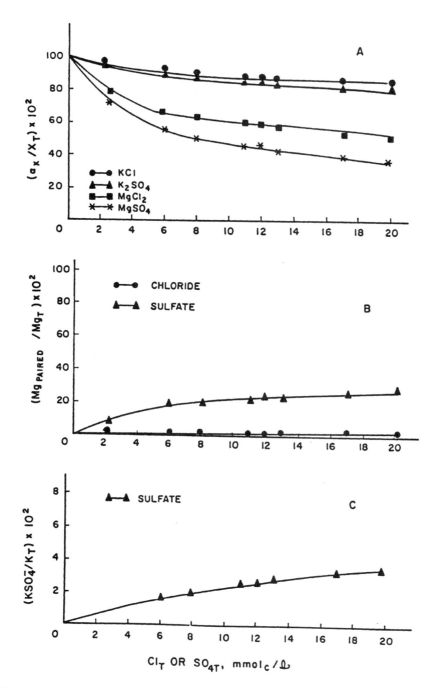

Figure 2.5. Influence of Cl^- and SO_4^{2-} concentration on (A) K^+ and Mg^{2+} activity, (B) Mg^{2+} pairing, and (C) K^+ pairing (from Evangelou and Wagner, 1987, with permission).

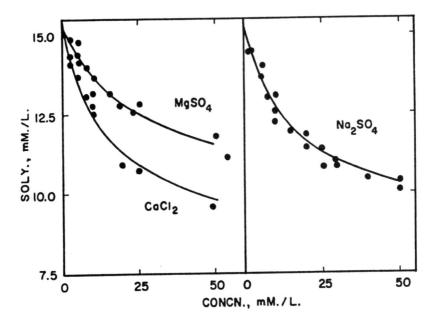

Figure 2.6. Gypsum ($CaSO_4 \cdot 2H_2Os$) solubility data demonstrating the common ion effect (from Tanji, 1969b, with permission).

rate of dissolution. Because of this, a new equilibrium point is established by precipitating $CaSO_4 \cdot 2H_2Os$. Experimental data show that $CaSO_4 \cdot 2H_2Os$ solubility in the presence of 50 mmol L^{-1} of the SO_4 salts ($MgSO_4$ or Na_2SO_4) decreases on average by approximately 33% in relationship to its solubility in distilled water (Fig. 2.6). This solubility suppression phenomenon is known as the *common ion effect*.

The data in Table 2.7 show selected minerals and their K_{sp} values, and the data in Table 2.8 contain the solubility of selected, environmentally important, minerals in distilled water in grams per liter.

TABLE 2.7. Solubility Product Constants[a] of Selected Minerals

Substance	Formula	K_{sp}
Aluminum hydroxide	$Al(OH)_3$	2×10^{-32}
Amorphous silica	$(SiO_2 + H_2O = H_2SiO_3)$	1.82×10^{-3}
Barium carbonate	$BaCO_3$	$1.0 \times 10^{-8.3}$
Barium carbonate	$BaCO_3$	5.1×10^{-9}
Barium chromate	$BaCrO_4$	1.2×10^{-10}
Barium iodate	$Ba(IO)_3$	1.57×10^{-9}
Barium oxalate	BaC_2O_4	2.3×10^{-8}
Barium sulfate	$BaSO_4$	1.3×10^{-10}

(continued)

TABLE 2.7. Continued

Substance	Formula	K_{sp}
Ca-phosphate (1)	$CaHPO_4 \cdot 2H_2O$	2.75×10^{-7}
Ca-phosphate (2)	$Ca_4H(PO_4)_3 \cdot 3H_2O$	1.26×10^{-47}
Ca-phosphate (3)	$Ca_5OH(PO_4)_3$	1.23×10^{-56}
Cadmium carbonate	$CdCO_3$	2.5×10^{-14}
Cadmium hydroxide	$Cd(OH)_2$	5.9×10^{-15}
Cadmium oxalate	CdC_2O_4	9×10^{-8}
Cadmium sulfide	CdS	2×10^{-28}
Calcium carbonate	$CaCO_3$	$1.0 \times 10^{-8.3}$
Calcium fluoride	CaF_2	4.9×10^{-11}
Calcium fluoride	CaF_2	$1.0 \times 10^{-10.4}$
Calcium hydroxide	$Ca(OH)_2$	$1.0 \times 10^{-5.1}$
Calcium monohydrogen phosphate	$CaHPO_4$	$1.0 \times 10^{-6.6}$
Calcium oxalate	CaC_2O_4	2.3×10^{-9}
Calcium sulfate	$CaSO_4$	1.2×10^{-6}
Cobalt carbonate	$CoCO_3$	1.0×10^{-12}
Copper carbonate	$CuCO_3$	$1.0 \times 10^{-9.9}$
Copper(I) bromide	$CuBr$	5.2×10^{-9}
Copper(I) chloride	$CuCl$	1.2×10^{-6}
Copper(I) iodide	CuI	1.1×10^{-12}
Copper(I) thiocyanate	$CuSCN$	4.8×10^{-15}
Copper(II) hydroxide	$Cu(OH)_2$	1.6×10^{-19}
Copper(II) sulfide	CuS	6×10^{-36}
Gypsum	$CaSO_4 \cdot 2H_2Os$	2.45×10^{-5}
Iron (III) hydroxide (αFe_2O_3)	$Fe(OH)_3$	2.0×10^{-43}
Iron carbonate	$FeCO_3$	$1.0 \times 10^{-10.3}$
Iron(II) hydroxide	$Fe(OH)_2$	8×10^{-16}
Iron(II) sulfide	FeS	6×10^{-18}
Iron(III) hydroxide	$Fe(OH)_3$	4×10^{-38}
Lanthanum iodate	$La(IO_3)_3$	6.2×10^{-12}
Lead carbonate	$PbCO_3$	3.3×10^{-14}
Lead chloride	$PbCl_2$	1.6×10^{-5}
Lead chromate	$PbCrO_4$	1.8×10^{-14}
Lead hydroxide	$Pb(OH)_2$	2.5×10^{-16}
Lead iodide	PbI_2	7.1×10^{-9}
Lead oxalate	PbC_2O_4	4.8×10^{-10}
Lead sulfate	$PbSO_4$	1.6×10^{-8}
Lead sulfide	PbS	7×10^{-28}
Magnesium ammonium phosphate	$MgNH_4PO_4$	3×10^{-13}
Magnesium carbonate	$MgCO_3$	1×10^{-5}
Magnesium carbonate	$MgCO_3$	1.0×10^{-5}
Magnesium hydroxide	$Mg(OH)_2$	1.8×10^{-11}
Magnesium oxalate	MgC_2O_4	8.6×10^{-5}
Manganese carbonate	$MnCO_3$	$1.0 \times 10^{-10.1}$
Manganese(II) hydroxide	$Mn(OH)_2$	1.9×10^{-13}

(continued)

TABLE 2.7. Continued

Substance	Formula	K_{sp}
Manganese(II) sulfide	MnS	3×10^{-13}
Mercury(I) bromide	Hg_2Br_2	5.8×10^{-23}
Mercury(I) chloride	Hg_2Cl_2	1.3×10^{-18}
Mercury(I) iodide	Hg_2I_2	4.5×10^{-29}
Nickel carbonate	$NiCO_3$	$1.0 \times 10^{-6.9}$
Quartz	$(SiO_2 + H_2O = H_2SiO_3)$	1.0×10^{-4}
Silver arsenate	Ag_3AsO_4	1.0×10^{-22}
Silver bromide	AgBr	5.2×10^{-13}
Silver carbonate	Ag_2CO_3	8.1×10^{-12}
Silver chromate	Ag_2CrO_4	1.1×10^{-12}
Silver cyanide	AgCN	7.2×10^{-11}
Silver iodate	$AgIO_3$	3.0×10^{-8}
Silver iodide	AgI	8.3×10^{-17}
Silver oxalate	$Ag_2C_2O_4$	3.5×10^{-11}
Silver sulfide	Ag_2S	6×10^{-50}
Silver thiocynate	AgSCN	1.1×10^{-12}
Siver chloride	AgCl	1.82×10^{-10}
Strontium carbonate	$SrCO_3$	$1.0 \times 10^{-8.8}$
Strontium oxalate	SrC_2O_4	5.6×10^{-8}
Strontium sulfate	$SrSO_4$	3.2×10^{-7}
Thallium(I) chloride	TlCl	1.7×10^{-4}
Thallium(I) sulfide	Tl_2S	1.0×10^{-22}
Zinc carbonate	$ZnCO_3$	1.0×10^{-7}
Zinc hydroxide	$Zn(OH)_2$	1.2×10^{-17}
Zinc oxalate	ZnC_2O_4	7.5×10^{-9}
Zinc sulfide	ZnS	4.5×10^{-24}

Source: Taken from L. Meites, *Handbook of Analytical Chemistry*, pp. 1–13. New York: Mc-Graw-Hill Book Company, Inc., 1963.

Table 2.8. Solubility of Salts in Water

Salt Type	Solubility (L^{-1})
KCl	238
NaCl	357
RbCl	770
LiCl	637
NH_4Cl	297
AgCl	0.0009 (exception)

In general, a similar behavior is expected from bromide salts, nitrate salts, perclorate (ClO_4^-) salts, fluoride salts, sulfate salts, hydroxides, and phosphates

$AgNO_3$	1220 (very soluble)
$CaCl_2 \cdot 2H_2O$	977
$MgCl_2 \cdot 6H_2O$	*(continued)*

Table 2.8. Continued

Salt Type	Solubility (L^{-1})
$BaCl_2$	375
$SrCl_2$	538

In general, a similar behavior is expected from bromide salts, nitrate salts, and perclorate (ClO_4^-) salts

$PbCl_2$	9.9
$PbBr_2$	4.5
$Pb(ClO_4)_2$	5000
CaF_2	0.016
MgF_2	0.076
BaF_2	1.25
SrF_2	0.11
PbF_2	0.64
AlF_3	Slightly soluble

Low solubility is also expected from the metal sulfides

$Ca(OH)_2$	1.85
$Mg(OH)_2$	0.009
$Ba(OH)_2$	56
$Sr(OH)_2$	4.1
$Pb(OH)_2$	0.15
$Cu(OH)_2$	Slightly soluble
$Cd(OH)_2$	0.0026
$Fe(OH)_2$	Significantly soluble
$Mn(OH)_2$	Significantly soluble
$Al(OH)_3$	Slightly soluble
$CaCO_3$	0.014
$MgCO_3$	1.79
$BaCO_3$	0.02
$SrCO_3$	0.01
$PbCO_3$	0.001
$CuCO_3$	Slightly soluble
$CdCO_3$	Slightly soluble
$MnCO_3$	Slightly soluble
$FeCO_3$	Slightly soluble

Low solubility is expected from phosphates

$AlCl_3$	700
AlF_3	Slightly soluble
$Al(NO_3)_3$	637
$MgSO_4 \cdot 7H_2O$	71
$CaSO_4 \cdot 2H_2O$	2.41
$SrSO_4$	0.1
$BaSO_4$	0.002
$PbSO$	0.042
$CuSO_4$	143
$Al_2(SO_4)_3$	313

[a]The values listed are mostly representative of low-temperature mineral solubility. Use these values only as a guide. The absolute solubility of minerals in fresh waters is a complex process.

ITERATION EXAMPLE

An example of an iteration procedure is presented below using I, K_{sp}, and K_{eq} of ion pairs for predicting mineral solubility. Consider

$$CaSO_4 \cdot 2H_2Os \Leftrightarrow Ca^{2+} + SO_4^{2-} \qquad K_{sp} = 2.45 \times 10^{-5} \qquad \text{(A)}$$

$$CaSO_4^0 \Leftrightarrow Ca^{2+} + SO_4^{2-} \qquad K_{eq} = 5.32 \times 10^{-3} \qquad \text{(B)}$$

and

$$(Ca^{2+})(SO_4^{2-}) = 2.45 \times 10^{-5} \qquad \text{(C)}$$

where the parentheses denote activity.

$$(Ca^{2+}) = (SO_4^{2-}) = Y \, \text{mol L}^{-1} \, \text{activity} \qquad \text{(D)}$$

Therefore,

$$Y = (2.45 \times 10^{-5})^{1/2} = 4.95 \times 10^{-3} \, \text{mol L}^{-1} \qquad \text{(E)}$$

First Iteration

a. Calculation of ionic strength (I):

$$I = 1/2 \sum m_i Z_i^2 \qquad \text{(F)}$$

where $m_i = Y$ in mol L^{-1} and Z denotes ion valence The only two chemical species (m_i) considered are Ca^{2+} and SO_4^{2-}.

$$I = 1/2 \, [(4.95 \times 10^{-3})2^2 + (4.95 \times 10^{-3})2^2] = 1.98 \times 10^{-2} \qquad \text{(G)}$$

b. Calculation of single-ion activity coefficients (Ca^{2+} and SO_4^{2-}) using the Guntelberg equation (see Table 2.1):

$$-\log \gamma = 0.5 \, Z^2 [I^{1/2}]/[1 + I^{1/2}] \qquad \text{(H)}$$

Substituting values into Equation H,

$$-\log \gamma = 0.5 \, (2)^2 [(1.98 \times 10^{-2})^{1/2}]/[1 + (1.98 \times 10^{-2})^{1/2}] \qquad \text{(I)}$$

and

$$\gamma = 0.566 \tag{J}$$

Considering (based on Eq. C) that $(Ca^{2+}) = (SO_4^{2-}) = (2.45 \times 10^{-5})^{1/2}$, a new dissociated concentration for $[Ca^{2+}]$ or $[SO_4^{2-}]$ would be estimated using the newly estimated γ value (Eq. J) as shown below:

$$[Ca^{2+}] = [SO_4^{2-}] = [2.45 \times 10^{-5}/(0.566)^2]^{1/2} = 8.74 \times 10^{-3} \text{ mol L}^{-1} \tag{K}$$

where the brackets denote concentration.

Second Iteration

a. Calculation of ionic strength (I):

$$I = 1/2 \sum m_i Z_i^2 \tag{L}$$

where $m_i = Y$ in mol L^{-1} and Z denotes ion valence.

$$I = 1/2 [(8.74 \times 10^{-3})2^2 + (8.74 \times 10^{-3})2^2] = 3.496 \times 10^{-2} \tag{M}$$

b. Calculation of single-ion activity coefficients (Ca^{2+} and SO_4^{2-}):

$$-\log \gamma = 0.5 \, Z^2[I^{1/2}]/[1 + I^{1/2}] \tag{N}$$

Substituting values into Equation N,

$$-\log \gamma = 0.5 \, (2)^2$$

$$[(3.496 \times 10^{-2})^{1/2}]/[1 + (3.496 \times 10^{-2})^{1/2}] \tag{O}$$

and

$$\gamma = 0.484 \tag{P}$$

Considering (based on Eq. E) that $(Ca^{2+}) = (SO_4^{2-}) = (2.45 \times 10^{-5})^{1/2}$, a new dissociated concentration for $[Ca^{2+}]$ or $[SO_4^{2-}]$ would be estimated using the newly estimated γ value (Eq. P) as shown below:

$$[Ca^{2+}] = [SO_4^{2-}] = [2.45 \times 10^{-5}/(0.484)^2]^{1/2} = 1.022 \times 10^{-2} \text{ mol L}^{-1} \tag{Q}$$

Third Iteration

a. Calculation of ionic strength (I):

$$I = 1/2 \sum [m_i Z_i]^2 \tag{R}$$

where $m_i = Y$ in mol L^{-1} and Z denotes ion valence.

$$I = 1/2 [(1.022 \times 10^{-2})2^2 + (1.022 \times 10^{-2})2^2] = 4.088 \times 10^{-2} \tag{S}$$

b. Calculation of single-ion activity coefficients (Ca^{2+} and SO_4^{2-}):

$$-\log \gamma = 0.5 \, Z^2[I^{1/2}]/[1 + I^{1/2}] \tag{T}$$

Substituting values into Equation T

$$-\log \gamma = 0.5 \, (2)^2 \, [(4.088 \times 10^{-2})^{1/2}]/[1 + (4.088 \times 10^{-2})^{1/2}] \tag{U}$$

and

$$\gamma = 0.46 \tag{V}$$

Considering (based on Eq. C) that $(Ca^{2+}) = (SO_4^{2-}) = (2.45 \times 10^{-5})^{1/2}$, a new dissociated concentration for $[Ca^{2+}]$ and $[SO_4^{2-}]$ would be estimated using the newly estimated γ value (Eq. V) as shown below:

$$[Ca^{2+}] = [SO_4^{2-}] = [2.45 \times 10^{-5}/(0.46)^2]^{1/2} = 1.076 \times 10^{-2} \text{ mol } L^{-1} \tag{W}$$

Note that the difference between the last two iterations in estimated $[Ca^{2+}]$ or $[SO_4^{2-}]$ is relatively insignificant (1.022×10^{-2} mol L^{-1} versus 1.076×10^{-2} mol L^{-1}), signifying that an answer has been found.

c. Calculation of $CaSO_4^0$ pairs:

$$CaSO_4^0 = (Ca^{2+}) \, (SO_4^{2-})/5.32 \times 10^{-3} \tag{X}$$

where $5.32 \times 10^{-3} = K_{eq}$. Since $(Ca^{2+}) = (SO_4^{2-}) = 4.95 \times 10^{-3}$ mol L^{-1}, by substituting activity values in Equation X, the concentration of $CaSO_4^0$ could be estimated:

$$(CaSO_4^0) = [CaSO_4^0] = (4.95 \times 10^{-3}) \, (4.95 \times 10^{-3})/5.32 \times 10^{-3}$$

$$= 4.61 \times 10^{-3} \text{ mol } L^{-1} \tag{Y}$$

Note that the γ value for a species with zero charge ($Z = 0$) is always approximately 1 (see Eq. H), and for this reason the activity of the $CaSO_4^0$ pair equals the concentration.

d. Summation of Ca and $CaSO_4^0$ in solution:

Pairs of $CaSO_4^0 = 4.61 \times 10^{-3}$ mol L^{-1} or 9.22 meq L^{-1}

Dissociated $Ca^{2+} = 1.076 \times 10^{-2}$ mol L^{-1} or 21.52 meq L^{-1}

Total Ca in solution $= CaSO_4^0$ pairs plus dissociated $Ca^{2+} = 30.74$ meq L^{-1}

Experimentally determined gypsum solubility is 30.60 meq L^{-1} (Tanji, 1969b) and Ca^{2+} activity is 4.95×10^{-3} mol L^{-1} (9.9 meq L^{-1}). The fact that there is agreement between experimentally determined gypsum solubility (30.60 meq L^{-1}) and estimated gypsum solubility (30.74 meq L^{-1}) (based on the iteration procedure used above and considering only the $CaSO_4^0$ pair) suggests that the $CaSO_4^0$ pair only contributes significantly to the solubility of gypsum.

2.1.4 Role of Hydroxide on Metal Solubility

When a salt is introduced to water (e.g., $AlCl_3s$), the charged metal (Al^{3+}) has a strong tendency to react with H_2O or OH^- and forms various Al–hydroxy species. Metal–hydroxide reactions in solution exert two types of influences on metal–hydroxide solubility, depending on the quantity of hydroxyl supplied. They either decrease or increase metal solubility. The solubility of a particular metal–hydroxide mineral depends on its K_{sp}, quantity of available hydroxyl, and solution pH of zero net charge. For example, aluminum (Al^{3+}) forms a number of hydroxy species in water as shown below:

$$Al_T = Al^{3+} + Al(OH)^{2+} + Al(OH)_2^+ + Al(OH)_3^0 + Al(OH)_4^- + Al(OH)_x^{3-x} \qquad (2.41)$$

where Al_T = total dissolved aluminum. The pH at which the sum of all Al–hydroxy species equals zero is referred to as the *solution pH of zero net charge*. For any particular metal (e.g., heavy metal), there appears to be a unique solution pH at which its solubility approaches a minimum. Below or above this solution pH, total dissolved metal increases (Fig. 2.7). This behavior is dependent on the *common-ion effect* (low pH) and the *ion-pairing or complexation effect* (high pH). An inorganic complex is an association of two oppositely charged ions, with both ions losing their individual hydration sphere and gaining one as a complex, hence, becoming a strong complex (Fig. 2.8).

Commonly, different metals exhibit different solution pH of zero net charge. For this reason, different metals exhibit minimum solubility at different pH values, which makes it difficult to precipitate effectively two or more metals, as metal–hydroxides, simultaneously. Thus metal–hydroxide solubility as a function of pH displays a U-shaped behavior. The lowest point in the U-shaped figure signifies the solution pH of zero net charge and is demonstrated below. Consider the solid $Fe(OH)_2s$,

$$Fe(OH)_2s \Leftrightarrow Fe^{2+} + 2OH^- \qquad (2.42)$$

with a K_{sp} of $10^{-14.5}$, and the solution complexation reaction

SOLUTION/MINERAL—SALT CHEMISTRY

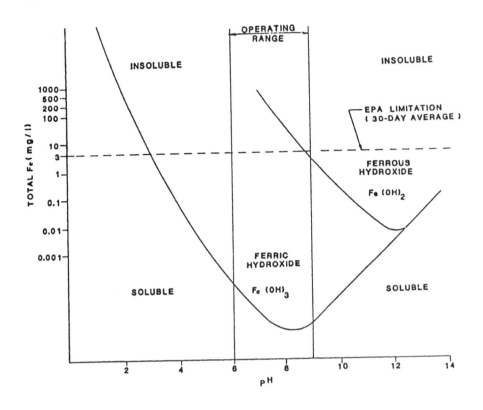

Figure 2.7. Solubility of Fe(OH)$_2$ and Fe(OH)$_3$ as a function of pH (from U.S. EPA, 1983).

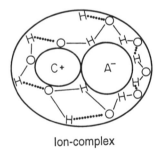

Ion-complex

Figure 2.8. Schematic of an ion-complex.

$$Fe(OH)_2s \Leftrightarrow 2OH^- = Fe(OH)_4^{2-} \tag{2.43}$$

with a K_{eq} of $10^{-4.90}$. Letting Fe_T be the total solubility of $Fe(OH)_2s$,

$$Fe_T = Fe^{2+} + Fe(OH)_4^{2-} \tag{2.44}$$

Based on the equilibrium expressions of Equations 2.42 and 2.43,

$$K_{sp} = (Fe^{2+})(OH^-)^2 = 10^{-14.5} \tag{2.45}$$

and

$$K_{eq} = Fe(OH)_4^{2-}/(OH^-)^2 = 10^{-4.90} \tag{2.46}$$

Rearranging and substituting Equations 2.45 and 2.46 into Equation 2.44 yields

$$Fe_T = K_{sp}/(OH^-)^2 + K_{eq}(OH^-)^2 \tag{2.47}$$

The pH of minimum solubility of $Fe(OH)_2s$ or solution pH of zero net charge can be obtained by differentiating Equation 2.47 and setting the derivative of Fe_T with respect to OH^- equal to zero. Therefore,

$$dFe_T/d(OH^-) = -2K_{sp}/(OH^-)^3 + 2K_{eq}(OH^-) \tag{2.48}$$

setting

$$dFe_T/d(OH^-) = 0 \tag{2.49}$$

then

$$OH^- = (2K_{sp}/2K_{eq})^{1/4} = [(2 \times 10^{-14.5})/(2 \times 10^{-4.90})]^{1/4}$$
$$= 10^{-2.80} \text{ mol L}^{-1} \tag{2.50}$$

Since pH $= 14 -$ pOH$^-$ (where pOH$^-$ denotes the negative log of OH$^-$), the pH of minimum solubility for $Fe(OH)_2s$ would be 11.21. The example above is only for demonstration purposes since only two of the many potentially forming Fe^{2+}–hydroxy species were employed. A graphical representation of the solubility of $Fe(OH)_2s$ (Eq. 2.47) and $Fe(OH)_3s$ as a function of pH are shown in Figure 2.7. The data in Figure 2.9 show the solubility of various heavy metals as a function of pH, whereas the data in Figure 2.10 show the decrease in metal–hydroxide solubility as pH increases (common ion effect). They do not, however, show the expected increase in metal–hydroxide solubility as pH increases.

The pH-dependent solubility behavior of metal–hydroxides and the corresponding solution pH of zero net charge can be demonstrated by deriving pH dependent solubility functions for all the metal–hydroxy species of a particular metal in solution.

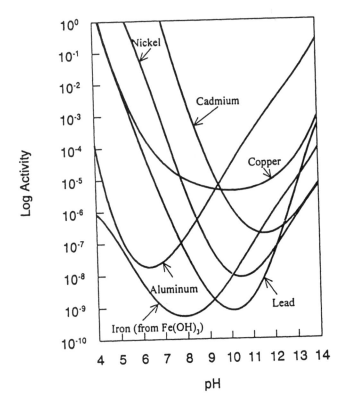

Figure 2.9. Solubility of various metal–hydroxides as a function of pH.

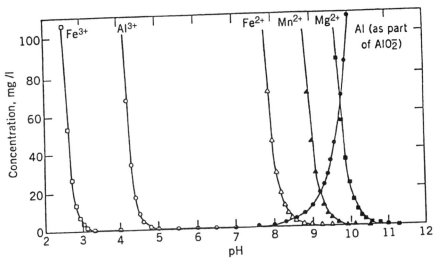

Figure 2.10. Solubility of various metal–hydroxides. Note that the figure exhibits only the common ion effect (from U.S. EPA, 1983).

These functions are then plotted as a function of pH. Functions of the Al–hydroxy species are derived below to demonstrate the process.

$$Al_T = Al^{3+} + Al(OH)^{2+} + Al(OH)_2^+ + \ldots \tag{2.51}$$

where Al^{3+} is assumed to be controlled by the solubility of $Al(OH)_3s$. The reaction describing the first Al–hydroxy species is

$$Al^{3+} + H_2O \overset{K_{eq1}}{\Longleftrightarrow} Al(OH)^{2+} + H^+ \tag{2.52}$$

$$K_{eq1} = \frac{(Al[OH]^{2+})\,(H^+)}{(Al^{3+})} = 10^{-5.1} \tag{2.53}$$

Rearranging Equation 2.53 to solve for $Al(OH)^{2+}$,

$$Al(OH)^{2+} = \frac{K_{eq1}(Al^{3+})}{(H^+)} \tag{2.54}$$

The reaction describing the second Al–hydroxy species is

$$Al(OH)^{2+} + H_2O \overset{K_{eq2}}{\Longleftrightarrow} Al(OH)_2^+ + H^+ \tag{2.55}$$

and

$$K_{eq2} = \frac{(Al[OH]_2^+)\,(H^+)}{Al(OH)^{2+}} = 10^{-9.9} \tag{2.56}$$

Rearranging Equation 2.56 to solve for $Al(OH)_2^+$

$$Al(OH)_2^+ = \frac{K_{eq2}(Al[OH]^{2+})}{(H^+)} \tag{2.57}$$

Substituting the $Al(OH)^{2+}$ term of Equation 2.57 with Equation 2.54 gives

$$Al(OH)_2^+ = \frac{K_{eq1}K_{eq2}(Al^{3+})}{(H^+)^2} \tag{2.58}$$

Finally, substituting Equations 2.54 and 2.58 into Equation 2.51 yields

$$Al_T = Al^{3+} + \frac{K_{eq1}(Al^{3+})}{(H^+)} + \frac{K_{eq1}K_{eq2}(Al^{3+})}{(H^+)^2} \tag{2.59}$$

A common denominator for Equation 2.59 is obtained by multiplying the first two terms on the right side of the equal sign with $(H^+)^2$ and (H^+), respectively. Thus,

$$Al_T = \frac{(H^+)^2(Al^{3+})}{(H^+)^2} + \frac{K_{eq1}(H^+)(Al^{3+})}{(H^+)^2} + \frac{K_{eq1}K_{eq2}(Al^{3+})}{(H^+)^2} \qquad (2.60)$$

or

$$Al_T = \frac{(Al^{3+})[(H^+)^2 + K_{eq1}(H^+) + K_{eq1}K_{eq2}]}{(H^+)^2} \qquad (2.61)$$

Taking the inverse of Equation 2.61,

$$\frac{1}{Al_T} = \frac{(H^+)^2}{(Al^{3+})[(H^+)^2 + K_{eq1}(H^+) + K_{eq1}K_{eq2}]} \qquad (2.62)$$

Equation 2.62 can be used to estimate the percentage of species as a function of pH as follows: Rearranging Equation 2.62 gives

$$\frac{(Al^{3+})}{Al_T} \times 100 = \frac{(H^+)^2}{[(H^+)^2 + K_{eq1}(H^+) + K_{eq1}K_{eq2}]} \times 100 \qquad (2.63)$$

To estimate the percentage of $Al(OH)^{2+}$ species, multiply both sides of Equation 2.62 with Equation 2.54:

$$\frac{Al(OH)^{2+}}{Al_T} \times 100 = \frac{(H^+)^2}{(Al^{3+})[(H^+)^2 + (H^+)(K_{eq1}) + (K_{eq1}K_{eq2})]} \cdot \frac{K_{eq1}(Al^{3+}]}{(H^+)} \times 100 \qquad (2.64)$$

Simplifying,

$$\frac{Al(OH)^{2+}}{Al_T} \times 100 = \frac{K_{eq1}(H^+)}{[(H^+)^2 + (H^+)K_{eq1} + K_{eq1}K_{eq2}]} \times 100 \qquad (2.65)$$

To estimate the percentage of $Al(OH)_2^+$ species, multiply both sides of Equation 2.62 with Equation 2.58:

$$\frac{Al(OH)_2^+}{Al_T} \times 100 = \frac{(H^+)^2}{(Al^{3+})[(H^+)^2 + (H^+)(K_{eq1}) + K_{eq1}K_{eq2}]} \cdot \frac{K_{eq1}K_{eq2}(Al^{3+})}{(H^+)^2} \times 100 \qquad (2.66)$$

Simplifying,

$$\frac{Al(OH)_2^+}{Al_T} \times 100 = \frac{K_{eq1}K_{eq2}}{[(H^+)^2 + (H^+)(K_{eq1}) + K_{eq1}K_{eq2}]} \times 100 \qquad (2.67)$$

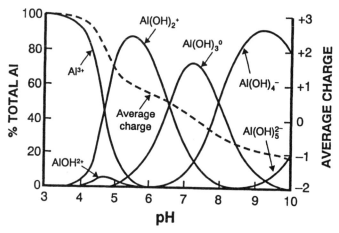

Figure 2.11. Relationship between pH and the distribution and average charge of soluble aluminum species (from Marion et al., 1976, with permission).

Equations 2.63, 2.65, and 2.67 can be plotted as a function of pH to produce a graphical solution. The aluminum speciation example above deals only with three Al species. However, any number of Al species can be calculated (Baes and Mesmer, 1976), as shown above, using the appropriate K_{eq} constants. The data in Figure 2.11 show six Al species and their average charge as a function of pH. The pH at which the $Al(OH)_3^0$ is at maximum is the solution pH of zero net charge. At this pH, one expects the $Al(OH)_3$ solid to exhibit its lowest solubility (see Section 2.1.5). The presence and actual behavior of these aluminum species in nature is of great interest because of their different toxicological effects.

SPECIAL NOTE

Minimum metal–hydroxide solubility or metal solubility at the solution pH of zero net charge can be approximated using the K_{eq} of the metal–hydroxide pair with zero charge ($M(OH)_n^0$. Consider,

$$Al(OH)_3s \overset{K_{sp}}{\Leftrightarrow} Al^{3+} + 3OH^- \tag{A}$$

$$Al^{3+} + 3OH^- \overset{K_{eq}}{\Leftrightarrow} Al(OH)_3^0 \tag{B}$$

Summing Reactions A and B,

$$\text{Al(OH)}_3\text{s} \overset{K_{sp}\cdot K_{eq}}{\Longleftrightarrow} \text{Al(OH)}_3^0 \tag{C}$$

and

$$(\text{Al(OH)}_3^0) = [\text{Al(OH)}_3^0] = K_{sp}\cdot K_{eq} \text{ mol L}^{-1} \tag{D}$$

Parentheses denote activity and brackets denote concentration of the species. The concentration of the Al(OH)_3^0 species represents approximately the lowest possible solubility point of the mineral and it is the product of two constants ($K_{sp}\cdot K_{eq}$). Thus, its magnitude is not in any way related to pH. Mineral solubility increases as pH increases above the solution pH of zero net charge because of increasing complexation effects, and mineral solubility also increases at pH values below the solution pH of zero net charge because of diminishing common-ion effects (Fig. 2A). All minerals are subject to the common-ion effect and many minerals are subject to the complexation or ion-pairing effect (Fig. 2B).

Consider the mineral $\text{CaSO}_4 \cdot 2\text{H}_2\text{O}$s. The common-ion effect can be demonstrated by expressing its solubility as

$$(\text{Ca}^{2+}) = K_{sp}/(\text{SO}_4^{2-}) \tag{E}$$

Equation E shows that when SO_4^{2-} activity (or concentration) increases (due perhaps to the addition of Na_2SO_4, a very water-soluble salt), Ca^{2+} activity (or concentration) decreases. It appears from Equation E that a continuous increase in SO_4^{2-} activity (or concentration) would suppress Ca^{2+} activity (or concentration) and would approach asymptotically zero. However, the latter does not occur because dissolved Ca (when under the influence of the common-ion effect) is controlled by the CaSO_4^0 pair. Consider

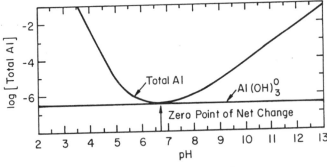

Figure 2A. Distribution of total dissolved Al and Al(OH)_3^0 pairs at an ionic strength 1 M and temperature of 25°C in solutions saturated with α-Al(OH)$_3$ (adapted from Baes and Mesmer, 1976, with permission).

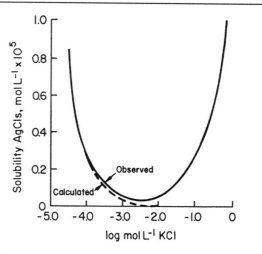

Figure 2B. Solubility of AgCls as a function of KCl concentration (from Skoog and West, 1976, with permission).

$$CaSO_4 \cdot 2H_2Os \overset{K_{sp}}{\Leftrightarrow} Ca^{2+} + SO_4^{2-} \tag{F}$$

ands

$$Ca^{2+} + SO_4^{2-} \overset{K_{eq}}{\Leftrightarrow} CaSO_4^0 \tag{G}$$

Summing Reactions F and G,

$$CaSO_4 \cdot 2H_2Os \overset{K_{sp} \cdot K_{eq}}{\Leftrightarrow} CaSO_4^0 \tag{H}$$

and

$$(CaSO_4^0) = [CaSO_4^0] = K_{sp} \cdot K_{eq} = (2.45 \times 10^{-5})(1.9 \times 10^2) = 4.65 \times 10^{-3} \text{ mol L}^{-1} \tag{I}$$

The concentration of the $CaSO_4^0$ pair is the product of two constants ($K_{sp} \cdot K_{eq}$), and therefore, is independent of the common-ion effect (Fig. 2C).

Congruent and Incongruent Dissolution

Congruent dissolution refers to the potential of a mineral to undergo dissolution but not form any secondary minerals by the dissolution products. For example, in the

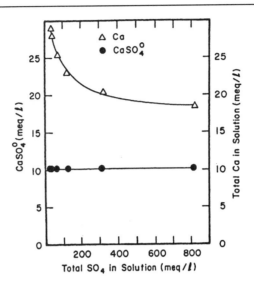

Figure 2C. Influence of SO_4 concentration on $CaSO_4^0$ and Ca^{2+} in a solution under equilibrium with $CaSO_4 \cdot 2H_2Os$.

case where $CaCO_3s$ is introduced to water the following reactions are hypothesized to take effect:

$$CaCO_3s \Leftrightarrow Ca^{2+} + CO_3^{2-} \tag{J}$$

$$CO_3^{2-} + 2H_2O \Leftrightarrow 2OH^- + CO_2 \tag{K}$$

and

$$Ca^{2+} + 2OH^- \Leftrightarrow Ca(OH)_2s \tag{L}$$

However, since the solubility of $CaCO_3s$ is much smaller than the solubility of $Ca(OH)_2s$ (Table 2.7), the latter does not form under the conditions described above, hence, there is congruent dissolution.

In the case where $Ca(OH)_2s$ is introduced to water,

$$Ca(OH)_2s \Leftrightarrow Ca^{2+} + 2OH^- \tag{M}$$

$$CO_2 + 2OH^- \Leftrightarrow HCO_3^- \tag{N}$$

and

$$Ca^{2+} + HCO_3^- \Leftrightarrow CaCO_3s + H^+ \tag{O}$$

Reactions M–O represent incongruent dissolution of $Ca(OH)_2s$ because the solubility of $CaCO_3s$ is much smaller than the solubility of $Ca(OH)_2s$ (Table 2.7). Therefore, introduction of $Ca(OH)_2s$ to water in equilibrium with atmospheric CO_2 leads to spontaneous formation of $CaCO_3s$. The well-known incongruent dissolution phenomena are those representing the dissolution of aluminosilicate minerals. For example, K-feldspars (orthoclase) undergo incongruent dissolution when exposed to water and carbonic acid to form kaolinite:

$$2KAlSi_3O_8s + 2H_2CO_3 + 9H_2O \rightarrow Al_2Si_2O_5(OH)_4s + 4H_4SiO_4 + 2K^+ + 2HCO_3^-$$
$$\text{(K-feldspars)} \qquad\qquad\qquad \text{(Kaolinite)} \tag{P}$$

Metal Hydrolysis

Charged metals (cations) in water behave as Lewis acids (willing to accept electrons). Water on the other hand, because it is willing to share its two unshared oxygen-associated pair of electrons, behaves as a Lewis base. Strong H_2O–metal (Lewis base–Lewis acid) interactions allow H^+ on the water molecule to dissociate, hence, low pH water is produced. The degree of dissociation of water interacting with a cation (M^{n+}) is described by the metal hydrolysis constant (Table 2A)

$$M^{n+} + H_2O \Leftrightarrow MOH^{(n-1)} + H^+ \tag{Q}$$

The larger the hydrolysis constant is the stronger the H_2O–metal interaction and the lower the solution pH.

Simultaneous Mineral Equilibria

An interesting case in mineral equilibria is the presence in a soil–water system of two minerals with a common ion. An example of such a case is barium sulfate ($BaSO_4$) plus calcium sulfate ($CaSO_4$). Which mineral would be controlling SO_4^{2-} in the system? Two conditions would need to be met in such a system; one is mass-balance while the second is charge balance. The mass-balance is given by

$$SO_4^{2-} = Ca^{2+} + Ba^{2+} \tag{R}$$

The charge balance is given by

$$H^+ + 2Ca^{2+} + 2Ba^{2+} = 2SO_4^{2-} + OH^- \tag{S_a}$$

Table 2A. **Solubility Products and Hydrolysis Constants of Metal Ions**[a]

Ion	$\log K_{sp}$	$\log K_{eq1}$ [a]	$\log K_{eq2}$
Be^{2+}	−21	−6.5	
Mg^{2+}	−10.8	−12	
Ca^{2+}	−5.0	−12.5	
Mn^{2+}	−12.5	−10.5	
Fe^{2+}	−14.8	−7	
Ni^{2+}	−15	−8	
Cu^{2+}	−19.5	−7.5	
Zn^{2+}	−17	−9.1	
Cd^{2+}	−14	−10	
Hg^{2+}	−25.5	−3.5	
Pb^{2+}	−18	−8	
Al^{3+}	−33.5	−5	−5.5
Fe^{3+}	−39	−2.9	−3.3
La3+	−20	−9	
Ti^{4+} [$(TiO_2)(OH)^2$]	−29	>−1	
Th^{4+}	−44	−4.1	

Source: From L. G. Sillen and A. E. Martell, *Stability Constants.* Special Publication No. 25, The Chemical Society, London, 1974.

Considering that the pH would be nearly neutral because of the small hydrolysis constants of the solution species involved (Ca^{2+}, Ba^{2+}, SO_4^{2-}) (Table 2A), no other solution species in significant concentration would be involved. Setting mass-balance equals to charge-balance,

$$SO_4^{2-} - Ca^{2+} - Ba^{2+} = -H^+ - 2Ca^{2+} - 2Ba^{2+} + 2SO_4^{2-} + OH^- \qquad (S_b)$$

and assuming that $H^+ \simeq OH^-$, then

$$SO_4^{2-} = Ca^{2+} + Ba^{2+} \qquad (T)$$

since

$$Ca^{2+} = K_{sp(CaSO_4)}/SO_4 \qquad (U)$$

and

$$Ba^{2+} = K_{sp(BaSO_4)}/SO_4 \qquad (V)$$

Substituting Equations U and V into Equation T and introducing K_{sp} values,

$$SO_4^{2-} = [(1.2 \times 10^{-6}) + (1.3 \times 10^{-10})]^{1/2} \tag{W}$$

Since the K_{sp} of $CaSO_4$ (1.2×10^{-6}) is at least 100 times greater than the K_{sp} of $BaSO_4$ (1.3×10^{-10}), $CaSO_4$ would be controlling the concentration of SO_4^{2-} in solution.

Solution Complexes and Mineral Solubility

Let us consider the mineral Ag_2Ss. Upon introducing water, Ag_2Ss undergoes dissolution until an equilibrium state is met:

$$Ag_2Ss \Leftrightarrow 2Ag^+ + S^{2-} \tag{X}$$

A mass-balance equation can be written as

$$1/2Ag^+ = S^{2-} \tag{Y}$$

However, since sulfide (S^{2-}) undergoes speciation when reacting with water, reactions for all potentially forming species would have to be written as

$$S^{2-} + H_2O \Leftrightarrow HS^- + OH^- \tag{Z}$$

$$HS^- + H_2O \Leftrightarrow H_2S + OH^- \tag{A'}$$

and

$$H_2O \Leftrightarrow H^+ + OH^- \tag{B'}$$

Complete mass and charge-balance equations can now be produced. The mass-balance equation is

$$Ag = 2S^{2-} + 2HS^- + 2H_2S \tag{C'}$$

and the charge-balance equation is

$$Ag^+ + H^+ = 2S^{2-} + HS^- + OH^- \tag{D'}$$

Considering that the K_{sp} of Ag_2Ss (6×10^{-50}) is too small to affect water pH,

$$OH^- \approx H^+ \tag{E'}$$

Setting the mass and charge-balance equations equal,

$$-Ag + 2S^{2-} + 2HS^- + 2H_2S = + Ag^+ - 2S^{2-} - HS^- \tag{F'}$$

Collecting terms and simplifying,

$$Ag^+ = 2S^{2-} + 3/2HS^- + H_2S \tag{G'}$$

We can now proceed to solve Equation G′ by substituting equations for the sulfide and OH^- species from reactions X, Z, A′, and B′. Therefore,

$$Ag^+ = \{(12 \times 10^{-50}) + 3/2(K_{eq1})(6 \times 10^{-50})(10^7) + (10^{14})K_{eq1}K_{eq2}(6 \times 10^{-50})\}^{1/3} \tag{H'}$$

and

$$Ag^+ = 2.54 \times 10^{-14} \text{ mol L}^{-1} \tag{I'}$$

where $6 \times 10^{-50} = K_{sp}$ of Ag_2Ss, $K_{eq1} = 8.3$ is the equilibrium constant for Reaction Z, $K_{eq2} = 1.8 \times 10^{-7}$ is the equilibrium constant for Reaction A′, and 10^7 is obtained by assuming OH^- equals 10^{-7} M and rearranging.

The above solution approach can be used for any problems involving mineral solubility and complexation. Complexes may either form by the species released from the particular mineral in question or by any other inorganic and/or organic ligands (e.g., chelators).

2.1.5 Solubility Diagrams

In addition to speciation diagrams on a percent species basis (Section 2.1.4), solubility diagrams can be plotted on a logarithmic scale of molar concentration. For example, in the case of $Al(OH)_3s$, a mineral which regulates aluminum in the solution of many soils, the solubility can be expressed as follows:

$$Al(OH)_3 \Leftrightarrow Al^{3+} + 3OH^- \tag{2.68}$$

with $K_{sp} = 1.9 \times 103^{-33}$. Using the equilibrium expression for the above reaction, when pH = 14 − pOH, substituting and solving for Al^{3+}, the logarithmic equation is

$$pAl^{3+} = -9.28 + 3\ pH \tag{2.69}$$

Similarly, for the reaction

$$Al^{3+} + H_2O \Leftrightarrow Al(OH)^{2+} + H^+ \tag{2.70}$$

$K_{eq} = 10^{-5.1}$. Using the equilibrium expression for the above reaction and solving for $Al(OH)^{2+}$, the logarithmic equation is

$$pAl(OH)^{2+} = -4.18 + 2pH \tag{2.71}$$

And finally, for the reaction

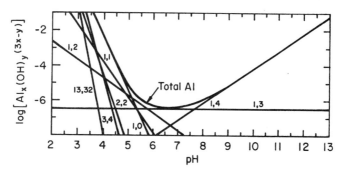

Figure 2.12. Distribution of hydrolysis products at ionic strength 1 M and temperature of 25°C in solutions saturated with α-Al(OH)$_3$ (gibbsite). (The first number denotes the number of Al atoms in the species; the second number denotes the number of OH molecules in the species. (adapted from Baes and Mesmer, 1976, with permission).

$$Al^{3+} + 4OH^- \Leftrightarrow Al(OH)_4^- + H^+ \tag{2.72}$$

$K_{eq} = 10^{-32.5}$. Using the equilibrium expression for the above reaction, substituting Al^{3+} by Equation 2.69, and considering that pH = 14 – pOH, substituting and rearranging the logarithmic equation yields

$$pAl(OH)_4^- = 14.22 - pH \tag{2.73}$$

One may plot Equations 2.69, 2.71, and 2.73 as $pAl_{hydroxy}$ species versus pH. This will produce three linear plots with different slopes. Equation 2.69 will produce a plot with slope 3, whereas Equations 2.71 and 2.73 will produce plots with slopes 2 and –1, respectively. The sum of all three aluminum species as a function of pH would give total dissolved aluminum. This is demonstrated in Figure 2.12, which describes the pH behavior of eight Al–hydroxy species. The following three points can be made based on Figure 2.12; (1) aluminum–hydroxide solubility exhibits a U-shaped behavior, (2) aluminum in solution never becomes zero, and (3) different aluminum species predominate at different pH values.

Aluminum–hydroxy species are known to exhibit different biological activity. For example, data have shown that some leafy plants are sensitive to Al^{3+} while others are known to be sensitive to Al(OH) monomers. It has also been shown that polymeric aluminum (more than one Al atom per molecule) is toxic to some organisms. Aluminum is one of the cations most difficult to predict in the soil solution. This is because it has the ability to form complex ions such as sulfate pairs and hydroxy–Al monomers and polymers.

The presence of sulfate in water can alter the solubilities of gibbsite and kaolinite, two minerals considered to control the concentration of aluminum in natural waters. Thus, minerals of lesser solubilities control the aluminum concentration in acid sulfate waters. These minerals most likely are: alunogen [Al$_2$(SO$_4$)$_3$·17H$_2$0s], with a K_{sp} of 10^{-7}; alunite [KAl$_3$(SO$_4$)$_2$(OH)$_6$s], with a K_{sp} of $10^{-85.4}$; jurbanite [Al(SO$_4$)(OH)·

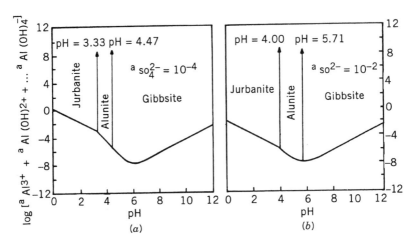

Figure 2.13. Solubility diagrams showing the pH range for the most stable Al-OH-SO_4 minerals for a potassium activity of 10^{-4} M and a sulfate activity of (a) 10^{-4} M and (b) 10^{-2} M (from Nordstrom, 1982b, with permission).

$5H_2Os]$, with a K_{sp} of $10^{-17.8}$; and basaluminite $[Al_4(SO_4)(OH)_{10}·5H_2Os]$, with a K_{sp} of $10^{-117.7}$. The pH stability range for these species at SO_4^{2-} activity of 10^{-4} mol L^{-1} is predicted to be between pH 0 and 3.3 for jurbanite and between pH 3.33 and 4.47 for alunite. Beyond pH 4.47, gibbsite $[Al(OH)_3s]$ appears to control aluminum in solution. This is demonstrated in Figure 2.13a. On the other hand, when SO_4^{2-} activity is 10^{-2} mol L^{-1}, between pH 0 and 4, jurbanite controls aluminum in solution, whereas between pH 4 and 5.7, alunite controls aluminum in solution. Beyond pH 5.7, gibbsite $[Al(OH)_3s]$ controls aluminum in solution (Fig. 2.13b).

2.2   SPECIFIC CONDUCTANCE

Specific conductance or electrical conductivity (EC) of a solution (e.g., natural water sample) is a function of total dissolved ions, type of ions, and their potential to form charged or noncharged pairs or complexes. The equation used to predict EC, in units of milimhos per centimeter (mmhos cm^{-1}), is

$$EC = AC/1000 \tag{2.72}$$

where A denotes the *limiting ionic conductance* in units of cm^2 (equiv-Ω)$^{-1}$, and C is concentration in equiv L^{-1}. Limiting ionic conductance values (A) for different ions are shown in Table 2.9. These data show that because A values between various ions differ, any two solutions with similar concentrations (in equivalents) will produce different EC values.

**TABLE 2.9.   Equivalent Ionic
Conductance of Different Ions Found
in Natural Water Systems**

Ion	A [cm^2 (equiv Ω)$^{-1}$]
H$^+$	349.8
K$^+$	73.6
Ca^{2+}	59.5
Mg^{2+}	53.1
OH$^-$	198.0
Cl$^-$	76.3
SO$_4^{2-}$	79.8
HCO$_3^-$	44.5
NO$_3^-$	71.4

Source: Tanji, 1969a.

Generally, the most commonly encountered ions in natural bodies of water, with
the exception of H$^+$ and OH$^-$, exhibit A values in the range of approximately 50–80
cm^2 (equiv-Ω)$^{-1}$. Hydrogen and OH$^-$ exhibit A values of 349.8 and 198.0 (equiv-Ω)$^{-1}$,
respectively. Because H$^+$ and OH$^-$ exhibit much greater A values than all other metallic
ions, and because ions in solution have a tendency to interact physically with each
other (form pairs or complexes), the relationship between solution EC and solution

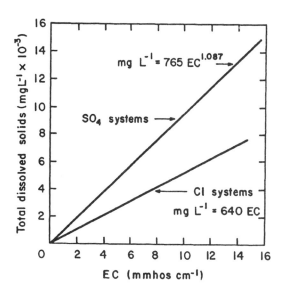

Figure 2.14.   Relationship between EC and dissolved solids for Cl$^-$-dominated and SO$_4^{2-}$-domi-
nated water systems (from Evangelou and Sobek, 1988, with permission).

salt content is pH-dependent. At low pH (< 3) (e.g., acid drainages), the contribution of H^+ to EC is significant. Additionally, ionic interactions also exhibit significant impact on the relationship between EC and salt concentration. The influence of ionic interactions on EC is demonstrated in Figure 2.14. Note that two solutions represented by different anions (Cl^- versus SO_4^{2-}) but the same EC differ greatly in total dissolved solids (Evangelou, 1995b).

Electrical conductivity is one of the most rapid and inexpensive measurements that can be made to assess water quality. It is a common practice to change EC values into milligrams per liter (mg L^{-1}) or parts per million (ppm) of dissolved solids. The equation for doing so is:

$$1 \text{ mmhos cm}^{-1}(EC) \times 10 = 10 \text{ meq L}^{-1} \times 64 \text{ mg (meq)}^{-1}$$

$$= 640 \text{ ppm dissolved solids} \qquad (2.73)$$

where 64 represents an average equivalent weight of salts commonly found in natural waters, and 10 represents a constant.

Example

Given: A water sample with an EC of 0.5 mmhos cm^{-1}
Find: Parts per million of dissolved solids
Solution: $0.5 \times 10 = 5 \text{ meq L}^{-1} \times 64 \text{ mg (meq)}^{-1} = 320 \text{ ppm}$

Equation 2.73 is only applicable to water samples with EC less than 1 mmhos cm^{-1}. For EC greater than 1 mmhos cm^{-1}, Equation 2.73 gives erroneous results. For example, according to Equation 2.73, a water sample with an EC of 2.3 mmhos cm^{-1} will have approximately 1472 ppm dissolved solids. If this water sample represented a sulfate water, its actual dissolved solids would be nearly 1900 ppm. For water samples with an EC above 1 mmhos cm^{-1} Equation 2.74 maybe more applicable.

$$ppm_{(dissolved\ solids)} = 640(EC)^{1.087} \qquad (2.74)$$

where EC is in mmhos cm^{-1} or dS m^{-1}.

2.3 ACIDITY–ALKALINITY

There are two kinds of acidity or alkalinity components in natural systems: (1) acidity or alkalinity dissolved in water, and (2) acidity or alkalinity present in the solid phase. The solid-phase acidity or alkalinity represents the reservoir of acidity or alkalinity encountered in water. By definition (commonly employed by commercial analytical laboratories), acidity refers to the amount of base added to 1 L of solution to bring the pH to 7 or 8.4. Alkalinity refers to the amount of acid added to 1 L of solution to bring the pH to 4.2. These definitions of acidity or alkalinity are limited in that they do not reveal the pH range at which maximum pH buffering capacity is exhibited. The pH range at which maximum pH buffering capacity is exhibited is related to particular chemical species in solution.

2.3.1 Alkalinity Speciation

There are two potential inorganic alkalinity sources in natural water systems. One is phosphate ($H_2PO_4^-/HPO_4^{2-}$) and the other is bicarbonate(HCO_3^-/H_2CO_3). The role of the $H_2PO_4^-/HPO_4^{2-}$ alkaline buffer is minimal, simply because its concentration in natural water systems is negligible. Bicarbonate plays a very important role in buffering large natural bodies of water. Two examples that show the role of bicarbonate and atmospheric CO_2 in controlling pH of rain and natural bodies of water are given below.

$$pCO_2 + H_2O \overset{K_{eq1}}{\Leftrightarrow} H_2CO_3 \tag{2.75}$$

where pCO_2 denotes partial pressure of CO_2, and

$$H_2CO_3 \overset{K_{eq2}}{\Leftrightarrow} H^+ + HCO_3^- \tag{2.78}$$

Summing Reactions 2.75 and 2.76,

$$pCO_2 + H_2O \overset{K_{eq1}K_{eq2}}{\Leftrightarrow} H^+ + HCO_3^- \tag{2.77}$$

and

$$K_{eq1}K_{eq2} = (H^+)(HCO_3^-)/pCO_2 \tag{2.78}$$

Since

$$H^+ = HCO_3^- \tag{2.79}$$

then

$$H^+ = (K_{eq1}K_{eq2}pCO_2)^{1/2} \tag{2.80}$$

Substituting values into Equation 2.80,

$$H^+ = [(10^{-1.47})(10^{-6.4})(10^{-3.52})]^{1/2} \tag{2.81}$$

TABLE 2.10. Carbon Dioxide Content of Water and pH

Pressure CO_2 (atm)	pH
0.0003	5.72
0.0030	5.22
0.0100	4.95
0.1000	4.45
1.0000	3.95

$$H^+ = 2.04 \times 10^{-6} \text{ mol L}^{-1} \tag{2.82}$$

$$pH = 5.69 \tag{2.83}$$

The actual measured pH of rainwater in equilibrium with atmospheric pCO_2 is somewhere around 5.6. Values of pH below 5.6 can either be due to an increased pCO_2 or to industrial emissions causing what is known as *acid rain*. As shown, the dissolution of CO_2 in water is an cause of acid water (Table 2.10). However, considering that soils and/or geologic systems are sources of HCO_3^- and CO_3^{2-}, when water contacts soil or

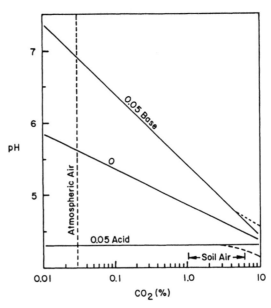

Figure 2.15. Dependence of pH in solution on the atmospheric pressure of CO_2 (expressed as percent by volume), and the addition of strong acid or base (from McBride, 1994, with permission).

TABLE 2.11. Composition of Soil Air[a]

Depth of Sample (ft)	Sandy Loam		Silty Clay Loam		Silty Clay	
	CO_2	O_2	CO_2	O_2	CO_2	O_2
1	0.8	19.9	1.0	19.8	1.7	18.2
2	1.3	19.4	3.2	17.9	2.8	16.7
3	1.5	19.1	4.6	16.8	3.7	15.6
4	2.1	18.3	6.2	16.0	7.9	12.3
5	2.7	17.9	7.1	15.3	10.6	8.8
6	3.0	17.5	7.0	14.8	10.3	4.6

[a]Given as percent by volume of air.

geologic environments, additional HCO_3^- is picked up and this shifts the pH value of the water upward.

The role of external alkalinity sources in regulating pH is demonstrated in Figure 2.15. It shows that addition of a base (e.g., an external source of OH^-) reduces to some degree the capacity of CO_2 to bring the pH below 4 as CO_2 approaches 10%. However, when CO_2 is less than 10%, the pH becomes highly sensitive to pCO_2 changes. Addition of external acidity, so that the pH becomes approximately 3.95 (Table 2.10), diminishes the capacity of CO_2 to have any pH control since only formation of H_2CO_3 (a weak acid) takes effect. Soil systems exhibit pCO_2 values higher than those encountered in the atmosphere (Table 2.11). For this reason, CO_2 is considered a cause of soil acidification, especially in the absence of an external HCO_3^- source. For example, the titration data in Figure 2.16 along with the data in Figure 2.15 show that CO_2 has the potential to drop the pH low enough (≈ 4.3) so that $Al(OH)_3$s dissolution would have an effect, a major environmental concern. However, the pH of soil resulting from pCO_2 could not get low enough to dissolve $Fe(OH)_3$s.

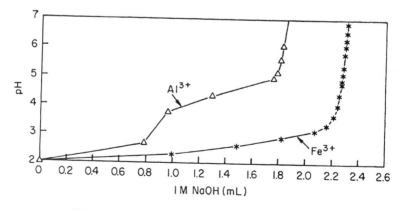

Figure 2.16. Potentiometric titration of Fe^{3+} and Al^{3+}.

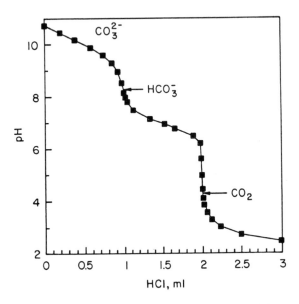

Figure 2.17. Potentiometric titration of Na_2CO_3 solution.

The maximum buffering capacity of the H_2CO_3–HCO_3^-–CO_3^{2-} system occurs at pH 6.4 and 10.3, because at pH 6.4 the H_2CO_3 is equal to HCO_3^- and at pH 10.3 the HCO_3^- is equal to CO_3^{2-}. These pH buffer relationships are shown in Figure 2.17. The same data can also be presented as carbonate species in percent versus pH (Fig. 2.18). The y intercepts at 50% of the species shown, when extrapolated to the x axis, are indicative of the pH values at which maximum buffering capacity occurs. To understand Figure 2.18 clearly, the following mathematical relationships should be understood. The total dissolved carbonate (C_T) is described by

$$C_T = H_2CO_3 + HCO_3^- + CO_3^{2-} \tag{2.84}$$

The percentage of any of the carbonate species in relationship to C_T can be expressed as

$$\% \ H_2CO_3 = (H_2CO_3/C_T) \times 100 \tag{2.85}$$

$$\% \ HCO_3^- = (HCO_3^-/C_T) \times 100 \tag{2.86}$$

$$\% \ CO_3^{2-} = (CO_3^{2-}/C_T) \times 100 \tag{2.87}$$

when

$$HCO_3^- = (H_2CO_3) \ (K_{eq1}/(H^+)) \tag{2.88}$$

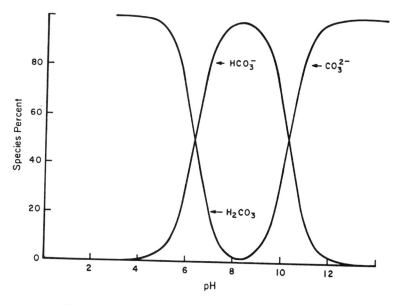

Figure 2.18. Speciation diagram of the carbonate system.

and

$$CO_3^{2-} = (K_{eq2}) (HCO_3^-)/(H^+) \qquad (2.89)$$

By substituting Equations 2.88 and 2.89 into Equation 2.84 and rearranging, the following equations are derived:

$$\% \ H_2CO_3 = [(H^+)^2/\{(H^+)^2 + K_{eq1}(H^+) + (K_{eq1}K_{eq2})\}] \times 100 \qquad (2.90)$$

$$\% \ HCO_3^- = [(H^+) (K_{eq1})/\{(H^+)^2 + K_{eq1}(H^+) + (K_{eq1}K_{eq2})\}] \times 100 \qquad (2.91)$$

and

$$\% \ CO_3^{2-} = [(K_{eq1}K_{eq2})/\{(H^+)^2 + K_{eq1}(H^+) + (K_{eq1}K_{eq2})\}] \times 100 \qquad (2.92)$$

where K_{eq1} and K_{eq2} are $10^{-6.4}$ and $10^{-10.3}$, respectively. Knowing the pH of a given water sample and its C_T (obtained by titration), % H_2CO_3, % HCO_3^-, and % CO_3^{2-} can be calculated as a function of pH by Equations 2.90–2.92, as shown in Figure 2.18.

2.3.2 Neutralization Potential

Generally, natural water contains two types of dissolved inorganic components, anions and cations. Anions are negatively charged, such as sulfate (SO_4^{2-}), chloride (Cl^-) and nitrate (NO_3^-), whereas cations are positively charged, such as calcium (Ca^{2+}), magnesium (Mg^{2+}), sodium (Na^+), and potassium (K^+). Beginning with the concept of electroneutrality,

$$\sum Z_i C_i - \sum Z_j A_j = 0 \tag{2.93}$$

where Σ denotes sum, Z_i and Z_j are the valency of cations and anions, respectively, and C_i and A_j denote the concentrations of anions and cations, respectively. For any water sample, the electroneutrality equation would be as follows:

$$[2Ca^{2+} + 2Mg^{2+} + Na^+ + K^+ + H^+] -$$

$$[2SO_4^{2-} + Cl^- + NO_3^- + OH^- + 2CO_3^{2-} + HCO_3^-] = 0 \tag{2.94}$$

Equation 2.94 can be used to calculate alkalinity by considering the following:

$$\text{Alkalinity} = 2CO_3^{2-} + HCO_3^- + OH^- - H^+ \tag{2.95}$$

Note that Equation 2.95 does not include the H_2CO_3 component, which is part of C_T (Eq. 2.84).

A more appropriate way of expressing alkalinity and/or acidity of acid waters emanating from natural environments is by defining the term commonly known as *neutralization potential* (NP):

$$NP = 2CO_3^{2-} + HCO_3^- + OH^- - HSO_4^-$$

$$- H^+ - 2Mn^{2+} - 2Fe^{2+} - 2Cu^{2+} - 3Al^{3+} \tag{2.96}$$

If the solution to Equation 2.96 gives a negative number, this value is considered to be acidity. However, if the solution gives a positive number, this value is considered to be alkalinity.

2.3.3 Alkalinity Contribution by CaCO3

Alkalinity of effluents in natural systems is often controlled by carbonate minerals. The most common carbonate mineral in the environment is limestone ($CaCO_3$). The factors that affect $CaCO_3$ solubility are pH and pCO_2. The equations that one needs to consider in estimating solubility and alkalinity are

$$CaCO_3 \overset{K_{eq1}}{\Longleftrightarrow} Ca_2^+ + CO_3^{2-} \tag{2.97}$$

$$CO_{2aq} \Longleftrightarrow k \cdot pCO_2 \tag{2.98}$$

$$CO_{2aq} + H_2O \overset{K_{eq2}}{\Longleftrightarrow} H^+ + HCO_3^- \qquad (2.99)$$

$$HCO_3^- \overset{K_{eq3}}{\Longleftrightarrow} H^+ + CO_3^{2-} \qquad (2.100)$$

where

$$k = \text{Henry's constant} = 10^{-1.5} \qquad (2.101)$$

$$pCO_2 = 10^{-3.5} \qquad (2.102)$$

$$K_{eq1} = 10^{-8.3} \qquad (2.103)$$

$$K_{eq2} = 10^{-6.4} \qquad (2.104)$$

$$K_{eq3} = 10^{-10.3} \qquad (2.105)$$

If

$$Ca^{2+} = K_{sp}/CO_3^{2-} \qquad (2.106)$$

then, based on Equations 2.98–2.102,

$$HCO_3^- = (CO_{2aq})(10^{-6.4})/H^+ \qquad (2.107)$$

$$CO_3^{2-} = (HCO_3^-)(10^{-10.3})/H^+ \qquad (2.108)$$

Substituting Equations 2.98 and 2.107 into Equation 2.108,

$$CO_3^{2-} = (pCO_2)(10^{-1.5})(10^{-6.4})(10^{-10.3})/(H^+)^2 \qquad (2.109)$$

Substituting Equation 2.109 into Equation 2.106 and taking logarithms on both sides of the equation,

$$\log Ca^{2+} = 9.9 - 2pH + \log 1/pCO_2 \qquad (2.110)$$

Because pH and pCO_2 are interdependent, another way to predict the solubility of $CaCO_3$ based on equilibrium considerations is as follows. Considering that

$$CO_{2aq} + H_2O \Longleftrightarrow H^+ + HCO_3^- \qquad (2.111)$$

and

$$HCO_3^- = (10^{-6.4})(CO_{2aq})/H^+ \qquad (2.112)$$

If

$$CO_{2aq} = pCO_2(10^{-1.5}) \qquad (2.113)$$

then

$$CO_{2aq} = (10^{-3.5})(10^{-1.5}) = (10^{-5}) \qquad (2.114)$$

By replacing CO_{2aq} in Equation 2.112 with the value estimated by Equation 2.114,

$$HCO_3^- = (10^{-11.4}/H^+) \qquad (2.115)$$

and

$$CO_3^{2-} = (10^{-10.3})(HCO_3^-)/H^+ \qquad (2.116)$$

Substituting Equation 2.115 into Equation 2.116 gives

$$CO_3^{2-} = (10^{-10.3})(10^{-11.4})/(H^+)^2 = (10^{-21.7})/(H^+)^2 \qquad (2.117)$$

Since

$$2Ca^{2+} + H^+ = 2CO_3^{2-} + HCO_3^- + OH^- \qquad (2.118)$$

and

$$Ca^{2+} = Ksp/CO_3^{2-} \qquad (2.119)$$

Then substituting Equation 2.117 into Equation 2.119 gives

$$Ca^{2+} = 10^{-8.3}/[(10^{-21.7}/(H^+)^2] \qquad (2.120]$$

or

$$Ca^{2+} = 10^{12.4}(H^+)^2 \qquad (2.121)$$

Substituting Equations 2.115, 2.117, and 2.121 into Equation 2.118 and rearranging gives

$$(H^+)^{13.7}(H^+)^4 + (H^+)^3 - 10^{-11.4}(H^+) = 10^{-21.4} \qquad (2.122)$$

Equation 2.122 can be solved by trial and error, giving $(Ca^{2+}) = 10^{-3.4}$, $(CO_3^{2-}) = 10^{-4.9}$, $(HCO_3^-) = 10^{-3}$, $(H_2CO_3) = 10^{-5}$, and $(H^+) = 10^{-8.4}$ (Garrels and Christ, 1965). By generating similar equations for various pCO_2 values, Figure 2.19 is produced. As shown from Equation 2.110, the solubility of $CaCO_3$ depends on two variables, pCO_2 and pH. When pCO_2 changes, pH and Ca^{2+} also change. The concen-

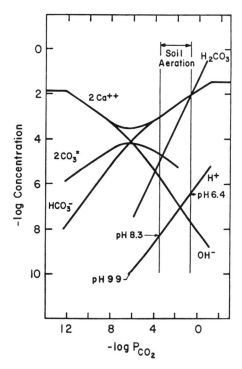

Figure 2.19. Stability diagram of $CaCO_3$ in natural water systems and soil or geologic solutions (from Jackson, 1975, with permission).

tration of Ca^{2+} reflects the level of alkalinity in water since mass balance reveals that any Ca^{2+} quantity (in equivalents) released from $CaCO_3$ must be accompanied by a similar quantity (in equivalents) of HCO_3^-. Figure 2.19 indicates that as pCO_2 increases, HCO_3^- also increases and parallels the increase of Ca^{2+}. When the pH decreases, however, OH^- and CO_3^{2-} decrease. However, the magnitude of the decreases of OH^- and CO_3^{2-} is small in comparison to the increases in HCO_3^-. Based on the above, alkalinity increases as pCO_2 increases, but pH decreases. Note that in the case of absence of $CaCO_3$, alkalinity is independent of pCO_2.

2.4 CHELATES

Chelates refer to organic molecules that have the potential to form inner-sphere complexes with divalent hard metals and heavy metals. The behavior of the complex ion is, in general, different from that of the noncomplexed ion. For example, copper hydroxide precipitates when sodium hydroxide is added to a solution of Cu^{2+}. If hydroxide is added to a solution containing strongly chelated Cu^{2+}, no precipitation takes place. *Chelation* is the process by which metals bond to ligands or functional

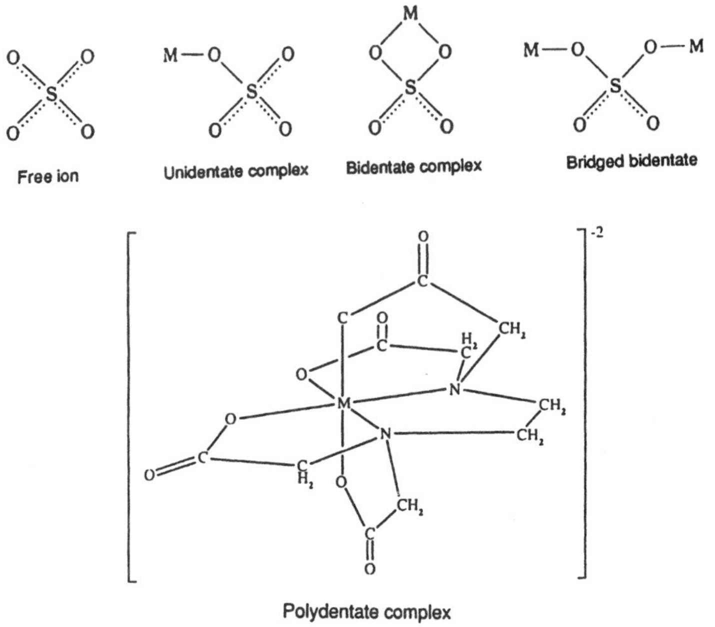

Figure 2.20. Schematic of ligands (from Skoog and West, 1976, with permission).

chelate groups. There are various coordination modes of metal–chelate complexes. These modes include *monodentates* (single point of chemical bonding), *bidentates* (double point of chemical bonding), and *polydentates* (three or more points of chemical bonding (Fig. 2.20).

The treatment of complex ion equilibria in solution is analogous to the treatment of weak acids. One of the best known chelators is the well-known EDTA (ethylene diamine tetraacetic acid). The metal stability constants for EDTA are very high, which indicates strong complexes. Various other compounds are available with high metal-stability constants for agricultural or environmental uses. Some of the more important ones are DPTA (diethylene triamine pentaacetic acid), CyDTA (cyclohexane diamine tetraacetic acid), EDDA [ethylene diamine di (O-hydroxyphenyl) acetic acid], or Chel-138.

Tables 2.12 and 2.13 list the logarithm of the stability constants for the complexes of these chelating agents with various metal ions. Note that with the exception of Chel-138, calcium and magnesium form rather stable complexes with these chelating agents; Fe^{3+} forms the most stable chelate of any metal listed. Generally, ferric iron is followed by Cu^{2+}, Zn^{2+}, Mn^{2+}, Fe^{2+}, Ca^{2+}, and Mg^{2+}. The weak acid properties of these chelating agents must be considered in any evaluation of their behavior. Because they are weak acids, the hydrogen ion tends to compete with the metal ions for association with the active groups.

TABLE 2.12. Log Stability Constants

Metal Ion	EDTA	DPTA	CyDTA	Chel-138
Mg^{2+}	8.69	9.0	10.3	2.3
Ca^{2+}	10.59	10.6	12.5	1.6
Cu^{2+}	18.80	21.0	21.3	>15
Zn^{2+}	16.50	18.1	18.7	9.26
Mn^{2+}	14.04	15.1	16.8	
Fe^{2+}	14.2	16.7		
Fe^{3+}	25.1	28.6		~30

To simplify notation, the anion form of the chelating agent is usually represented by the letter Y. Thus, $EDTA^{4-} = Y^{4-}$ and the acid formed is $H_4EDTA = H_4Y$. The disassociation of hydrogen ions takes place in steps:

$$H_4Y \Leftrightarrow H^+ + H_3Y^- \tag{2.123}$$

and

$$K_1 = (H^+)(H_3Y^-)/(H_4Y) \tag{2.124}$$

$$H_3Y^- \Leftrightarrow H^+ + H_2Y^{2-} \tag{2.125}$$

and

$$K_2 = (H^+)(H_2Y^{-2}/(H_3Y^{1-}) \tag{2.126}$$

$$H_2Y^{2-} \Leftrightarrow H^+ + HY^{3-} \tag{2.127}$$

TABLE 2.13. Formation Constants for EDTA Complexes

Cation	K_{eq}	$\log K_{eq}$	Cation	K_{eq}	$\log K_{eq}$
Ag^+	2.1×10^7	7.32	Cu^{2+}	6.3×10^{18}	18.80
Mg^{2+}	4.9×10^8	8.69	Zn^{2+}	3.2×10^{16}	16.50
Ca^{2+}	5.0×10^{10}	10.70	Cd^{2+}	2.9×10^{16}	16.46
Sr^{2+}	4.3×10^8	8.63	Hg^{2+}	6.3×10^{21}	21.80
Ba^{2+}	5.8×10^7	7.76	Pb^{2+}	1.1×10^{18}	18.04
Mn^{2+}	6.2×10^{13}	13.79	Al^{3+}	1.3×10^{16}	16.13
Fe^{2+}	2.1×10^{14}	14.33	Fe^{3+}	1.3×10^{25}	25.1
Co^{2+}	2.0×10^{16}	16.31	V^{3+}	7.9×10^{25}	25.9
Ni^{2+}	4.2×10^{18}	18.62	Th^{4+}	1.6×10^{23}	23.2

and

$$K_3 = (H^+)(HY^{3-})/(H_2Y^{2-}) \tag{2.128}$$

$$HY^{3-} \Leftrightarrow H^+ + Y^{4-} \tag{2.129}$$

and

$$K_4 = (H^+)(Y^{4-})/(HY^{3-}) \tag{2.130}$$

Considering that

$$C_T = H_4Y + H_3Y^- + H_2Y^{-2} + HY^{-3} + Y^{-4} \tag{2.131}$$

taking the inverse of the above equation,

$$1/C_T = 1/\{H_4Y + H_3Y^- + H_2Y^{-2} + HY^{-3} + Y^{-4}\} \tag{2.132}$$

Replacing the EDTA species in solution as a function of H^+ and the various dissociation constants (K_i) gives

$$1/C_T = 1/\{(H^+)^4 + K_1(H^+)^3 + K_1K_2(H^+)^2 + K_1K_2K_3(H^+)^3$$
$$+ K_1K_2K_3K_4(H^+)^4\} \tag{2.133}$$

For simplicity, Equation 2.133 is written as

$$1/C_T = 1/\left\{(H^+)^4 + \sum_{i=1, j=4}^{i=4, j=1} K_i(H^+)^j\right\} \tag{2.134}$$

and rearranging

$$\alpha_0 = H_4Y/C_T = (H^+)^4/\left\{(H^+)^4 + \sum_{i=1, j=4}^{i=4, j=1} K_i(H^+)^j\right\} \tag{2.135}$$

$$\alpha_1 = H_3Y^{1-}/C_T = K_1(H^+)^3/\left\{(H^+)^4 + \sum_{i=1, j=4}^{i=4, j=1} K_i(H^+)^j\right\} \tag{2.136}$$

$$\alpha_2 = H_2Y^{2-}/C_T = K_1K_2(H^+)^2/\left\{(H^+)^4 + \sum_{i=1, j=4}^{i=4, j=1} K_i(H^+)^j\right\} \tag{2.137}$$

$$\alpha_3 = HY^{3-}/C_T = K_1K_2K_3(H^+)/\left\{(H^+)^4 + \sum_{i=1, j=4}^{i=4, j=1} K_i(H^+)^j\right\} \tag{2.138}$$

and

$$\alpha_4 = Y^{4-}/C_T = K_1K_2K_3K_4 \left/ \left\{ (H^+)^4 + \sum_{i=1, j=4}^{i=4, j=1} K_i(H^+)^j \right\} \right. \tag{2.139}$$

The components α_i denote the mole fraction of the species present in solution at any given pH. A plot of α_i versus pH is given in Figure 2.21.

The EDTA–metal complexes are rather unique because they always form in 1:1 ratio regardless of the charge on the cation. The EDTA–metal formation constant can be described by the reaction

$$M^{n+} + Y^{4-} \Leftrightarrow MY^{(n-4)+} \tag{2.140}$$

and the equilibrium expression is

$$K_{eq} = (MY^{[n-4]+})/(M^{n+})(Y^{4-}) \tag{2.141}$$

Equation 2.140 shows that formation of the $MY^{(n-4)+}$ complex is Y^{4-} dependent. The species Y^{4-} is pH dependent. Figure 2.21 shows that the species Y^{4-} begins to form above pH 7 and approaches maximum at pH 12. Because of this pH dependency of the Y^{4-} species, conditional formation constants are used to estimate EDTA-metal complexes in solution. Considering that

$$\alpha_4 = (Y^{4-})/(C_T) \tag{2.142}$$

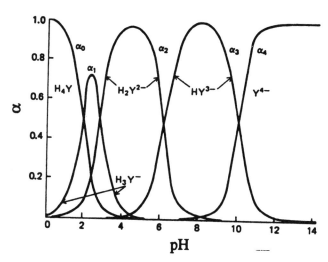

Figure 2.21. Composition of EDTA solutions as a function of pH (from Skoog and West, 1976, with permission).

then

$$Y^{4-} = (\alpha_4)(C_T) \qquad (2.143)$$

By replacing the species Y^{4-} in Equation 2.141 with Equation 2.143

$$K_{con} = \alpha_4 K_{eq} = (MY^{[n-4]+})/(M^{n+})C_T \qquad (2.144)$$

where K_{con} = conditional formation constant. Therefore, by knowing α_4 at any given pH (Eq. 2.139), $MY^{(n-4)+}$ can be estimated for any pH

$$[MY^{[n-4]+}] = (K_{con})(M^{n+})(C_T) \qquad (2.145)$$

or

$$[MY^{[n-4]+}] = (\alpha_4 K_{eq})(M^{n+})(C_T) \qquad (2.146)$$

Using Equation 2.146, the lowest possible pH at which the total metal in solution can be complexed by an equamolar EDTA solution can be predicted (Fig. 2.22).

Other competing reactions, however, must also be considered to determine whether or not a particular metal chelate would be stable. For example, when two metals are present in a solution system (e.g., Fe^{2+} and Cu^{2+}) in the presence of EDTA, speciation calculations can be carried out as follows: Consider

$$[FeY^{[2-4]+}] = (\alpha_4 K_{eqFe})(Fe^{2+})(C_T) \qquad (2.147)$$

and

$$[CuY^{[2-4]+}] = (\alpha_4 K_{eqCu})(Cu^{2+})(C_T) \qquad (2.148)$$

Note that

$$(Y_T^{-4}) = Y^{4-} + FeY^{(2-4)+} + CuY^{(2-4)+} \qquad (2.149)$$

Substituting all the terms of Equation 2.149 by Equations 2.143, 2.147, and 2.148 and taking the inverse gives

$$1/(Y_T^{4-}) = 1/\{(\alpha_4)(C_T) + (\alpha_4 K_{eqFe})(Fe^{2+})(C_T) + (\alpha_4 K_{eqCu})(Cu^{2+})(C_T)\} \qquad (2.150)$$

and by multiplying Equation 2.150 by Equation 2.143, 2.147, or 2.148, the following three equations describing the distribution of the metal–EDTA species as well as the Y^{4-} species are produced:

$$\alpha_{0'} = Y^{4-}/(Y_T^{4-}) = (\alpha_4)(C_T)/\{[1 + (K_{eqFe})(Fe^{2+}) + (K_{eqCu})(Cu^{2+})]\} \qquad (2.151)$$

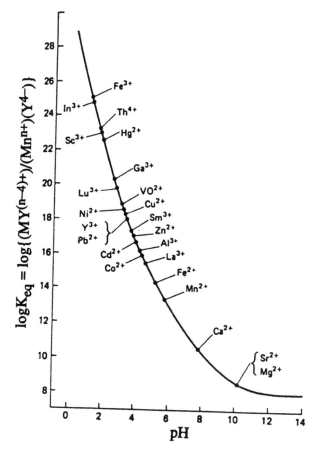

Figure 2.22. Minimum pH needed for 100% complexation of various cations with equamolar EDTA (from Reilley and Schmid, 1958, with permission).

$$\alpha_{Fe} = FeY^{(2-4)+}/(Y_T^{-4}) = (K_{eqFe})(Fe^{2+})/\{[1 + (K_{eqFe})(Fe^{2+}) + (K_{eqCu})(Cu^{2+})]\}$$

and

$$(2.152)$$

$$\alpha_{Cu} = CuY^{(2-4)+}/(Y_T^{-4}) = (K_{eqCu})(Cu^{2+})1/\{1 + (K_{eqFe})(Fe^{2+}) + (K_{eqCu})(Cu^{2+})]\}$$

$$(2.153)$$

The various chelating agents described above are all manmade. There are also many natural chelating agents. In fact, nature abounds with chelating agents. The main structural requirement is that the compound have groups capable of associating with the metal ion on adjacent carbon atoms. Such groups include amine and carboxyl.

Other groups which can associate with metal ions include $R_2CH\text{-}OH$, $R_2C=O$, and $RCH_2\text{-}SH$. Thus, organic acids, amino acids, amines and sugars can all be chelating agents if they meet the steric requirements for ring formation.

PROBLEMS AND QUESTIONS

1. Given a 10-mM KCl solution, calculate the following parameters using the appropriate equations given in this chapter:
   a. Ionic strength.
   b. Activity coefficients (γ) of K^+ and Cl^-.
   c. Activity of K^+ and Cl^-.
   d. Discuss the difference between activity and concentration.

2. Given a 10-mM $MgCl_2$ solution, calculate the following parameters using the appropriate equations given in this chapter:
   a. Ionic strength.
   b. Activity coefficients (γ) of Mg^{2+} and Cl^-.
   c. Activity of Mg^{2+} and Cl^-.
   d. Discuss the difference between the activity of divalent and monovalent ions (see answers to problem 1).
   e. Using the equilibrium constant of the $MgCl^+$ pair (Table 2.4), calculate its activity and then its concentration.
   f. What can you conclude about the potential of Mg^{2+} to form pairs with the Cl^-?

3. An aquifer was found to contain the mineral barite ($BaSO_4$s, s = solid, K_{sp} = 1.3 $\times 10^{-10}$).
   a. Calculate the expected barium (Ba^{2+}) activity in the aquifer's water when in equilibrium with $BaSO_4$s.
   b. Calculate the expected barium (Ba^{2+}) concentration in the aquifer's water when in equilibrium with $BaSO_4$s.
   c. Explain whether the calculated Ba^{2+} concentration is above or below the allowed maximum contaminant level (1 ppm).

4. A company manufactures calcium oxide (CaO) for commercial purposes (e.g., it is used to remove sulfur from the smokestacks of coal-fired plants), but a portion of this oxide is rejected because of low quality. The management decides to discard it in an adjacent river and quickly CaO becomes $Ca(OH)_2$s as follows:

$$CaO + H_2O \rightarrow Ca(OH)_2s \ (s = solid)$$

and $Ca(OH)_2$s reaches an equilibrium state with the river water:

$$Ca(OH)_2s = Ca^{2+} + 2OH^-; \quad K_{sp} = 10^{-5.43}$$

a. Calculate the maximum potential pH of the river water in equilibrium with $Ca(OH)_2s$.

b. Explain whether the pH of the river water is going to increase, decrease, or remain the same if the CaO discharge practice is discontinued.

c. Consider also

$$Ca(OH)_2s = CaOH^+ + OH^-, \quad K_{eq} = 10^{-4.03}$$

Calculate the approximate total Ca concentration in the river water contributed by the $Ca(OH)_2$.

5. Cyanide (CN^-) is a complexing agent used in the processing of various metal ores. Using the speciation procedure outlined in Sections 2.1.4 and 2.4 (the pK_a of hydrogen cyanide (HCN) is 9.2 (HCN $\Leftrightarrow H^+ + CN^-$).

a. Produce two equations that would allow you to calculate the concentration of HCN and CN^- as a function of pH.

b. Plot these data similarly to Figure 2.18.

c. Since HCN is gaseous and toxic, point out the range of pH and the values at which HCN will persist.

PART II
Soil Minerals and
Surface Chemical Properties

3 Soil Minerals and
Their Surface Properties

3.1 COMPOSITION AND STRUCTURE OF SOIL MINERALS

The major mineral groups commonly found in soil include: (1) aluminosilicates, (2) oxides, and (3) organic matter. Through their surface electrochemical properties, these soil minerals control adsorption, transformation, and release behavior of chemical constituents (e.g., nutrients and contaminants) to water or soil solution. Soil-surface electrochemical properties vary between soil types and depend on factors such as parent material, climate, and vegetation (Table 3.1). Generally, the overall makeup of soil is (Fig. 3.1)

1. Inorganic mineral matter (defined as soil material made up mostly of oxygen, silicon, and aluminum—many other metals in small quantities may be included)
2. Organic mineral matter (defined as soil material having derived mostly from plant residues and made up mostly of carbon, oxygen, and hydrogen)
3. Solutes (refers to the portion of soil composed of water and mostly dissolved salts (plant nutrients)
4. Air (refers to the gaseous portion of soil composed of the same gases found in the atmosphere (oxygen, nitrogen, and carbon dioxide) but in different proportions)

Soil is a highly complex, highly variable biomolecular sieve with an array of physical and chemical properties. The physical properties include:

1. Macroporosity (composed of pores with diameter greater than 200 μm)
2. Microporosity (composed of pores with diameter less than 200 μm)

100

TABLE 3.1. The 11 Soil Orders

Name of Order	Derivation of Order Name	Character of the Soils
Entisol	Nonsense syllable "ent," from "recent"	Negligible differentiation of horizons in alluvium, frozen ground, desert sand, and so on, in all climates
Vertisol	L. *verto*, turn, invert	Clay-rich soils that hydrate and swell when wet, and crack on drying. Mostly in subhumid to arid regions
Inceptisol	L. *inceptum*, beginning	Soils with only slight horizon development. Tundra soils, soils on new volcanic deposits, recently deglaciated areas, and so on
Aridisol	L. *aridus*, dry	Dry soils; salt, gypsum, or carbonate accumulations common
Mollisol	L. *mollis*, soft	Temperate grassland soils with a soft, organic-enriched, thick, dark surface layer
Spodosol	Gr. *spodos*, wood ash	Humid forest soils. Mostly under conifers, with a diagnostic iron- or organic-enriched B horizon and commonly also an ash-gray leached A horizon
Alfisol	Syllables from the chemical symbols Al and Fe	Clay-enriched B horizon, young soils commonly under deciduous forests
Ultisol	L. *ultimus*, last	Humid temperate to tropical soils on old land surfaces, deeply weathered, red and yellow, clay-enriched soils
Oxisol	F. *oxide*, oxide	Tropical and subtropical lateritic and bauxitic soils. Old, intensely weathered, nearly horizonless soils
Histosol	Gr. *histos*, tissue	Bog soils, organic soils, peat, and muck. No climatic distinctions
Andisols	Modified from Ando	From volcanic ejecta, dominated by alophane or humic complexes

Source: Hausenbuiller, 1985; Buol et al., 1997.

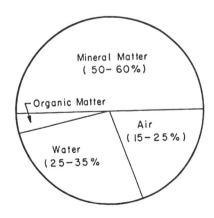

Figure 3.1. Average composition of soil material.

3. Physical stability (refers to bonding strength between soil particles forming aggregates)
4. External/internal surface and its geometry (defined as magnitude of soil's specific surface in square meters per gram and depth or width of clay's internal surface in nanometers)

The chemical properties include:

1. Permanent charge (defined as cation exchange capacity, CEC, which is independent of pH)
2. Variable charge (defined as pH-dependent CEC)
3. Point of zero charge, PZC (defined as the pH at which the net surface charge is zero or CEC minus anion exchange capacity, AEC, equals zero)
4. Inner-sphere/outer-sphere surface complexes (defined as strong surface complexes or inner-sphere complexes, as opposed to weak surface complexes or outer-sphere complexes)
5. Hydrophobic–hydrophilic potential (defined as the potential of soil to adsorb water; hydrophobic = does not like water; hydrophilic = likes water)
6. pH buffering (defined as the potential of soil to resist pH changes)

These physicochemical properties play a major role in regulating surface and groundwater chemistry or quality and nutrient availability to plants or soil organisms. In this chapter, soil makeup and the physical and chemical properties of soil are discussed.

3.2 ALUMINOSILICATE MINERALS

The approximate elemental composition of the earth's inorganic mineral surface is reported in Table 3.2. Elemental composition alone, however, cannot justify the unique properties of soil and how such properties influence the soil and water environment. The component that explains many of the physical and chemical properties of soil is the molecular arrangement of elements, forming structures with unique physicochemical properties. Soil mineral structures are briefly discussed below.

The inorganic minerals of soil are classified into (a) *primary minerals* and (b) *secondary minerals* (Table 3.3). Primary minerals are minerals with the chemical composition and structure obtained during the crystallization process of molten lava, whereas secondary minerals are those that have been altered from the original structure and chemical composition by weathering, a process referred to as the geomorphic cycle (Fig. 3.2). Generally, the size of soil mineral particles varies from clay-sized colloids (< 2 μm) to gravel (< 2 mm) and rocks.

Aluminosilicates or phyllosilicates are inorganic crystalline structures which make up a large part of the < 0.2 mm soil-sized particles. These minerals, commonly referred to as clay minerals, consist of Si–O tetrahedrons, in which one silicon atom (Si^{4+}) is

TABLE 3.2. Approximate Elemental Composition of the Earth's Outer-Surface Layer

Element	Percent
O	46.5
Si	27.6
Al	8.1
Fe	5.1
Ca	3.6
Mg	2.1
Na	2.8
K	2.6
Ti	0.6
P	0.12
Mn	0.09
S	0.06
Cl	0.05
C	0.04

centered between four oxygen atoms (Fig. 3.3), and Al–O octahedrons, in which one aluminum atom (Al^{3+}) is centered between six oxygen atoms ($-O^{2-}$) (OH molecules may also be included depending on mineral type) (Fig. 3.4). The cations of Si^{4+} and Al^{3+} are referred to as coordinating cations. The simple Si–O tetrahedral and Al–O octahedral structures form sheets by sharing oxygen atoms (Fig. 3.5). The substitution of an aluminum atom for silicon in the tetrahedron or the substitution of divalent cations (e.g., Fe^{2+} and Mg^{2+}) for aluminum in the octahedral sheet is a common occurrence.

The amount of substitution in the tetrahedral and octahedral sheets and the ratio of octahedral to tetrahedral sheets are the primary differentiating characteristics between the many clay minerals (Fig. 3.6). For example, clays that have one tetrahedral sheet and one octahedral sheet are known as 1:1 clay minerals (e.g., kaolin group) (Fig. 3.7); clays that have two tetrahedral sheets and one octahedral sheet are known as 2:1 clay minerals (e.g., smectite group) (Fig. 3.8) or mica and vermiculite (Fig. 3.9), while clays that have two tetrahedral sheets and two octahedral sheets are known as 2:2 clay minerals (e.g., chlorite) (Fig. 3.10). These sheet arrangements give rise to various mineral surface identities such as magnitude (specific surface), functional groups, and interactions with solution species.

Substitution of a given coordinating cation by a cation with lower valence in a mineral gives rise to *permanent negative charge* or cation exchange capacity, CEC. Location of this substitution (e.g., tetrahedral sheet or octahedral sheet) gives rise to clay minerals with unique physicochemical properties. For example, in the case of vermiculite, a 2:1 clay mineral, most of the coordinating-cation substitution takes place in the tetrahedral sheet (Table 3.4), which limits the mineral's potential to expand its

TABLE 3.3. Selected Common Primary and Secondary Minerals[a]

Name	Chemical Formula
	Primary Minerals
Quartz	SiO_2
Muscovite	$KAl_2(AlSi_3O_{10})(OH)_2$
Biotite	$K(Mg, Fe)_3(AlSi_3O_{10})(OH)_2$
Feldspars	
Orthoclase	$KAlSi_3O_8$
Microcline	$KAlSi_3O_8$
Albite	$NaAlSi_3O_8$
Amphiboles	
Tremolite	$Ca_2Mg_5Si_8O_{22}(OH)_2$
Pyroxenes	
Enstatite	$MgSiO_3$
Diopside	$CaMg(Si_2O_6)$
Rhodonite	$MnSiO_3$
Olivine	$(Mg, Fe)_2SiO_4$
Tourmaline	$(Na, Ca)(Al, Fe^{3+}, Li, Mg)_3Al_6(BO_3)_3(Si_6O_{18})(OH)_4$
	Secondary Minerals
Clay minerals[b]	
Kaolinite	$Si_4Al_4O_{10}(OH)_8$
Montmorillonite	$M_x(Al, Fe^{2+}, Mg)_4Si_8O_{20}(OH)_4$ (M = interlayer metal cation)
Vermiculite	$(Al, Mg, Fe^{3+})_4(Si, Al)_8O_{20}(OH)_4$
Chlorite	$[M\ Al\ (OH)_6](Al, Mg)_4(Si, Al)_8O_{20}(OH)_9F)_4$
Allophane	$Si_3Al_4O_{12} \cdot nH_2O$
Goethite	$FeOOH$
Hematite	$\alpha\text{-}Fe_2O_3$
Maghemite	$\gamma\text{-}Fe_2O_3$
Ferrihydrite	$Fe_{10}O_{15} \cdot 9H_2O$
Gibbsite	$Al(OH)_3$
Pyrolusite	$\beta\text{-}MnO_2$
Dolomite	$CaMg(CO_3)_2$
Calcite	$CaCO_3$
Gypsum	$CaSO_4 \cdot 2H_2O$

Source: From Sparks (1995).

[a]Adapted from *Mineralogy: Concepts, Descriptions, Determinations* by Berry, Mason, and Dietrich. Copyright © 1959 by W.H. Freeman and Company and Hurlbut, C. S., Jr., and Klein, C. (1977). *Manual of Mineralogy*, 19th ed. Copyright © 1977, John Wiley & Sons, Inc. Reprinted by permission of John Wiley & Sons, Inc.

[b]Formulas are for the full-cell chemical formula unit.

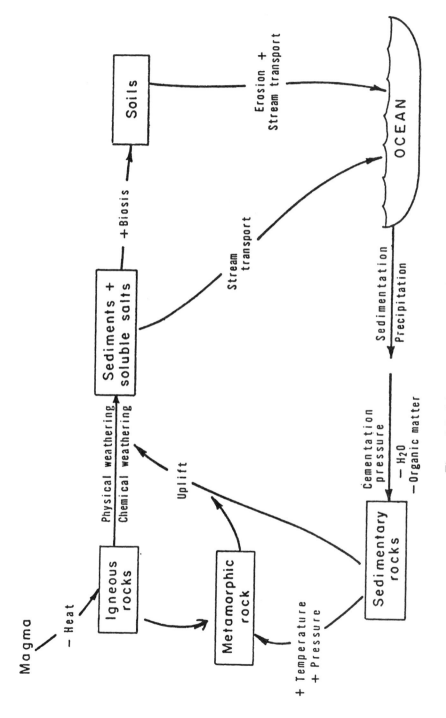

Figure 3.2. Schematic of the geomorphic cycle.

Figure 3.3. Schematic of a silicon tetrahedron made up of four oxygen atoms surrounding one much smaller silicon atom. In a clay crystal, each oxygen atom may be part of a two-silicon tetrahedral, of a one-silicon tetrahedron and a two-aluminum octahedral, or of a one-silicon tetrahedron and a hydrogen atom (from Taylor and Ashcroft, 1972, with permission).

Figure 3.4. Schematic of an aluminum octahedron made up of six oxygen atoms forming an octahedral structure around one aluminum atom. The six oxygens of the octahedron satisfy the three valence bonds of the central aluminum, leaving each oxygen with one and one-half unbalanced valence bonds, which are satisfied by a second aluminum, a silicon, or one hydrogen (from Taylor and Ashcroft, 1972, with permission).

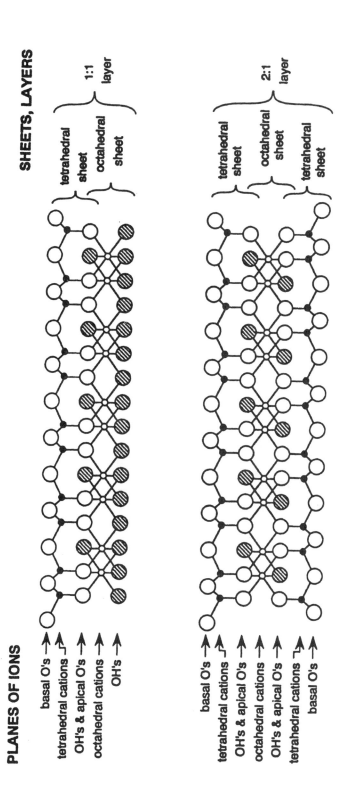

Figure 3.5. Compositional arrangement of silica tetrahedral and alumina octahedral (from Schulze, 1989, with permission).

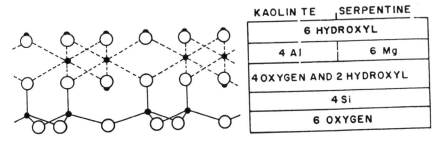

KAOLIN TE	SERPENTINE
6 HYDROXYL	
4 Al	6 Mg
4 OXYGEN AND 2 HYDROXYL	
4 Si	
6 OXYGEN	

STRUCTURAL COMPARISON OF TWO 1:1 LAYER SILICATE MINERALS

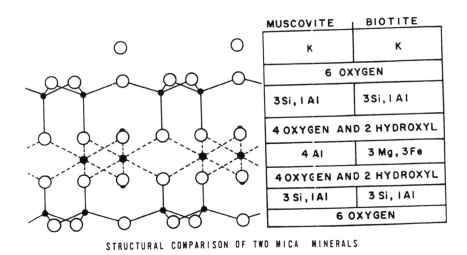

MUSCOVITE	BIOTITE
K	K
6 OXYGEN	
3 Si, I Al	3 Si, I Al
4 OXYGEN AND 2 HYDROXYL	
4 Al	3 Mg, 3 Fe
4 OXYGEN AND 2 HYDROXYL	
3 Si, I Al	3 Si, I Al
6 OXYGEN	

STRUCTURAL COMPARISON OF TWO MICA MINERALS

Figure 3.6. Structural comparisons of 1:1 and 2:1 layer mineral silicates.

interlayer space (Table 3.5), identified as the space between two tetrahedral sheets occupied by charge balancing cations (Fig. 3.11). The reason for limiting interlayer expansion is the strong interaction between the charge-balancing cations and the two tetrahedral sheets owing to their (1) physical proximity to the exchangeable cations and (2) limited number of O^{2-} sharing the excess negative charge produced by the coordinating-cation substitution.

In the case of smectites (e.g., bentonite and montmorillonite, also 2:1 clay minerals), some coordinating-cation substitutions take place in the octahedral sheet (Table 3.4). Because the octahedral location of the coordinating-cation substitution is far removed from the interlayer spacing where the cations balancing this excess negative charge reside, and because more O^{2-} in the octahedral sheet shares this excess negative charge,

Pyrophyllite	Talc	Illite	Vermiculite	Beidellite	Saponite	Montmorillonite	Nontronite
No Interlayer Cations		0-1 K & Water	Mg and Water	Replaceable Cations and Water			
6 OXYGEN							
4 Si	4 Si	3 Si 1 Al	3 Si 1 Al	3 Si 1 Al	4 Si, Al	4 Si, Al	4 Si, Al
4 OXYGEN AND 2 HYDROXYL							
4 Al	6 Mg	4 Al···	4 Al·	4 Al	6 Mg	4 Al, Mg	4 Fe
4 OXYGEN AND 2 HYDROXYL							
4 Si	4 Si	3 Si 1 Al	3 Si 1 Al	3 Si 1 Al	4 Si, Al	4 Si, Al	4 Si, Al
6 OXYGEN							

STRUCTURAL COMPARISON OF SEVERAL 2:1 LAYER SILICATE MINERALS

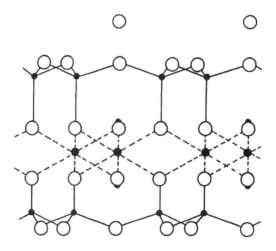

Figure 3.6. Continued

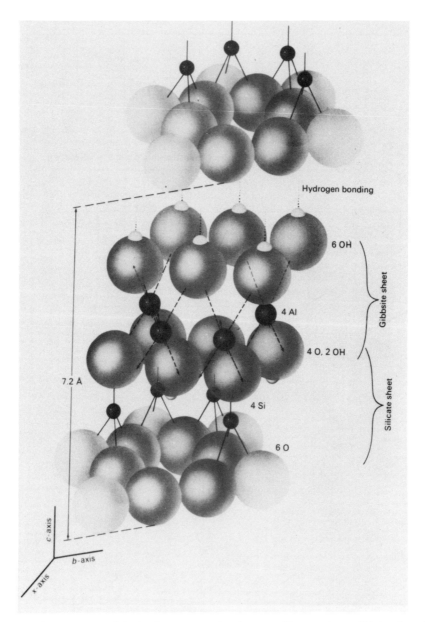

Figure 3.7. Schematic of the kaolin structure showing one silicate and one gibbsite sheet in each layer, which has been expanded along the c-axis to show bonding. The basic unit is repeated along the two horizontal axes to form layers. Adjacent layers are held together by hydrogen bonding (from Taylor and Ashcroft, 1972, with permission).

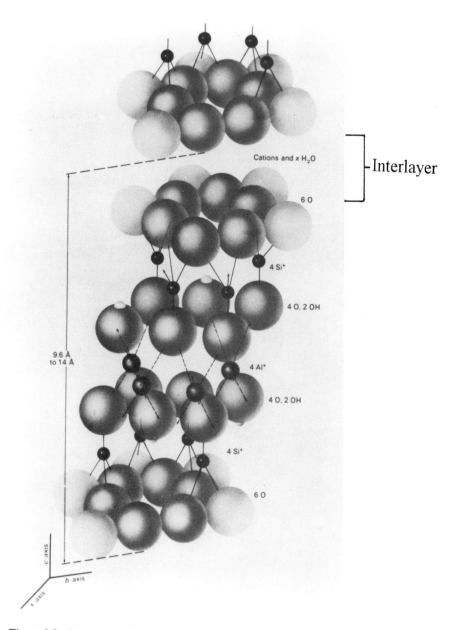

Figure 3.8. Schematic of the smectite structure showing one gibbsite sheet between two silicate sheets. The basic unit is repeated many times in the horizontal directions to produce layers. The basic unit with the 9.6 Å c-axis spacing expands to 14 Å when water enters between layers. The exchangeable cations located between the layers produce the counter charge for the isomorphous substitution. It occurs in the layers marked with an asterisk (from Taylor and Ashcroft, 1972, with permission).

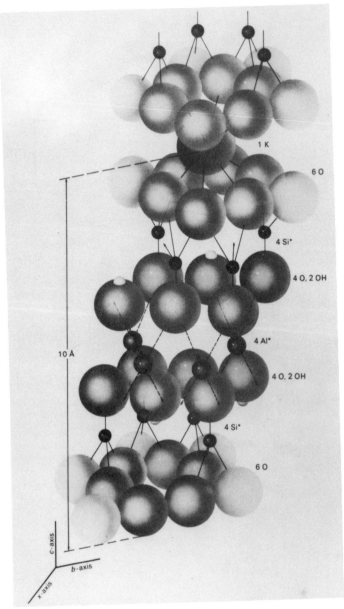

Figure 3.9. Schematic of the illite structure showing one gibbsite sheet between two silicate sheets. Potassium holds the layers together in a 12-fold coordination, preventing expansion. Isomorphous substitution can occur in the layers marked with an asterisk (from Taylor and Ashcroft, 1972, with permission).

TABLE 3.4. Clay Minerals that Commonly Occur in Soils

Clay Name	Formula	Negative Charge			Formula Weight	Calculated CEC	Observed CEC	Cause
		In IV[a]	In VI[b]	Total				
Kaolinite	$(Si_4)^T(Al_4)^O O_{10}(OH)_8$	0	0	0	516	0	5	Minute substitution
Muscovite mica	$K_2(Si_6Al_2)^T(Al_4)^O O_{20}(OH)_4$	−2	0	−2	796	251	40	Blocked by K
Vermiculite	$K(Si_6Al_2)^T(Al_4)^O O_{20}(OH)_4$	−2	0	−2	757	264	132	1 K Remains
Montmorillonite	$(Si_{7.6}Al_{0.4})^T(Al_{3.7}Mg_{0.3})^O O_{20}(OH)_4$	−0.4	−0.3	−0.7	719	97	100	—

[a]Or T (tetrahedral) sheet.
[b]Or O (octahedral) sheet.

113

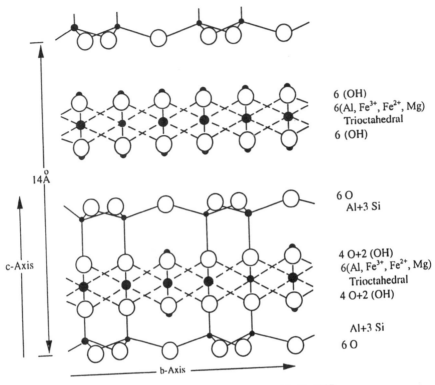

Chlorite   for example, Al Mg$_5$(OH)$_{12}$ (Al2Si$_6$)Al2Si$_6$)Al Mg$_5$O$_{20}$(OH)$_4$

Figure 3.10. Schematic of the chlorite structure showing two gibbsite sheets between two silicate sheets (from Jackson, 1964, with permission).

the interaction between the charge-balancing cations and the two tetrahedral sheets is rather weak, leading to a relatively large interlayer expansion (Table 3.5).

Minerals such as the kaolin group (1:1) and chlorite (2:2) do not exhibit significant expansion potential between layers (a layer is a mineral structure composed of combinations of sheets) (Table 3.5, Figs. 3.7 and 3.10).

TABLE 3.5. Clay Minerals and Some of Their Properties

Property	Kaolinite	Muscovite	Vermiculite	Montmorillonite
Minimum c axis spacing (Å)	7	10	10	10
Maximum c axis spacing (Å)	7	10+	16	20 Ca >100 Na
CEC (cmol/100 g)	5	40	130	100
Type of lattice	1:1	2:1	2:1	2:1

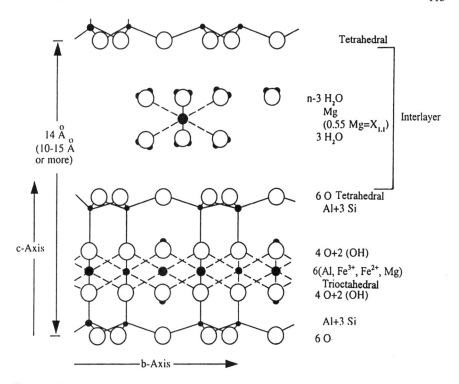

Figure 3.11. Schematic of the vermiculite structure showing one brucite sheet between two silicate sheets. Isomorphous substitution occurs mostly in the tetrahedral sheet with some in the octahedral sheet (from Jackson, 1964, with permission).

It follows that, generally, degree of cation substitution controls magnitude of permanent CEC and location of substitution (e.g., tetrahedral versus octahedral) controls expansion and specific surface of 2:1 clay minerals (Table 3.6). Additional information on clay surfaces and their behavior is presented in Chapter 4.

TABLE 3.6. Clay Minerals and Some of Their Surface Properties

Mineral	CEC ($cmol_c kg^{-1}$)	Specific Surface ($m^2 g^{-1}$)	Layer Type
Kaolinite	3–15	10–20	1:1
Montmorillonite	60–100	600–800	2:1
Vermiculite	110–160	600–800	2:1
Illite	20–40	40–180	2:1
Chlorite	10–30	70–150	2:2
Mica	20–40	70–120	2:1

SOIL MINERAL TERMS AND DEFINITIONS

Brief Definitions and Discussion

I. Crystal. A crystal is made up of ions with a certain arrangement which repeats itself indefinitely.

II. Ions. Ions are made up of a positive nucleus composed of protons and neutrons which are surrounded by negatively charged electrons. In general, when the positive charges (protons) exceed the negative charges, the ions are referred to as cations (K^+, Na^+, Ca^{2+}, Mg^{2+}, Al^{3+}), but when the negative charges (electrons) exceed the positive charges (protons), the ions are referred to as anions (Cl^-, F^-).

III. Quasi-crystal. A quasi-crystal is made up of ions with a certain arrangement which does not repeat itself indefinitely.

IV. Amorphous (noncrystal). An amorphous material is made up of ions without any repeating arrangement.

V. Makeup of crystals. Soil mineral crystals are made out of ions. However, not all crystals are made out of ions. Exceptions include sulfur crystals (which are made of S atoms), diamonds (which are made out of carbon atoms), and asphalt crystals (which are made out of large organic molecules arranged in a repeating fashion).

VI. How does one understand clays? Most substances in nature are understood by their reactivity with other substances. For example,

$$CaCO_3 + HCl \rightarrow CO_{2gas} + CaCl_2 \tag{A}$$

They also are understood by their dissolution in water. For example,

$$CaCl_2 + H_2O \rightarrow Ca^{2+} + 2Cl^- \tag{B}$$

Clays are generally insensitive to such approaches. They don't react readily with acids or bases, and they don't go into solution when reacted with water. Their reactions are characterized as being slow and incomplete. Important information about clay structure was not available until 1930 with the introduction of X rays (Dixon and Weed, 1989).

General Rules of Bonding in Soil Mineral Structures

Electronegativity relates to ionization potential for cations, or electron affinity for anions. Pauling's general rule on bonding states that ions of closer electronegativity have a greater tendency to form covalent bonds, (NaCl = 2.1) < (CaS = 1.5) < (CuS = 0.5) < (CS = 0) (Table 3A). Ionic radii and electronegativities permit the formulation of some specific rules about chemical bond formation.

TABLE 3A. Partial List of Electronegativities and Percentages of Ionic Character of Bonds with Oxygen

Ion	Electronegativity	Ionic Character (%)
Ca^{2+}	0.7	89
K^+	0.8	87
Na^+	0.9	83
Li^+	1.0	82
Ba^{2+}	0.9	84
Ca^{2+}	1.0	79
Mg^{2+}	1.2	71
Be^{2+}	1.5	63
Al^{3+}	1.5	60
B^{3+}	2.0	43
Mn^{2+}	1.5	72
Zn^{2+}	1.7	63
Sn^{2+}	1.8	73
Pb^{2+}	1.8	72
Fe^{2+}	1.8	69
Fe^{3+}	1.9	54
Ag^+	1.9	71
Cu^+	1.9	71
Cu^{2+}	2.0	57
Au^+	2.4	62
Si^{4+}	1.8	48
C^{4+}	2.5	23
P^{5+}	2.1	35
N^{5+}	3.0	9
Se	2.4	
S	2.5	
O	3.5	
I	2.5	
Cl	3.0	
F	4.0	

I. For any given cation and two different anions, the larger anion forms a stronger covalent bond, for example, MgS > MgO; S^{2-} 1.85 Å crystalline radius > O^{2-} 1.40 Å crystalline radius.

II. For any given anion and two different cations, the smaller cation forms a stronger covalent bond or is more covalently bonded, for example, MgO > BaO; Mg^{2+} 0.65 Å crystalline radius < Ba^{2+} 1.35 Å crystalline radius. Data on cation size is presented in Table 3B.

TABLE 3B. Ion Sizes and Ionic Hydration

Ion	Ionic Radii (Å)	
	Not Hydrated	Hydrated
Li^+	0.68	10.03
Na^+	0.98	7.90
K^+	1.33	5.32
NH_4^+	1.43	5.37
Rb^+	1.49	3.6
Cs^+	1.65	3.6
Mg^{2+}	0.89	10.8
Ca^{2+}	1.17	9.6
Sr^{2+}	1.34	9.6
Ba^{2+}	1.49	8.8
Al^{3+}	0.79	—
La^{3+}	1.30	—

III. For any two given ions of similar size but different charge, the one with the highest charge (z) is the most covalently bonded (e.g., Ca–O > Na–O in Na_2O; Ca, $z = 2$; Na, $z = 1$).

IV. Ions of metals in the middle of the periodic table form more covalent character bonds with anions than do ions of metals in the first two or three groups of the periodic table (e.g., CdS > CaS).

Rules Explaining Cation–Anion Coordination

I. Coordination Number. The number of ions surrounding the ion of opposite charge in a mineral. The coordination number refers to a specific cation. For example, in the case of NaCl, the cation (Na^+) coordinates six anions (Cl^-).

II. Coordinating Cation. The cation that is surrounded by the coordinated anions.

III. Hedra or "Hedron." The number of planes or surfaces created by the anion coordination is called "hedra." The number of "hedra" or surfaces is coordination number dependent.

IV. The Radium Ratio (R_r). It is defined as the ratio of the cation crystalline radius to the anion crystalline radius (R_c/R_a). The driving force which explains the radium ratio's role on coordination is that of the closest packing. In general,

A. If R_c/R_a (R_r) is between 0.41 and 0.73, the cation would most likely coordinate six anions (e.g., Na/Cl, $R_r = 0.54$).

B. If R_c/R_a (R_r) is between 0.73 and 1, the cation would most likely coordinate eight anions (e.g., Cs/Cl, $R_r = 0.92$).

C. If R_c/R_a (R_r) is between 0.22 and 0.41, the cation would most likely coordinate four anions (e.g., Zn^{2+}/S^{2-}, $R_r = 0.40$).

Clay mineral structures are composed of the cations Si^{4+}, Al^{3+}, Ca^{2+}, Mg^{2+}, $Fe_{2+,3+}$, $Mn^{3+,2+}$, Cu^{2+}, and the anions O^{2-} and OH^-. A summary of potential cation coordination numbers is presented in Table 3C.

V. Tetrahedron. A solid that is geometrically characterized by having four planes, formed by four O^{2-} coordinated by a Si^{4+} (Fig. 3.3).

VI. Octahedron. A solid that is geometrically characterized by having eight planes, formed by six O^{2-} coordinated by an Al^{3+} ion (Fig. 3.4).

VII. Isomorphism. It is a term that refers to two compounds with similar structure but different chemical formulas. The ions that differ are of the same size (ionic radii do not differ more than 15%), but not necessarily the same charge. For example, Mg_2SiO_4 (fosterite) versus Fe_2SiO_4 (faylite).

VIII. Polymorphism. It implies that compounds have similar formula but different structure. For example, aragonite ($CaCO_3$) versus calcite (also $CaCO_3$).

TABLE 3C. Comparison of Observed Coordination Number with Numbers Predicted from Geometric Crystalline Radius Ratios

Ion	Radius Ratio[a]	Coordination Predicted from Ratio	Observed Coordination Number	Theoretical Limiting Radius Ratios
Ca^{2+}	1.19	12	12	
				1.00
K^+	0.95	8	8–12	
Sr^{2+}	0.80	8	8	
				0.73
Ca^{2+}	0.71	6	6, 8	
Na^+	0.69	6	6, 8	
Fe^{2+}	0.53	6	6	
Mg^{2+}	0.47	6	6	
				0.41
Al^{3+}	0.36	4	4, 6	
Si^{4+}	0.30	4	4	
				0.22
S^{6+}	0.21	3	4	
B^{3+}	0.16	3	3, 4	

[a]Ionic radius/radius of O^-. Radius of $O^- = 1.40$ Å.

IX. Phyllosilicates. Minerals (clays) made out of O^{2-} coordinated by metal cations and stacked in a certain sequence (Fig. 3.6).

A. SHEET. Combinations of planes of ions form sheets (Fig. 3.5).

B. LAYERS. Combinations of sheets form layers (Fig. 3.5).

C. INTERLAYERS. The zones between the layers where the basal oxygens of two layers meet (Figs. 3.8 and 3.11).

D. UNIT CELL. The total assemblage of a layer plus interlayer material or smallest repeating three-dimensional array of ions in a crystal (Table 3.3).

E. LAYER CHARGE. The magnitude of charge per formula unit or structure or the difference between the sums of cationic and anionic charges in a mineral per unit cell.

F. SILOXANE CAVITY. Phylosilicate functional group made up of six planar tetrahedral O^{2-} (see Chapter 4).

G. SILANOL. Phylosilicate tetrahedral -OH edge functional group (see Chapter 4).

H. ALUMINOL. Phylosilicate octahedral -OH edge functional group (see Chapter 4).

How Are Layer Silicates Differentiated?

I. Number and sequence of tetrahedral and octahedral sheets
    A. 1:1—One tetrahedral to one octahedral (kaolin, halloysite, dickite, nacrite) (Fig. 3.7).
    B. 2:1—Two tetrahedral to one octahedral (mica, pyrophylite, talk, montmorillonite, vermiculite) (Fig. 3.8).
    C. 2:2—Two tetrahedral to two octahedral (chlorite) (Fig. 3.10).

II. The layer charge per unit cell of structure
    A. Pyrophylite and talc exhibit zero charge per unit cell
    B. Mica exhibits −1 charge per unit cell
    C. Vermiculite exhibits −0.6 to −0.9 charge per unit cell
    D. Montmorillonite exhibits −0.3 to −0.6 charge per unit cell

III. The type of interlayer bond and interlayer cations
    A. Kaolin—Hydrogen bond (Fig. 3.7)
    B. Pyrophylite and talc—Van der Waals attraction force between basal planes (Fig. 3.6)
    C. Mica—Unhydrated K^+ shared between adjacent siloxane cavities (Fig. 3.9)
    D. Vermiculite—Commonly, hydrated Mg^{2+} shared between adjacent siloxane cavities (Fig. 3.11)

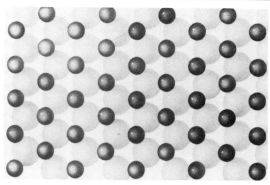

Figure 3A. Schematic of a brucite sheet showing the relative locations of all Mg^{2+} by removing the top layer of oxygen. Minerals may contain any combination of Al^{3+} and Mg^{2+}, while Fe^{3+} or Fe^{2+} may also substitute isomorphously (from Taylor and Ashcroft, 1972, with permission).

E. Smectite—Commonly, hydrated Na^+ or Ca^{2+} shared between adjacent siloxane cavities (Fig. 3.8)

IV. The type of cations in the octahedral sheet

   A. For example, Al^{3+} in the octahedral sheet forms kaolinite, but Mg^{2+} in the octahedral sheet forms antigorite

V. Number of positions occupied in the octahedral sheet

   A. Triochahedral—All available cation-coordinating positions in the octahedral sheet are occupied by a divalent metal which acts as the coordinating cation (Fig. 3A).

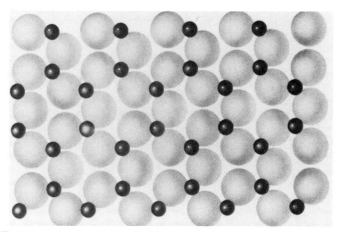

Figure 3B. Schematic of a portion of gibbsite sheet showing the relative locations of all Al^{3+} by removing the top layer of oxygen. Minerals may contain any combination of Al^{3+} and Mg^{2+}, while Fe^{3+} or Fe^{2+} may also substitute isomorphously (from Taylor and Ashcroft, 1972, with permission).

B. Dioctahedral—Only two of the available three cation-coordinating positions are occupied in the octahedral sheet by a trivalent metal which
acts as the coordinating cation (Fig. 3B)

Clay mineral term refers to layer silicates, however, in soil science it is loosely
used and it is often taken to represent any soil material with an effective diameter
of less than 2 μm.

Clay Mineral Groups in Soil

I.   Kaolin group
II. Mica group
III. Smectite group
IV. Vermiculite group
V.   Chlorite group

I.   Kaolin group
    A. Minerals and structural composition
        1. Consists of one tetrahedral and one octahedral sheet
        2. Kaolinite is an important mineral in soils of temperate and tropical
           climates
        3. Hydrated halloysite has the same structure as kaolinite with water
           (single layer of H_2O molecules) sandwiched between the mineral
           layers (10-Å spacing). This mineral is commonly encountered in
           tropical soils
        4. Nacrite—more regular layers and larger in size (a polymorph of
           kaolinite)
        5. Dickite—same as nacrite, a polymorph of kaolinite
        6. Antigorite—Al^{3+} has been substituted by Mg^{2+} in the octahedral
           position
    B. Some general properties of the kaolin group
        1. Particle size for kaolin ranges from 0.1 to 1 μm in diameter. For
           nacrite, the most ordered, the size of the crystal approaches the
           range of 1 mm
        2. Cation exchange capacity (CEC)—specific surface. CEC varies
           from 1 to 10 meq/100 g or $cmol_c$ kg^{-1}, while specific surface var-
           ies from 10 to 20 m^2 g^{-1}. CEC is pH dependent and therefore not
           due to isomorphous substitution
        3. Hydration and plasticity—small between layers because they are
           nonexpanding due to hydrogen bonding
    C. Some general information on kaolinitic soils

    1. Kaolinitic soils—Southeastern United States

    2. Low CEC—require specific management practices

    3. Usually low water holding capacity

    4. Resistant to weathering

    5. Physical problems (e.g., compaction and water infiltration) not severe

II. Mica group

  A. Some general information on noncharged 2:1 layer minerals (pyrophylite)

    1. Consists of two tetrahedral and one octahedral sheet

    2. Serves as an excellent example for understanding the structure of 2:1 soil clay minerals because of no isomorphous substitution

    3. Layers are held together by van der Waals forces (nonexpanding)

    4. Exchange capacity very low—only due to broken bonds at the edges of the crystals

  B. Mica and some of its properties

    1. Similar to pyrophylite—except that one out of four Si^{4+} in the tetrahedral is replaced by aluminum (Al^{3+})

    2. High negative charge

    3. Interlayer K^+ is tightly held in 12 coordination and perfectly fits in the hexagonal siloxane cavities. Muscovite is dioctahedral; biotite is trioctahedral

    4. Layer charge is high but the K ions are held tightly and therefore nonexchangeable

    5. High charge (−1 per unit formula); CEC = 20–40 meq/100 g or $cmol_c$ kg^{-1}; specific surface 70–120 m^2 g^{-1}

III. Smectite group

  A. Minerals and structural composition

    The following substitutions in the pyrophylite give rise to the smectite group: Mg^{2+} for Al^{3+} or Fe^{3+} for Al^{3+} in octahedral, and Al^{3+} for Si^{4+} in tetrahedral

    1. Montmorillonite

      a. Mg^{2+} replaces Al^{3+} in the octahedral position (1 out of six Al^{3+} in octahedral are replaced by Mg^{2+})

      b. Negative charge taken care by exchangeable cations

    2. Beidellite

      a. Al^{3+} replaces Si^{4+} in the tetrahedral position; Extra Al in octahedral

    3. Nontronite

      a. Ferric iron (Fe^{3+}) replaces Al^{3+} in octahedral

        b. Some replacement of Si^{4+} by Al^{3+} in the tetrahedral

    4. Saponite

        a. Mg^{2+} replaces Al^{3+} in octahedral

  B. General properties of the smectite group

    1. Crystal size varies from 0.01 to 0.1 μm

    2. Irregularities in structure causes breakdown of crystals to smaller size

    3. CEC = 80–120 meq/100 g and specific surface 600–800 m^2g^{-1}

    4. Hydration and plasticity high because of expanding interlayers (> 18 Å)

    5. Soils high in montmorillonite possess high CEC and high H_2O holding capacity but exhibit slow intake of H_2O, "puddle" easily; swell and shrink and thus subject to dispersion–flocculation phenomena

IV. Vermiculite group

  A. Minerals and structural composition

    1. They are formed by alteration of micas through K^+ replacement by exchangeable Mg^{2+}

    2. Interlayer spacing expands from 10 to 14–15 Å because Al^{3+} substitutes Si^{4+} in the tetrahedral position

    3. Mg^{2+} and Fe^{2+} also substitute Al^{3+} in the octahedral position

    4. Net charge = –0.7 per unit formula weight

    5. CEC = 120–150 meq/100 g or $cmol_c$ kg^{-1}

    6. Specific surface = 600–800 m^2 g^{-1}

  B. General properties of vermiculitic soil

    1. Swells less than smectitic soil because of its higher charge in the tetrahedral sheet

    2. Mineral collapses to 10 Å with K^+ and NH_4^+

    3. Possesses higher elasticity and plasticity than kaolin and mica

V. Chlorite group

  A. Minerals and structural composition

    1. Gibbsite [$Al(OH)_3$; Fig. 3B] and/or brucite [$Mg(OH)_2$; Fig. 3A] substitute the exchangeable cations in the interlayer

    2. CEC = 10–40 meq/100 g

    3. Surface area = 70–150 m^2g^{-1}

Primary Mineral Classification Based on Structural Arrangement

I. Independent tetrahedra (Fig. 3C)

  A. Olivines, Fe, Mg (SiO_4)

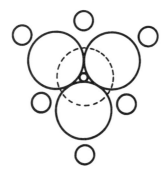

Figure 3C. Individual tetrahedra held together by divalent cations, for example, olivine (smaller spheres represent coordinating cations).

    1. Isomorphous series end-member
       a. Fayalite (Fe_2SiO_4)
       b. Forsterite (Mg_2SiO_4)
    2. Compact, physically strong
    3. Chemically reactive because Fe^{2+} and Mg^{2+} are readily exposed at crystal faces
    4. Alter to serpentine, magnetite, goethite, magnesite, opal
  II. Chains of tetrahedra
    A. Pyroxenes-Single chains (Fig. 3D)
       1. Augite-Ca(Mg, Fe, Al)(Si, Al)$_2$ O$_6$
    B. Amphiboles—Double chains (Fig. 3E)
       1. Hornblende—(Na, Ca)$_2$ (Mg, Fe, Al)$_5$ (Si, Al)$_5$ (Si, Al)$_8$ O$_{22}$ (OH)$_2$
    C. Weathering—proceeds parallel to chains

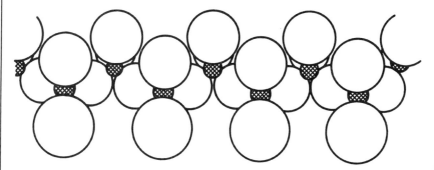

Figure 3D. Tetrahedra unichain silicates, for example, pyroxenes (enstatite, diopside) (smaller spheres represent coordinating cations).

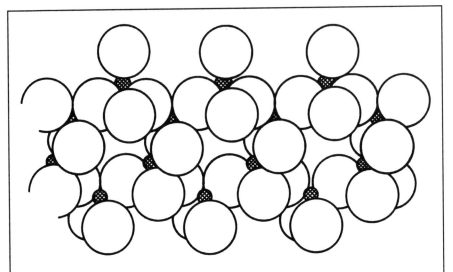

Figure 3E. Chain of linked hexagonal rings of tetrahedral units (duochain), for example, amphibole (tremulite, hornblende) (smaller spheres represent coordinating cations).

III. Three-dimensional network of tetrahedra
    A. Each oxygen bonded to two Si atoms
    B. Al^{3+} substituting for Si^{4+}, $1/3 > (Al/Al + Si) > 1/4$
        1. Feldspars
            a. Plagioclase feldspars
                i. $(Na, Ca) (Al, Si)_4 O_8$
            b. Potassic feldspars
                ii. Orthoclase, microcline ($KAlSi_3 O_8$)
        2. Three-dimensional network of tetrahedra with no Al^{3+} substituting
            a. Quartz—SiO_2 (Fig. 3F)
IV. Two-dimensional sheets of tetrahedra and octahedra
    A. Substitution in octahedral–tetrahedral sheets, "charge balancing" cations present
        1. Mica
    B. No substitution in either octahedral or tetrahedral sheets; no charge-balancing cations present
        1. Pyrophylite (Dioctahedral form), talc (Trioctahedral form)
    C. Substituting of Al for Si in tetrahedral sheet
        2. Muscovite (Dioctahedral form), biotite (Trioctahedral form)

Additional information on primary minerals is given in Tables 3D and 3E.

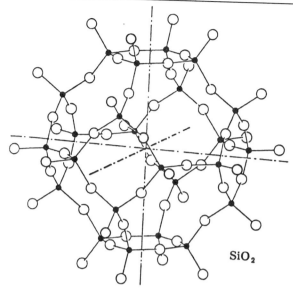

Figure 3F. Continuous framework structure of SiO_2 demonstrating the arrangement of SiO_4 tetrahedra. Open circles represent O^- and small solid circles represent Si^{4+} (from Berry and Mason, 1959, with permission).

TABLE 3D. Primary Minerals as Sources of Essential Nutrients

Primary Mineral Class	Representative Minerals	Chemical Formulae	Essential Elements
Feldspars	Orthoclase	$KAlSi_3O_8$	K, Na, Ca
	Albite	$NaAlSi_3O_8$	
	Anorthite	$CaAl_2Si_2O_8$	
Pyroxenes	Enstatite	$MgSiO_3$	Ca, Mg, Fe
	Diopside	$Ca, MgSi_2O_6$	
	Augite	$Ca, (Mg, Fe, Al)(Si, Al)_2O_6$	
Amphibole	Hornblende	$(Na, Ca)_2(Mg, Fe, Al)_5(Si, Al)_8O_{22}(OH)_2$	Na, Ca, Mg, Fe
Olivine	Forsterite	Mg_2SiO_4	Mg, Fe
	Fayalite	Fe_2SiO_4	
Apatite	Apatite	$Ca_{10}(F, OH, Cl)_2 (PO_4)_6$	Ca, P, Cl
Tourmaline	Tourmaline	$Na(Mg, Fe)_3Al_6(BO_3)_3Si_6O_{18}(OH)_4$	Na, Mg, Fe, B
Sulfides[a]	Pyrite	FeS_2	Fe, S
Mica	Muscovite	$K_2Al_2Si_8Al_4O_{20}(OH)_4$	K, Fe, Mg
	Biotite	$K_2Al_2Si_6(Fe, Mg)_6O_{20}(OH)_4$	

[a]Trace elements occur as impurities in the primary minerals, especially the sulfides.

TABLE 3E. Classification of Silicate Minerals Based on Structural Arrangements in Silicate Crystal Lattices

Silicate Structural Groups	Structural Arrangement	Formula of Group	Number of O Ions Shared	Average Si:O Ratio (Unsubstituted)	Mineral Series Example (Isomorphous Series)[a]	
					Series Name	Idealized Formula with Substitution Involved
S1	Independent silica tetrahedra (unitetrahedral silicates) (Fig. 3C)	$(SiO_4)_n^{-4}$	0	SiO_4^{-4}	Olivine	$(Mg, Mn, Fe)_2SiO_4$
S2	Linked pairs of tetrahedra (duotetrahedral silicates) (Fig. 3G)	$(Si_2O_7)_n^{-6}$	1	$SiO_{3.5}^{-3}$	Melilite	$(Ca,Na)_2(MgAl)(SiAl)_2O_7$
S3	Closed rings of tetrahedra (cyclosilicates):					
	(a) trigonal rings;	$(Si_3O_9)_n^{-6}$	2	SiO_3^{-2}	Benitoite	$Ba\,TiSi_3O_9$
	(b) hexagonal rings (Figs. 3H and 3I)	$(Si_6O_{18})_n^{-12}$	2	SiO_3^{-2}	Beryl	$BeAlSi_6O_{18}$
S4	Infinite chains of tetrahedra (unichain silicates); pyroxene family (Fig. 3D)	$Si_2O_7^+$ $(SiO_3)_n^{-2}$	2	SiO_3^{-2}	Diopside	$CaMg(SiO_3)_2$

S5	Infinite double chains of tetrahedra (duochain silicates); amphibole family (Fig. 3E)	$Si_6O_{18^+}$ $(Si_4O_{11})_n^{-6}$	2 and 3	$SiO_{2.75}^{-1.5}$	Tremolite	$(OH)_2Ca_2Mg_5\,Si_8O_{22}$
S6	Infinite layers of tetrahedra (layer silicates): (a) hexagonal linkage, 1:1 layers, kaolin family; (b) hexagonal linkage, 2:1 layers, pyrophyllite family	$Si_6O_{18} + (Si_2O_5)_n^{-2}$ $Si_6O_{18} + (Si_2O_5)_n^{-2}$	3	$SiO_{2.5}^{-1}$	Kaolinite Montmorillonite Muscovite	$(OH)_4Al_2Si_2O_5$ $(OH)_2Al_{1.67}Mg_{0.33}$ $Si_4O_{10} \times H_2O$ $(OH)_2KAl_2(AlSi_3)O_{10}$
S7	Infinite framework of tetrahedra (framework silicates) (Fig. 3F):	$\left.\begin{array}{l}Si_3O_9\\Si_6O_{18}\end{array}\right\} + (SiO_2)_n$	4	SiO_2^{-0}	Quartz	SiO_2
	(a) trigonal and hexagonal linkage, silica family; (b) 4-tetragonal, 2-hexagonal	$\left.\begin{array}{l}(Si_4O_{12})_4\\(SiO_{18})_2\end{array}\right\} + (Si_4O_8)_n$	4	SiO_2^{-0}	Orthoclase	$K(Al,\,Si)_4O_8$

[a]One example only. There are many isomorphous series in each structural group. Each isomorphous series contains several individual minerals or "species."
Source: Adapted from M. L. Jackson. Crystal chemistry of soils. I. The fundamental structural groups and families of silicate minerals. *Soil Sci. Soc. Amer. Proc.* 1948.

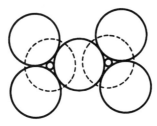

Figure 3G. Pair of tetrahedra linked by sharing one oxygen, for example, melilite (Ca, Na)$_2$(Mg, Al)(Si, Al)$_2$O$_7$ (smaller spheres represent coordinating cations).

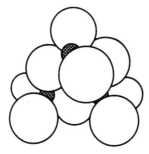

Figure 3H. Closed rings composed of three tetrahedra sharing two oxygens [(Si$_3$O$_9$)$_n^{6-}$] and held together in crystals by cations, for example, benitoite (BaTiSi$_3$O$_9$) (smaller spheres represent coordinating cations).

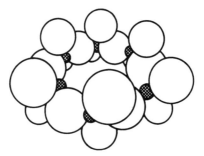

Figure 3I. Closed rings composed of six tetrahedra sharing two oxygens [(Si$_6$O$_{18}$)$_n^{12-}$] and held together in crystals by cations, for example, beryl (BeAlSi$_6$O$_{18}$) (smaller spheres represent coordinating cations).

3.3   METAL–OXIDES

Iron and manganese are commonly found in soils and often in large quantities. Iron commonly occurs in soil in the 2+ and 3+ oxidation states while manganese occurs in the 2+, 3+, and 4+ oxidation states. Iron of the iron–oxide form is in the 3+ oxidation state and manganese of the manganese–oxide form is the 3+ or 4+ oxidation states. The solubility of these two oxides is extremely low in the pH range typically encountered in most soils.

Iron– and Mn–oxides exhibit a structure similar to that of $Al(OH)_3$ or gibbsite, another oxide mineral commonly encountered in soils in large quantities. The manganese oxides as a group are almost exclusively in octahedral arrangement. They differ from gibbsite, however, by having oxygens at the corners of the octahedron instead of hydroxyls (Fig. 3.12). The most common Fe–oxides, hematite (Fe_2O_3) and goethite (FeOOH), also have the gibbsite octahedral arrangement. Metal oxides are present in soils as:

1. Free oxides
2. Clay mineral coatings
3. Clay edges

Metal–oxides exhibit charge due to protonation and deprotonation of the oxygen coordinated to the metal (Fe^{3+}, Mn^{3+}, Mn^{4+}, Al^{3+}, and Si^{4+}). This charge is known as variable charge or *pH-dependent charge*. On the average, soil pH-dependent charge may vary from 1 mmol/100 g soil (milliequivalents, meq per 100 g) to 30 meq/100 g. The pH-dependent charge can be positive or negative depending on the pH, and the specific pH at which the positive charges on the surface of the oxide equal the negative charges on the surface of the oxide (see Section 3.5).

3.4   SOIL ORGANIC MATTER

Soil organic matter (SOM) is made of two major groups of compounds: The nonnitrogenous compounds, which are mainly carbohydrates, and the nitrogenous com-

Gibbsite

Figure 3.12. Diagram showing the compositional arrangement of gibbsite (from Schulze, 1989, with permission).

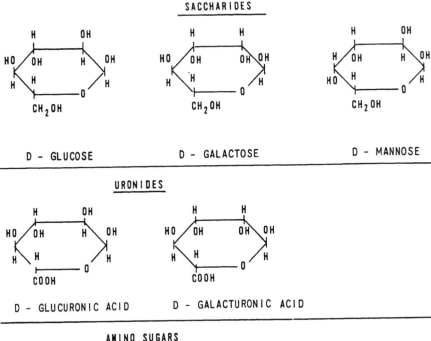

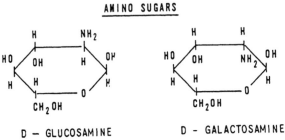

Figure 3.13. Chemical composition and structure of saccharides, chlorides, amino sugars, amino acids, amides, and a heterocyclic amino acid.

pounds principally derived from proteins. Some of the simple compounds making up soil organic material are shown in Figure 3.13. The degradation steps of plant residues leading to formation of soil organic matter are briefly described below.

Cellulose is one of the first plant-cell polysaccharides (carbohydrate) to be attacked by microorganisms at the early stages of decomposition. It is composed of glucose units bound together into a long linear chain, b, of 1–4 linkages. The first decomposition step is enzymatic hydrolysis of the linear chain by an extracellular enzyme called cellulase which breaks cellulose to cellobiose, the repeating unit of cellulose (Fig. 3.14). Cellobiose is then broken down intracellularly by the enzyme b-glucosidase. Although the by-products differ to some extent by the type of organism consuming

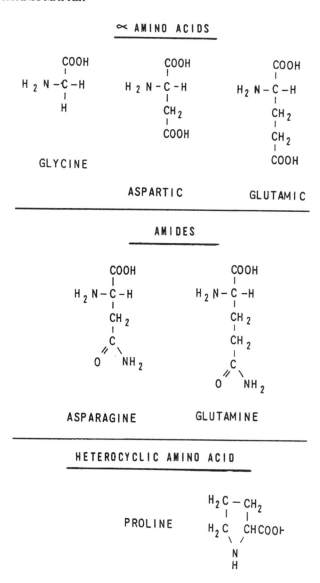

Figure 3.13. Continued

the cellulose, those usually found are formic acid, lactic acid, succinic acid, ethyl alcohol, butelene glycol, and hydrogen, and about two thirds of the original cellulose is converted into carbon dioxide.

Hemicelluloses, the second group of plant-cell constituents, are water-insoluble polysaccharides. During the first stage of attack, hemicelluloses disappear more rapidly than cellulose. Later, hemicelluloses slow down in degradation owing to their heterogeneity. Fungi, bacteria, and actinomycetes are able to break down hemicelluloses by

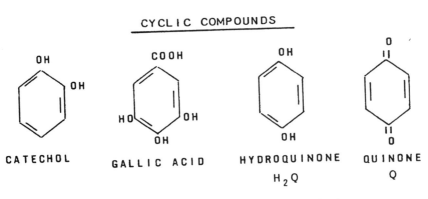

Figure 3.14. Schematic of cellobiose, a repealing unit of cellulose.

the enzyme hemicellulase into water-soluble, simple sugars which are then absorbed by the microbial cells, where oxidation takes place and various other organic acids are formed.

The third plant cell constituent, lignin, has as its basic unit a phenyl–propane type of structure. Organisms capable of breaking down lignin are the fungi, such as basidiomycetes, cladosporium, helminthosporium, and humicola. Eventually, lignin is broken down into vanillin and vanillic acid or other methoxylated aromatic structures by the fungi. In the final stages, low-molecular-weight organic acids are produced.

Generally therefore, as organic matter decomposes, the carbon and nitrogen are partially used by microorganisms, partially escape as gas (carbon dioxide), or are transformed into other substances. The plant material takes on a black coloration due to the oxidation of sugars, forming quinones (Fig. 3.15). The volume decreases because of the diminished strength of the plant fibers resulting from the microbial digestion of their wall structures. The final product is humus, the completely decomposed organic material. Seventy to eighty percent of the organic matter, by weight, found in most soils consists of humic substances (Schnitzer, 1986).

CYCLIC COMPOUNDS

CATECHOL   GALLIC ACID   HYDROQUINONE   QUINONE

H_2Q Q

Figure 3.15. Cyclic compound produced by oxidation of simple sugars.

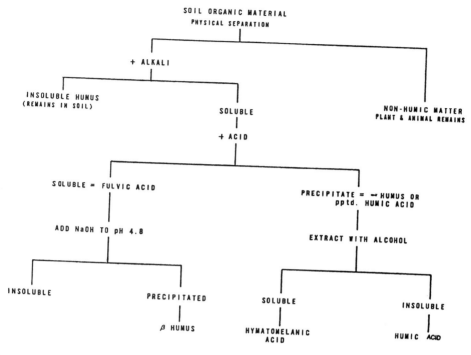

Figure 3.16. Soil organic matter fractionation procedure.

Generally, humic substances are defined as condensed polymers of aromatic and aliphatic compounds produced by decomposition of plant and animal residues and by microbial synthesis. They are amorphous, dark-colored, and hydrophilic, with a wide range in molecular weight from a few hundreds to several thousands. Furthermore, humic substances contain a large number of nonidentical functional groups, with different pK_a values, and are partitioned into three main fractions based upon their solubility behavior (Fig. 3.16) (Evangelou, 1995b):

1. Humic acid—soluble in dilute alkali, but precipitates in acid solution
2. Fulvic acid—soluble both in alkali and acid solutions
3. Humin—insoluble both in alkali and acid solutions

3.4.1 Humic Substances

The basic structure of humic substances consists of aromatic rings of the di- or trihydroxy-phenol type bridged by $-O-$, $-CH_2-$, $-NH-$, $-N=$, $-S-$, and other groups and contains both free OH groups and the double linkage of quinones (Stevenson, 1982). Some investigators proposed structural formulae for humic and fulvic acids, but to this day none have proven satisfactory. A model used to demonstrate the general structure and function of humic and fulvic acids is shown in Figure 3.17. Humic

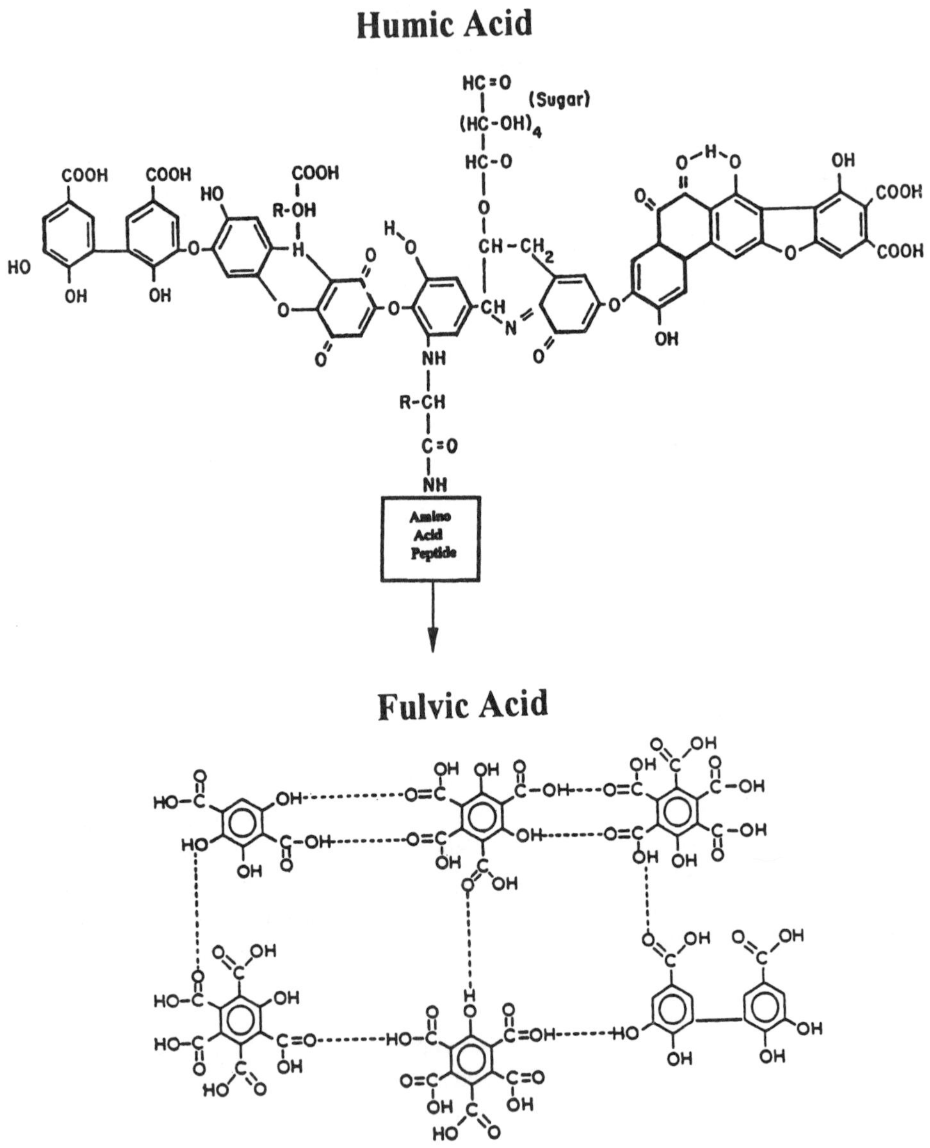

Humic Acid

Fulvic Acid

Figure 3.17. Model of soil humic and fulvic acids (from Stevenson, 1982, and Schnitzer and Khan, 1972, with permission).

substances generally exhibit similar elemental- and functional-group composition in spite of the origin and environmental conditions of their formation. However, humic substances formed under different climatic regimes and natural soils contain different ranges of aromaticity and functional group contents. The aromatic component of soil humic substances has been reported to range from 35 to 92% (Hatcher et al., 1981).

Generally, humic and fulvic acids have a similar structure, but they differ in molecular weight and elemental and functional group contents. Fulvic acid is lower in molecular weight and contains more oxygen-containing functional groups and elemental oxygen, but less nitrogen and carbon per unit weight than humic acid (Schnitzer and Khan, 1972; Sposito et al., 1976; Schnitzer, 1991). Furthermore, almost all oxygen in fulvic acid can be accounted for by functional groups (COOH, OH, C=O), whereas a high portion of oxygen in humic acid occurs as a structural component of the nucleus (e.g., in ether or ester linkage) (Stevenson, 1985).

Functional groups of humic substances are fractioned into three clusters of different dissociation constants:

1. Very weak cluster, which is assumed to include phenolic–hydroxyl residues, carboxyl residues, and nitrogenous bases of very high pK_a's
2. Weak cluster, which is assumed to include various aliphatic and aromatic-carboxyl residues and nitrogenous bases of medium pK_a's
3. Strong cluster, which is assumed to include the carboxyl residues of low pK_a of salicylate or phthalate

Humic substances exhibit:

1. Polyfunctionality due to the broad range of functional group reactivity
2. Molecular negative charge due to proton dissociation from various functional groups
3. Hydrophilicity due to formation of strong hydrogen bonds with water molecules
4. Structural lability due to intermolecular association and molecular conformation changes in response to changes in pH, redox potential (Eh), electrolyte concentration, and functional-group binding

3.4.2 Reaction Among Humic Substances, Clays, and Metals

Heavy metal and soil colloid (clay, humic substances, or combination) interactions are explained on the basis of ion exchange, surface adsorption, or chelation reactions. The potential of humic substances to form complexes with heavy metals results from oxygen-containing functional groups such as carboxyl (COOH), hydroxyl (OH), and carbonyl (C=O). The extent of heavy metal retention by mixtures of soil colloids (organics plus clays) varies with ionic strength, pH, type of clay minerals, type of functional groups, and type of competing cations.

Commonly, the amount of metal ions adsorbed by the solid surface increases with increasing pH for humic substances, clays, or clay—humic acid mixtures. Metal-ions adsorbed in acid media increase with pH until the threshold value required for partial dissolution of solid and formation of soluble metal–humate complexes is exceeded (Fig. 3.18). Metal–organic complexes experience three types of interactions, which

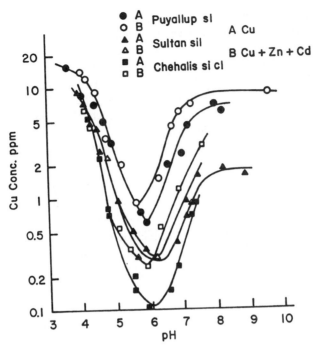

Figure 3.18. Concentration of Cu as a function of pH in three different soil samples (from Kuo and Baker, 1980, with permission).

determine the apparent solubility of metals as a function of pH. These interactions, as summarized by Evangelou (1995b), are:

1. Protons (H^+) compete with cations for organic binding sites
2. Hydroxyl ion (OH^-) competes with humic substances for the cationic metal-ion
3. Soft metals compete with hard metals for organic functional groups

Commonly, humic substances form a strong complex with clay. The most likely mechanisms of clay–humic complex formation are:

1. By anion and ligand exchange to clay edges
2. By cation or water bridges to basal clay surfaces
3. By H-bonding to the siloxane or gibbsite sheet
4. By van der Waals forces
5. By trapping in the crystal pores
6. By adsorption in interlayer spaces

The relative contribution of the various mechanisms to forming clay–humic complexes are different for different clay minerals.

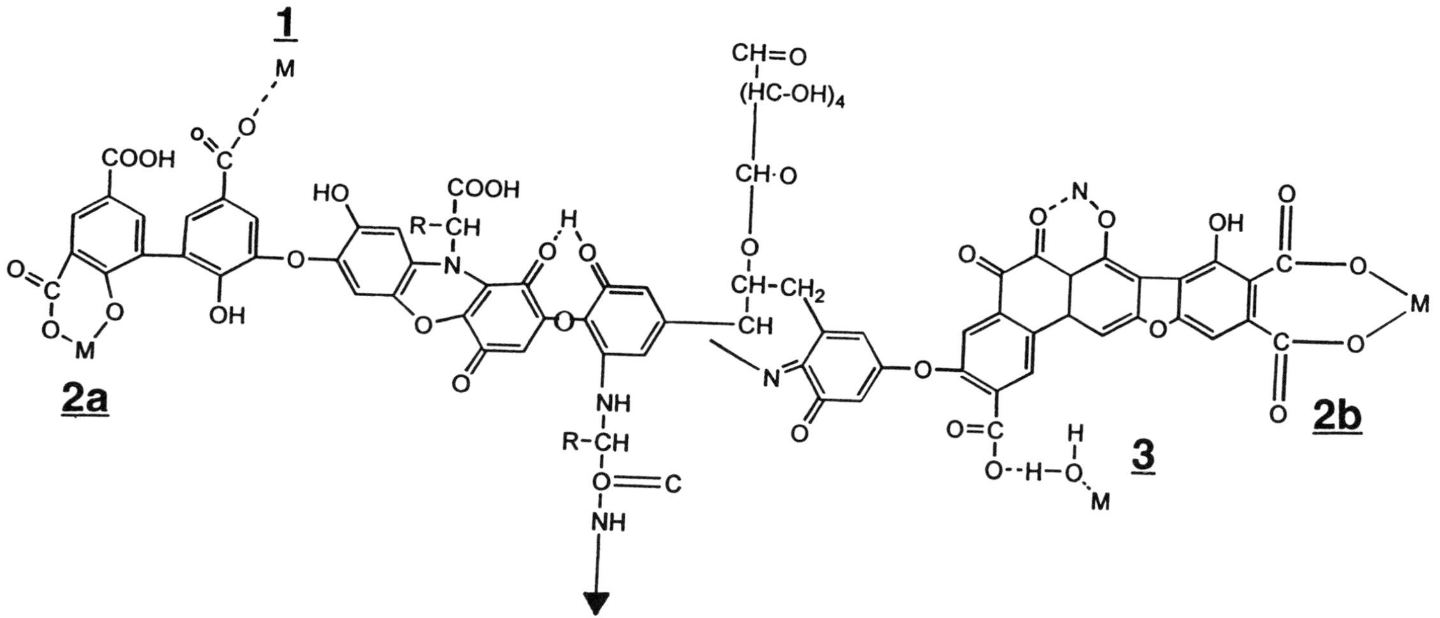

Figure 3.19. Schematic presentation of metal–humic substances complexation (adapted and modified from Stevenson, 1982, with permission). **1**, Electrostatic interaction; **2**, inner-sphere complexation; **3**, weak water bridging.

139

3.4.3   Mechanisms of Complex Formation

The adsorption reaction that occurs between metallic ions and the charged surfaces of clay-organics may involve formation of either relatively weak outer-sphere complexes, or strong inner-sphere complexes.

An outer-sphere complex is a relatively weak electrostatic association between a charged surface and a hydrated cation. This cation can easily be exchanged by other cations also capable of forming outer-sphere complexes. An inner-sphere complex is a relatively strong complex between a charged surface and an unhydrated cation. This cation can only be effectively exchanged by cations which are capable of forming inner-sphere complexes. Studies have shown that humic substances interact with metal ions in a number of ways. The schematic presentation of humic substances–metal complexes can be seen in Figure 3.19 (Evangelou, 1995b).

Interaction **1** denotes electrostatic forces between humic substances (negatively charged) and metal ions (positively charged). It is a relatively weak interaction (outer-sphere complex) and the cation can be readily exchanged by other weakly bonding cations,

Interaction **2** denotes chelation or inner-sphere complexation. For this interaction to take place, the ligand must contain at least two donor atoms capable of being positioned within the ligand forming a ring. This type of interaction commonly appears to have two modes: in the first mode the metal is bound in both phenolic OH and COOH

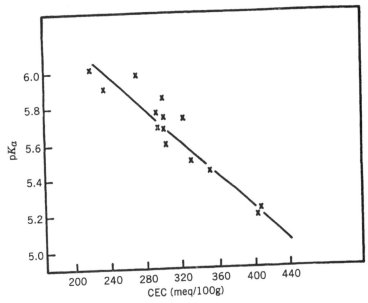

Figure 3.20. Relationship between charge (CEC) and pK_a (from Posner, 1966, with permission).

groups (**2a**); in the second mode, the metal is bound only by COOH groups (**2b**) (Schnitzer, 1969; Gamble et al., 1970),

Interaction **3** shows metal–humic substances interacting through water bridging. It is a weak interaction involving metals with high hydration energy (Bohn et al., 1985).

It is often difficult for one to distinguish the potential of organic material to form inner-sphere complexes with heavy metals, but it is known that such complexes represent a significant fraction of the total charge of SOM. The total charge of SOM is related to the pK_a (Fig. 3.20). The greater the pK_a value is, the lower the charge (CEC). Organic material contribution to soil CEC has been reported to vary from 36 $cmol_c kg^{-1}$ at pH 2.5 to 213 $cmol_c kg^{-1}$ at pH 8. Furthermore, it has been reported that the apparent pK_a of organic matter increased from 4 to 6 when trivalent metals (Al^{3+}) were complexed (Thomas and Hargrove, 1984). This suggests, according to Figure 3.20, that upon Al^{3+} complexation, CEC decreases by more than half.

3.5 CLAY MINERAL SURFACE CHARGE

There are two types of charges at the surface of mineral particles: (1) permanent charge and (2) variable charge. Permanent charge is due to isomorphous substitutions, whereas variable charge is caused by dissociation of mineral-edge hydroxyls.

3.5.1 Permanent Structural Charge

As pointed out in Section 3.2, when coordinating cations of higher valence are replaced by cations of lower valence, a deficit of internal positive charge results or, conversely, a net negative charge on the mineral surface is generated. Correspondingly, a negative electrical potential is created at the mineral surface. To maintain electrical neutrality in the system, cations from solution are absorbed onto the negatively charged mineral surface (Fig. 3.21). Figure 3.21 is only a representation of exchangeable cations and surface negative charge, because the latter is of a delocalized nature and the internal charge deficit is averaged over the whole surface. For this reason, as the distance increases from the particle surface to the bulk solution, the concentration of counterions decreases at some exponential rate (Fig. 3.22). Thus, a large portion of the adsorbed cations remain at some distance from the mineral surface partially or nearly fully hydrated. The electrical force acting on these cations is pulling them toward the particle surface, whereas the force of diffusion is pulling them away from the surface. The net interaction between these two forces allows the absorbed ions to extend outward to a point where the surface's attractive electrical force equals that of diffusion, a repulsive force. This theory of ion adsorption and distribution was first described by Gouy and Chapman, independently, in 1910 and is now known as the Gouy–Chapman diffuse double layer theory. The complete derivation of the Gouy–Chapman model has been outlined by several authors (Singh and Uehara, 1986; Stumm and Morgan, 1981; Van Olphen, 1971; Gast, 1977; Babcock, 1963).

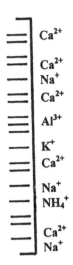

Figure 3.21. Schematic of cation adsorption by a negatively charged surface.

Application of the electric double layer theory to soil minerals at a quantitative level is difficult because soil mineral surfaces at the microscopic scale are not well defined, that is, they are neither perfectly spherical nor flat, as the double layer requires. However, application of the double layer theory at a qualitative level is appropriate because it explains much of the behavior of soil minerals in solution, for example, dispersion, flocculation, soil permeability, and cation and/or anion adsorption. When equilibrium between the counterions at the surface (near the charged surface) and the equilibrium solution is met, the average concentration of the counterions at any

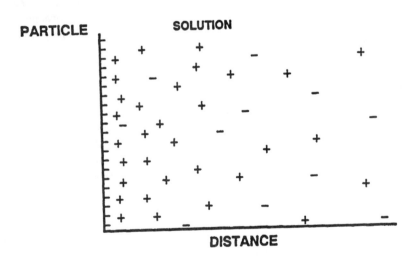

Figure 3.22. Distribution of charged species with respect to distance from a charged surface.

distance x from the charged surface can be expressed as a function of the average electrical potential, ψ, by the Boltzman equation. In the case of cations,

$$n_+ = n_{+0}e^{z_+ F\psi_x/\kappa T} \tag{3.1}$$

and in the case of anions

$$n_- = n_{-0}e^{z_- F\psi_x/\kappa T} \tag{3.2}$$

and

$$\rho = zF(n_+ - n_-) \tag{3.3}$$

where

n_+ = concentration of positive counterions
n_{+0} = concentration of positive counterions in the equilibrium solution
n_- = concentration of negative counterions
n_{-0} = concentration of negative counterions in the equilibrium solution
z_+ = valence of positive counterion (cation)
z_- = valence of negative counterion (anion)
F = Faraday's constant (96,490 C eq^{-1})
ψ_x = electrical potential at any distance x from the charged surface (V)
R = universal gas constant (8.314 Jmol^{-1} °K)
ρ = charge density at any point
T = absolute temperature
κ = Boltzman constant (1.38 × 10^{-23} J °K^{-1})

When ψ is negative, $n_+ > n_-$, whereas when ψ_x is positive, $n_- > n_+$. The surface charge density ρ and the electrical point ψ at any point are related by Poisson's equation:

$$d^2\psi/dx^2 = -(4\pi/\varepsilon)\rho \tag{3.4}$$

where $d^2\psi/dx^2$ describes variation in the electric field strength, $-d\psi/dx$, with distance and ε is the dielectric constant. By combining Equations 3.1–3.4, and considering that $\sinh x = (e^x - e^{-x})/2$, the fundamental differential equation for the double layer is obtained:

$$d^2\psi/dx^2 = k^2/\sinh (zF\psi/RT)(zF/RT) \tag{3.5}$$

The component k is the inverse of the thickness of the double layer, which extends from the solid surface to the point where the local potential is that of the bulk solution, and is given by (Stumm and Morgan, 1970)

$$k = [(8\pi ne^2 z^2/\varepsilon\kappa T]^{1/2} \tag{3.6}$$

where e = electron charge (1.6×10^{-19} C), n = number of ion pairs (cm^{-3}), and ε = relative dielectric permittivity (dimensionless, ε = 80 for water).

Also, for $\psi^0 < 25$ mV, Equation 3.5 reduces to

$$d^2\psi/dx^2 = k^2\psi \tag{3.7}$$

and upon integrating twice with the appropriate boundary conditions (Stumm and Morgan, 1981),

$$\psi = \psi_0 e^{-kx} \tag{3.8}$$

Equation 3.7 points out that the variation in the electric field strength ($-d\psi/dx$) is related to the second power of the inverse of the thickness of the double layer times ψ, while Equation 3.8 shows that ψ decays exponentially with respect to distance (x) from the surface (Fig. 3.23). A plot of $\ln(\psi/\psi_0)$ versus x produces a straight line with slope k, which is the inverse of the double layer thickness. The assumption $\psi_0 < 25$ mV is not applicable to all soil minerals or all soils. Commonly, clay minerals possess more than 25 mV in surface electrical potential, depending on ionic strength. The purpose of the assumption was to demonstrate the generally expected behavior of charged surfaces.

The total surface charge, σ, can be obtained by solving for ρ using the Poisson equation (Equation 3.4) and integrating with respect to x by considering as boundary conditions for x, 0 and ∞.

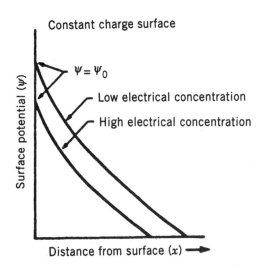

Figure 3.23. Schematic of electric potential distribution for two electrolyte concentrations as a function of distance from a constant charge surface (from Dixon and Weed, 1977, with permission).

$$\sigma = -\int_0^\infty \rho dx = \varepsilon/4\pi \int_0^\infty (d^2\psi/dx^2)dx$$

$$= -\varepsilon/4\pi[d\psi/dx]_x = 0 \tag{3.9}$$

Since, from the first integration of Equation 3.5,

$$d\psi/dx = (RT/zF)\, 2k\, \sinh\, (zF\psi/RT) \tag{3.10}$$

substituting Equation 3.6 into Equation 3.10 and then into Equation 3.9 gives

$$\sigma = [(2/\pi)n\varepsilon\kappa T)]^{1/2}\sinh(zF\psi_0/2RT) \tag{3.11}$$

Equation 3.11 shows that the surface charge is directly related to the surface electrical potential (ψ_0) and number of ion pairs per cubic centimeter (n) or ionic strength (I).

Note that the only terms in Equation 3.6 that are variable are the valence and concentration of the ions in the bulk solution. In further dissecting Equation 3.6, one finds that under constant temperature, $8\pi ne^2z^2/\varepsilon\kappa$ is a constant and can be defined as λ. Therefore,

$$k = \lambda^{1/2}\,(I)^{1/2} \tag{3.12a}$$

or

$$k^{-1} \cong 3.0 \times 10^{-8}\,(I)^{1/2}\;(\text{cm}) \tag{3.12b}$$

where I = ionic strength of the solution. For a small ψ_0 ($\psi_0 < 25$ mV), Equation 3.11 reduces to

$$\sigma = (\varepsilon k/4\pi)\psi_0 \tag{3.13a}$$

or

$$\sigma = (\lambda^{1/2}I^{1/2}\varepsilon/4\pi)\psi_0 \tag{3.13b}$$

Equations 3.12 and 3.13 reveal that when I increases, k increases but ψ_0 decreases, leading to a constant σ. The term σ is related to the experimental CEC. Since the concentration and valence of the charged solution constituents dictate the distance that the diffuse layer could extend into the bulk solution, when I increases (either by increasing ion concentration or by increasing ion valence), the drop in potential occurs a short distance from the solid surface, thereby creating a thinner diffuse double layer (Fig. 3.23).

It follows that the CEC of a constant charge mineral should be independent of pH, salt concentration, or type of metal cation. Charged surfaces however, respond to ion valence according to the Schulze–Hardy rule. It states that ions of higher valence are

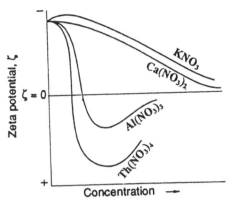

Figure 3.24. Electrical potential effects of various electrolytes (from Taylor and Ashroft, 1972, with permission).

more strongly adsorbed by a surface than ions of lower valence. The behavior is graphically demonstrated in Figure 3.24. It shows that cations with higher valence are more effective in suppressing the surface electrical potential or zeta potential (defined as ψ at the point of slippage or boundary between water moving with the charged-particle surface and bulk or immobile water; see Chapter 9) than cations of lower valence. Furthermore, cations with high valence have a tendency to induce surface-charge reversal.

3.5.2 Variable Charge

Metal oxides and other nonsilicate or silicate minerals possess variable charge or pH-dependent charge. Variable charge arises from the protonation and deprotonation of functional groups at surfaces. Figure 3.25 represents the terminal edge of an iron oxide crystal under three pH conditions. In the acid condition, an excess of adsorbed H^+ results in a net positive charge at the oxygen and hydroxyl functional groups. High pH conditions induce oxygen deprotonation with the surface gaining a net negative charge. At near neutral pH, the positive and negative charges are equal. This pH value is referred to as *point of zero charge* (PZC). Generally, the PZC represents the pH of maximum particle agglomeration and lowest potential mineral solubility (Parks and DeBruyn, 1962).

As noted, metal oxides are not the only minerals with variable charge behavior. Kaolinite, an aluminosilicate, may attain as much as 50% of its negative charge by deprotonation of terminal oxygens and hydroxyls at high pH (pH~7).

In variable charge or pH-dependent charge minerals, the surface potential, ψ_0, remains constant and is not affected by the concentration of ions in solution. In the case of permanent charge minerals, however, ψ_0, varies with the concentration of salt in solution (Fig. 3.26). The relationship between ψ_0 and surface charge is given by the Gouy–Chapman model, as previously demonstrated.

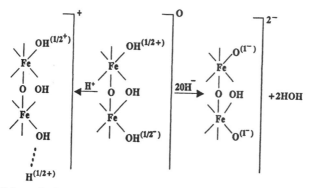

Figure 3.25. Schematic showing the zero point of charge of an iron oxide surface and the charge generated as a function of added hydrogen (H^+) or hydroxyls (OH^-) (from Singh and Uehara, 1986, with permission).

$$\sigma = [(2/\pi)n\varepsilon\kappa T)]^{1/2}\sinh(zF\psi_0/2RT) \qquad (3.14)$$

where

σ = total surface charge (C cm^{-2})

n = number of ion pairs (cm^{-3})

ε = dielectric constant of solution

ψ_0 = surface electric potential

κT = Boltzman's constant times absolute temperature

z = valence

e = electron charge

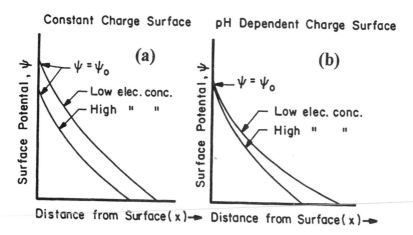

Figure 3.26. Surface electrical potential as a function of distance from the surface of a constant charge or pH-dependent charge.

Equation 3.14 reveals that as surface electrical potential (ψ_0) increases, total surface charge (σ) increases. The surface electrical potential ψ_0 can be approximated by the Nernst equation:

$$\psi_0 = (RT/F)\ln C/C_0 \tag{3.15}$$

where

    C = concentration of potential determining ion in solution
    C_0 = concentration of potential determining ion at $\psi_0 = 0$
    F = Faraday's constant
    T = temperature

Note that variably charged soil mineral surfaces do not exactly obey the Nernst equation. The assumption is only valid at approximately one pH unit above or below the PZC or at approximately 25 mV of surface electrical potential. The assumption is only used for demonstration purposes. Since H^+ and OH^- are considered to be potential determining ions (PDIs), Equation 3.15 can be rewritten as

$$\psi_0 = (RT/F)\ln H^+/H_0^+ \tag{3.16}$$

or

$$\psi_0 = 59(-pH + pH_0) \text{ mV at } 25°C \tag{3.17}$$

where pH_0 is the pH at which σ_0 and ψ_0 equal zero, and represents the PZC. Equation 3.14 reveals that the CEC of soil composed of variably charged minerals is dependent on pH, salt concentration, and ion valence. It follows from Equation 3.17 that when pH decreases, ψ_0 decreases. Furthermore, as ψ_0 decreases, σ (or CEC) also decreases. On the other hand, as salt concentration and/or ion valence increase, σ also increases (Equation 3.14).

    In summary, constant charge minerals (variable surface potential) differ from variably charged (constant surface potential) minerals because for constant charge minerals the electrolyte components (e.g., cations) represent potential determining ions, thus the surface electrical potential (ψ_0) decreases as electrolyte concentration increases (observe Fig. 3.26a and note the difference in ψ_0 at the two electrolyte levels). For the variably charged minerals, the surface potential ψ_0 remains constant when electrolyte concentration increases (observe Fig. 3.26b and note that there is no difference in ψ_0 at the two electrolyte levels). This implies that the particular electrolyte is indifferent with respect to the surface. The only time ψ_0 would change is when the electrolyte contains potential determining ions. In the case of a variably charged surface potential, determining ions include those capable of undergoing reversible inner-sphere complexation with the surface. Such ions commonly include H^+ and/or OH^-.

3.5.3 Mixtures of Constant and Variably Charged Minerals

Figure 3.27 demonstrates that soil systems are mixtures of variable and constant charge minerals. It appears that as pH rises above 3, the positive charge or anion exchange capacity (AEC) of the minerals decrease, but the CEC remains unaffected up to pH 5 and increases significantly above pH 5. Therefore, above pH 5 the potential of soil to adsorb cations (e.g., Ca^{2+}, Mg^{2+}, Mn^{2+}, and Fe^{2+}), increases, whereas the potential of soil to adsorb anions (e.g., SO_4^{2-}) decreases. Since natural soils are mixed systems with respect to their charge, that is they contain both variable plus constant charge minerals,

$$\sigma_T = \sigma_c + \sigma_v \tag{3.18}$$

where σ_T = total charge, σ_c = constant charge, and σ_v = variable charge.

Considering that σ_v is given by Equation 3.14, substituting its ψ_0 by Equation 3.17 and collecting terms gives

$$\sigma_v = [(2/\pi)n\varepsilon\kappa T)]^{1/2}\sinh 1.15z\,(pH_0 - pH) \tag{3.19}$$

For a 1:1 indifferent electrolyte and for pH values within one unit of pH_0, $\sinh(pH_0 - pH) \approx 1.15(pH_0 - pH)$ and Equation 3.19 can be rewritten as

$$\sigma_v = \Lambda\,n^{1/2}(pH_0\text{-}pH) \tag{3.20}$$

where $\Lambda = 1.67 \times 10^{-6}$ esu cm^{-1} (Uehara and Gillman, 1980) and

$$\sigma_T = \sigma_c + 1.67 \times 10^{-6}n^{1/2}(pH_0 - pH) \tag{3.21}$$

Equation 3.21 points out two zero points of charge, when $\sigma_T = 0$ and when $\sigma_v = 0$. The zero point of charge of the mixture is

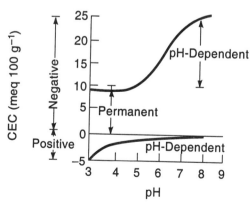

Figure 3.27. Mineral cation exchange capacity (CEC) behavior as a function of pH (from Bohn et al., 1985, with permission).

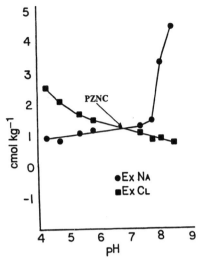

Figure 3.28. Mineral cation and anion exchange capacity as a function of pH.

$$\sigma_c + \sigma_v = 0 \tag{3.22}$$

Substituting the term σ_v in Equation 3.22 by Equation 3.20 and rearranging,

$$\sigma_c = -n^{1/2}(1.67 \times 10^{-6})[pH_0 - pH] \tag{3.23}$$

and

$$pH_{znc} = pH_0 + \sigma_c/n^{1/2}(1.67 \times 10^{-6}) \tag{3.24}$$

The term pH_{znc} in Equation 3.24 represents the point of zero net charge (PZNC). It is the pH value at which the cation exchange capacity equals the anion exchange capacity (Fig. 3.28). Equation 3.24 shows that pH_{znc} varies with ionic strength (n), whereas pH_0 or PZC is an intrinsic property of the mineralogically heterogeneous soil (Uehara and Gillman, 1980).

3.5.4 Relevant Soil Charge Components

The relationship between pH and surface charge of soils is rather difficult to predict. This is especially true in soil systems where clay minerals represent a mixture of constant and variable charge. Surface charge generation on variably charged colloid surfaces takes place because of specific adsorption or inner-sphere complex formation of H^+ or OH^-. However, specific adsorption is not limited to H^+ and OH^-. Other ions such as HPO_4^{2-}, Fe^{3+}, and Al^{3+}, may also specifically or strongly adsorb and may influence the surface charge of soils (Wann and Uehara, 1978; Singh and Uehara, 1986; Uehara and Gillman, 1981; Sposito, 1984a). Wann and Uehara (1978) and Singh and

TABLE 3.7. Examples of Oxides and pH of Point Zero Charge

Oxide	pH_{pzc}
Aluminum oxide	9.1
Aluminum trihydroxide	5.0
Iron oxide	6–8
Manganese oxide	2–4.5
Silicon oxide	2
Kaolinite	4.5
Montmorillonite	2.5

Uehara (1986) presented evidence that weak specific adsorption also has an influence on variable charge surfaces.

Ions forming outer-sphere complexes differ with respect to their potential to influence surface charge of variably charged minerals (e.g., metal-oxides), when compared to ions forming inner-sphere complexes (e.g., HPO_4^{2-} and $FeOH^{2+}$). Generally, cations forming outer-sphere complexes (e.g. Ca^{2+}) shift the PZC to lower pH values while cations forming inner-sphere complexes (e.g. Al^{3+}) shift the PZC to higher pH values (Singh and Uehara, 1986). Anions forming outer-sphere complexes (e.g. SO_4^{2-}) shift the PZC to higher pH values, whereas anions forming inner-sphere complexes (e.g., HPO_3^{2-}) shift the PZC to lower pH values (Sposito, 1981a; Singh and Uehara, 1986). Table 3.7 shows the PZC of various minerals commonly found in natural soil.

Wann and Uehara (1978) reported that in iron-rich soil, the use of $CaCl_2$ as background electrolyte shifted the PZC to lower pH values than when the background electrolyte was NaCl. They attributed this shift to the ability of Ca^{2+} to undergo low-affinity specific adsorption or form outer-sphere complexes, thus increasing the negative charge of the mineral's surface by displacing protons. The same authors also reported a decrease in the PZC of an oxisol from pH 4.7 to 3.5 in the presence of 1500 ppm P as PO_4^{3-} with $CaCl_2$ as the background electrolyte (Fig. 3.29). The data in Figure 3.29 reveal that the potential of the soil to adsorb cations increases as a function of pH, phosphate added, and cation valence. The higher the valence is, the higher the potential of the charged mineral surface to adsorb cations (note the difference in the slope between the two lines in Fig. 3.29).

The role of pH on metal adsorption by soils is also demonstrated in Figure 3.30. These data show that as pH increases, lead (Pb) adsorption increases. However, Pb adsorption appears to be greater for the A soil horizon than the B soil horizon. Since the A soil horizon contains a greater amount of organic matter, this increase in metal adsorption is most likely due to the organic matter's high charge density and greater pH dependence.

Attempts to model soil–metal adsorption behavior using the double-layer model have not been adequate (Van Raij and Peech, 1972). The reason is that in the

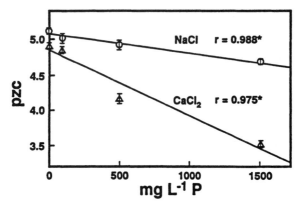

Figure 3.29. Relationship between PZC_{pH} (pH_0) and phosphorus (P) levels (from Wann and Uehara, 1978, with permission).

Gouy–Chapman double layer model, it is assumed that ions behave as point charges and can approach the surface without limit. This leads to predicting extremely high concentrations of metals at high surface potentials. Because of this, Stern proposed a model (Van Raij and Peech, 1972 and references therein) where ions are given a finite size and the surface charge is assumed to be balanced by the charge in solution, which

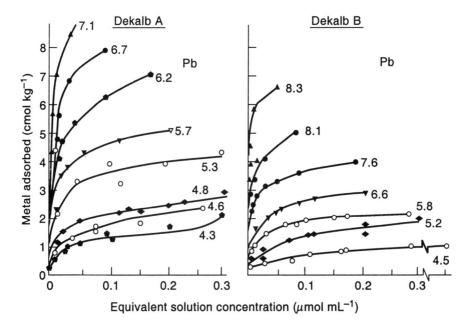

Figure 3.30. Adsorption isotherms for lead (Pb) by a soil's A and B horizons adjusted at various pH levels by $Ca(OH)_2$ (from Harter, 1983, with permission).

is distributed between two layers. The charge closer to the mineral surface, called the Stern layer, is represented by σ_1; the charge in the other layer (diffuse layer) is represented by σ_2 (Fig. 3.31). Thus,

$$\sigma_T = \sigma_1 + \sigma_2 \tag{3.25}$$

where

$$\sigma_1 = N_1 Ze/1 + (N_A q/M)\exp[-(Z\varepsilon\psi_\delta + \phi/\kappa T)] \tag{3.26}$$

and

$$\sigma_2 = (2ne\kappa T/\pi)^{1/2}\sinh(ze\psi_\delta/2\kappa T) \tag{3.27}$$

where
N_1 = number of available sites
N_A = Avogadro's number
q = is the density of the solvent
M = molecular weight of solvent
ε = relative dielectric permittivity (dimensionless, $\varepsilon = 80$ for water)
ψ_δ = electric potential at the boundary between the Stern layer and the diffuse layer
ϕ = energy of specific adsorption and all other terms as previously defined

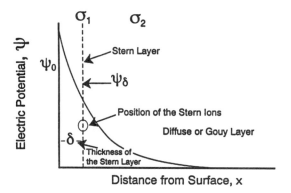

δ = Specific adsorption potential
ψ_0 = Surface potential
ψ_δ = Stern potential
σ_1 = Stern layer charge
σ_2 = Diffuse layer charge

Figure 3.31. Diagram of the Stern model (from Van Olphen, 1977, with permission).

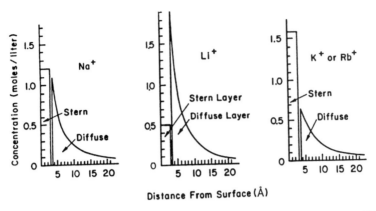

Figure 3.32. Calculated cation distribution near a mineral surface (from Shainberg and Kemper, 1966, with permission).

The data in Figure 3.32 show the theoretical estimated concentration of Na^+, Li^+, and K^+ or Rb^+ between the Stern layer and the diffuse layer based on the Stern model. The distribution appears to be consistent, as expected, with the hydration energy of the cation. The greater the heat of hydration (see Chapter 4, Table 4.1) is, the greater the concentration of the cation in the diffuse layer in relationship to the Stern layer. The Stern model has been the basis for many variations of the model recently also known as the surface complexation model (Goldberg, 1992).

3.6 SOIL–MINERAL TITRATIONS

Soil or soil–mineral titrations are often used to establish surface acidity composition and acid–base behavior. Soil or soil–mineral surfaces are complex in nature owing to their large variation in functional group content and behavior. For example, the data in Figure 3.33 show that soil surface acidity is made up mostly by Al and a smaller quantity of H^+. The titration behavior of such soil would depend on amount of Al present, affinity by which this Al is adsorbed by the surface, degree of surface Al hydroxylation, and finally the pK_a values of the surface-associated H^+. Commonly, two types of titrations are employed to evaluate soil or soil–mineral surfaces: (1) conductimetric titration and (2) potentiometric titration.

3.6.1 Conductimetric Titration

Conductimetric titration denotes change in specific conductance of any given clay or soil suspension as a function of base or acid added. An ideal conductimetric titration curve is shown in Figure 3.34. The first slope of the curve (left-hand side) is attributed to easily dissociated hydrogen (very low pK_a surface-functional groups) (Kissel et al.,

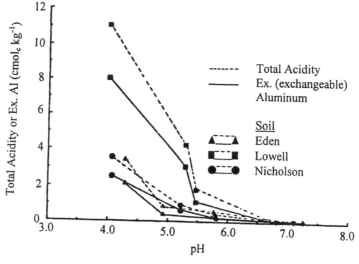

Figure 3.33. Makeup of surface acidity in two Kentucky soils (from Lumbanraja and Evangelou, 1991, with permission).

1971). The second and third slopes are attributed to Al^{3+} and interlayer hydroxy-aluminum (Kissel and Thomas, 1969; Rich, 1970). Conductimetric titrations of two soil samples at three initial pH values are shown in Figure 3.35. None of the soil samples contain dissociated protons. This was concluded because the lines with the

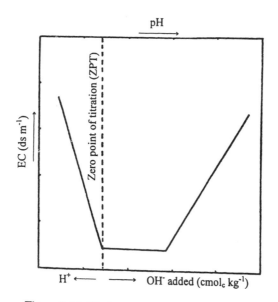

Figure 3.34. Ideal conductimetric titration plot.

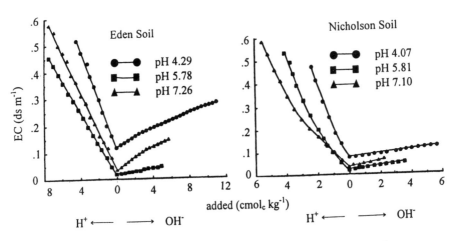

Figure 3.35. Conductimetric titration plots for various Kentucky soils.

negative (downward) slope (near the y-axis) represent HCl titrations. If strong acid groups were present, at the lowest pH values of the titration a line with negative slope would appear if KOH was the titrant. The absence of such line suggests that the difference between total acidity and Al^{3+}, shown in Figure 3.33, represents weak acid groups on the clay surfaces.

Some soils produce conductimetric titration lines that parallel the x axis; they represent adsorbed Al^{3+}. When the slopes of the titration lines become positive, they represent surface-adsorbed Al–hydroxy species (Kissel and Thomas, 1969, Kissel et al., 1971, and Rich, 1970). For example, at all three initial pH values, the Nicholson soil (Fig. 3.35) exhibits titration lines that nearly parallel the x axis. On the other hand, the Eden soil exhibits a line nearly parallel to the x axis only for the sample with an initial pH of 5.78. One expects this behavior to be exhibited by the sample with initial pH of 4.3. For the samples with initial pH 4.3 or 7.3 the titration lines exhibit positive slopes, suggesting neutralization of surface-adsorbed Al–hydroxy species. The overall data in Figure 3.35 show that most surface acidity of the soils is dominated by aluminum under various degrees of hydroxylation.

3.6.2 Potentiometric Titration

Potentiometric titration denotes a change in the pH of any given clay or soil suspension as a function of base or acid added. Generally, three types of potentiometric titration curves are produced (Fig. 3.36). The first type, represented by Figure 3.36a, shows a common crossover point for all three potentiometric curves, representing three different concentrations of an indifferent electrolyte (i.e., $NaNO_3$). The crossover point of the titrations is known as the point of zero salt effect (PZSE). The intercept of the dotted line with the titration lines is known as the pH of zero titration (PZT). For a pure oxide,

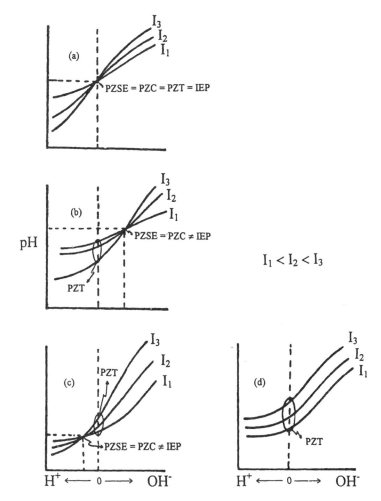

Figure 3.36. Ideal potentiometric titration plots under various concentrations (I_1–I_3) of an indifferent electrolyte.

the PZSE is identical to PZC or isoelectric point (IEP) (Sposito, 1981a). The PZSE of any given mineral may also represent

1. pH of minimum mineral solubility
2. pH of maximum particle settling rate
3. pH at which AEC equals CEC, also known as point of zero net charge (PZNC). This is most likely to occur when inner-sphere complexes are not affected by pH and ionic strength (Sposito, 1984a)

The second type of potentiometric titration curves are shown in Figures 3.36b and c. Figure 3.36b shows that the crossover point of the same colloid system as in Figure

3.36a shifts to higher pH. This implies that a cation has been adsorbed by the colloid in an inner-sphere mode. In the case of Figure 3.36c, the crossover point shifts to a lower pH than that in Figure 3.36a. This implies that an anion has been adsorbed by the colloid in an inner-sphere mode. Finally, the third type of potentiometric titration curves are shown in Figure 3.36d, which reveals no crossover point. This behavior is often associated with permanent charge minerals (Sposito, 1981a; Uehara and Gilman, 1980).

Experimental potentiometric titration curves are shown in Figures 3.37 and 3.38. The data in Figure 3.37 show the PZSE of an iron oxide by the crossover point of the various background salt titrations at approximately pH 6.8. The titration data in Figure 3.38 show that a given soil's acid–base behavior could be dependent on the soil's initial pH (PZT). The Eden soil exhibited a PZSE when the PZT was near 4. At the PZT of approximately 6, no definite trend in variable charge behavior was apparent because the titration graphs were displaced to the right upon increasing the concentration of the background electrolyte, demonstrating titration behavior of a permanently charged soil. Finally, for the PZT of about pH 7, the soil exhibits variable charge behavior with a PZSE of approximately 7.4.

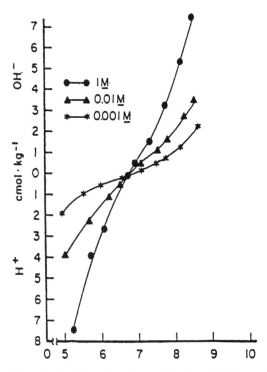

Figure 3.37. Potentiometric titrations of an iron oxide (hematite) formed under laboratory conditions (the point at which all three lines converge is representative of the PZC) (from Evangelou, 1995b, with permission).

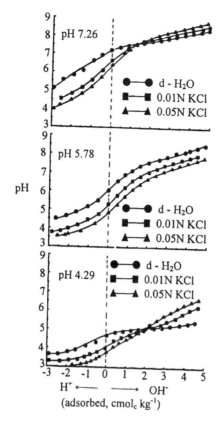

Figure 3.38. Potentiometric titrations of Kentucky soil (from Lumbanraja and Evangelou, 1991, with permission).

The apparent shift in the PZSE of a soil by changing initial pH suggests removal of exchangeable bases from clay surfaces or buildup of Al–hydroxy species on the soil's surfaces (Uehara and Gillman, 1981). Parker et al. (1979) cautioned that a PZSE does not necessarily reflect the adsorption or desorption of a potential determining ion (H^+ or OH^-). Potentiometric titrations also involve exchange reactions and/or dissolution reactions. Therefore, a drop in the PZSE could also be caused by the dissolution of Al during soil acidification, whereas an increase in PZSE could be caused by the formation of Al–hydroxy species on the soil's surfaces. Hendershot and Lavkulich (1983) demonstrated that illite did not exhibit a PZSE when PZT was in the range of 6–7; however, it did exhibit a PZSE near pH 4 when coated with Al. They also reported that soil samples behaved similarly.

3.6.3 Soil Acidity

Generally speaking, soil acidity is separated into three types:

1. Soluble and exchangeable acidity (Al^{3+} plus H^+)
2. Titratable acidity (soluble and exchangeable Al^{3+} plus H^+ and nonexchangeable Al–hydroxy or Fe–hydroxy polymers)
3. Total acidity, which refers to titratable acidity up to pH 8.2. Titratable acidity includes H_3O in the pH range < 4, Al^{3+} in the pH range of 4–5.6, "strong" aluminum hydroxy in the pH range 5.6–7.6, and "weak" aluminum–hydroxy in the pH range greater than 7.6 (Thomas and Hargrove, 1984)

A process that leads to low soil pH and soil acidification is hydrolysis. The metal most commonly associated with soil acidification is Al^{3+}. Aluminum ions on mineral surfaces hydrolyze to produce H^+, which in turn attacks the clay surfaces to produce more acidity. The process is demonstrated below:

$$-Al^{3+} + nH_2O \Leftrightarrow -Al(OH)_n^{(3-n)} + nH^+ \tag{3.28}$$

If, for whatever reason, available OH^- in soil increases (e.g., by liming), Al–OH monomers form large Al–hydroxy polymers. This leads to the formation of double Al–hydroxy rings $[Al_{10}(OH)_{22}]^{8+}$, leading to triple rings $[Al_{13}(OH)_{30}]^{9+}$, and so on. The continuous increase in OH^- relative to Al decreases the charge per Al from 0.8^+ to 0.33^+. When the charge contributed per Al finally reaches zero, formation of crystalline aluminum hydroxide takes effect (Hsu and Bates, 1964).

This implies that Al in soils is found in various stages of hydroxylation, hence at various degrees of positive charge (e.g., 8^+ to 18^+). For this reason, various other anions, besides OH^-, may be found associated with Al. Furthermore, because polymerization and production of crystalline $Al(OH)_3$ takes time, titratable acidity, which represents a quick procedure, may not necessarily reflect soil-available acidity. A similar problem persists with KCl-extractable soil Al. Polymeric aluminum is strongly adsorbed by soil mineral surfaces and, for this reason, such Al may not be extractable with metal salts (e.g., KCl).

Soil aluminum has drawn a great deal of research interest because of its chemical complexity and agricultural importance. In the range of 1–3 ppm, aluminum in soil is highly toxic to plants. It has been shown that as Al concentration in the soil solution increases, plant productivity decreases. For this reason, Al^{3+} complexing agents such as SOM (Fig. 3.39) support plant growth even at pH values at which aluminum is expected to induce high plant toxicity (Fig. 3.40). Although it is known that plant toxicity is induced by soluble aluminum, the species responsible for this toxicity is not well known. For some plants, it has been shown that total dissolved aluminum and relative root growth give poor correlation (Fig. 3.41), while correlation between the activity of Al^{3+} species in solution and relative root growth appears to be significantly improved (Fig. 3.42).

Soil acidity is commonly neutralized by $CaCO_3$ or $Mg(CO_3)_2$. The general reactions that explain soil acidity neutralization by $CaCO_3$ are as follows:

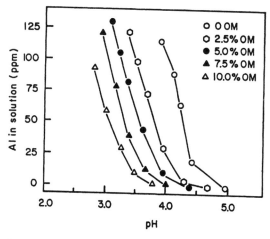

Figure 3.39. Aluminum concentration in solution at various amounts of H peat and base (after Hargrove and Thomas, 1981, with permission).

$$CaCO_3 + H_2O \Leftrightarrow Ca^{2+} + HCO_3^- + OH^- \qquad (3.29)$$

and from Equation 3.28,

$$OH^- + H^+ \rightarrow H_2O \qquad (3.30)$$

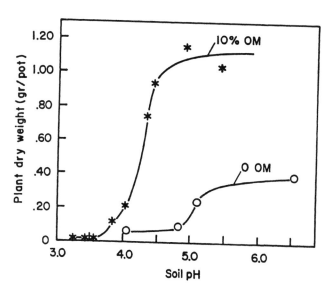

Figure 3.40. Dry weight of plants as a function of organic matter content and soil pH (from Hargrove and Thomas, 1981, with permission).

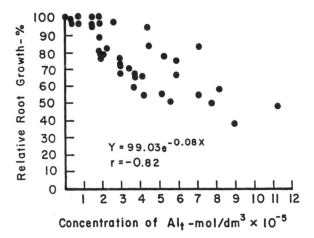

Figure 3.41. Relationship between total dissolved aluminum and relative root growth (from Pavan et al., 1982, with permission).

The overall reaction of acid soil with $CaCO_3$ is as follows:

$$2Al\text{-soil} + 3CaCO_3 + 3H_2O \rightarrow 3Ca\text{-soil} + 2Al(OH)_3 + 3CO_2 \qquad (3.31)$$

When all soil acidity has converted to Ca-soil, the pH would be about 8.3. Commonly, only exchangeable acidity is easily neutralized. A significant fraction of the titratable acidity may remain intact owing to its extremely weak form or high apparent pK_a values. Soil acidity is for the most part produced in soils under high rainfall regimes (tropical or temperate regions).

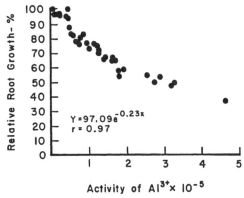

Figure 3.42. Relationship between Al^{3+} activity in solution and relative root growth (from Pavan et al., 1982, with permission).

3.7   SOIL AND SOIL SOLUTION COMPONENTS

There are mainly three forms of chemical constituents in soil material (1) pore–solution constituents, (2) surface-adsorbed constituents, and (3) potentially decomposable–soluble solids. Pore–solution constituents represent those dissolved in the pore water and involve a number of cations and anions. A range of concentration values most commonly encountered in soils is: K^+, 1–10 mg L^{-1}; Na^+, 1–5 mg L^{-1}; Ca^{2+}, 20–200 mg L^{-1}; Mg^{2+}, 2–50 mg L^{-1}; Si, 10–50 mg L^{-1}; SO_4, 60–300 mg L^{-1}; F^-, 0.1–0.1 mg L^{-1}; Cl^-, 50–500 mg L^{-1}; Mn^{2+}, 0.1–10 mg L^{-1}; Cu^{2+}, 0.03–0.3 mg L^{-1}; Al < 0.01, Fe < 0.005, Zn < 0.005, P, 0.002–0.03 mg L^{-1}; Mo, 0.001–0.01 mg L^{-1}. These concentrations reflect soils under temperate climatic zones. Under arid environments, salt concentrations close to those of seawater are sometimes common.

Surface-adsorbed constituents are those composed of minerals or nutrients (e.g., Ca^{2+} Mg^{2+}, K^+, and Na^+), heavy metals (e.g., Al^{3+}, Fe^{2+}, Mn^{2+}, Cu^{2+}, and Pb^{2+}) and/or H^+. Potentially decomposable–soluble solids include pyrite, carbonates, metal oxides, and primary minerals such as feldspars.

Soil CEC is composed of two types of constituents: (1) weak Lewis acid metals, commonly referred to as "bases" (e.g., Ca^{2+}, Mg^{2+}, K^+, and Na^+) and (2) relatively strong Lewis acid metals, H^+, and heavy metals, depending on the nature of the sample, for example, geologic waste or natural soil (Al^{3+}, H^+, Fe^{2+}, Fe^{3+}, and Mn^{2+}). The term percent base saturation is commonly used to describe the percent of the sum of exchangeable "bases" relative to the CEC near pH 7 or at pH 7 (CEC_7). The equation for percent base saturation is given as

$$\%\text{base saturation} = (\text{exchangeable bases}/CEC_7) \times 100 \qquad (3.32)$$

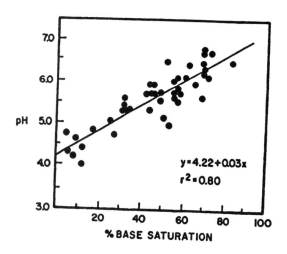

Figure 3.43. The relationship between soil suspension pH and percent base saturation (from Magdoff and Bartlett, 1985, with permission).

An empirical relationship between pH and percent base saturation is shown in Figure 3.43. This relationship appears to be linear, but this does not imply any mechanistic molecular meaning, because soils or clay minerals contain many different functional sites. The general behavior of these data, however, is of practical value. For example, a sample with base saturation of approximately 20% will exhibit a pH of approximately 5, while a pH of approximately 7 suggests a percent base saturation of 100%.

3.8 ROLE OF SOIL–MINERALS IN CONTROLLING WATER CHEMISTRY

Soil modifies water chemistry or quality through the processes of

1. Surface-exchange hydrolysis
2. Dispersion by monovalent metal ions
3. Soil's catalytic role in many chemical and/or electrochemical reactions
4. Precipitation reactions of heavy metals through hydroxylation
5. Oxidation reactions of organics and inorganics
6. Hydrolysis reactions of organics and inorganics
7. Condensation reactions of organics
8. Physical adsorption of metals and metalloids
9. Chemical reactions with metalloids
10. Soil-dissolution reactions

Overall, soil systems behave as complex biomolecular sieves. It is the purpose of this book to elucidate these soil processes in the following chapters.

PROBLEMS AND QUESTIONS

1. Explain (a) coordination, (b) coordination number, (c) tetrahedral, and (d) octahedral.

2. What is the driving force in cation coordination?

3. Name the various clay mineral groups present in soil.

4. What are 2:1, 1:1, and 2:2 clay minerals? How do they differ?

5. Explain how and why the surface charge properties of various 2:1 clay minerals might differ?

6. What is a clay interlayer? How does it form? What is its role in ion adsorption?

7. What are primary minerals and how do they differ from secondary clay minerals?

8. Explain isomorphous substitution and its practical significance.

9. Name the various cation exchange sites on clay minerals.

10. What is soil organic matter? How does it form?

11. Explain the two major groups of organic compounds found in soil.

12. What are (a) outer-sphere complexes and (b) inner-sphere complexes? What is their practical significance in soil?

13. Explain the potential reactivity of soil organic matter with (a) cations and (b) clay minerals. Explain the practical significance of this reactivity.

14. How does humic acid differ from fulvic acid? What is the practical significance of this difference in soil?

15. Consider a surface characterized by the following reactions:

$$SOH_2^+ \Leftrightarrow SOH + H^+ \qquad pK_a = 4.0$$

$$SOH \Leftrightarrow SO^- + H^+ \qquad pK_a = 8.0$$

where S denotes charged mineral surface.
a. Plot net surface charge as a function of pH.
b. What is the pH of zero point of charge?
c. Describe briefly (2–3 sentences maximum) how would you determine the pK_a values of this surface?

16. Explain the relationship between variably charged soils and surface electric potential and the relationship between constant charge soils and surface electric potential. Define the role of potentially determining ions in variably and constant charged soils. Discuss the practical meaning of the above.

17. Ten grams of soil were displaced with 250 mL of 1 M ammonium acetate and made to a final volume of 1 L. Analysis of the final 1-L solution showed 20 mg L^{-1} Ca, 2 mg L^{-1} Mg, 1 mg L^{-1} K, and 0.5 mg L^{-1} Na. Estimate exchangeable cations in meq/100 g soil.

18. After the 1-M ammonium acetate extraction in problem 17, the sample was rinsed twice with distilled water and the NH_4^+ concentration of the second rinse was determined to be 180 mg L^{-1}. The rinsed moist sample (60% moisture by weight) was then displaced with 250 mL of 1 M KCl solution and made to a final volume of 1 L. Analysis of the final 1-L solution showed 30 mg L^{-1} NH_4^+. Estimate the CEC (in meq/100 g) of the soil.

19. Based on your calculations in problems 17 and 18, estimate the percent base saturation of the sample and discuss its practical significance.

20. Explain how one may determine the PZSE. How is it related to the PZC? What is the practical significance of the PZC?

21. Explain when and why PZC equals the IEP of a mineral.

22. Explain the difference between the PZNC and the PZSE. When would one expect PZNC to be the same as PZSE? How would you determine that PZNC and PZSE differ?

23. Explain what type of soil mineral would not show an apparent PZSE.

24. Calculate the change in ionic strength when the thickness of the double layer is suppressed by 50% (see Eq. 3.12).

25. Give two reasons for the observed decrease in CEC of soil organic matter when in the presence of some soluble Al.

26. Give an explanation for the constant CEC of soil organic matter when in the presence of soluble sodium.

27. Name the various functional groups of (a) clay mineral surfaces and (b) soil organic matter. Explain which of these functional groups exhibits constant charge or variable charge behavior and discuss the practical significance of this behavior.

28. Based on the double layer theory, explain the potential effect of temperature on soil CEC.

29. Based on the classical double layer theory, name two assumptions that are now known not to be valid when one tries to predict ion adsorption on the basis of surface electrical potential.

30. How did the Stern model get around the limitations of the double-layer or Gouy–Chapman model?

31. On what basis does the Stern model distinguish the potential location of ions near a surface under the influence of an electrical potential?

32. Based on the classical double layer theory, name all parameters affecting CEC and explain how they affect it.

4 SORPTION AND EXCHANGE REACTIONS

4.1 SORPTION PROCESSES

Mineral solubility and precipitation were discussed in Chapter 2 as techniques for predicting the release of ionic constituents to water or soil solutions or the removal of ionic constituents from water or soil solutions. In this chapter, a second source/sink for ionic constituents (e.g., contaminants or nutrients) is presented and the mechanisms controlling this sink are referred to as *adsorption* or *sorption*. Both terms denote the removal of solution chemical species from water by mineral surfaces (e.g., organics, metal oxides, and clays) and the distinction between the two terms is based on the mechanism(s) responsible for this removal. In adsorption, a chemical species may be adsorbed by a surface either electrostatically or chemically (electron sharing), whereas in sorption, a chemical species may accumulate on a mineral's surface either through adsorption, hydrophobic interactions, and/or precipitation.

Mineral surfaces may catalyze ion precipitation via a number of mechanisms such as simultaneous adsorption of cations and anions. When the solution activity of two ionic species (anion and cation) reaches the saturation point with respect to a given mineral, surface precipitation follows. Another potential surface precipitation mechanism could be based on the inability of a mineral's surface to exclude counterions. For example, assuming that a surface attracts a cation (i.e., Ca^{2+}), it encounters water with low dielectric constant (the dielectric constant of free water is 80 while that of surface-adsorbed water is less than 20), which increases the force of attraction between oppositely charged ions. In this case, failure of the mineral's surface to repulse the associated anion (i.e., SO_4^{2-}), allows $CaSO_4 nH_2O$ precipitation to occur.

Upon reacting with a surface, a chemical species may form either an inner- or an outer-sphere complex (Fig. 4.1). These two terms are defined in Chapter 3. When a mineral's surface and a particular contaminant react covalently (*chemisorption*) or through short-range electrostatic interactions (*physical adsorption*), three reaction modes may be identified: (1) monodentate, (2) bidentate, and (3) polydentate (Fig. 4.2). Generally, a monodentate is represented schematically by a single point of attachment, a bidentate is represented by two points of attachment, and a polydentate is represented by more than three points of attachment. The first two modes, monodentate and bidentate, are common to inorganic and organic mineral surfaces, while the polydentate mode is common to organic surfaces. A prerequisite to polydentate

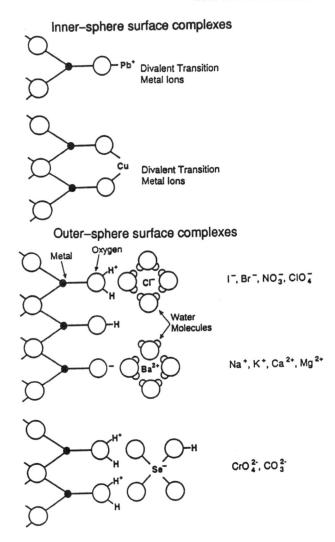

Figure 4.1. Schematic demonstrating inner- and outer-sphere complexes between inorganic ions and hydroxyl groups of an oxide surface (adapted from Hayes, 1987).

complexes is the potential of organic surfaces to undergo reconfiguration during complexation.

The term specific adsorption, often encountered in the scientific literature, describes adsorption of chemical constituents by a surface owing to some unique characteristics between the surface and the adsorbing species which allows a high degree of selectivity (e.g., good fit) between an adsorbing species and a surface site (steric effect). Chemisorption and physical adsorption are distinguished by the magnitude of the heat of adsorption. The higher the heat of adsorption is, the stronger the bond. Chemisorption

Inner-sphere surface complexes

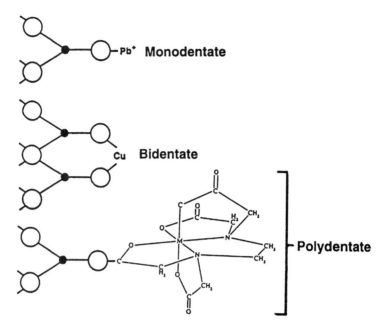

Figure 4.2. Schematic of inner-sphere monodentate and bidentate metal complexes between metal ions and hydroxyl groups of an oxide surface (adapted from Hayes, 1987).

exhibits higher heat of adsorption (> 20 kcal mol^{-1}) than physical adsorption (< 10 kcal mol^{-1}) (McBride, 1994 and references therein).

4.1.1 Surface Functional Groups

The mineral surface components responsible for adsorption are the so-called functional groups. In the case of minerals with permanent charge (e.g., aluminosilicates), the functional group is the siloxane or ditrigonal cavity (Figs. 4.3 and 4.4); in the case of clay–mineral edges, the surface functional groups are –OH species capable of dissociating hydrogen (H^+). There are three such potential functional groups on clay minerals. The first group is the –Al–OH (octahedral) or *aluminol* with a pK_a of around 5, the second is the *silanol* (–Si–OH) (tetrahedral) with a pK_a of around 9, and the third is the intermediate –Si–Al–OH$_2$, or Lewis-acid site (OH is shared between a tetrahedral sheet and an octahedral sheet), with an apparent pK_a of around 6–7 (see the meaning of pK_a in Chapter 1) (Fig. 4.5). Overall, these functional groups are distributed on three different clay–mineral site types. The first site type is known as planar and is represented by mostly exposed siloxane cavities. The second site type is known as edge and is represented by aluminol, silanol, and intermediate functional

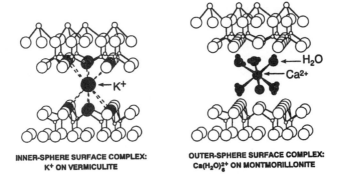

INNER-SPHERE SURFACE COMPLEX:
K⁺ ON VERMICULITE

OUTER-SPHERE SURFACE COMPLEX:
Ca(H₂O)₆²⁺ ON MONTMORILLONITE

Figure 4.3. Schematic of inner- and outer-sphere complexes between metal cations and the siloxane ditrigonal cavities of 2:1 clay minerals (from Sposito, 1984a, with permission).

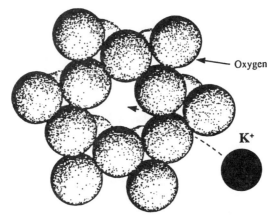

Figure 4.4. Schematic of the siloxane ditrigonal cavity showing the fit of the K^+ ion in the cavity, thus producing an inner-sphere complex (from Sposito, 1984a, with permission) (average inside diameter of siloxane cavity is 0.26 nm while K^+ diameter is 0.133 nm).

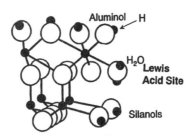

Kaolinite Surface Hydroxyls

Figure 4.5. Schematic of the kaolinite surface hydroxyls. In addition to the basal OH groups, the schematic is showing the aluminol groups, the silanol groups, and the Lewis acid sites contributing to water adsorption (from Sposito, 1984a, with permission).

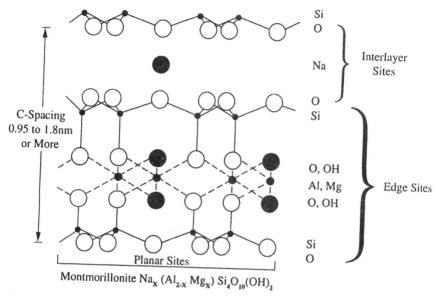

Figure 4.6. Schematic structure of montmorillonite showing the planar sites, the interlayer sites, and edge sites (adapted from Jackson, 1964).

−OH groups. The third site type is known as interlayer and is represented by the space between adjacent siloxane cavities (Fig. 4.6).

The Henderson–Hasselbalch equation (see Chapter 1) can be used to explain functional surface-group behavior. Consider the functional group −Al–OH. Upon dissociation it produces −Al–O$^-$, which attracts cations and forms surface-metal complexes such as −Al–O · nH$_2$OM, where M denotes any metal cation. According to the Henderson–Hassalbalch equation, a surface functional group with a pK_a of 5 signifies that at pH 5 half of the surface is dissociated and, therefore, its negative charge is one half of the potential maximum. At approximately 2 pH units above the pK_a, all the surface groups are dissociated and thus the negative surface charge approaches maximum. At approximately 2 pH units below the pK_a, all surface groups are protonated and thus the negative surface charge approaches zero. Any further decrease in pH may induce surface-charge reversal by forming −Al–OH$_2^+$, which allows surfaces to form complexes with anions (e.g., Cl$^-$ and SO$_4^{2-}$) (Fig. 4.1). The functional groups −Al–OH and −Al–Si–OH are known to exhibit surface-charge reversal potential within natural pH ranges, whereas the −Si–OH does not exhibit this characteristic.

It follows that the method by which pH regulates charge and metal adsorption by a surface (S) can be described as a competitive process. For example,

$$-S-O-H + K^+ \Leftrightarrow -S-O\cdots K + H^+ \qquad (4.1)$$

the symbols −O⋯K signify an outer-sphere surface oxide–potassium complex. In such a case (Reaction 4.1), the affinity of −S–O$^-$ for H$^+$ is much greater than the affinity of

–S–O⁻ for K⁺. In essence, –O–H, an inner-sphere complex, is much stronger than –O⋯K; the latter's formation potential is pK_a or pH dependent. Under these conditions, relatively weak surface-metal complexes (e.g., outer-sphere complexes) begin to form approximately 2 pH units below the pK_a, and their formation potential increases as pH increases.

When –S–O⁻ forms an inner-sphere complex with a heavy metal (M⁺), the reaction is expressed by

$$\text{–S–O–H} + M^+ \Leftrightarrow \text{–S–O–M} + H^+ \tag{4.2}$$

Considering that –O–H may be a weaker complex than –O–M, formation of the latter would be relatively independent of pH. The latter complex would involve a strong bond (e.g., chemisorption). The same explanation applies to anion adsorption. For example, phosphate (PO_4) adsorption by oxides may take place in an outer- or inner-sphere mode of the monodentate or bidentate type (Fig. 4.7).

These mechanisms also apply to all soil metal–oxides and SOM (soil humic substances). Soil organic matter functional groups include enolic and alcoholic OH with pK_a near 9, as well as carboxylic OH with a pK_a around 4. The positive charge on organic matter is mostly due to amino groups ($-NH_2^+$) exhibiting a wide range of pK_a values. However, SOM–NH_2 groups exhibit a short life span owing to their rapid degradation by soil microbes. The dominant surface charge in SOM is mostly negative owing to the resistance of charge-contributing molecules to microbial degradation. Thus, because of its electron configuration, –O⁻ behaves as a Lewis base and participates in the makeup of most functional groups in soil systems.

Specific examples of clay–mineral structures and the role of ligands (complexing structures) are discussed below. There are two types of 2:1 clay minerals—those that have the potential to swell significantly (e.g., smectite) and those with limited swelling potential (e.g., vermiculite) (Figs. 4.6, 4.8, and 4.9). These two mineral groups do not differ greatly in their overall structure. In the case of smectite (e.g., montmorillonite), most of the isomorphous substitution is in the octahedral sheet; in vermiculite, most of the isomorphous substitution is in the tetrahedral sheet. Therefore, in vermiculite the close proximity of the siloxane cavity (located in the tetrahedral sheet; see Chapter 3) to the site of isomorphous substitution (also in the tetrahedral sheet) allows greater

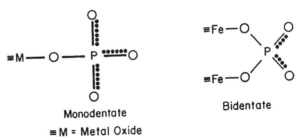

Monodentate

≡M = Metal Oxide

Bidentate

Figure 4.7. Schematic of adsorption of phosphate by an iron oxide in an inner-sphere bidentate or monodentate mode.

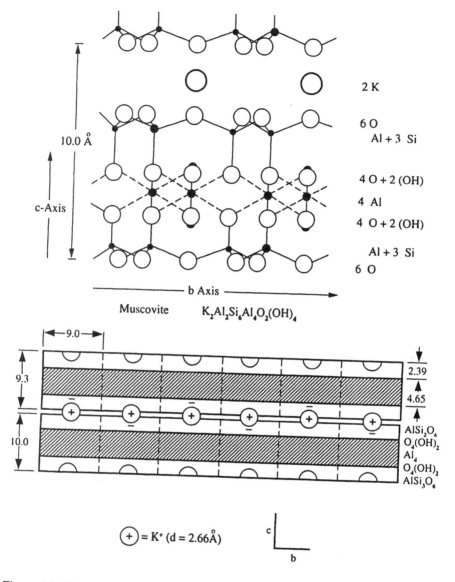

Figure 4.8. Schematic of K–vermiculite. The numbers on the lower portion of the diagram represent the distances (Å) of the various sheets (from Jackson, 1964, with permission).

attraction between the siloxane cavity and the interlayer cations than in montmorillonite. In the case of montmorillonite, the large distance between the siloxane cavity (located in the tetrahedral sheet) and the site of isomorphous substitution (located in the octahedral sheet) diminishes the force of attraction between the siloxane cavity and the interlayer cations.

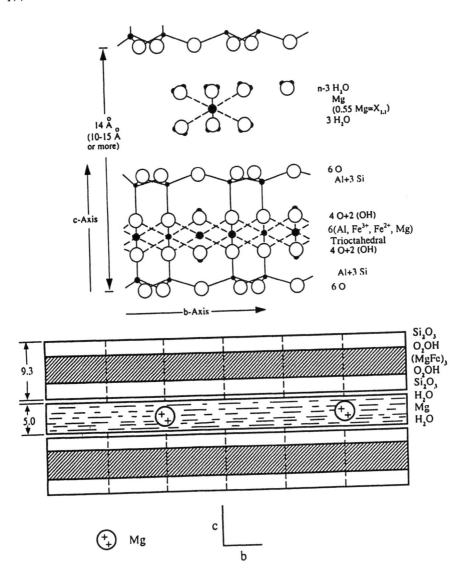

Figure 4.9. Schematic of Mg–vermiculite. The numbers on the lower portion of the diagram represent the distances (Å) of the various sheets (top from Jackson, 1964, with permission).

A large attraction force between interlayer cations and adjacent siloxane cavities allows some cations with certain hydration energy to dehydrate. If the dehydrated cation radius is smaller than the inside diameter of the siloxane cavity, the mineral could collapse and an inner-sphere complex would form (e.g., K-vermiculite) (Fig. 4.3). When vermiculite contains a relatively strongly hydrated cation such as Ca^{2+} or

Mg^{2+} (Table 4.1) its interlayer spacing would be approximately 5 Å or 0.05 nm wide (1 Å = 10^{-8} cm) (Fig. 4.9). When K^+ replaces the interlayer Ca^{2+}, or Mg^{2+}, the low hydration energy of K^+ and its good fit in the siloxane cavity would allow vermiculite to collapse and bring the 5-Å (0.05 nm) interlayer gap to near 0 Å. Thus, an inner-sphere complex would form and such K^+ is often referred to as "fixed K^+". Vermiculite is also known to form inner-sphere complexes with NH_4^+.

Because of the higher affinity for higher-valence cations by the interlayer siloxane cavity, relatively low pH values allow the interlayer to become loaded with Al–hydroxy polymers. Complete occupation of the interlayer spacing by octahedrally coordinated aluminum–hydroxy polymers produces chlorotized vermiculite or montmorillonite (Fig. 4.10). Such minerals are referred to as 2:2 clay minerals. Under these conditions, the interlayer space is limited to approximately 4 Å (0.04 nm) and Al is nearly permanently lodged in the interlayer, diminishing mineral CEC. However, in most cases chlorotization is partial. Thus, Al–hydroxy islands are formed (Fig. 4.11). In such cases, the interlayer gap is also held at approximately 4 Å and many interlayer

TABLE 4.1. Crystallographic Radii and Heats and Entropies of Ion Hydration at 25°C

Ion	Crystallographic Ion Radius (mm)	Heat of Hydration, ΔH (kJ mol^{-1})	Entropy of Hydration, ΔS (J mol^{-1} °K^{-1})
H^+	—	−10	109
Li^+	0.060	−506	117
Na^+	0.095	−397	87.4
K^+	0.133	−314	51.9
Rb^+	0.148	−289	40.2
Cs^+	0.169	−225	36.8
Be^{2+}	0.031	−2470	—
Mg^{2+}	0.065	−1910	268
Ca^{2+}	0.099	−1580	309
Ba^{2+}	0.135	−1290	159
Mn^{2+}	0.080	−1830	243
Fe^{2+}	0.076	−1910	272
Cd^{2+}	0.097	−1790	230
Hg^{2+}	0.110	−1780	180
Pb^{2+}	0.120	−1460	155
Al^{3+}	0.050	−4640	464
Fe^{3+}	0.064	−4360	460
La^{3+}	0.115	−3260	368
F^-	0.136	−506	151
Cl^-	0.181	−377	98.3
Br^-	0.195	−343	82.8
I^-	0.216	−297	59.8
S^{2-}	0.184	−1380	130

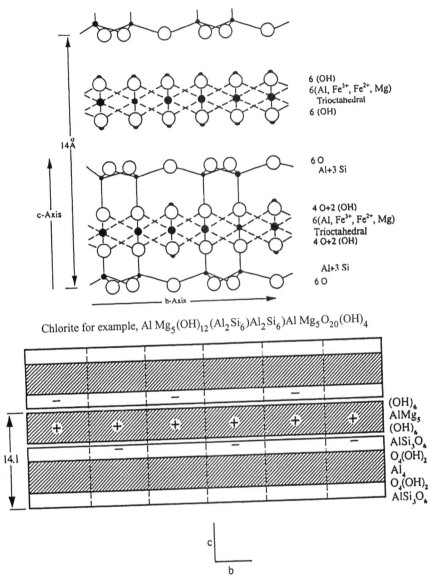

Chlorite for example, $Al\, Mg_5(OH)_{12}(Al_2Si_6)Al_2Si_6)Al\, Mg_5O_{20}(OH)_4$

Figure 4.10. Schematic structure of chlorite showing the various isomorphous substitutions and *d*-spacing (top from Jackson, 1964, with permission).

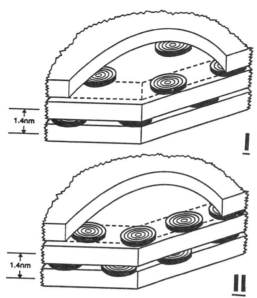

Figure 4.11. Schematic showing the distribution of aluminum–hydroxy polymers in the interlayer space of 2:1 clay minerals. (I) A uniform distribution; (II) an "atol" distribution (from Dixon and Jackson, 1962, with permission).

siloxane cavities remain free of aluminum. The pathway to these siloxane cavities is highly tortuous (channelized) and only cations with low hydration energy (e.g., K^+ and NH_4^+) are able to reach them through diffusion.

Vermiculite is also known for the so-called *wedge effect* (Fig. 4.12), which may explain certain cation replaceability behavior in the interlayer. For example, cations such as Ca^{2+} may expand the outermost portion of the interlayer, but because Ca^{2+} is held strongly in such sites, K^+ would be trapped deeper in the interlayer. On the other

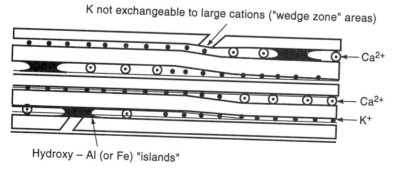

Figure 4.12. Schematic demonstration of the potential role of an aluminum–hydroxy polymer on cation exchangeability (from Rich, 1968, with permission).

hand, K ions adjacent to Ca^{2+} may be replaced by diffusing cations, owing to the propping effect induced by hydrated Ca^{2+}.

4.2 ADSORPTION–SORPTION MODELS

Equilibrium between solution and adsorbed or sorbed phases is a condition commonly used to evaluate adsorption or sorption processes in soils or soil–clay minerals. As previously stated, equilibrium is defined as the point at which the rate of the forward reaction equals the rate of the reverse reaction. Two major techniques commonly used to model soil adsorption or sorption equilibrium processes are (1) the Freundlich approach and (2) the Langmuir approach. Both involve adsorption or sorption isotherms. A sorption isotherm describes the relationship between the dissolved concentration of a given chemical species (adsorbate) in units of micrograms per liter (μg L^{-1}), milligrams per liter (mg L^{-1}), microequivalents per liter (μequiv L^{-1}), or millimoles per liter (mmol L^{-1}), and the sorbed quantity of the same species by the solid phase (adsorbent) in units of adsorbate per unit mass of adsorbent (solid) (e.g., μg kg^{-1}, mg kg^{-1}, μeq kg^{-1}, or mmol kg^{-1}) at equilibrium under constant pressure and temperature. Sorption isotherms have been classified into four types, depending on their general shape (Fig. 4.13):

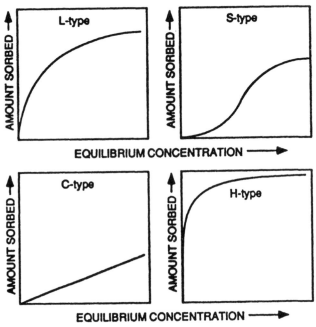

Figure 4.13. Classification of adsorption isotherms (adapted from Sposito, 1948a).

1. L-type: describes high-affinity adsorption between the adsorbate and adsorbent and usually indicates chemisorption (e.g., phosphate–soil interactions)

2. S-type: describes adsorbate–adsorbate interactions on the adsorbent, often referred to as clustering, and/or the interaction of the adsorbate with solution ligands. When ligand saturation is reached, adsorption proceeds (e.g., aluminum–fulvic acid–clay interactions)

3. C-type: describes partitioning, which suggests interaction between a generally hydrophobic adsorbate with a hydrophobic adsorbent (e.g., pesticide–organic matter interactions)

4. H-type: describes strong chemisorption interactions, which is basically an extreme case of the L-type isotherms (e.g., phosphate–iron oxide interactions)

4.2.1 Freundlich Equilibrium Approach

Soils are multicomponent systems consisting of solid, liquid, and gaseous phases. These three phases are constantly in a dynamic state, trying to maintain equilibrium. Any type of perturbation in one phase influences the other two phases so that a new equilibrium state is approached. An equilibrium process that has been extensively investigated in soil systems employing the Freundlich equation involves sorption. Consider the reaction

$$-S + 1/nC \Leftrightarrow -SC \qquad (4.3)$$

where $-S$ denotes a heterogeneous soil mineral surface, C denotes a chemical substance (e.g., organic contaminant in solution, mg L^{-1}) in equilibrium with $-S$, $1/n$ is an empirical linearization constant, and $-SC$ denotes the adsorbate–surface complex in milligrams of adsorbate per unit mass of soil (e.g., per 100 g of soil). The equilibrium expression for Reaction 4.3 is given by

$$K_D = -SC/C^{1/n} \qquad (4.4)$$

where K_D is the distribution coefficient. By rearranging Equation 4.4 to solve for $-SC$,

$$-SC = K_D[C^{1/n}] \qquad (4.5)$$

A diagram of $-SC$ versus C will produce a curvilinear plot, as shown in Figure 4.14. Taking logarithms on both sides of Equation 4.5 gives

$$\log\text{–}SC = \log K_D + 1/n \log C \qquad (4.6)$$

When plotted as log $-SC$ versus log C, Equation 4.6 produces a linear plot with log K_D as the y intercept and $1/n$ as the slope. Figure 4.15 shows the linearized form of the Freundlich plot of the data in Figure 4.14. The linearized Freundlich plot has no particular molecular mechanistic interpretation; it simply represents an empirical approach for predicting the distribution of a constituent (e.g., herbicide) between the

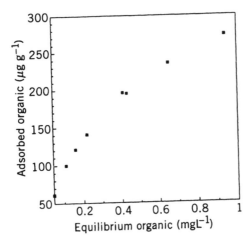

Figure 4.14. Example of a Freundlich isotherm.

solution phase and the solid phase. Such a model can be used, for example, to predict the leachability of a particular herbicide in soil. Generally, for soils with similar $1/n$, the greater the K_D, the lesser is the potential of the herbicide to leach.

In some sorption cases $1/n$ equals 1. When this condition is met, a plot of $-SC$ versus C will produce a straight line with K_D as slope (C-type isotherm; Fig. 4.16). This type of isotherm best describes soil sorption of hydrophobic organics (e.g., chlorinated hydrocarbons) (Fig. 4.17). The linearity of such data can be explained by the schematic model in Figure 4.18. In this model, (linear partition model), the hydrophobic organic contaminant distributes itself linearly between hydrophobic organic matter adsorbed on an inorganic mineral particle and solution. Linearity

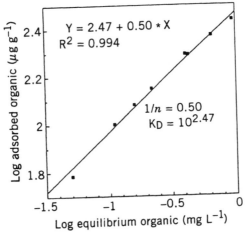

Figure 4.15. Linearized form of data in Figure 4.14.

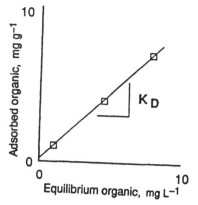

Figure 4.16. Linear Freundlich plot.

is obtained because no single specific interaction takes place between the adsorbent (organic matter) and the adsorbate (contaminant), and thus no saturation is attained.

Since SOM is most often the main component responsible for sorption of organic chemicals, e.g., contaminants (Fig. 4.19), researchers use octanol (an eight-carbon alcohol) to simulate organic matter–hydrophobic contaminant sorption phenomena.

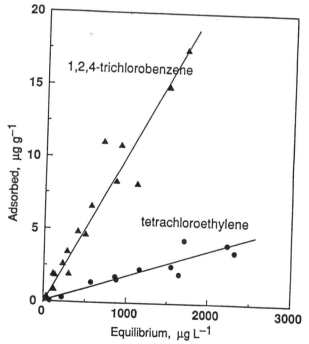

Figure 4.17. Sorption of trichlorobenzene and tetrachloroethylene by two Michigan soils (adapted from Weber et al., 1992).

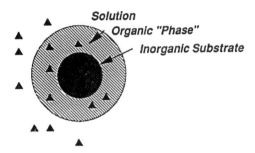

Figure 4.18. Linear "partition" model.

They carry out this process by extrapolating the octanol–water partition coefficient (K_{ow}) of a particular organic contaminant to the SOM–water partition coefficient (K_{oc}) and finally to the soil–water partition coefficient (K_D) as follows. Consider that

$$K_{ow} = C_o/C_w \qquad (4.7a)$$

where C_o denotes concentration of contaminant in octanol and C_w denotes concentration of contaminant in water, and

$$K_{oc} = C_{oc}/C_w \qquad (4.7b)$$

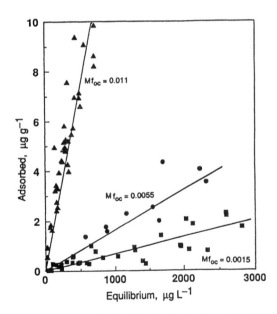

Figure 4.19. Sorption of tetrachloroethylene by three Michigan subsoils (adapted from Weber et al., 1992).

where C_{oc} denotes concentration of contaminant in organic carbon and C_w denotes concentration of contaminant in water. Consider also that the mass fraction of organic carbon in soil (M_{foc}) is given by the equation

$$M_{foc} = \text{carbon mass/soil mass} \tag{4.7c}$$

Then

$$-SC = K_D[C] = M_{foc}K_{oc}[C] \tag{4.8}$$

and

$$K_{oc} = K_D/M_{foc} \tag{4.9}$$

The magnitude of K_{oc} for a particular contaminant can be obtained from the octanol–contaminant partition coefficient K_{ow} by the following relationship (Means et al., 1987):

$$\log K_{oc} = 1.00 \log K_{ow} - 0.317 \tag{4.10}$$

Therefore, by knowing the organic matter content (M_{foc}) of any given soil and introducing K_{oc} from Equation 4.10 into Equation 4.8, one may produce the adsorption coefficient (K_D) of the given hydrophobic organic chemical for any given soil. However, Equation 4.10 does not necessarily apply to all organics under all soil conditions. Therefore, in the case of untested organic contaminants, sorption evaluation is needed under a wide range of soil conditions (e.g., soil pH, ionic strength, solution composition, and soil type).

4.2.2 Langmuir Equilibrium Approach

The Langmuir equilibrium approach was developed in 1918 by Langmuir to describe vapor adsorption on a homogeneous surface. It incorporates several assumptions when employed to model adsorption of chemical species in soil–solution suspensions. These assumptions are that

1. The number of surface adsorption sites are fixed
2. Adsorption includes a single monolayer
3. Adsorption behavior is independent of surface coverage
4. All adsorption sites are represented by similar types of functional groups
5. The isotherm displays L-type behavior

Mathematically, a Langmuir-type reaction can be described as follows:

$$-S + C \Leftrightarrow -SC \tag{4.11}$$

where the terms are defined as in Section 4.2.1. The equilibrium expression for Reaction 4.11 is

$$K_L = [-SC]/[-S][C] \qquad (4.12)$$

Solving for –SC by rearranging gives

$$-SC = K_L[-S][C] \qquad (4.13)$$

Considering that

$$-S = [S_T] - [-SC] \qquad (4.14)$$

where S_T denotes total number of adsorption sites, by substituting Equation 4.14 into Equation 4.13 and rearranging,

$$-SC = K_L S_T C/[1 + K_L C] \qquad (4.15)$$

When $K_L C <<< 1$,

$$-SC = K_L S_T C \qquad (4.15a)$$

When plotted as –SC versus C, Equation 4.15 produces a curvilinear line approaching asymptotically S_T (Fig. 4.20). When $K_L C <<< 1$ (Eq. 4.15a), a plot of –SC versus C would produce a straight line with slope $K_L S_T$ and the y intercept would be zero. Upon rearranging Equation 4.15 such that

$$-SC = S_T C/[1/K_L + C] \qquad (4.16)$$

or

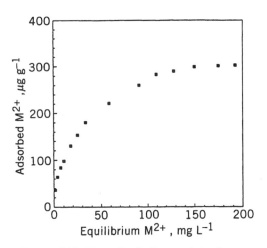

Figure 4.20. Example of a Langmuir isotherm.

$$-SC = S_T C/[K_{1/L} + C] \qquad (4.17)$$

and assuming that $-SC = 1/2 S_T$, it can be shown that

$$K_{1/L} = C \qquad (4.18)$$

Thus, $K_{1/L}$ denotes the concentration of adsorbate (e.g., contaminant) at which 50% of the total adsorption sites are occupied by the contaminant. Therefore, the larger $K_{1/L}$ is, the lower the affinity of the adsorbent (surface) for the adsorbate (contaminant). The magnitude of $K_{1/L}$ allows one to predict the potential of a particular surface to keep a contaminant from being released to solution. One can rearrange Equation 4.16 to produce a linear equation:

$$C/-SC = 1/K_L S_T + C/S_T \qquad (4.19)$$

A plot of $C/-SC$ versus C will produce a straight line with slope $1/S_T$ and y intercept of $1/K_L S_T$ (Fig. 4.21). Such a plot will allow one to determine the so-called adjustable parameters, S_T and K_L. Note that both the Freundlich equation and the Langmuir equation produce a surface adsorption affinity constant, but the latter (Langmuir) also produces the often-sought adsorption maximum (S_T).

Equation 4.19 could be modified to produce a competitive interaction such as

$$-SB + C \Leftrightarrow -SC + B \qquad (4.20)$$

where B is the competing cation. By substituting the term C with the term C/B, and considering that K_L represents a unitless competitive selectivity coefficient, $K_{C/B}$, Equation 4.19 can be expressed for the competitive form by

$$(C/B)/(-SC) = 1/K_{C/B} S_T + (C/B)/S_T \qquad (4.21)$$

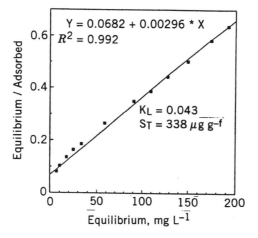

Figure 4.21. Linearized form of the Langmuir data in Figure 4.20.

A plot of $(C/B)/(-SC)$ versus C/B will produce a straight line with slope $1/S_T$ and y intercept of $1/K_{C/B}S_T$.

Researchers and field practitioners often employ two-site models to describe Langmuir-type behavior. The expression is as follows:

$$-SC = K_L S_T C/[1 + K_L C] + K'_L S'_T C/[1 + K'_L C] \qquad (4.22)$$

However, the fit of experimental data to a one- or two-site model does not necessarily provide any mechanistic meaning. It only serves as a tool to model reactions for predictive purposes.

4.2.3 Surface Complexation Models

Currently, an advanced approach to treating experimental adsorption data is through the use of surface complexation models based on a molecular description of the electric double layer (Chapter 3). These models include (1) the constant-capacitance model (CCM), (2) the triple-layer model (TLM), (3) the Stern variable surface charge–variable surface potential model (VSC–VSP), (4) the generalized two-layer model, and (5) one pK_a model (Goldberg, 1992). These models are fairly complex to use and require various degrees of computing power (MINTEQA2/PRODEFA2). Considering that the purpose of this book is to introduce the reader to the principles of soil and water chemistry, no detailed discussion is carried out on these models. For further information about these models, see Goldberg (1992). Some discussion on the CCM model is presented here because it has been tested extensively by Goldberg (1992) and it deals with inner-sphere complexes, an important component for predicting the behavior of heavy metals and oxyanions such as arsenic (AsO_2, AsO_4), and selenium (SeO_3, SeO_4) in soils.

The constance-capacitance model (CCM) was developed by the research groups of Stumm and Schindler (Goldberg, 1992 and references therein). It explicitly defines surface species, chemical reactions, equilibrium constant expressions, and surface activity. The surface functional group is defined as SOH which signifies a surface hydroxyl bound to a metal ion S (Al or Fe) of an oxide mineral, or aluminol–silanol group on the clay edge. Furthermore, the CCM model is characterized by the following four assumptions (Goldberg, 1992):

1. Surface complexes are of the inner-sphere type
2. Anion adsorption takes place by ligand exchange
3. An indifferent background electrolyte is employed to maintain constant ionic strength
4. The relationship between surface charge and surface potential is linear (see also Chapter 3)

Two types of mineral surface constants are considered by the CCM. One is a protonation–deprotonation constant and the other is an adsorbate complexation constant. The model is based on the consideration that the formation of an inner-sphere complex is controlled by a chemical potential (e.g., potential of surface, S, and contaminant to react chemically) and an electrical potential (determined by specific adsorption or inner-sphere complexes of electrical potential determining ions H^+ and OH^-) which controls the physical attraction–repulsion of the contaminant by the surface. For example, in the case of a deprotonation reaction,

$$-SH \Leftrightarrow S^- + H^+ \tag{4.23}$$

where $-S$ denotes surface. The intrinsic conditional equilibrium constant $[K_{-(int)}]$ expression describing Reaction 4.23 is

$$K_{-(int)} = ([-S^-][H^+]/[-SH])\exp[-F\psi/RT] \tag{4.24}$$

where

F = Faraday's constant (96,490 C eq^{-1})
ψ = surface electrical potential (V)
R = universal gas constant (8.314 J mol^{-1} °K^{-1})
T = absolute temperature (°K)

In the case of surface adsorption of a particular chemical species, for example, a metal (M^{m+}), the reaction is

$$-SH + M^{m+} \Leftrightarrow -SM^{(m-1)} + H^+ \tag{4.25}$$

where m denotes metal charge. The intrinsic conditional equilibrium constant $[K_{-(int)}]$ expression describing Reaction 4.25 is

$$K_{m(int)} = ([-SM^{(m-1)}][H^+]/[-SH][M^{m+}])\exp[(m-1)F\psi/RT] \tag{4.26}$$

Equation 4.26, as noted above, has two terms—a chemical complexation term, $[-SM^{(m-1)}][H^+]/[-SH][M^{m+}]$, and an electrical potential term, $\exp[(m-1)F\psi/RT]$. The electrical potential term is directly dependent on pH (see Eq. 4.27b). Therefore, by regulating the magnitude of the physical attraction between a metal and a surface, the electrical potential indirectly affects the magnitude of surface complexes that could form. Recall that before a chemical complex or inner-sphere complex between two reactants is formed, they must physically collide successfully at a given frequency. When a metal–ion (e.g., contaminant) and a surface are oppositely charged, physical collisions maximize, and, thus, inner-sphere complex formation maximizes. When a metal–ion (e.g., contaminant) and a surface are similarly charged, physical collisions minimize, and inner-sphere complex formation also minimizes.

According to the Nernst equation, the surface electrical potential (ψ) can be approximated as a function of H^+ activity or pH as follows:

$$\psi = (RT/F)\ln H^+/H_0^+ \tag{4.27a}$$

or

$$\psi = 59(-pH + pH_o)mV \text{ at } 25° \tag{4.27b}$$

where H^+ denotes hydrogen activity in solution, H_0^+ denotes hydrogen activity in solution at the PZC of the mineral surface (see Chapter 3), and the other terms are as previously defined. Therefore, based on the Nernst equation (Eq. 4.27b), when pH equals pH_o or $H^+ = H_0^+$, ψ equals zero, and Equation 4.24 reduces to

$$K_{-(int)} = [-S^-][H^+]/[-SH] \qquad (4.28a)$$

or, in the case of the surface-metal complex (Equation 4.26),

$$K_{m(int)} = [-SM^{(m-1)}[H^+]/[-SH][M^{m+}] \qquad (4.28b)$$

It follows that the classical Langmuir equation (see Eq. 4.12), when used to describe metal-ion adsorption by a charged surface,

$$K_m = [-SM/[-S][M] \qquad (4.29)$$

is only useful for experimental data representing a particular single pH value. For Equation 4.29 to be applicable to a wide range of pH values, the electrical term described by Equation 4.27 has to be introduced. This is demonstrated below, where adsorption is expressed on a surface-fraction basis (θ).

ADSORPTION ON A SURFACE-FRACTION BASIS

Consider the reaction

$$-S^- + M^+ \Leftrightarrow -SM \qquad (A)$$

where the meaning of the terms is as previously defined. The equilibrium expression for Reaction A can be expressed as

$$K_m = -SM/[-S^-][M^+] \qquad (B)$$

Expressing the terms $-SM$ and $-S^-$ on a surface-fraction basis, θ,

$$\theta = [-SM]/[[S_T]] \qquad (C)$$

where

$$[S_T] = [-SM] + [-S^-] \qquad (D)$$

and $-S^-$ on a surface-fraction basis is $1 - \theta$, then

$$[1 - \theta] = ([S_T] - [-SM])/[S_T] \qquad (E)$$

By substituting Equations C and E into Equation B

$$K_m = \theta/([1 - \theta][M^+]) \qquad (F)$$

Considering that surface coverage by a metal (–SM) is given in units of θ (see box above), and $1 - \theta$ represents the uncovered surface (–S) (see Equation E), then based on Equation F and the pH-dependent surface electrical potential, ψ (Equation 4.27),

$$\theta/1 - \theta = K_m M^{m+} \{\exp- [(m - 1)F\psi/RT]\} \qquad (4.30)$$

where K_m denotes the intrinsic conditional equilibrium constant. Substituting the Nernst equation (Eq. 4.27) for ψ gives

$$\theta/1 - \theta = K_m M^{m+} [(H_0^+/H^+)]^{(m-1)} \qquad (4.31)$$

Taking logs on both sides of Equation 4.31 and considering a metal cation with valence 2 gives

$$\log(\theta/1 - \theta) = \log K_m + \log M^{m+} - (pH_0 - pH) \qquad (4.32)$$

Equation 4.32 reveals that plotting percent surface coverage ($\theta \times 100$) versus pH will produce a sigmoidal plot. Figure 4.22 (percent metal adsorbed versus pH) shows, as predicted by Equation 4.32, a sigmoidal plot characterized by a narrow pH range where adsorption sharply increases (adsorption edge). This only applies to a single type of surface complex. When more than one type of surface complex form, the number of equations and constants needed to predict such complexes increases. Furthermore, in such systems, a plot of percent adsorption versus pH may not necessarily produce an ideal sigmoidal plot.

As a general rule, adsorption by soil mineral surfaces exhibiting variable charge is predicted by the pK_a value of the adsorbing compound (conjugated acid) or in the case of metal cations, their hydrolysis constant. In the latter, the pH position of the adsorption edge is generally related to the metal's acid–base properties. The greater

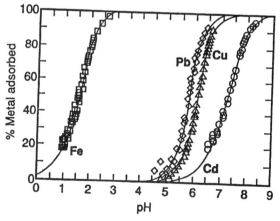

Figure 4.22. Relationship between metal surface coverage and pH (from Schindler et al., 1976, with permission).

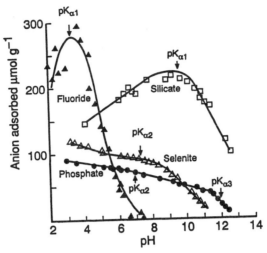

Figure 4.23. Relationship between pH and anion adsorbed by goethite (from Hingston et al., 1972, with permission).

the hydrolysis constant is, the lower the adsorption pH (Fig. 4.22). In the case of weak conjugated acids (e.g., oxyanions), the adsorption maximum is exhibited around the pH closest to the pK_a of the acid (Figs. 4.23 and 4.24). The length of the pH adsorption plateau, where surface adsorption is maximized, is related to the number of pK_a values of the weak conjugated acid under consideration (Fig. 4.23).

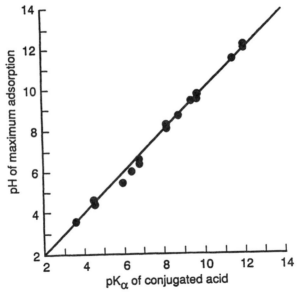

Figure 4.24. Relationship between pK_a of conjugated acid and pH of maximum adsorption by two oxides, goethite and gibbsite (adapted from Hingston et al., 1972).

4.3 EXCHANGE REACTIONS

Cation exchange in soils or clay minerals involves replacement of a given cation on a given mineral surface by another cation. Exchange equations are commonly used to evaluate ion availability to plant roots and/or release of metals to soil water (e.g., heavy metals to groundwater or surface water). There are two major types of cation-exchange reactions in soil systems—homovalent and heterovalent cation exchange.

4.3.1 Homovalent Cation Exchange

Homovalent cation exchange refers to exchanging cations with similar valence. For example, with Ca^{2+}–Mg^{2+} exchange, the reaction can be expressed as follows:

$$ExMg + Ca^{2+} \Leftrightarrow ExCa + Mg^{2+} \tag{4.33}$$

Based on Reaction 4.33, the cation-exchange selectivity coefficient (K_{Ca-Mg} or K_G) can be expressed as

$$K_{Ca-Mg} = [ExCa/ExMg][(Mg^{2+})/(Ca^{2+})] \tag{4.34}$$

where ExMg and ExCa denote exchangeable cations in units of centimoles per kilogram (cmol kg^{-1}) or millimoles per hundred grams (mmol/100 g), Ex denotes the soil exchanger with a charge of 2–, and Mg^{2+} and Ca^{2+} denote solution concentration in units of millimoles per liter (mmol L^{-1}). Equation 4.34 can be used to solve for ExCa:

$$ExCa = K_{Ca-Mg}(CEC)(CR_{Ca})[1 + CR_{Ca} K_{Ca-Mg}]^{-1} \tag{4.35}$$

where CEC denotes cation exchange capacity in millimoles per hundred grams, CR_{Ca} denotes calcium adsorption ratio and is expressed by

$$CR_{Ca} = [Ca^{2+}]/[Mg^{2+}] \tag{4.36}$$

where the brackets denote concentration in millimoles or moles per liter in the solution phase, and

$$CEC = ExCa + ExMg \tag{4.37}$$

A plot of ExCa versus CR_{Ca} will produce a curvilinear line, asymptotically approaching CEC. The pathway of such line from ExCa = 0 to ExCa = CEC depends on K_{Ca-Mg} and CEC (see the section entitled Relationship Between (CR_{Ca} and ExCa).

An important component of homovalent exchange is the magnitude of the exchange selectivity coefficient. Commonly, homovalent cation-exchange reactions in soils or soil minerals exhibit a selectivity coefficient somewhere around 1 (Table 4.2). This value signifies that the soil mineral surface does not show any particular adsorption preference for either of the two cations. However, for a mineral where the K_{Ca-Mg} is

TABLE 4.2. Thermodynamic Equilibrium Constants of Exchange (K_{eq}) and Standard Enthalpy of Exchange (ΔH^0ex) Values for Binary Exchange Processes on Soils and Soil Components[a]

Exchange Process	Exchanger	K_{eq}	ΔH^0ex (kJ mol^{-1})	Reference
Ca–Na	Soils	0.42–0.043		Mehta et al. (1983)
Ca–Na	Calcareous soils	0.38–0.09		Van Bladel and Gheyi (1980)
Ca–Na	World vermiculite	0.98	39.38	Wild and Keay (1964)
Ca–Na	Camp Berteau montmorillonite	0.72		
Ca–Mg	Calcareous soils	0.89–0.75		Van Bladel and Gheyi (1980)
Ca–Mg	Camp Berteau montmorillonite	0.95		Van Bladel and Gheyi (1980)
Ca–K	Chambers montmorillonite	0.045	16.22	Hutcheon (1966)
Ca–NH$_4$	Camp Berteau montmorillonite	0.035	23.38	Laudelout et al. (1967)
Ca–Cu	Wyoming bentonite	0.96	−18.02	El-Sayed et al. (1970)
Na–Ca	Soils (304 K)	0.82–0.16		Gupta et al. (1984)
Na–Li	World vermiculite	11.42	−23.15	Gast and Klobe (1971)
Na–Li	Wyoming bentonite	1.08	−0.63	Gast and Klobe (1971)
Na–Li	Chambers montmorillonite	1.15	−0.47	Gast and Klobe (1971)
Mg–Ca	Soil	0.61		Jensen and Babcock (1973)
Mg–Ca	Kaolinitic soil clay (303 K)	0.65	6.96	Udo (1978)
Mg–Na	Soils	0.75–0.053		Mehta et al. (1983)
Mg–Na	World vermiculite	1.73	40.22	Wild and Keay (1964)
Mg–NH$_4$	Camp Berteau montmorillonite	0.031	23.38	Landelout et al. (1967)
K–Ca	Soils	5.89–323.3		Deist and Talibudeen (1967a)
K–Ca	Soil	12.09		Jensen and Babcock (1973)
K–Ca	Soils	19.92–323.3	−3.25 to −5.40	Deist and Talibudeen (1967b)
K–Ca	Soil	0.46	−15.90	Jardine and Sparks (1984b)
K–Ca	Soils	0.64–6.65	−3.25 to −5.40	Goulding and Talibudeen (1984)
K–Ca	Soils	6.42–6.76	−16.28	Ogwada and Sparks (1986b)
K–Ca	Soil silt	0.86		Jardine and Sparks (1984b)

(continued)

TABLE 4.2. Continued

Exchange Process	Exchanger	K_{eq}	$\Delta H^0 ex$ (kJ mol^{-1})	Reference
K–Ca	Soil clay	3.14		Jardine and Sparks (1984b)
K–Ca	Kaolinitic soil clay (303 K)	16.16	−54.48	Udo (1978)
K–Ca	Clarsol montmorillonite	12.48		Jensen (1972b)
K–Ca	Danish kaolinite	32.46		Jensen (1972b)
K–Mg	Soil	5.14		Jensen and Babcock (1973)
K–Na	Soils	4.48–6.24		Deist and Talibudeen (1967a)
K–Na	Wyoming bentonite	1.67	−2.53	Gast (1972)
K–Na	Chambers montmorillonite	3.41	−4.86	Gast (1972)

Source: Adapted from Sparks (1955).
[a]The exchange studies were conducted at 298 K; exceptions are noted. $K_{eq}/2$ approximately equals K_G.

greater than 1, the surface prefers Ca^{2+}, and for a mineral where the K_{Ca-Mg} is less than 1, the surface prefers Mg^{2+} (Table 4.2).

Cation preference is commonly demonstrated through fractional isotherms. Fractional isotherms are plots of equivalent or mole fraction in the exchange phase versus equivalent or mole fraction in the solution phase (e.g., in the case of Ca^{2+}–Mg^{2+},

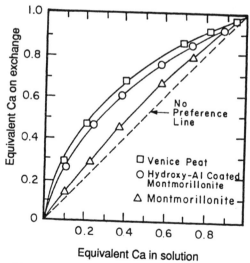

Figure 4.25. Relationships between equivalent fraction of solution phase versus equivalent fraction of exchange phase for Ca^{2+}–Mg^{2+} exchange in various soil minerals (from Hunsaker and Pratt, 1971, with permission).

ExCa/CEC vs. $Ca^{2+}/[Ca^{2+} + Mg^{2+}]$ or ExMg/CEC vs. $Mg^{2+}/[Ca^{2+} + Mg^{2+}]$). Such an isotherm is shown in Figure 4.25. The diagonal broken line shows no preference ($K_{Ca-Mg} = 1$). A line above the diagonal line reveals that the surface prefers Ca^{2+}, while any line below the nonpreference line reveals that the surface prefers Mg^{2+}.

RELATIONSHIP BETWEEN CRCa AND ExCa

In the case of Ca^{2+}–Mg^{2+} exchange, the reaction can be expressed as follows:

$$ExMg + Ca^{2+} \Leftrightarrow ExCa + Mg^{2+} \tag{A}$$

Equation (A) can be used to solve for exchangeable Ca^{2+}, ExCa

$$ExCa = K_{Ca-Mg}(CEC)(CR_{Ca})[1 + CR_{Ca} K_{Ca-Mg}]^{-1} \tag{B}$$

with all terms as previously defined. Note that K_{Ca-Mg} is also known as K_G. Taking the limits for Equation B at CR_{Ca} approaching zero (0) and CR_{Ca} approaching infinity (∞)

$$\text{limit ExCa} = 0; \quad \text{limit ExCa} = CEC$$
$$CR_{Ca} \to 0 \quad CR_{Ca} \to \infty$$

The derivative of Equation B with respect to CR_{Ca} is

$$dExCa/dCR_{Ca} = K_{Ca-Mg} CEC/[1 + K_{Ca-Mg}CR_{Ca}]^{-2} \tag{C}$$

and

$$\text{limit } dExCa/dCR_{Ca} = K_{Ca-Mg}CEC; \quad \text{limit } dExCa/dCR_{Ca} = 0$$
$$CR_{Ca} \to 0 \quad CR_{Ca} \to \infty$$

The limits of Equations B and C at zero and infinity imply that a plot of Equation B in terms of CR_{Ca} versus ExCa (with ExCa being the dependent variable) at constant CEC and K_{Ca-Mg} will give a curvilinear line approaching the CEC asymptotically (Fig. 4A). A plot of CR_{Ca} versus ExCa/CEC would also produce a curvilinear line asymptotically approaching 1. The pathway from ExCa/CEC = 0 to ExCa/CEC = 1 depends only on K_{Ca-Mg} or K_G (Fig. 4B). Similar conclusions apply to all other homovalent cation-exchange reactions (e.g., $K^+ - NH_4^+$ and $K^+ - Na^+$).

The CR_{Ca} versus ExCa/CEC relationship allows us to predict the quantity of Ca^{2+} adsorbed by knowing the solution-phase composition (CR_{Ca}). The practicality of this relationship is that it is relatively simple to analyze the solution phase in the laboratory as opposed to the exchange phase. Knowing the composition of the solution phase, the exchange phase is easily predicted based on the relationship discussed above. It is necessary to know the exchange phase because the majority of the cations reside on it.

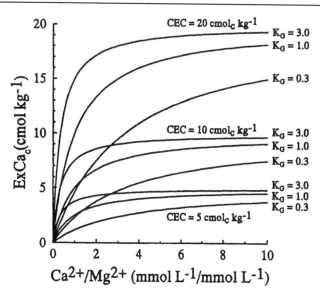

Figure 4A. Relationship between CR_{Ca} and ExCa representing various hypothetical soils with different CEC and K_G values.

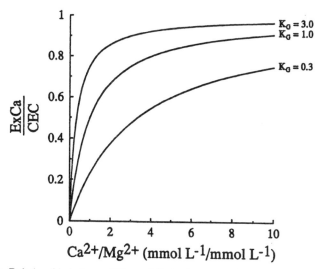

Figure 4B. Relationship between CR_{Ca} and Ca-load representing various hypothetical soils with different CEC and K_G values.

NONPREFERENCE HOMOVALENT ISOTHERMS

For a homovalent ion-exchange reaction (e.g., Ca^{2+}–Mg^{2+}), when $K_{Ca-Mg} = 1$, it follows from Equation 4.35 that

$$ExCa/CEC = [Ca^{2+}/Mg^{2+}]/[1 + Ca^{2+}/Mg^{2+}] \quad (D)$$

It can also be shown that

$$[ExCa/CEC] = [Ca^{2+}]/[Ca^{2+} + Mg^{2+}] \quad (E)$$

and, therefore,

$$[Ca^{2+}]/[Ca^{2+} + Mg^{2+}] = [Ca^{2+}/Mg^{2+}]/[1 + Ca^{2+}/Mg^{2+}] \quad (F)$$

Consider

$$[Ca^{2+}]/[Ca^{2+} + Mg^{2+}] = f[Ca^{2+}/Mg^{2+}] \quad (G)$$

and

$$f = [Ca^{2+} \cdot Mg^{2+}]/[(Ca^{2+})^2 + Ca^{2+} \cdot Mg^{2+}] \quad (H)$$

Substituting the right-hand side of Equation H for f in Equation G gives

$$[Ca^{2+}]/[Ca^{2+} + Mg^{2+}] = \{[Ca^{2+} \cdot Mg^{2+}]/[(Ca^{2+})^2 + Ca^{2+} \cdot Mg^{2+}]\}[Ca^{2+}/Mg^{2+}] \quad (I)$$

Dividing the numerator and the denominator of Equation I by $(Mg^{2+})^2 \cdot Ca^{2+}$ produces Equation F. Therefore, a plot of $[ExCa/CEC]$ versus $[Ca^{2+}]/[Ca^{2+} + Mg^{2+}]$ (see Eq. E) produces a straight line with slope 1 (nonpreference isotherm). Furthermore, upon considering Equation F and introducing cation-solution activities, it appears that since γ_{Ca} approximately equals γ_{Mg} (where γ_i equals the single-ion activity coefficient for species i in solution), the relationship of $ExCa/CEC$ versus $[Ca^{2+}]/[Ca^{2+} + Mg^{2+}]$ is shown to be independent of ionic strength.

4.3.2 Heterovalent Cation Exchange

Heterovalent cation-exchange reactions in soils or soil minerals involve exchange of a monovalent cation with a divalent cation or vice versa. Cations with valence higher than 2 (i.e., Al^{3+}) are also involved in heterovalent exchange, but to a lesser extent owing to the low pH needed for the Al^{3+} species to persist in the solution and exchange phase. The equation that is most commonly used to describe heterovalent cation exchange is the Gapon exchange equation. For example, for $Na^+ - Ca^{2+}$ exchange,

$$ExCa_{1/2} + Na^+ \Leftrightarrow ExNa + 1/2\,Ca^{2+} \quad (4.38)$$

Based on Reaction 4.38, the Gapon exchange selectivity coefficient (K_G) can be expressed as

$$K_G = [ExNa/ExCa_{1/2}][[Ca^{2+}]^{1/2}/[Na^+]] \tag{4.39}$$

where ExNa and $ExCa_{1/2}$ denote exchangeable cations in units of centimoles per kilogram ($cmol_c\ kg^{-1}$) or per hundred grams (meq/100 g), Ex denotes the soil exchanger with a charge of 1– and Na^+ or Ca^{2+} denote solution cations in terms of concentration in units of millimoles per liter ($mmol\ L^{-1}$). Equation 4.39 can be rearranged to solve for ExNa:

$$ExNa = K_G\ (CEC)(SAR)[1 + SAR\ K_G]^{-1} \tag{4.40}$$

where CEC is in centimoles per kilogram or milliequivalents per hundred grams,

$$SAR = [Na^+]/[Ca^{2+}]^{1/2} \tag{4.41}$$

where the brackets denote concentration in millimoles or moles per liter in the solution phase, and

$$CEC = ExCa_{1/2} + ExNa \tag{4.42}$$

A plot of SAR versus ExNa (see Eq. 4.40) will produce a curvilinear line, asymptotically approaching CEC (see section entitled Relationship Between SAR and ExNa).

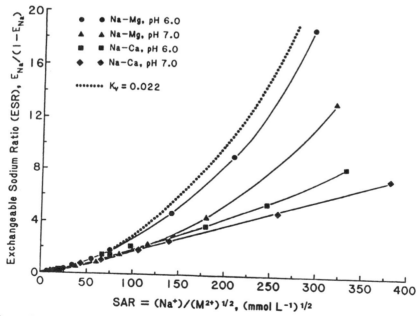

Figure 4.26. Relationship between SAR and exchangeable sodium ration (ExNa/ExCa$_{1/2}$ or ExNa/ExMg$_{1/2}$ for soil minerals under different pH values (data taken from Fletcher et al., 1984).

Rearranging Equation 4.40 gives

$$ExNa/ExCa_{1/2} = K_G\{[Na^+]/[Ca^{2+}]^{1/2}\} \qquad (4.43)$$

Theoretically, a plot of SAR versus $ExNa/ExCa_{1/2}$ or exchangeable sodium ratio (ESR) will produce a straight line with slope equal to K_G. The average magnitude of K_G for soils of the arid west is approximately 0.015 $(mmol\ L^{-1})^{-1/2}$. However, the experimental K_G appears to be dependent on pH, salt concentration, and clay mineralogy. Furthermore, the experimental K_G does not appear to be constant across the various sodium loads. Commonly, as sodium load increases, K_G also increases. Furthermore, as pH increases, K_G decreases (Fig. 4.26).

An important component of heterovalent exchange reactions is the absolute magnitude of K_G. Commonly, monovalent–divalent cation-exchange reactions (e.g., Na^+–Ca^{2+}) exhibit a K_G somewhere around 0.5 when the solution concentration units are in moles per liter (note, a K_G obtained with solution concentration units in millimoles per liter, when multiplied by the square root of 1000, gives the K_G in units of moles per liter to the negative half power). Therefore, the K_G of heterovalent cation exchange depends on solution concentration units, while the K_G of homovalent cation-exchange reactions is independent of solution units. Commonly, the K_G of Na^+–Ca^{2+} exchange in soils varies from 0.50 to 2 $(mol\ L^{-1})^{-1/2}$, depending on soil mineralogy and pH. In the case of K^+–Ca^{2+} exchange, the K_G may vary from approximately 1 to approximately 300 $(mol\ L^{-1})^{-1/2}$, depending also on mineralogy and pH (Table 4.2). Variation in K_G (e.g., K^+–Ca^{2+}) due to mineralogy is in the order of illite > smectite > organic matter (Fig. 4.27).

There are generally two potential uses for cation-exchange selectivity coefficients: (1) predicting solution-exchange phase chemistry and (2) evaluating the behavior of surface-functional groups. When a selectivity coefficient, representing two heterovalent (or homovalent) cations, remains constant for a given fraction of the exchange isotherm, it (selectivity coefficient) can be used to predict the distribution of the two cations between the solution and exchange phases in that portion of the isotherm.

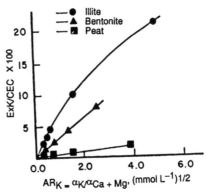

Figure 4.27. Relationship between AR_K and K-load on the exchange for various soil minerals (from Salson, 1962, with permission).

Furthermore, any variation in the magnitude of a given cation-selectivity coefficient with respect to cation-exchange load may signify variation in surface-functional groups. However, because the Gapon equation is not consistent with the thermodynamics of chemical reactions (it considers that reactions occur on an equivalent basis), it cannot be used for evaluating surface-functional groups. For this reason, Vanselow (1932) produced an exchange equation consistent with chemical reactions by using reactants and products in units of moles, which are consistent with the thermodynamics of chemical reactions. Nevertheless, the Gapon equation is widely used to predict the distribution of cations between the solution and exchange phases because of its simplicity.

RELATIONSHIP BETWEEN SAR AND ExNa

An equation that is most commonly used to describe heterovalent cation exchange, such as Na^+–Ca^{2+} exchange, is the Gapon exchange equation. For example, for the Na^+–Ca^{2+} system,

$$ExCa_{1/2} + Na^+ \Leftrightarrow ExNa + 1/2\ Ca^{2+} \tag{A}$$

Equation A can be used to solve for ExNa:

$$ExNa = K_G(CEC)(SAR)[1 + SAR\ K_G]^{-1} \tag{B}$$

with all terms as previously defined for heterovalent exchange reactions. Taking the limits for Equation B at SAR approaching zero (0) and SAR approaching infinity (∞):

$$\text{limit ExNa} = 0; \qquad \text{limit ExNa} = CEC$$
$$SAR \to 0 \qquad\qquad SAR \to \infty$$

The derivative of Equation B with respect to SAR is

$$dExNa/dSAR = K_G CEC/[1 + K_G SAR]^{-2} \tag{C}$$

and

$$\text{limit } dExNa/dSAR = K_G CEC; \qquad \text{limit } dEXNa/dSAR = 0$$
$$SAR \to 0 \qquad\qquad SAR \to \infty$$

The limits of Equations B and C at zero and infinity imply that a plot of Equation B in terms of SAR versus ExNa (with ExNa being the dependent variable) at a constant CEC and K_G will give a curvilinear function approaching the CEC asymptotically (Fig. 4C). Upon normalizing Equation B by dividing both sides by CEC:

$$\text{ExNa/CEC} = K_G(\text{SAR})[1 + \text{SAR } K_G]^{-1} \qquad \text{(D)}$$

A plot of SAR versus ExNa/CEC will produce a curvilinear line, asymptotically approaching 1. The pathway of such line from ExNa/CEC = 0 to ExNa/CEC = 1 depends only on K_G. This is demonstrated in Figure 4D. This relationship has some practical utility. Commonly, up to ExNa/CEC = 0.2 or exchangeable sodium percentage (ESP) of 20%, the relationship between ExNa/CEC versus SAR is nearly linear and, strictly fortuitously for some soils when ESP equals 15, SAR also equals 15 (solution units are in millimoles per liter). Furthermore, at an SAR or ESP 15, agricultural soils with mostly 2:1 mineralogy swell enough to stop water infiltration and gaseous exchange, rendering soil unsuitable for crop production (see Chapter 11).

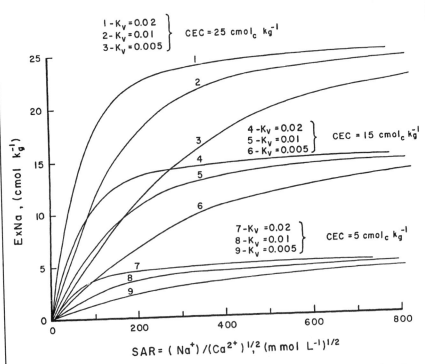

Figure 4C. Theoretical relationships between SAR and ExNa representing various hypothetical soils with different CEC and K_G values (from Evangelou and Phillips, 1987, with permission).

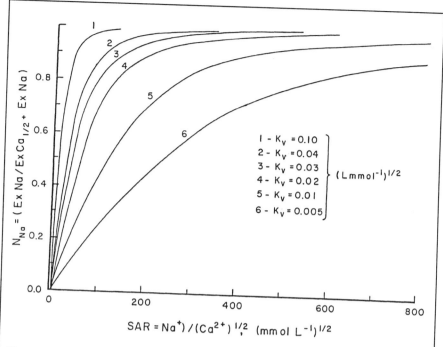

Figure 4D. Theoretical relationships between SAR and Na-load representing various hypothetical soils with different CEC and K_G values (from Evangelou and Phillips, 1987, with permission).

4.3.3 The Vanselow Equation

A binary cation-exchange reaction at equilibrium in soil or soil minerals involving Na^+ and Ca^{2+} can be written as

$$1/2\ Ex_2Ca + Na^+ \Leftrightarrow ExNa + 1/2\ Ca^{2+} \qquad (4.44a)$$

where Ex denotes an exchanger phase taken to have a charge of negative one (-1) and Na^+ and Ca^{2+} denote solution species. The correct thermodynamic expression for the Na^+–Ca^{2+} exchange reaction is

$$ExCa + 2Na^+ \Leftrightarrow ExNa_2 + Ca^{2+} \qquad (4.44b)$$

where Ex is 2– and the K'_{eq} is

$$K'_{eq} = (a_{ExNa})^2(a_{Ca})/(ExCa)(a_{Na})^2 \qquad (4.44c)$$

The thermodynamic exchange equilibrium constant K_{eq} for Reaction 4.44a is

$$K_{eq} = (^aExNa)(^aCa^{1/2})/(^aExCa^{1/2})(^aNa) \qquad (4.45)$$

where ^aCa and ^aNa denote activity of solution phase Na^+ and Ca^{2+}, respectively, and ^aExNa and ^aExCa denote activity of exchange phase Na^+ and Ca^{2+}, respectively. Equation 4.44c is related to Equation 4.45 by $(K'_{eq})^{1/2} = K_{eq}$. However, because of convenience and consistency with the Gapon equation, Equation 4.45 will be used for the rest of the discussion. Solution activity a_i is defined by the equation

$$a_i = \gamma_i \cdot c_i \tag{4.46}$$

where γ_i = solution activity coefficient of species i and c_i = solution concentration of species i. When solution ionic strength (I) approaches zero, solution phase a_i is set to 1, hence $\gamma_i = 1$. For mixed-electrolyte solutions, when $I > 0$, the single-ion activity concept introduced by Davies (see Chapter 9) is employed to estimate γ_i.

The activity component of the adsorbed or solid phase is defined by employing the mole-fraction concept introduced by Vanselow (1932). According to Vanselow (1932), for a heterovalent binary exchange reaction such as Na^+–Ca^{2+}, assuming that the system obeys ideal solid-solution theory, the activity term (a_{Exi}) is defined by

$$a_{ExNa} = X_{Na} = \frac{ExNa}{ExNa + Ex_2Ca} \tag{4.47}$$

and

$$a_{Ex_2Ca} = X_{Ca} = \frac{Ex_2Ca}{ExNa + Ex_2Ca} \tag{4.48}$$

where X_{Na} and X_{Ca} denote adsorbed mole fraction of Na^+ and Ca^{2+}, respectively, and Ex denotes exchange phase with a valence of 1–. For a system where ideal solid-solution behavior is not obeyed,

$$a_{Ex_i} = f_i Ex_i \tag{4.49}$$

where f_i denotes adsorbed-ion activity coefficient. In the mole-fraction concept, the sum of exchangeable Na^+ (ExNa) and exchangeable Ca^{2+} (Ex_2Ca) is expressed in moles per kilogram of soil. Hence, the denominator of Equations 4.47 and 4.48 is not a constant even though the sum of exchangeable Na^+ and exchangeable Ca^{2+} in units of charge equivalents is a constant. Equivalent fractions (E_i) for Na^+ and Ca^{2+} are defined by

$$E_{Na} = \frac{ExNa}{ExNa + 2Ex_2Ca} \tag{4.50}$$

and

$$E_{Ca} = \frac{2Ex_2Ca}{ExNa + 2Ex_2Ca} \tag{4.51}$$

For the binary system above, the CEC of the soil is

$$CEC = ExNa + 2Ex_2Ca \tag{4.52}$$

In exchange systems such as the one above, it is assumed that any other cations (e.g., exchangeable K^+ and/or H^+) are present in negligible quantities and do not interfere with Na^+–Ca^{2+} exchange, or H^+ is tightly bound to the solid surface, giving rise only to pH-dependent charge.

Based on the concepts above, a working equilibrium exchange expression for Reaction 4.44a can be given as

$$K_V = \{(X_{Na})/(X_{Ca^{1/2}})\}\{(a_{Ca^{1/2}})(a_{Na})\} \tag{4.53}$$

where X_{Na} and X_{Ca} denote mole fractions on the exchange phase for Na^+ and Ca^{2+}, respectively, a_{Na} and a_{Ca} denote solution activity of Na^+ and Ca^{2+}, respectively, and K_V is the Vanselow exchange selectivity coefficient. Commonly, the magnitude of K_V is taken to represent the relative affinity of Na^+ with respect to Ca^{2+} by the clay surface. When K_V equals 1, the exchanger shows no preference for either Na^+ or Ca^{2+}. When $K_V > 1$, the exchanger prefers Na^+ and when $K_V < 1$, the exchanger prefers Ca^{2+}.

Cation preference in heterovalent exchange is demonstrated through *fractional isotherms*. Fractional isotherms are plots of equivalent fraction in the solution phase versus equivalent fraction of the exchange phase (e.g., in the case of Na^+–Ca^{2+}, $Na^+/[Na^+ + 2Ca^{2+}]$ versus E_{Na}, ExNa/CEC, or $2Ca^{2+}/[Na^+ + 2Ca^{2+}]$) versus E_{Ca} or $2Ex_2Ca/CEC$. The nonpreference line for a heterovalent exchange shown in Figure 4.28 is the solid line ($K_V = 1$). A line above the nonpreference line reveals that the

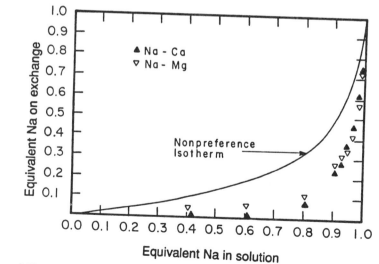

Figure 4.28. Relationship between Na-load on the exchange phase versus Na-equivalent fraction in the solution phase for Na^+–Ca^{2+} exchange in montmorillonite (from Sposito and LeVesque, 1985, with permission).

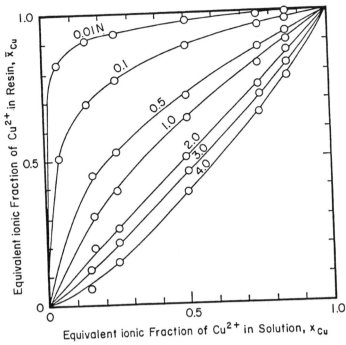

Figure 4.29. Relationships between Cu^{2+}-load on the exchange phase versus Cu^{2+}-equivalent fraction in the solution phase for Na^+–Cu^{2+} exchange on cation exchange resin (from Subba and David, 1957, with permission).

surface prefers Na^+, while a line below the nonpreference line reveals that the surface prefers Ca^{2+}. Heterovalent fractional cation preference isotherms differ from homovalent cation preference isotherms in that (a) the nonpreference line in heterovalent exchange is never the diagonal line and (b) preference in heterovalent exchange is ionic-strength dependent (Fig. 4.29). The latter is also known as the square root effect.

A number of researchers have carried out various studies involving binary heterovalent exchange on various clay minerals. For example, Sposito and Mattigod (1979) showed that for the exchange reactions of Na^+ with trace metal cations (Cd^{2+}, Co^{2+}, Cu^{2+}, Ni^{2+}, and Zn^{2+}) on Camp Berteau montmorillonite, K_V was constant and independent of exchanger composition up to an equivalent fraction of trace metal cations of 0.70. This indicates that the cationic mixture on the exchanger phase, up to an equivalent fraction of trace metal cations of 0.70, behaves as an ideal mixture (obeys ideal solid-solution theory). Van Bladel et al. (1972) studied Na^+–Ca^{2+} exchange on the same kind of mineral and found that there was a more pronounced selectivity of clay for Ca^{2+} ions at the calcium-rich end of the isotherm. Levy and Hillel (1968) reached the same conclusion studying Na^+–Ca^{2+} exchange on montmorillonitic soils. It appears that the magnitude of K_V in soil minerals is variable and meticulous experiments are needed to quantify it. In general, it can be said that the selectivity coefficient (K_V) for a binary exchange reaction depends primarily on ionic strength and on two dimensionless parameters, one being a measure of the proportion of cations in the soil-absorbing complex and the other a measure of their proportions in the soil-solution phase.

4.3.4 Relationship Between K_V and K_G

Evangelou and Phillips (1987 and 1988) carried out a detailed characterization of the relationship between K_V and K_G. Their findings are summarized below. The relationship between K_G and K_V is described by

$$K_G = K_V [1 + E_{Na}]^{1/2} [2(1 - E_{Na})^{1/2}]^{-1} \tag{4.54}$$

Equation 4.54 shows that even if K_V is constant across the entire exchange isotherm, K_G is exchangeable Na-load dependent. By taking the limit of Equation 4.57 at the E_{Na} of 0.60, it can be shown that $K_V = K_G$, and when the monovalent cation approaches an equivalent fraction of 1, $K_G = \infty$ (Figs. 4.30 and 4.31). In summary, similar conclusions would be reached on the behavior of a heterovalent cation exchange up to an equivalent monovalent fraction load of 0.20 employing K_V or K_G.

There are many other types of exchange equations, such as the Gaines and Thomas, the Kerr, and the Krishnamoarthy–Overstreet (Table 4.3). Generally, however, the Capon and Vanselow equations are the most widely used.

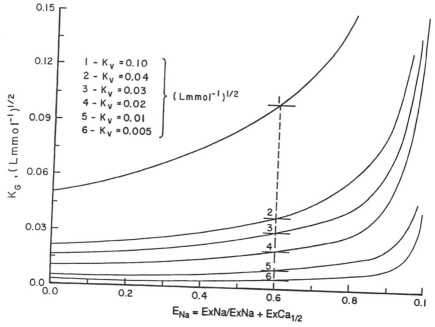

Figure 4.30. Theoretical relationship between Na-load and K_V or K_G (from Evangelou and Phillips, 1987, with permission).

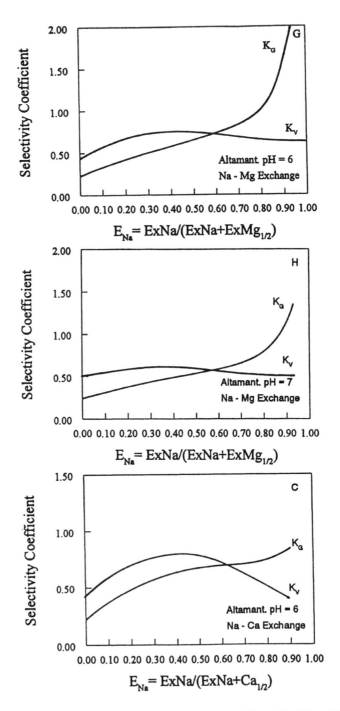

Figure 4.31. Experimental relationship between Na-load and K_V or K_G (adapted from Evangelou and Phillips, 1988).

TABLE 4.3. Cation Exchange Selectivity Coefficients for Homovalent (K–Na) and Heterovalent (K–Ca) Exchange

Selectivity Coefficient	Homovalent Exchange[a]	Heterovalent Exchange[b]
Kerr	$$K_K = \frac{\{\text{K-soil}\}\,[[\text{Na}^+]^c}{\{\text{Na-soil}\}\,[\text{K}^+]}$$	$$K_K = \frac{\{\text{K-soil}\}^2\,[\text{Ca}^{2+}]}{\{\text{Ca-soil}\}\,[\text{K}^+]^2}$$
Rothmund and Kornfeld[e]	$$K_F = \left(\frac{\{\text{K-soil}\}}{\{\text{Na-soil}\}}\right)^n \frac{[\text{Na}^+]}{[\text{K}^+]}$$	$$K_F = \left(\frac{\{\text{K-soil}\}}{\{\text{Ca-soil}\}}\right)^n \frac{[\text{Ca}^{2+}]}{[\text{K}^+]^2}$$
Vanselow[d]	$$K_V = \frac{\{\text{K-soil}\}\,[\text{Na}^+]}{\{\text{Na-soil}\}\,[\text{K}^+]}$$	$$\left[\frac{\{\text{K-soil}\}^2\,[\text{Ca}^{2+}]}{\{\text{Ca-soil}\}\,[\text{K}^+]^2}\right]$$
	$$\text{or } K_V = K_K$$	$$\left[\frac{1}{\{\text{K-soil}\} + \{\text{Ca-soil}\}}\right]$$
		or
		$$K_K\left[\frac{1}{\{\text{K-soil}\} + \{\text{Ca-soil}\}}\right]$$
Krishnamoorthy–Overstreet	$$K_{KO} = \frac{\{\text{K-soil}\}\,[\text{Na}^+]}{\{\text{Na-soil}\}\,[\text{K}^+]}$$	$$K_{KO} = \left[\frac{\{\text{K-soil}\}^2\,[\text{Ca}^{2+}]}{\{\text{Ca-soil}\,[\text{K}^+]^2}\right]$$
	$$\text{or } K_{KO} = K_K$$	$$\left[\frac{1}{\{\text{K-soil}\} + 1.5\,\{\text{Ca-soil}\}}\right]$$
Gaines–Thomas[d]	$$K_{GT} = \frac{\{\text{K-soil}\}\,[\text{Na}^+]}{\{\text{Na-soil}\}\,[\text{K}^+]}$$	$$K_{GT} = \left[\frac{\{\text{K-soil}\}^2\,[\text{Ca}^{2+}]}{\{\text{Ca-soil}\,[\text{K}^+]^2}\right]$$
	$$\text{or } K_{GT} = K_K$$	$$\left[\frac{1}{2[2\{\text{Ca-soil}\} + \{\text{K-soil}\}]}\right]$$
Gapon	$$K_G = \frac{\{\text{K-soil}\}\,[\text{Na}^+]}{\{\text{Na-soil}\}\,[\text{K}^+]}$$	$$K_G = \frac{\{\text{K-soil}\}\,[\text{Ca}^{2+}]^{1/2}}{\{\text{Ca}_{1/2}\text{-soil}\}\,[\text{K}^+]}$$

Source: Table adapted from Sparks (1995).

[a]The homovalent exchange reaction (K-Na exchange) is Na-soil + $K^+ \Leftrightarrow$ K-soil + Na^+.

[b]The heterovalent exchange reaction (K-Ca exchange) is Ca-soil + $2K^+ \Leftrightarrow 2$K-soil + Ca^{2+}, except for the Gapon convention where it would be $Ca_{1/2}$-soil + $K^+ \Leftrightarrow$ K-soil + 1/2 Ca^{2+}.

[c]Brackets denote concentration in the solution phase in moles per liter; braces denote concentration in the exchanger phase in moles per kilogram.

[d]Vanselow (1932) and Gaines and Thomas (1953) originally expressed both solution and exchanger phases as activity. For simplicity, in this table they represent concentrations.

[e]Walton (1949) units expressed as concentration in the solution phase in moles per liter and exchange phase in moles per kilogram.

4.3.5 Ion Preference

Ion preference or selectivity is defined as the potential of a charged surface to demonstrate preferential adsorption of one ion over another. Such ion preference is described by the *lyotropic series* and an example, from highest to lowest preference, is: $Ba^{2+} > Pb^{2+} > Sr^{2+} > Ca^{2+} > Ni > Cd^{2+} > Cu^{2+} > Co^{2+} > Mg^{2+} > Ag^+ > Cs^+ > Rb^+ > K^+ > NH_4^+ > Na^+ > Li^+$ (Helfferich, 1972). Note, however, that this lyotropic series is not universally applicable. The series often depends on the nature of the adsorbing surface (Sullivan, 1977). Some rules of cation selectivity are listed below:

I. Selectivity Rules Based on the Physical Chemistry of Cations

A. Generally, for two cations with the same valence (e.g. Na^+ vs. K^+), the cation with the smaller hydrated radius, or least negative heat of hydration (Tables 4.1 and 4.4), is preferred (e.g., K^+).

B. Generally, for two cations with different valence (e.g., Na^+ and Ca^{2+}), the cation with the higher valence, Ca^{2+}, due to its greater polarization potential, is preferred (polarization is the distortion of the electron cloud around the adsorbing cation due to the electric field of the charged surface). Larger anions show greater polarization potential than smaller anions. For this reason, larger anions would be preferred by a positively charged surface.

C. For two cations with the same valence, but one a stronger acid than the other (e.g., Cu^{2+} versus Ca^{2+}), the cation with stronger acid behavior or higher hydrolytic constant would be preferred if the surface behaved as a relatively strong base.

TABLE 4.4. Ion Sizes and Ionic Hydration

	Ionic Radii (Å)	
Ion	Not Hydrated	Hydrated
Li^+	0.68	10.03
Na^+	0.98	7.90
K^+	1.33	5.32
NH_4^+	1.43	4.4
Rb^+	1.49	3.6
Cs^+	1.65	3.6
Mg^{2+}	0.89	10.8
Ca^{2+}	1.17	9.6
Sr^{2+}	1.34	9.6
Ba^{2+}	1.49	8.8
Al^{3+}	0.79	—
La^{3+}	1.30	—

D. For any cation with any valence, when in the presence of two different anions, the anion with the highest potential to form neutral pairs with the cation controls the latter's adsorption potential, assuming that the anions do not react with the surface. For example, Ca^{2+} in the presence of Cl^- exhibits greater adsorption potential than Ca^{2+} in the presence of SO_4^{2-} due to the latter's greater potential to form neutral $CaSO_4^0$ pairs.

E. For any heavy-metal cation, when in the presence of two different anions, the anion with the highest potential to form surface complexes controls the metal's adsorption potential. For example, Ni^{2+} in the presence of $Ca(NO_3)_2$ exhibits greater adsorption potential than Ni^{2+} in the presence of $CaSO_4$ owing to the sulfate's potential to react with the surface and produce sites with high specificity for Ca^{2+}.

II. Selectivity Rules Based on the Physical Chemistry of Surfaces

A. A surface that has the potential to form inner-sphere complexes with certain monovalent cations (e.g., K^+ or NH_4^+) shows stronger preference for such cations than any other cation (e.g., vermiculite–K^+ or vermiculite–NH_4^+ versus vermiculite–Na^+ or vermiculite–Ca^{2+}).

B. A surface that does not have the potential to form inner-sphere complexes prefers cations with higher valence. This preference depends on the magnitude of the surface-electrical potential. For example, a surface with high electrical potential shows lower preference for monovalent cations (e.g., Na^+) in the presence of a divalent cation (e.g., Ca^{2+}) than a surface with low electrical potential.

C. Surfaces that exhibit pH-dependent electrical potential show various degrees of selectivity for the same cation. For example, kaolinite or kaolinitic soils at high pH (high negative surface electrical potential) shows increasing preference for divalent cations than monovalent cations, while at low pH (low negative electrical potential), kaolinite, or kaolinitic soils show increasing preference for monovalent cations than divalent cations.

D. Surfaces that have the potential to undergo conformational changes, such as humic acids, prefer higher-valence cations (e.g., Ca^{2+}) rather than lower-valence cations (e.g., K^+).

E. Surfaces that exhibit weaker acid behavior (high pK_a) show a stronger preference for heavy metals than hard metals in comparison to surfaces with stronger acid behavior (low pK_a). For example, illite or kaolinite show a stronger preference for Cu^{2+} or Cd^{2+} than montmorillonite.

NONPREFERENCE HETEROVALENT ISOTHERMS

In the case of the K_V equation for K^+–Ca^{2+} exchange,

$$K_V = [X_K/X_{Ca^{1/2}}][a_{Ca^{1/2}}/a_K] \tag{A}$$

where $AR_K = a_K/a_{Ca}^{1/2}$. Upon rearranging and solving for ExK,

$$ExK = CEC \cdot K_V AR_K/[4 + (K_V AR_K)^2]^{1/2} \tag{B}$$

Assuming that for the nonpreference isotherm, $K_V = 1$,

$$ExK/CEC = AR_K/[4 + (AR_K)^2]^{1/2} \tag{C}$$

Using a number of AR_K values, as $I \to 0$, to cover the entire exchange isotherm, equivalent fractional loads for K^+ can be estimated. A plot of $(K^+)/[(K^+) + (2Ca^{2+})]$ vs. ExK/CEC will produce a curvilinear line representing the nonpreference isotherm. Upon introducing γ_i values in the $K^+/[K^+ + 2Ca^{2+}]$ term such that

$$[(K^+)/\gamma_K]/\{(K^+)/\gamma_K + 2[(Ca^{2+})/\gamma_{Ca}]\} \tag{D}$$

it can be shown that as I increases, γ_K and γ_{Ca} decrease disproportionally, with γ_{Ca} decreasing significantly more than γ_K (see Chapter 2). Therefore, as I increases, the term $[(K^+)/\gamma_K]/\{(K^+)/\gamma_K + 2[(Ca^{2+})/\gamma_{Ca}]\}$ would decrease, suggesting that the nonpreference isotherm, $K^+/[K^+ + 2Ca^{2+}]$ vs. ExK/CEC, would be shifted upward, making it ionic-strength dependent (Fig. 4.29). Sposito et al. (1981) derived a nonpreference heterovalent (monovalent–divalent) exchange isotherm showing analytically its dependency on ionic strength. The equation is

$$E_{Ca} = 1 - \{1 + 2/\beta TMC[1/(1 - E_{Cas})^2 - (1 - E_{Cas})]\}^{-1/2} \tag{E}$$

where E_{Ca} = equivalent fraction of Ca^{2+} on the solid phase, TMC = total metal concentration (TMC = K + 2Ca), E_{Cas} = equivalent Ca^{2+} concentration in the solution phase, and $\beta = \gamma_K^2/\gamma_{Ca}$, where γ_K and γ_{Ca} represent single-ion activity coefficients for K^+ and Ca^{2+}, respectively.

4.3.6 Adsorbed-Ion Activity Coefficients

An experimentally variable K_V with respect to exchangeable Na^+ load (Eq. 4.53) can be transformed to the thermodynamic exchange constant (K_{eq}) as follows:

$$K_{eq} = K_V(f_{Na}/f_{Ca}^{1/2}) \tag{4.55}$$

where f_{Na} and f_{Ca} denote adsorbed-ion activity coefficients for Na^+ or Ca^{2+}, respectively. The equations producing f_{Na} and f_{Ca} are

$$\ln f_{Na} = (1 - E_{Na}) \ln K_V - \int_{E_{Na}}^{1} \ln K_V \, dE_{Na} \tag{4.56}$$

and

$$\frac{1}{2} \ln f_{Ca} = -E_{Na} \ln K_V + \int_0^{E_{Na}} \ln K_V \, dE_{Na} \tag{4.57}$$

where E_{Na} denotes the Na–equivalent fraction (ExNa/CEC) on the exchange phase. For a detailed discussion of Eqs. 4.55 and 4.56 refer to Evangelou and Phillips (1989), Sparks (1995), and references therein. Adsorbed-ion activity coefficients denote the relative strength of the interaction between adsorbing cation and surface. However, although adsorbed-ion activity coefficients reveal a great deal about the functional nature of any given mineral surface, their practical utility in soils has been limited.

EXAMPLE ON ADSORBED-ION ACTIVITY COEFFICIENTS

The equation employed to calculate the thermodynamic exchange constant (K_{eq}) from the K_V of Na⁺–Ca²⁺ exchange (see Reaction 4.38 for stoichiometry), for example, is as follows:

$$\ln K_{eqV} = \int_0^1 \ln K_V dE_{Na} = \ln K_{eq} \tag{A}$$

The $\ln K_{eq}$ constant can be calculated from Equation A by first producing a plot of $\ln K_V$ versus E_{Na} as shown in Figure 4E. Recall that K_V can be determined from Equation 4.53 employing the experimental solution and exchange data for Na⁺ and Ca²⁺ covering the entire exchange isotherm. The plot in Figure 4E is then separated into a number of "trapezoids" (listed 1–5). Summation of the areas (ln K_V times E_{Na}) of all five trapezoids produces ln K_{eq}.

Technically, ln K_{eq} can be estimated through Equation A by integrating under the entire ln K_V curve in Figure 4E. However, because there is no mathematical function available (describing the relationship between ln K_V and E_{Na}) to take the integral, the process is a little more time consuming because it requires a more practical approach. This approach involves cutting up carefully the entire figure (Fig. 4E) along its borders and then taking the weight of this cut-paper sample. Knowing the relationship of paper weight per unit area (ln K_V vs. E_{Na}), the area under the entire ln K_V plot can be estimated. Using this approach, the area in Figure 4E was estimated to be 2.035 or ln K_{eq} = 2.035. Additional approaches for integrating under the curve (E_i vs. ln K_V) of complex exchange data include computer graphics or analytical approaches.

For the direction and the stoichiometry of the exchange reaction used in this text (Reaction 4.38), the equations for estimating adsorbed-ion activity coefficients according to Argersinger et al. (1950) are

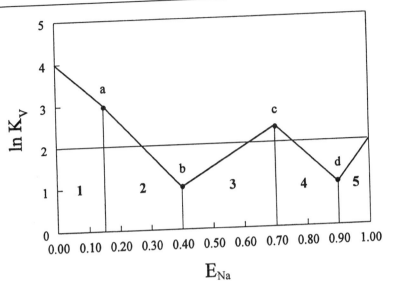

Figure 4E. Hypothetical data of Na^+ equivalent fraction on the exchange phase versus ln K_V. The line parallel to the x axis represents ln K_{eq}.

$$-\ln f_{Na} = (1 - E_{Na}) \ln K_V - \int_{E_{Na}}^{1} \ln K_V dE_{Na} \qquad (B)$$

and

$$-1/2 \ln f_{Ca} = - (E_{Na}) \ln K_V + \int_{0}^{E_{Na}} \ln K_V dE_{Na} \qquad (C)$$

An example is given below showing how to estimate adsorbed-ion activity coefficients for points **a**, **b**, **c**, and **d** in Figure 4E. For point **a**, integrating under $E_{Na} = 0$ to $E_{Na} = 0.14$ and solving Equation C gives $f_{Ca} = 0.86$, whereas for adsorbed Na^+ on point **a**, integrating under $E_{Na} = 0.14$ to $E_{Na} = 1$ and solving Equation B gives $f_{Na} = 0.36$. For point **b**, integrating under $E_{Na} = 0$ to $E_{Na} = 0.4$ and solving Equation C gives $f_{Ca} = 0.29$, while for adsorbed Na^+ on point **b**, integrating under $E_{Na} = 0.40$ to $E_{Na} = 1$ and solving Equation B gives $f_{Na} = 1.53$. For point **c**, integrating under $E_{Na} = 0$ to $E_{Na} = 0.7$ and solving Equation C gives $f_{Ca} = 1.537$, and for adsorbed Na^+ on point **c**, integrating under $E_{Na} = 0.70$ to $E_{Na} = 1$ and solving Equation B gives $f_{Na} = 0.78$. Finally, for point **d**, integrating under $E_{Na} = 0$ to $E_{Na} = 0.9$ and solving Equation C gives $f_{Ca} = 0.14$, while for adsorbed Na^+ on point **d**, integrating under $E_{Na} = 0.90$ to $E_{Na} = 1$ and solving Equation B gives $f_{Na} = 1.05$.

Multiplying the K_V corresponding to points **a**, **b**, **c**, and **d** in Figure 4E with the ratio of $f_{Na}/f_{Ca}^{1/2}$ at each corresponding point produces the K_{eq} obtained by solving Equation A. For example, ln K_V at point **a** is given as 3, by taking the antilog $K_V = 20.08$, multiplying it by $f_{Na}/f_{Ca}^{1/2}$ at point **a**, 0.36/0.93, we get 7.77 vs. 7.65 obtained by solving graphically Equation A. The small difference between the two K_V values are due possibly to an error in estimating graphically the area under the curve in Figure 4E. When we perform the same calculation at point **b**, where $K_V = 2.72$ (ln $K_V = 1$), multiplying it by $f_{Na}/f_{Ca}^{1/2}$ at point **b**, 1.53/0.53, gives ln $K_V = 2.061$ or $K_V = 7.85$.

4.3.7 Quantity–Intensity Relationships

Availability of nutrients such as K^+ and Ca^{2+} to plants in soil systems is related to the quantity and form of these nutrients in the solid phase. The quantity and form of the nutrients are related to their chemical potential (Beckett, 1964). However, it is not possible to directly measure the chemical potential of an ion in the solid phase, but it is possible to measure the difference in the chemical potential between two ions in the solution phase at equilibrium with the solid phase. The latter can then be related to the chemical potential difference of the two ions in the solid phase (Beckett, 1964, 1972; Nye and Tinker, 1977).

An approach often used to predict the relative chemical potential of K^+ in soils is the quantity–intensity (Q/I) relationship. A typical Q/I plot for a binary cation system is shown in Figure 4.32 with the following components:

ΔExK = Quantity factor (Q) and represents changes [gains (+) or losses (−) of exchangeable K^+]

CR_K = intensity factor (I) or concentration ratio for K^+

ExK^0 = labile or exchangeable K^+

ExK_s = specific K^+ sites

CR_K^0 = equilibrium concentration ratio for K^+ [$K^+/(Ca^{2+})^{1/2}$]

PBC_K = linear potential buffering capacity for K^+

This relationship (Fig. 4.32) implies that the ability of a soil system to maintain a certain concentration of a cation in solution is determined by the total amount of the cation present in readily available forms (exchangeable and soluble) and the intensity by which it is released to the soil solution.

The Q/I relationship can be demonstrated as follows: Consider the exchange reaction given by Reaction 4.38 and replacing Na^+ by K^+, the Gapon expression is

$$K_G = [(ExK)/(ExCa_{1/2})][(Ca^{2+})^{1/2}/(K^+)] \qquad (4.58)$$

and

$$[ExK] = [ExCa_{1/2}]\, K_G(K^+)/(Ca^{2+})^{1/2} \qquad (4.59)$$

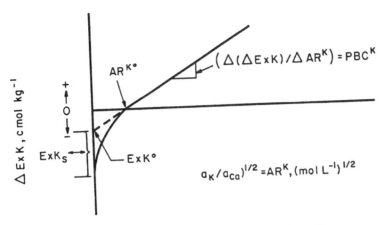

Figure 4.32. Ideal K quantity/intensity (Q/I) relationship.

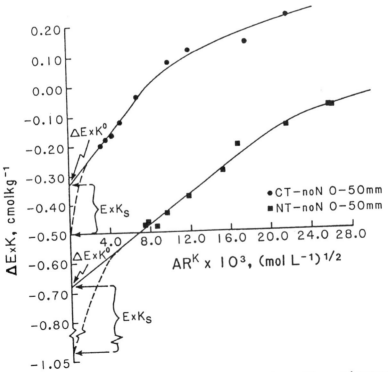

Figure 4.33. Potassium Q/I plots of Maury silt loam soil under no-tillage and conventional tillage conditions (from Evangelou and Blevins, 1988).

Equation 4.59 can be further written as

$$\text{ExK} = K_G[\text{ExCa}_{1/2}]\text{CR}_K \tag{4.60}$$

where

$$\text{CR}_K = K/(\text{Ca}^{2+})^{1/2} \tag{4.61}$$

For most agricultural soils under K-fertilization, $\text{ExK} <<< \text{ExCa}_{1/2}$ and Equation 4.59 then becomes a linear equation where ExK is the dependent variable, CR_K is the independent variable, and $K_G[\text{ExCa}_{1/2}]$ is the slope. This slope can also be expressed as the derivative of ExK with respect to CR_K:

$$d[\text{ExK}]/d\text{CR}_K = K_G[\text{ExCa}_{1/2}] \tag{4.62}$$

The right-hand side of Equation 4.62 is the PBC_K (Fig. 4.32).

$$\text{PBC}_K = K_G[\text{ExCa}_{1/2}] \tag{4.63}$$

Experimental data of Q/I plots are shown in Figure 4.33. These plots show how different soil-management practices affect the potential of soil to supply K^+ to plants. Beckett (1972) pointed out that based on Q/I, the potential of a soil to deliver K^+ to plants should be dependent of CR_K. Furthermore, he pointed out that each plant species should exhibit a critical CR_K ratio (CCR_K) (Beckett, 1972). Below that ratio, a particular plant species would respond to added K. Note that for a number of different soils, the same CR_K would represent different quantities of exchangeable K. In general, the CCR_K for most plants, according to Beckett, is around 0.005 $(\text{mol L}^{-1})^{1/2}$.

Q/I JUSTIFICATION

Consider an equilibrium state following increases or decreases in [ExK] by addition or removal of K^+ from the soil, keeping in mind that $[\text{ExK}] <<< [\text{ExCa}_{1/2}]$. Then Equation 4.58 can be rearranged as follows:

$$\pm\Delta[\text{ExK}] = K_G\{[\text{ExCa}_{1/2}] \pm [(\Delta\text{ExK})]\}\text{CR}_{K'} - [\text{ExK}^0] \tag{A}$$

where

$$\text{CR}_{K'} = [K \pm \Delta(K)]/[Ca \pm \Delta(Ca)]^{1/2} \tag{B}$$

and $\text{ExK}^0 = \text{ExK}$ in soil at equilibrium. The ($\pm$) in Equations A and B indicate that for (+), K^+ is being added and for (–), K^+ is being removed. After the addition or removal of K^+ (still assuming that $[\text{ExK}] <<< [\text{ExCa}_{1/2}]$), Equation A is still a linear equation where $(K_G[\text{ExCa}_{1/2}] \pm \Delta[\text{ExK}])$ is the slope and ExK^0 is the constant of the linear equation. Equation 4.63 now becomes

$$PBC_{K'} = slope = K_G([ExCa_{1/2}] \pm \Delta[ExK]) \qquad (C)$$

It follows from Equation A that as $CR_{K'} \to 0$, $\Delta[ExK] \to ExK^0$. The value of ExK^0 in Figure 4.32 indicates variable K^+ (labile or exchangeable). On the other hand, when $\Delta[ExK] \to 0$, $CR_{K'} \to CR_K^0$. The value of CR_K^0 represents the equilibrium concentration ratio of K^+ for the soil sample. The straight line section of the Q/I plot of Figure 4.32 represents the linear buffering capacity of soil ($PBC_{K'}$), as long as K_G remains constant. For most agricultural soils under K–fertilization, the contribution of ExK and/or $\Delta[ExK]$ to the CEC is insignificant; that is, $[ExCa_{1/2}] >>> \Delta[ExK]$ and Equation C reduces to

$$PBC_{K'} = PBC_K = CEC \; K_G \qquad (D)$$

When ExK represents a significant portion of the soil's CEC,

$$ExK = K_G [CEC - ExK]CR_K \qquad (E)$$

where

$$[CEC - ExK] = [ExCa_{1/2}] \qquad (F)$$

and

$$d[ExK]/dCR_K = K_G \cdot CEC/(1 + K_G \cdot CR_K)^2 = PBC_K \qquad (G)$$

Equation G points out that for a given K_G value, a given CEC, and a relatively small CR_K, the product of $K_G \cdot CR_K$ is near zero and Equation G reduces to Equation D. However, when using the same K_G value, the same CEC, and significantly larger CR_K values, so that the denominator of Equation G is significantly > 1, the PBC_K better fits a curvilinear function asymptotically approaching some upper limit.

4.3.8 Ternary Exchange Systems

Laboratory studies commonly consider that clays and/or soils are two-ion exchange systems (binary). Field soils, however, are at least three-ion exchange systems (ternary) (Curtin and Smillie, 1983; Adams, 1971). It is therefore assumed that the data of binary exchange reactions can be employed to predict ternary exchange reactions. For this assumption to be valid, one has to accept that binary-exchange selectivity coefficients are independent of exchanger-phase composition. However, Chu and Sposito (1981) have shown at a theoretical level that one cannot predict exchange–phase/solution– phase interactions of a ternary system solely from data obtained from binary exchange. These researchers argued that experimental data obtained from systems with three ions are necessary.

Farmers often use NH_4 and K salts as fertilizer sources. Even though applied NH_4 has a short life span in agricultural soils (1–3 wk or more depending on rates of nitrification), the K–NH_4–Ca exchange interaction controls the distribution of these cations between the exchange and solution phases during that period. Thus, the availability of K and NH_4 in the solution phase would be affected by all ions present, or the relative chemical potential of adsorbed cations would depend on the number and type of cations present in the soil system. This can be demonstrated by the Q/I relationship of three cation (ternary) exchange systems (Lumbanraja and Evangelou, 1990).

The Q/I plot of a ternary (three-ion) soil system can be explained by the Vanselow (1932) exchange expression (Evangelou and Phillips, 1987, 1988 and references therein). Note that for exchangeable K^+ fractional loads <0.20, the situation encountered in most agricultural soils, the Vanselow and the Gapon equations are indistinguishable ($K_G = 1/2K_V$). For a ternary system such as $K-NH_4-Ca$, the equations that describes PBC_K at constant CR_{NH4} (Lumbanraja and Evangelou, 1990) as CR_K approaches the critical potassium concentration ratio (CCR_K) [0.002 − 0.005 (mol $L^{-1})^{1/2}$] are

$$PBC_K = CEC \cdot K_{V1}/[4 + (K_{V2} CR_{NH4})^2]^{1/2} \qquad (4.64)$$

and at constant CR_K as CR_{NH4} approaches values in the range of 0.002 to 0.005 (mol $L^{-1})^{1/2}$

$$PBC_{NH4} = CEC \cdot K_{V2}/[4 + (K_{V1}CR_K)^2]^{1/2} \qquad (4.65)$$

Equation 4.64 demonstrates that when $CR_K \rightarrow CCR_K$, PBC_K will depend on the CEC, K_{V1}, K_{V2} and on the magnitude of the constant CR_{NH4} value, where K_{V1} is the Vanselow K^+-Ca^{2+} exchange selectivity coefficient and K_{V2} is the Vanselow $NH_4^+-Ca^{2+}$ exchange selectivity coefficient. Assuming the CEC remains constant and for a constant CR_{NH4} value, PBC_K will depend directly on K_{V1} and inversely on K_{V2}. A high K_{V1}, which indicates a high affinity for K, manifests itself as an extremely high slope on the low portion of the Q/I plot. Furthermore, in accordance with Equation G (Q/I Justification section) (Evangelou et al., 1994), an increase in CR_K or CR_{NH4} (Equation 4.64) would suppress PBC_K (upper portion of CR_K or Q/I plot) assuming CEC, K_{V1}, and K_{V2} are constant over the appropriate activity ratio range. A similar discussion applies to Equation 4.65, which represents change in exchangeable NH_4^+ at a constant CR_K (Figures 4.34 and 4.35).

In contrast to the PBC_K for the $K-NH_4-Ca$ ternary exchange system (i.e., Equation 4.64) the PBC_K for the $K-Ca$ binary system, as $CR_K \rightarrow CCR_K$, is described by Equation D in the previous boxed section (Q/I Justification) which for convenience is shown below using K_{V1}

$$PBC_K = (1/2)(K_{V1})(CEC) \qquad (4.66)$$

Equation 4.66 demonstrates that, when CR_K approaches CCR_K, PBC_K will depend on the product of K_{V1} and CEC only. Assuming that CEC remains constant, the PBC_K is dependent on K_{V1} only. A high K_{V1} indicates a high affinity for the K^+ and manifests itself as an extremely high slope on the low portion ($CR_K \rightarrow 0$) of the Q/I plot. Furthermore, an increase in CR_K would have a suppressing effect on PBC_K (upper portion of Q/I plot) for a constant CEC and K_{V1} (Evangelou et al., 1994). The same applies to the NH_4-Ca exchange system.

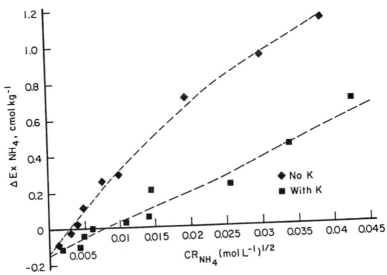

Figure 4.34. Ideal ammonium Q/I plots in the presence or absence of K.

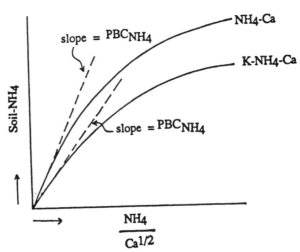

Figure 4.35. Experimental ammonium Q/I plots in the presence or absence of K.

4.3.9 Influence of Anions

Ion exchange equilibria are affected by the type of anions used (Babcock and Schultz, 1963; Rao et al., 1968), owing to the pairing properties of divalent ions (Tanji, 1969b; Adams, 1971). Tanji (1969b) demonstrated that approximately one third of the dissolved $CaSO_4^0$ of a solution in equilibrium with gypsum ($CaSO_4 \cdot 2H_2O$) is in the $CaSO_4^0$ pair form. Magnesium behavior is similar to Ca^{2+} because Mg^{2+} pairing ability with SO_4^{2-} is approximately equal to that of Ca^{2+} (Adams, 1971). By investigating Na^+–Ca^{2+} exchange reactions in Cl^- or SO_4^{2-} solutions, Babcock and Schultz (1963) and Rao et al. (1968) found that exchange selectivity coefficients for the two cations were different. But they demonstrated that this difference was nearly eliminated when solution-phase activities and $CaSO_4^0$ were considered (Fig. 4.36).

Evangelou (1986) investigated the influence of Cl^- or SO_4^{2-} anions on the Q/I relationships for K^+ in several soils. The data from these studies demonstrated that correcting the solution data for single-ion activity considering ion pairing and single-ion solution activity did not compensate for the experimental differences in the Q/I relationships between the chloride and sulfate systems (Figs. 4.37 and 4.38). Evangelou (1986) concluded that these differences may be due to competitive interactions of $CaCl^+$ with Ca^{2+} and/or KSO_4^- with SO_4^{2-} for exchange sites and possibly SO_4^{2-} specific adsorption onto the soil's solid surfaces (Fig. 4.39). Sposito et al. (1983) have

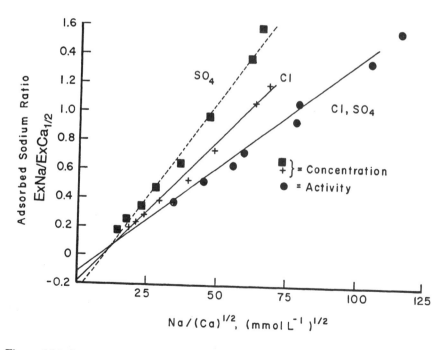

Figure 4.36. Relationship between $Na/Ca^{1/2}$ in solution (in concentration and activity units) and Na-load on the exchange phase (from Rao et al., 1968, with permission).

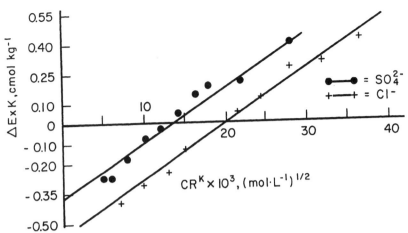

Figure 4.37. Relationship between CR_K and ΔExK with SO_4^{2-} or Cl^- as the balancing cation in solution (from Evangelou, 1986, with permission).

shown that external clay surfaces adsorbed charged pairs of $CaCl^+$ and $MgCl^+$ in $Na^+-Ca^{2+}-Cl^-$ and $Ma^+-Mg^{2+}-Cl^-$ clay suspension systems.

The influence of anion on K^+ rate of adsorption, mobility, and retention were investigated by Sadusky and Sparks (1991) on two Atlantic coastal plain soils. They reported that the type of anion had little effect (if any) on the rate of K^+ adsorption, but had an effect on the amount of K^+ adsorbed. They found that K^+ adsorbed in the presence of a particular accompanying anion was of the order $SiO_3 > PO_4 > SO_4 > Cl$

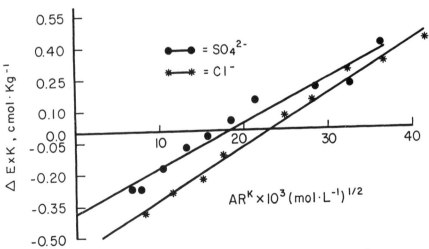

Figure 4.38. Relationship between activity ratio K^+ (AR_K) and ΔExK with SO_4^{2-} or Cl^- as the balancing anion in solution (from Evangelou, 1986, with permission).

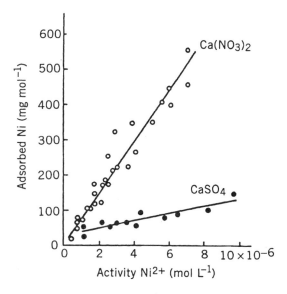

Figure 4.39. Relationship between activity of Ni^2 in solution and adsorbed Ni with CaSO$_2$ or Ca(NO$_3$)$_2$ as the background electrolyte (from Mattigod et al., 1981, with permission).

> ClO$_4$. These findings strongly imply that exchange reactions in soils are affected by the anion present in the soil. For a review of this material, see also Evangelou et al. (1994).

4.3.10 Exchange Reversibility

It is assumed that the shape and form of cation exchange is identical between the adsorption and desorption mode (Fig. 4.40a). However, Nye and Tinker (1977) suggested that the desorption process may be affected by hysteresis (Fig. 4.40b). The graphical relationship presented in Figure 4.40b shows that the difference between the quantity of adsorbed X$^+$ and the quantity of desorbed X$^+$(X$_d$) is the quantity of X$^+$ fixed, X$_f$ (Arnold, 1970). In this case, the term "fixed" denotes exchange-reaction nonreversibility under the conditions described by the experiment. The term "fixed" often seen in the literature denotes nonextractability of the cation by concentrated salt solutions (Quirk and Chute, 1968; Barshad, 1954; Lurtz, 1966; Opuwaribo and Odu, 1978). Arnold (1970) points out that an isotherm demonstrating a hysteresis effect can exhibit two different shapes (linear and curvilinear). If the desorption isotherm is linear, it indicates that the (X) is being desorbed from low-affinity sites only. If the desorption isotherm is curvilinear, the cation (X) is being desorbed from high- (Stern layer) and low- (diffuse layer) affinity sites. Relative high–low (Stern layer–diffuse layer) affinity sites of monovalent cations are schematically demonstrated in Figure 4.41).

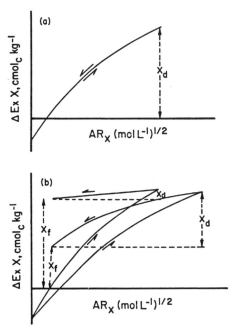

Figure 4.40. Relationship between AR_X and ΔExX for the forward and reverse reactions. See text for further explanation (from Lumbanraja and Evangelou, 1994, with permission).

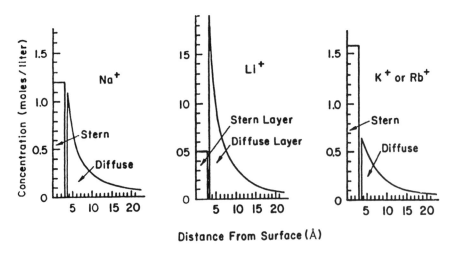

Figure 4.41. Schematic distribution of monovalent cations between the Stern layer (high-affinity sites) and the diffuse layer (low-affinity sites) as predicted by the double layer theory (from Shainberg and Kemper, 1966, with permission).

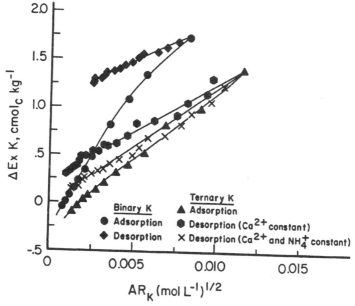

Figure 4.42. Relationship between AR_K and ΔExK for the forward and reverse K–Ca exchange in the presence and absence of NH_4^+ (from Evangelou et al., 1994, with permission).

The hysteresis effect appears to be affected by the length of the equilibration period. Nye and Tinker (1977) point out that a true hysteresis effect would persist no matter what length of time was given for a true equilibrium to establish. In contrast, if the difference between an adsorption and a desorption isotherm is eliminated by extending the equilibration time, this would be considered a relaxation effect (Everett and Whitton, 1952).

Examples of K^+ and NH_4^+ adsorption–desorption plots for the two-cation system (K–Ca; NH_4–Ca) and the three-cation system (K–NH_4–Ca) are shown in Figure 4.42. These data show that there is a significant hysteresis or relaxation effect in the 24-hr desorption process. This hysteresis–relaxation effect in the case of K–Ca exchange is more pronounced without added NH_4^+ than with added NH_4^+ in the desorbing solution.

4.3.11 Thermodynamic Relationships

From the K_{eq} constant, estimated graphically from the Vanselow exchange selectivity coefficient, a number of thermodynamic parameters can be estimated. For example,

$$\Delta G^0 = -RT \ln K_{eq} \qquad (4.67)$$

where ΔG^0 is the standard Gibbs free energy of exchange and all other terms are as previously defined (see Chapter 3). Substituting values for the constants in Equation 4.67 converts to

$$\Delta G^0 = -1.364 \log K_{eq} \qquad (4.68)$$

For any given reaction, where reactants and products are at the standard state, a negative ΔG^0 denotes that the reaction will move spontaneously from left to right. When ΔG^0 is zero, it implies that the system is in equilibrium and $K_{eq} = 1$. Finally, for any given reaction where reactants and products are at the standard state, a positive ΔG^0 denotes that the reaction will not move spontaneously from left to right. Such a reaction, however, could become spontaneous by changing the concentrations of the reactants and/or products, thus giving a negative ΔG^0 for the reaction. Finally, the more negative the value of ΔG^0 is, the greater the value of K_{eq} (e.g., greater adsorption potential, greater solubility, greater ion preference). In the case of exchange reactions, ΔG^0 denotes ion preference between two competing cations for the same surface. In the case of K–Ca, it signifies that Ca^{2+} is the surface cation and K^+ is the solution cation. If ΔG^0 is negative, the surface prefers K^+, but if ΔG^0 is positive, the surface prefers Ca^{2+}. Note that this preference is independent of ionic strength.

Also, from K_{eq}, the standard enthalpy of exchange, ΔH^0 can be calculated (Table 4.2) using the van't Hoff equation:

$$\ln(K_{eqT_1}/K_{eqT_2}) = (\Delta H^0/R) - \{(1/T_2) - (1/T_1)\} \qquad (4.69)$$

where T_1 and T_2 denote two different temperatures. From the classical thermodynamic relationship,

$$\Delta G^0 = \Delta H^0 - T\Delta S^0 \qquad (4.70)$$

the standard entropy of exchange, ΔS^0 can be calculated by rearranging Equation 4.70:

$$\Delta S^0 = (\Delta H^0 - \Delta G^0)/T \qquad (4.71)$$

Based on Reaction 4.70 one may observe the requirements of ΔS^0 and ΔH^0 in order for $K_{eq} > 1$, or for the reactants in their standard state to be converted spontaneously to their products in their standard state. If ΔH^0 is negative and ΔS^0 is positive, the reaction will be spontaneous. If, on the other hand, ΔH^0 is positive and ΔS^0 is negative, the reaction will not be spontaneous. Finally, if ΔH^0 and ΔS^0 have the same sign, the reaction may or may not be spontaneous, depending on the magnitude of ΔS^0, ΔH^0, and temperature (see Daniels and Alberty, 1975).

Enthalpy (ΔH^0) is defined as the heat given out or taken up during a reaction. If heat is given out during a reaction, it is referred to as an exothermic reaction and ΔH^0 is negative. If heat is taken up during a reaction, it is referred to as an endothermic reaction and ΔH^0 is positive. A negative ΔH^0 denotes stronger bonds within the products, whereas a positive ΔH^0 denotes stronger bonds within the reactants.

Entropy (ΔS^0) is defined as the tendency of a given system to come apart. For a given ΔH^0, the greater ΔS^0 is, the greater the tendency of a reaction to become spontaneous because of the increased tendency of the system to come apart or go to disarray. Exchange reactions on clay surfaces involving hard cations are known to be mostly electrostatic processes and entropy of exchange plays a significant role in determining cation preference with respect to size and valence.

PROBLEMS AND QUESTIONS

1. To understand the behavior of Pb in soils, you need to understand how Pb partitions itself between the solid and solution phases. Consider the following lead adsorption experiment. To each of 8 bottles, 10.18 g of soil and 200 mL of Pb solution was added. The initial and equilibrium Pb concentrations are given below:

Initial Pb Concentration $(mg\ L^{-1})$	Equilibrium Pb Concentration $(mg\ L^{-1})$
3.15	0.05
5.20	0.11
6.32	0.16
7.40	0.22
10.39	0.41
10.36	0.43
12.61	0.65
14.84	0.94

a. Should you fit a Langmuir or Freundlich isotherm? Why?

b. Fit the appropriate isotherm and calculate the adjustable parameters.

c. Assume there was a field test for Pb that measured solution Pb concentration. The procedure is to add 10.18 g of soil to 100 mL of test solution. If the measured solution Pb concentration was 0.50 mg L^{-1}, how much lead is on the solid phase? What is the total lead concentration of the soil? **NOTE:** total lead = solid phase lead + solution phase lead.

2. The following arsenate (AsO_4^{3-}) adsorption data was obtained by adding 20.42 g of kaolinite to 200 mL of arsenate solution,

Initial Concentration $(mg\ L^{-1})$	Equilibrium Concentration $(mg\ L^{-1})$
4.88	1.20
10.09	3.56
15.36	6.78
20.11	10.1
30.87	17.6
40.62	25.0
51.78	33.4
80.96	58.4
117.24	90.7
138.29	109
157.91	128
180.53	150
205.73	175
222.83	192

a. Should you fit a Langmuir or Freundlich isotherm? Why?

b. Fit the appropriate isotherm and calculate the adjustable parameters.

3. Compare the lead and arsenate isotherms. Although the same soils were not used for these experiments, it is still possible to make qualitative comparisons. Does arsenate or lead adsorb more to soils? Why?

4. Use the data below to perform the following calculations:

a. Plot the selectivity coefficients K_G and K_V as a function of fractional NH_4 load and describe your observations

b. Plot the same data as cNH_4/cK vs. $ExNH_4/(ExK + ExNH_4)$ and describe your observations

c. Plot $cNH_4/cK + cNH_4$ vs. $ExNH_4/(ExK + ExNH_4)$ and describe ion preference

d. Plot $^cNH_4/^cK$ vs. $ExNH_4/ExK$; estimate the selectivity coefficient from the slope(s) of the line and compare these values to those estimated in a.

Experimental Data on NH₄–K Exchange at 297 K on Vermiculite Clay

| | mM | | | mol$_c$ kg^{-1} | |
pH	$cNH_4{}^a$	cK	$ExNH_4{}^b$	ExK	ΣEx_i
8.07	0.085	8.647	0.003	0.378	0.381
8.02	0.207	8.500	0.007	0.375	0.382
8.01	0.363	8.227	0.011	0.365	0.378
7.74	0.854	7.920	0.028	0.343	0.371
7.54	1.613	7.183	0.051	0.302	0.353
7.37	2.330	6.483	0.073	0.259	0.322
7.20	4.760	4.420	0.147	0.164	0.311
6.98	7.860	1.480	0.209	0.060	0.269
6.94	8.690	0.816	0.238	0.035	0.269
6.97	9.073	0.411	0.237	0.022	0.260
6.94	9.250	0.221	0.252	0.015	0.267
6.91	9.470	0.092	0.253	0.011	0.264

^{a}The term c_i denotes concentration of cation i in solution at equilibrium.
^{b}The term Ex_i denotes concentration of adsorbed cation i at equilibrium; ΣEx_i denotes total adsorbed cations.

5. Use the data below to perform the following calculations:

a. Plot the selectivity coefficients K_G and K_V as a function of fractional NH_4 load and describe your observations

b. Plot the same data as cNH_4/cCa vs. $ExNH_4/(ExCa_{1/2} + ExNH_4)$ and describe your observations

c. Plot $cNH_4/2cCa + cNH_4$ vs. $ExNH_4/(ExCa_{1/2} + ExNH_4)$ and include the nonpreference isotherm; describe ion preference

d. Plot $cNH_4/cCa^{1/2}$ vs. $ExNH_4/ExCa_{1/2}$; estimate the selectivity coefficient from the slope(s) of the line and compare these values to those estimated in **a**.

Experimental Data on NH_4–Ca Exchange at 297 K on Vermiculite Clay

	mM			mol kg^{-1}	
pH	$cNH_4{}^a$	cCa	$ExNH_4{}^b$	ExCa	ΣEx_i
7.68	0.130	3.890	0.002	0.627	0.627
7.72	0.331	3.783	0.004	0.542	0.546
7.72	0.516	3.653	0.006	0.479	0.485
7.70	0.879	2.917	0.009	0.402	0.411
7.41	1.563	2.867	0.012	0.352	0.364
7.22	2.370	2.767	0.018	0.310	0.328
7.09	4.810	1.887	0.020	0.255	0.275
7.08	8.243	0.476	0.083	0.185	0.268
6.88	8.923	0.252	0.136	0.113	0.219
6.91	9.397	0.127	0.190	0.067	0.257
6.91	9.490	0.070	0.224	0.043	0.267
6.89	9.597	0.031	0.229	0.024	0.253

[a]The term c_i denotes concentration of cation i in solution at equilibrium.
[b]The term Ex_i denotes concentration of adsorbed cation i at equilibrium; ΣEx_i denotes total adsorbed cations.

6. If the selectivity coefficient for K–Ca exchange is 5, calculate exchangeable K for a soil with $K/Ca^{1/2}$ of 0.1 and CEC of 25 meq/100 g. What would happen if the CEC is doubled?

PART III
Electrochemistry and Kinetics

5 Redox Chemistry

5.1 REDOX

Many metals and metalloids in soil–water systems have the potential to change oxidation states. Generally, an oxidation state is characterized by the net charge of the elemental chemical species. For example, iron may be present in soil as iron(II) (Fe^{2+}) or Iron(III) (Fe^{3+}). Similarly, manganese maybe present as manganese(II) (Mn^{2+}), manganese(III) (Mn^{3+}), or manganese(IV) (Mn^{4+}). In oxidized soil–water systems, the environment is considered electron deficient, and the higher oxidation states of iron or manganese (e.g., Fe^{III}, Mn^{III}, or Mn^{IV}) are most stable. In reduced soil–water systems, the environment is considered electron rich, and the lowest oxidation states of manganese or iron (e.g., Mn^{II} or Fe^{II}) are most stable. It turns out that the lower oxidation states of Fe and Mn are more soluble than their higher oxidation states (see Chapter 2). Mineral solubility is not always directly related to the lower oxidation state of redox-sensitive metals. For example, reduced metals such as Fe^{II} and Mn^{II} would be soluble if no other ion is present to react with and form insoluble minerals. At intermediate reduced environments, sulfur would be present in the sulfate form (SO_4) and its potential to precipitate Fe^{2+} as $FeSO_4$ would be limited owing to the high K_{sp} of ferrous sulfate. In strongly reduced environments, sulfur would be reduced to S^{-II} and FeS, a very insoluble mineral, would form.

The terms *oxidation* and *reduction* with respect to chemical processes in soil–water systems refers to potential electron-transfer processes. Under oxidation, a chemical element or molecular species donates electrons (e^-), whereas under reduction a chemical element or molecular species accepts electrons. The potential of an atom of any given element to react depends on the affinity of its nucleus for electrons and the strong tendency of the atom to gain maximum stability by filling its outer electron shell or comply with the octet rule. The octet rule states that to gain maximum stability an atom must have eight electrons in its outer shell or outermost energy level.

In general, the nature of ions in soil–water systems depends on type and oxidation state(s) of the elements involved (Table 5.1). The oxidation state of elements in nature can be estimated following the three rules stated below (Stumm and Morgan, 1981):

TABLE 5.1. Oxidation States

Nitrogen Compounds		Sulfur Compounds		Carbon Compounds	
Substance	Oxidation States	Substance	Oxidation States	Substance	Oxidation States
NH_4^+	N = −3, H = +1	H_2S	S = −2, H = +1	HCO_3^-	C = +4
N_2	N = 0	$S_8(s)$	S = 0	HCOOH	C = +2
NO_2^-	N = +3, O = −2	SO_3^{2-}	S = +4, O = −2	$C_6H_{12}O_6$	C = 0
NO_3^-	N = +5, O = −2	SO_4^{2-}	S = +6, O = −2	CH_3OH	C = −2
HCN	N = −3, C = +2, H = +1	$S_2O_3^{2-}$	S = +2, O = −2	CH_4	C = −4
SCN^-	S = −1, C = +3, N = −3	$S_4O_6^{2-}$	S = +2.5, O = −2	C_6H_5COOH	C = −2/7
		$S_2O_6^{2-}$	S = +5, O = −2		

Source: Reproduced from Stumm and Morgan (1981).

1. For a monoatomic substance, its oxidation state is equal to its electronic charge. For example, in the case of metal iron, its oxidation state is zero because the electronic charge of iron in its ground state is zero.

2. For a compound formed by a number of atoms, the oxidation state of each atom is the charge remaining on the atom when each shared pair of electrons is completely assigned to the more electronegative ion. When an electron pair is shared by two atoms with the same electronegativity, the charge is divided equally between the two atoms. For example, in the case of FeS_2, the S_2 is assigned two negative charges, Fe possesses two positive charges, and each sulfur is assigned a single negative charge [$S(I^-)$].

3. For a molecule with no charge, the sum of the oxidation states of the elements involved is zero, whereas for molecules with charge, it is equal to the difference in the algebraic sum of the charge of each of the elements involved.

As pointed out in Chapter 1, generally, two types of bonds may form in nature: (1) ionic bonds and (2) covalent bonds. Ionic bonds, which are common to salts (e.g., NaCl, KCl, and $NaNO_3$), are a product of weak electrostatic forces between ions with complete outer electron shells, explaining the high solubility of such salts. Covalent bonds are strong bonds that occur between atoms because of their strong tendency to complete their outer electron orbital shell or fulfill the octet rule by sharing electrons (e.g., Cl_2; each chlorine atom provides an electron). Thus, reactions and types of reactions occur because of electron configurations. There are a number of elements/atoms (listed in Table 1.5, Chapter 1), with any number of electron configurations, depending on their redox state. In other words, a particular atom of an element may react ionically or covalently, depending on its redox state. For example, sulfur in its +6 oxidation state exists as sulfate (SO_4^-), generally a soluble ion. However, sulfur in its −2 oxidation state exists as S^{2-}, generally very insoluble because of its tendency to react covalently with most metals forming insoluble metal sulfides.

TABLE 5A. Soil and Sediment Processes Causing pH Changes

1. *Nitrification*

$$NH_4^+ + 1.5\,O_2 \rightarrow NO_2^- + 2H^+ + H_2O$$

2. *Denitrification*

$$H_2O + NO_3^- + 2e^- \rightarrow NO_2^- + 2OH^-$$
$$2\,H_2O + NO_2^- + 3e^- \rightarrow 1/2\,N_2 + 4OH^-$$
$$1\,1/2\,H_2O + NO_2^- + 2e^- \rightarrow 1/2\,N_2O + 3OH^-$$

3. *Oxidative metabolism (enzymatically available organic matter + $O_2 \rightarrow CO_2 + H_2O$)*

$$H_2O + CO_2 \Leftrightarrow H_2CO_3$$

4. *Other electron acceptors for microbiological respiration*

$$2\,H_2O + MnO_2 + 2e^- \rightarrow Mn^{2+} + 4OH^-$$
$$Fe(OH)_3 + e^- \rightarrow Fe^{2+} + 3OH^-$$
$$6\,H_2O + SO_4^{2-} + 8\,e^- \rightarrow H_2S + 10\,OH^-$$

5. *Sulfur oxidation*

$$H_2O + S + 1\,1/2\,O_2 \rightarrow SO_4^{2-} + 2H^+$$

Therefore, oxidation–reduction processes in nature control the behavior of elements or substances. During oxidation–reduction, the potential for reactions to take effect changes because the redox status of elements changes. A summary of soil–water mineral–ion properties known to be affected by redox chemistry is listed below:

1. Mineral solubility [$Mn(II)(OH)_2$ soluble vs. $Mn(IV)O_4$ insoluble]
2. Soil–water pH (reducing conditions increase pH, oxidizing conditions decrease pH; Table 5A)
3. Mineral surface chemistry [PZC of δ–$MnO_2 = 2.8$ or PZC of β–$MnO_2 = 7.2$ vs. $Mn(II)(OH)_2$ a very soluble mineral under natural pH with very high PZC]
4. Availability/presence of certain chemical species [Fe^{2+} vs. Fe^{3+} or $HAs(III)O_2$ vs. $HAs(V)O_4^{2-}$]
5. Persistence or toxicity of chemical species [$N(III)O_2^-$ persistence very low vs. $N(V)O_3^-$ persistence high; As(III) high toxicity vs. As(V) low toxicity]
6. Salt content or electrical conductivity of solution [S(II) insoluble vs. $S(VI)O_4$ soluble]
7. Volatility of chemical species [$S(VI)O_4^{2-}$ nonvolatile vs. $H_2S(II)$ volatile]

5.2 REDOX-DRIVEN REACTIONS

Redox is characterized by two types of reactions: (a) *oxidation*, denoting electron (e^-) donation,

$$\text{Simple sugar} + 1/4H_2O \rightarrow 1/4CO_2 + H^+ + e^- \tag{5.1}$$

and (b) *reduction*, denoting electron (e^-) reception,

$$H^+ + 1/4O_2 + e^- \Leftrightarrow 1/2H_2O \tag{5.2}$$

In nature however, there is no such thing as free electrons. For a redox reaction to proceed, both oxidation and reduction would have to be coupled. This is demonstrated by summing Reactions 5.1 and 5.2, which shows that the electron released by the oxidizing species (Reaction 5.1) is accepted by the reducing species (Reaction 5.2):

$$\text{Simple sugar} + 1/4O_2 \rightarrow 1/4CO_2 + 1/4H_2O \tag{5.3}$$

These equations show how simple sugar (reduced carbon), when in the presence of atmospheric oxygen (O_2), is converted to carbon dioxide (CO_2) and water via electron transfer.

SOME THERMODYNAMIC RELATIONSHIPS

Consider an equilibrium state between species A (reactant) and species B (product):

$$A \overset{K_{eq}}{\Leftrightarrow} B \tag{A}$$

If reactant A and product B are introduced at unit activity (α), one of three conditions would be met:

1. $\Delta G^0 = 0$, reaction at equilibrium
2. ΔG^0 = negative, reaction will move from left to right to meet its equilibrium state
3. ΔG^0 = positive, reaction will move from right to left in order to meet its equilibrium state (for the definition of ΔG or ΔG^0, see Chapter 7).

From classical thermodynamics,

$$\Delta G_A = \Delta G_A^0 + RT \ln \alpha_A \tag{B}$$

and

$$\Delta G_B = \Delta G_B^0 + RT \ln \alpha_B \tag{C}$$

furthermore, when $\alpha_A = 1$ or $\alpha_B = 1$, ΔG in Equations B and C equals ΔG^0.
At equilibrium,

$$\Delta G_{products} = \Delta G_{reactants} \tag{D}$$

---

and by substituting Equations B and C into Equation D,

$$\Delta G_B^0 + RT \ln \alpha_B = \Delta G_A^0 + RT \ln \alpha_A \qquad \text{(E)}$$

Collecting terms,

$$\Delta G_B^0 - \Delta G_A^0 = -RT \ln \alpha_B + RT \ln \alpha_A \qquad \text{(F)}$$

Setting

$$\Delta G_r^0 = (\Delta G_B^0 - \Delta G_A^0) \qquad \text{(G)}$$

where ΔG_r^0 = Gibb's free energy of reaction, since

$$\Delta G_r^0 = -RT \ln K_{eq} \qquad \text{(H)}$$

By substituting Equation H into Equation F and rearranging,

$$RT \ln K_{eq} = RT \ln \alpha_B - RT \ln \alpha_A \qquad \text{(I)}$$

Dividing Equation I by RT and raising both sides to base e gives

$$K_{eq} = \alpha_B / \alpha_A \qquad \text{(J')}$$

and the ΔG of a reaction is related to the K_{eq} by the equation

$$\Delta G_r = \Delta G_r^0 + RT \ln K_{eq} \qquad \text{(J'')}$$

Note, however, that an equilibrium state is truly met only when electrochemical potentials of products and reactants are equal. Therefore,

$$\Delta G_{products} + \psi_{products} \Leftrightarrow \Delta G_{products} + \psi_{products} \qquad \text{(K)}$$

where ψ = electrical potential. Replacing the ΔG terms of Equation K by Equations B and C gives

$$\Delta G_B^0 + RT \ln \alpha_B + \psi_B = \Delta G_A^0 + RT \ln \alpha_A + \psi_A \qquad \text{(L)}$$

and rearranging Equation L,

$$(\Delta G_B^0 - \Delta G_A^0) + (\psi_B - \psi_A) = RT \ln \alpha_A - RT \ln \alpha_B \qquad \text{(M)}$$

Setting

$$\Delta G_r^0 = (\Delta G_B^0 - \Delta G_A^0) \qquad \text{(N)}$$

and considering that

$$\Delta G_r^0 = - RT \ln K_{eq1} \quad \text{or} \quad \Delta G_r^0 = - 1.364 \log K_{eq1} \qquad (O)$$

by substituting Equation O into Equation M and rearranging, gives

$$RT \ln K_{eq1} = RT \ln \alpha_B - RT \ln \alpha_A + (\psi_B - \psi_A) \qquad (P)$$

Dividing Equation P by RT and raising it to base e gives

$$K_{eq1} = \alpha_B/\alpha_A \, e^{\Delta\psi/RT} \qquad (Q)$$

Note, at equilibrium $\Delta\psi$ is set to 0 and

$$K_{eq1} = \alpha_B/\alpha_A \qquad (R)$$

Thus, Equation J is indistinguishable from Equation R:

$$K_{eq1} = K_{eq} = \alpha_B/\alpha_A \qquad (S)$$

5.3 REDOX EQUILIBRIA

To understand redox equilibria, one needs to understand how two half-cells intercon-
nected by a salt bridge (e.g., KCl, an electrical conductor) can attain an electron
equilibrium state. A schematic of a salt bridge is shown in Figure 5.1. In this figure,
half-cell I is composed of a platinum wire as well as ions of Cu^+ and Cu^{2+}. Cell II is
also made of a container, as is cell I, a platinum wire, hydrogen gas (H_2) at 1 atm
pressure, and hydrogen ions (H^+). Cell II is known as the standard hydrogen electrode
(SHE). The entire system is maintained at 25°C and the components making up cell
II are by convention set at activity one ($\Delta G^0 = 0$). Furthermore, the difference in
electrical potential between the platinum wire and the cell's solution is assumed to be
zero. If switch S is closed and an electron gradient between cells II and I exists,
electrons will move from cell II to cell I and the following reactions will take effect:

$$1/2H_{2(gas)} \Leftrightarrow H^+ + e^- \qquad (5.4)$$

$$Cu^{2+} + e^- \Leftrightarrow Cu^+ \qquad (5.5)$$

If switch S is opened, an electrical potential difference between the two cells would
be recorded by the voltmeter, assuming there is an electron gradient between the two
cells. Considering that cell II has its electrical potential set at zero, the voltmeter's
measurement reflects the electrical potential of cell I. Since electrons flow from high-
to low-electron potential, upon opening switch S, the voltmeter may record a positive
or a negative electrical potential (Eh). By convention, Eh is positive if electron activity
in cell I is less than in cell II, and Eh is negative if electron activity in cell I is greater

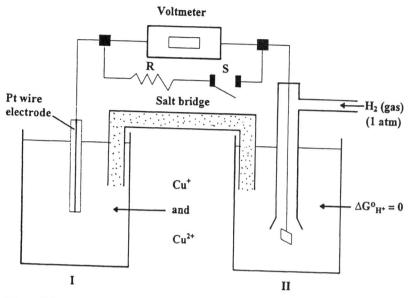

Figure 5.1. Schematic of a redox cell discussed in text (adapted from Drever, 1982).

than in cell II. Using this concept, one may measure the electrical potential of soil in comparison to SHE.

5.3.1   Redox as Eh and the Standard Hydrogen Electrode (SHE)

A reaction demonstrating the conversion of an oxidized species to a reduced species without forming any precipitate is shown below:

$$Fe^{3+} + e^- \Leftrightarrow Fe^{2+} \tag{5.6}$$

Reaction 5.6 is known as a half-cell reaction; to move from left to right, there should be an electron donor. This electron donor is the hydrogen half-cell reaction:

$$1/2H_{2(gas)} \Leftrightarrow H^+ + e^- \tag{5.7}$$

By convention $\alpha_{H_{2(gas)}}$, α_{H^+}, and α_{e^-} in the SHE are set to 1 ($\Delta G^0 = 0$) and the overall reaction is

$$Fe^{3+} + 1/2H_{2(gas)} \Leftrightarrow Fe^{2+} + H^+ \tag{5.8}$$

This reaction's redox level can be expressed in volts of Eh, which is obtained by

$$Eh = -\Delta G/nF \tag{5.9}$$

where F is Faraday's constant (23.06 kcal per volt gram equivalent) and n is the number of electrons (representing chemical equivalents) involved in the reaction. From Reactions J″ (in the boxed section of this chapter) and 5.9,

$$-\Delta G/nF = -\Delta G^0/nF - RT/nF \ \{\ln[(Fe^{2+})/(Fe^{3+})]\} \tag{5.10}$$

where T is temperature and R is the universal gas constant (1.987×10^{-3} kcal deg^{-1} mol^{-1}) and

$$Eh = E^0 - 2.303 \ RT/nF\{\log[(Fe^{2+})/(Fe^{3+})]\} \tag{5.11}$$

or

$$Eh = E^0 + 0.059 \ \{\log[(Fe^{3+})/(Fe^{2+})]\} \tag{5.12}$$

where E^0 is the standard electrode potential for Reaction 5.6. Equation 5.12 can be generalized for any redox reaction at 25°C:

$$Eh = E^0 + \{0.059/n\} \ \{\log(\text{activity of oxidized species/activity of reduced species})\} \tag{5.13}$$

Considering that

$$\Delta G_r^0 = \Delta G_{Fe^{2+}}^0 - \Delta G_{Fe^{3+}}^0 = -20.30 - (-2.52) = -17.78 \ \text{kcal mol}^{-1} \tag{5.14}$$

and from Equation 5.9,

$$E^0 = 17.78/\{(1)(23.06)\} = 0.77 \ \text{V} \tag{5.15}$$

or

$$Eh = 0.77_{Fe^{3+}Fe^{2+}} + 0.059\{\log[(Fe^{3+})/(Fe^{2+})]\} \tag{5.16}$$

This reaction shows that Eh is related to the standard potential ($E^0 = 0.77$ V) and the proportionality of Fe^{3+} to Fe^{2+} in the solution phase. A number of E^0 values representing various reactions in soils are given in Tables 5.2 and 5.3. Note that Equation 5.16 is analogous to the Henderson–Hasselbalch equation. A plot of Eh versus ($Fe^{3+}/Fe^{3+} + Fe^{2+}$) would produce a sigmoidal line with midpoint E^0 (Fig. 5.2) (recall that the Henderson–Hasselbalch equation gives the pK_a at the titration midpoint; see Chapter 1). To the left of E^0 (midpoint in the x axis; $E^0 = 0.77$ V) the reduced species (e.g., Fe^{2+}) predominates, whereas to the right of E^0 (midpoint in the x axis), the oxidized species (e.g., Fe^{3+}) predominates.

5.3.2 Redox as pe and the Standard Hydrogen Electrode (SHE)

In the case of the redox reaction

$$Fe^{3+} + e^- \Leftrightarrow Fe^{2+} \tag{5.17}$$

TABLE 5.2.   Reduction Potentials of Selected Half-Cell Reactions in Soil–Water Systems at 25°C

Reaction	E^o (V)
$F_2 + 2e^- = 2F^-$	+2.87
$Cl_2 + 2e^- = 2Cl^-$	+1.36
$NO_3^- + 6H^+ + 5e^- = 1/2N_2 + 3H_2O$	+1.26
$O_2 + 4H^+ + 4e^- = 2H_2O$	+1.23
$NO_3^- + 2H^+ + 2e^- = NO_2^- + H_2O$	+0.85
$Fe^{3+} + e^- = Fe^{2+}$	+0.77
$SO_4^- + 10H^+ + 8e^- = H_2S + 4H_2O$	+0.31
$CO_2 + 4H^+ + 4e^- = C + 2H_2O$	+0.21
$N_2 + 6H^+ + 6e^- = 2NH_3$	+0.09
$2H^+ + 2e^- = H_2$	+0
$Fe^{2+} + 2e^- = Fe$	−0.44
$Zn^{2+} + 2e^- = Zn$	−0.76
$Al^{3+} + 3e^- = Al$	−1.66
$Mg^{2+} + 2e^- = Mg$	−2.37
$Na^+ + e^- = Na$	−2.71
$Ca^{2+} + 2e^- = Ca$	−2.87
$K^+ + e^- = K$	−2.92
$1/2N_2O + e^- + H^+ = 1/2N_2 + 1/2 H_2$	+1.76
$NO + e^- + H^+ = 1/2N_2O + 1/2H_2O$	+1.58
$1/2NO_2^- + e^- + 3/2H^+ = 1/4N_2O + 3/4H_2O$	+1.39
$1/5NO_3^- + e^- + 6/5H^+ = 1/10N_2 + 3/5H_2O$	+1.24
$NO_2^- + e^- + 2H^+ = NO + H_2O$	+1.17
$1/4NO_3^- + e^- + 5/4H^+ = 1/8N_2O + 5/8H_2O$	+18.9
$1/6NO_2^- + e^- + 4/3H^+ = 1/6NH_4^+ + 1/3H_2O$	+1.12
$1/8NO_3^- + e^- + 5/4H^+ = 1/8NH_4^+ + 3/8H_2O$	+0.88
$1/2NO_3^- + e^- + H^+ = 1/2NO_2^- + 1/2H_2O$	+0.83
$1/6NO_3^- + e^- + 7/6H^+ = 1/6NH_2OH + 1/3H_2O$	+0.67
$1/6N_2 + e^- + 4/3H^+ = 1/3NH_4^+$	+0.27
$1/2O_3 + e + H^+ = 1/2O_2 + 1/2H_2O$	+2.07
$OH + e^- = OH^-$	+1.98
$O_2^- + e^- + 2H^+ = H_2O_2$	+1.92
$1/2H_2O_2 + e^- + H^+ = H_2O$	+1.77
$1/4O_2 + e^- + H^+ = 1/2H_2O$	+1.23
$1/2O_2 + e^- + H^+ = 1/2H_2O_2$	+0.68
$O_2 + e^- = O_2^-$	−0.56

Source: Reproduced from Stumm and Morgan (1981) and Sparks (1995).

TABLE 5.3. Electrode Potentials of Phenol, Acetic Acid, Ethanol, and Glucose

Redox Couple	E^0 (V)
Phenol–CO_2	
$\quad 6CO_2 + 28H^+ + 28e^- = C_6H_5OH + 11H_2O$	0.102
Acetic acid–CO_2	
$\quad 2CO_2 + 8H^+ + 8e = CH_3COOH + 2H_2O$	0.097
Ethanol–CO_2	
$\quad 2CO_2 + 11H^+ + 11e^- = CH_2CH_2OH + 3H_2O$	0.079
Glucose–CO_2	
$\quad 6CO_2 + 24H^+ + 24e^- = C_6H_{12}O_6 + 6H_2O$	−0.014

Source: Reproduced from Stumm and Morgan (1981).

the half-cell of the SHE can be omitted because its $\Delta G^0 = 0$, and

$$K_{eq} = [(Fe^{2+})/(Fe^{3+})(e^-)] \tag{5.18}$$

rearranging

$$1/e^- = K_{eq} [(Fe^{3+})/(Fe^{2+})] \tag{5.19}$$

and taking logarithms on both sides of Equation 5.19,

$$pe = \log K_{eq} + \{\log(Fe^{3+})/(Fe^{2+})\} \tag{5.20a}$$

or

$$pe = -pe^0 + \{\log(Fe^{3+})/(Fe^{2+})\} \tag{5.20b}$$

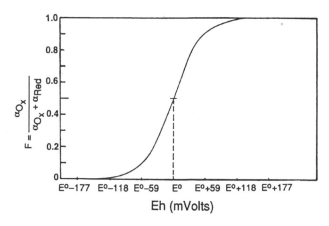

Figure 5.2. Relationship between redox electrical potential, Eh, and relative fraction of oxidized species (adapted from Kokholm, 1981).

where K_{eq} (e^0) is the equilibrium constant of the Fe^{3+} (Reaction 5.17):

$$pe^0 = -\log K_{eq} = -17.78 \text{ kcal mol}^{-1}/1.364 = -13.03 \qquad (5.21)$$

Substituting $\log K_{eq} = 13.03$ into Equation 5.20b gives

$$pe = 13.03 + \log [(Fe^{3+})/(Fe^{2+})] \qquad (5.22)$$

Assuming that $(Fe^{3+})/(Fe^{2+})$ is set to 1, Equation 5.22 reduces to

$$pe = 13.03 \qquad (5.23)$$

TABLE 5.4. Equilibrium Constants of Redox Processes Pertinent in Soil–Water Systems (25°C)

Reaction	pe^0 ($e \log K_{eq}$)	pe^0 (W)[a]
(1) $1/4O_2(g) + H^+ + e^- = 1/2 H_2O$	+20.75	+13.75
(2) $1/5NO_3^- + 6/5 H^+ + e^- = 1/10N_2(g) + 3/5H_2O$	+21.05	+12.65
(3) $1/2MnO_2(s) + 1/2HCO_3^- (10^{-3}) + 3/2H^+ + e^- = 1/2MnCO_3(s) + H_2O$	—	+8.9[b]
(4) $1/2NO_3^- + H^+ + e^- = 1/2NO_2^- + 1/2H_2O$	+14.15	+7.15
(5) $1/2NO_3^- + 5/4H^+ + e^- = 1/8NH_4^+ + 3/8H_2O$	+14.90	+6.15
(6) $1/6NO_2^- + 4/3H^+ + e^- = 1/6NH_4^+ + 1/3H_2O$	+15.14	+5.82
(7) $1/2CH_3OH + H^+ + e^- = 1/2CH_4(g) + 1/2H_2O$	+9.88	+2.88
(8) $1/4CH_2O + H^+ + e^- = 1/4CH_4(g) + 1/4H_2O$	+6.94	-0.06
(9) $FeOOH(s) + HCO_3^- (10^{-3}) + 2H^+ + e^- = FeCO_3(s) + 2H_2O$	—	-0.8[b]
(10) $1/2CH_2O + H^+ + e^- = 1/2CH_3OH$	+3.99	-3.01
(11) $+1/6SO_4^{2-} + 4/3H^+ + e^- = 1/6S(s) + 2/3H_2O$	+6.03	-3.30
(12) $1/8SO_4^{2-} + 5/4H^+ + e^- = 1/8H_2S(g) + 1/2H_2O$	+5.25	-3.50
(13) $1/8SO_4^{2-} + 9/8H^+ + e^- = 1/8HS^- + 1/2H_2O$	+4.25	-3.75
(14) $1/2S(s) + H^+ + e^- = 1/2H_2S(g)$	+2.89	-4.11
(15) $1/8CO_2(g) + H^+ + e^- = 1/8CH_4(g) + 1/4H_2O$	+2.87	-4.13
(16) $1/6N_2(g) + 4/3H^+ + e^- = 1/3NH_4^+$	+4.68	-4.68
(17) $1/2(NADP^+) + 1/2H^+ + e^- = 1/2(NADPH)$	-2.0	-5.5
(18) $H^+ + e^- = 1/2H_2(g)$	0.0	-7.00
(19) Oxidized ferredoxin + e^- = reduced ferredoxin	-7.1	-7.1
(20) $1/4CO_2(g) + H^+ + e^- = 1/24$ (glucose) + $1/4H_2O$	-0.20	-7.20
(21) $1/2HCOO^- + 3/2H^+ + e^- = 1/2CH_2O + 1/2H_2O$	+2.82	-7.68
(22) $1/4CO_2(g) + H^+ + e^- = 1/4CH_2O + 1/4H_2O$	-1.20	-8.20
(23) $1/2CO_2(g) + 1/2H^+ + e^- = 1/2HCOO^-$	-4.83	-8.33

Source: Reproduced from Stumm and Morgan (1981).

[a]Values for pe^0(W) apply to the electron activity for unit activities of oxidant and reductant in neutral water, that is, at pH = 7.0 for 25°C.

[b]These data correspond to $(HCO_3^-) = 10^{-3} M$ rather than unity and so are not exactly pe^0(W); they represent typical aquatic conditions more nearly than pe^0(W) values do.

Equation 5.23 demonstrates that as long as Fe^{3+} and Fe^{2+} are present in the system at equal activities, pe is fixed at 13.03. If electrons are added to the system, Fe^{3+} will become Fe^{2+}, and if electrons are removed from the system, Fe^{2+} will become Fe^{3+}. Equation 5.22 can be generalized for any redox reaction at 25°C,

$$pe = (1/n)K_{eq} + (1/n)\{log(\text{activity of oxidized species})/(\text{activity of reduced species})\}$$
(5.24)

TABLE 5.5. Equilibrium Constants for a Few Redox Reactions (25°C)

Reaction	Log K_{eq}	E^0 (V)
$Na^+ + e^- = Na(s)$	−47	−2.71
$Zn^{2+} + 2e^- = Zn(s)$	−26	−0.76
$Fe^{2+} + 2e^- = Fe(s)$	−14.9	−0.44
$Co^{2+} + 2e^- = Co(s)$	−9.5	−0.28
$V^{3+} + e^- = V^{2+}$	−4.3	−0.26
$2H^+ + 2e^- = H_2(g)$	0.0	0.00
$S(s) + 2H^+ + 2e^- = H_2S$	+4.8	+0.14
$Cu^{2+} + e^- = Cu^+$	+2.7	+0.16
$AgCl(s) + e^- = Ag(s) + Cl^-$	+ 3.7	+0.22
$Cu^{2+} + 2e^- = Cu(s)$	+11.4	+0.34
$Cu^+ + e^- = Cu(s)$	+8.8	+0.52
$Fe^{3+} + e^- = Fe^{2+}$	+13.0	+0.77
$Ag^+ + e^- = Ag(s)$	+13.5	+0.80
$Fe(OH)_3(s) + 3H^+ + e^- = Fe^{2+} + 3H_2O$	+17.1	+1.01
$IO_3^- + 6H^+ + 5e^- = 1/2I_2(s) + 3H_2O$	+104	+1.23
$MnO_2(s) + 4H^+ + 2e^- = Mn^{2+} + 2H_2O$	+43.6	+1.29
$Cl_2(g) + 2e^- = 2Cl^-$	+46	+1.36
$Co^{3+} + e^- = Co^{2+}$	+31	+1.82
$1/2NiO_2 + e^- + 2H^+ = 1/2Ni^{2+} + H_2O$	29.8	+1.76
$PuO_2^+ + e^- = PuO_2$	26.0	+1.52
$1/2PbO_2 + e^- + 2H^- = 1/2Pb^{2+} + H_2O$	24.8	+1.46
$PuO_2 + e^- + 4H^+ = Pu^{3+} + 2H_2O$	9.9	+0.58
$1/3HCrO_4^- + e^- + 4/3H^+ = 1/3Cr(OH)_3 + 1/3H_2O$	18.9	+1.12
$1/2AsO_4^{3-} + e^- + 2H^+ = 1/2AsO_2^- + H_2O$	16.5	+0.97
$Hg^{2+} + e^- = 1/2Hg_2^{2+}$	15.4	+0.91
$1/2MoO_4^{2-} + e^- + 2H^+ = 1/2MoO_2 + H_2O$	15.0	+0.89
$1/2SeO_4^{2-} + e^- + H^+ = 1/2SeO_3^{2-} + 1/2H_2O$	14.9	+0.88
$1/4SeO_3^{2-} + e^- + 3/2H^+ = 1/4Se + 3/4H_2O$	14.8	+0.87
$1/6SeO_3^{2-} + 4/3H^+ = 1/6H_2Se + 1/2H_2O$	7.62	+0.45
$1/2VO_2^+ + e^- + 1/2H_3O^+ = 1/2V(OH)_3$	6.9	+0.41
$PuO_2 + e^- + 3H^+ = PuOH^2 + H_2O$	2.9	+0.17

Source: Reproduced from Stumm and Morgan (1981) and Sparks (1995).

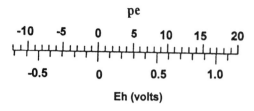

Figure 5.3. Relationship between pe and Eh (adapted from Drever, 1982).

A number of half-cell redox reactions pertinent in soil–water systems are given in Tables 5.4 and 5.5. Note that at 25°C, pe = 16.9 Eh and Eh = 0.059 pe. Graphically, the relationship between pe and Eh is shown in Figure 5.3.

5.3.3 Redox as Eh in the Presence of Solid Phases

The discussion in Sections 5.3.1 and 5.3.2 was based on the electrochemical cell shown in Figure 5.1, which included a platinum and a hydrogen electrode. This electrode pair gave us an understanding of pe and Eh, but one should keep in mind that the electrode potential in cell I is independent of the SHE and platinum electrode. The redox potential is determined only by the redox couple under consideration (e.g., Fe^{3+}, Fe^{2+}) when both members are present in solution or in contact with solution. In natural soil–water systems, more often than not, several redox pairs maybe present (e.g., Fe^{3+}, Fe^{2+}; SO_4^{2-}, H_2S; NO_3^-, NO_2^-). In such situations, each pair will produce its own pe or Eh, but the three values may or may not be the same. When a chemical equilibrium between the three redox pairs is met, the three pe or Eh values will be the same (Drever, 1982). In nature, redox couples may also involve solution and solid phases, for example, Fe^{2+}, $Fe(II)(OH)_3$; Mn^{2+}, $Mn(III)(OOH)$; $Fe(II)CO_3^\circ$, $Fe(II)(OH)_3$; their role in determining redox potential is demonstrated below.

Consider a reaction where two solid phases are involved. Thus, in addition to electron transfer (redox process), one of the solid phases decomposes while the other one forms. For example,

$$FeCO_{3(s)} + 3H_2O \Leftrightarrow 3\,Fe(OH)_{3(s)} + HCO_3^- + 2H^+ + e^- \qquad (5.25)$$

Reaction 5.25 is a half-cell reaction. In order for this half-cell reaction to move from left to right, there should be an electron sink. In soils, a most common electron sink is O_2, although other soil minerals, as we shall see later in this chapter, can act as electron sinks (see also Chapter 6). The equilibrium expression (K_{eq}) for Reaction 5.25 is

$$K_{eq} = \{[Fe(OH)_{3(s)}]\,(HCO_3^-)\,(H^+)^2\,(e^-)\}/\{(FeCO_{3(s)})\,(H_2O)^3\} \qquad (5.26)$$

In this case, H^+ activity cannot be set to 1 because the two solids, $Fe(OH)_{3(s)}$ and $FeCO_{3(s)}$, cannot persist at such low pH. Unit activity, however, is assigned to all other pure phases, $Fe(OH)_{3(s)}$, HCO_3^-, $FeCO_{3(s)}$, and H_2O. The above reaction's redox level can be expressed in units of volts of Eh obtained from the relationship

$$Eh = -\Delta G/nF \qquad (5.27)$$

where F is Faraday's constant (23.06 kcal per volt gram equivalent) and n is the number of electrons (representing chemical equivalents) involved in the reaction. From the classical thermodynamic relationships (outlined in Reactions A to J'' in the boxed section) and Reaction 5.26 [considering that activity of the pure phases ($Fe(OH)_{2(s)}$, $Fe(OH)_{3(s)}$), H_2O, and HCO_3^- is set to 1],

$$-\Delta G/nF = -\Delta G^0/nF - RT/nF\{\ln(H^+)^2\} \qquad (5.28)$$

where R is the universal gas constant (1.987×10^{-3} kcal deg^{-1} mol^{-1}). If ΔG_r^0 (Garrels and Christ, 1965) is as follows:

$$FeCO_{3(s)} = -161.06 \text{ kcal mol}^{-1}$$

$$H_2O = -56.69 \text{ kcal mol}^{-1}$$

$$Fe(OH)_{3(s)} = -166.0 \text{ kcal mol}^{-1}$$

$$HCO_3^- = -140.31 \text{ kcal mol}^{-1}$$

and

$$\Delta G_{reaction}^0 = \Delta G_{products}^0 - \Delta G_{reactants}^0 \qquad (5.29)$$

then

$$\Delta G_r^0 = -166.0 + (-140.31) - (-161.06) - (3x - 56.69) = 24.82 \text{ kcal mol}^{-1} \quad (5.30)$$

For natural systems, a more accurate representation is $HCO_3^- = 10^{-3}$ M. From Equation 5.27, $E^0 = -24.82$ kcal mol^{-1}/$\{(1)(23.06)\} = -1.08$ V. By replacing the $-\Delta G/nF$ terms of Equation 5.28 with Eh terms (see Equation 5.27),

$$Eh = E^0 - RT/nF\{\ln(H^+)^2\} \qquad (5.31)$$

and

$$Eh = -1.08 - RT/nF\{\ln(H^+)^2\} \qquad (5.32)$$

Equation 5.32 can be converted to base 10:

$$Eh = -1.08 - 2.303\ RT/nF\{\log(H^+)^2\} \qquad (5.33)$$

Using R, standard temperature (25°C), $n = 1$, and Faraday's constant gives

$$Eh = -1.08 - 0.059 \{\log(H^+)^2\} \tag{5.34}$$

or

$$Eh = -1.08 + 0.118 \, pH \tag{5.35}$$

A plot of pH versus Eh will give a straight line with slope 0.118 and y intercept of -1.08. Equation 5.35 shows that as long as $FeCO_{3(s)}$ and $Fe(OH)_{3(s)}$ are present in the system, E^0 is fixed at -1.08 V. If electrons are added to the system, $Fe(OH)_{3(s)}$ will become $FeCO_{3(s)}$, and if electrons are removed from the system, $FeCO_{3(s)}$ will become $Fe(OH)_{3(s)}$.

5.3.4 Redox as pe in the Presence of Solid Phases

For the previous reaction,

$$FeCO_{3(s)} + 3H_2O \Leftrightarrow Fe(OH)_{3(s)} + HCO_3^- + 2H^+ + e^- \tag{5.36}$$

Its equilibrium expression is

$$K_{eq} = \{[Fe(OH)_{3(s)}](HCO_3^-)(H^+)^2(e^-)\}/\{(FeCO_{3(s)})(H_2O)^3 \tag{5.37}$$

If the activity of the pure phases $[Fe(OH)_{3(s)}, FeCO_{3(s)}]$, H_2O as well as HCO_3^- are set to 1, taking negative logarithms on both sides of Equation 5.37 and rearranging gives

$$pe = pK_{eq} - 2 \, pH \tag{5.38}$$

or

$$pe = pe^0 - 2 \, pH \tag{5.39}$$

If ΔG_r^0 (Garrels and Christ, 1965) is as follows:

$$FeCO_{3(s)} = -161.06 \text{ kcal mol}^{-1}$$

$$H_2O = -56.69 \text{ kcal mol}^{-1}$$

$$Fe(OH)_{3(s)} = -166.0 \text{ kcal mol}^{-1}$$

$$HCO_3^- = -140.31 \text{ kcal mol}^{-1}$$

and

$$\Delta G_{reaction}^0 = \Delta G_{products}^0 - \Delta G_{reactants}^0 \tag{5.40}$$

then

$$\Delta G_r^0 - 166.0 + (-140.31) + (0) - (-161.06) - (3x - 56.69) = 24.82 \text{ kcal mol}^{-1}$$
$$(5.41)$$

From Equation O in the boxed section,

$$\log K = -\Delta G_r^0/1.364 = -24.82/1.364 = -18.20 \qquad (5.42)$$

Substituting -18.20 into Equation 5.38 and rearranging,

$$18.20 = \text{pe} + 2 \text{ pH} \qquad (5.43)$$

A number of half-cell redox reactions pertinent in soil–water systems are given in Table 5.4. Equation 5.43 shows that as long as $FeCO_{3(s)}$ and $Fe(OH)_{3(s)}$ are present in the system, pe^0 is fixed at 18.20. If electrons are added to the system, $Fe(OH)_{3(s)}$ will become $FeCO_{3(s)}$, and if electrons are removed from the system, $FeCO_{3(s)}$ will become $Fe(OH)_{3(s)}$.

5.4   STABILITY DIAGRAMS

Commonly, redox processes in nature are quantified through the use of stability diagrams, assuming an equilibrium state, which are bounded by the upper and lower stability limits of water. For the upper stability limit of water, the reaction is

$$2H_2O \Leftrightarrow O_{2(g)} + 4H^+ + 4e^- \qquad (5.44)$$

and

$$K_{eq} = (H_2O)^2/(pO_{2[g]})(H)^4(e^-)^4 \qquad (5.45)$$

where pO_2 is the partial pressure of $O_{2(g)}$. Taking logs on both sides of Equation 5.45, assuming $\alpha_{H_2O} = 1$ and $pO_{2(g)} = 1$, and rearranging gives

$$\text{pe} = - \text{pH} + 1/4 \log K_{eq} \qquad (5.46)$$

and

$$1/4 \log K_{eq} = -\{\Delta G_r^0/[(4)(1.364)]\} = 20.88 \qquad (5.47)$$

and the redox potential (pe) is described by the equation

$$\text{pe} = -\text{pH} + 20.88 \qquad (5.48)$$

Multiplying both sides of Equation 5.48 by 0.059 gives

$$Eh = -0.059 \text{ pH} + 1.23 \qquad (5.49)$$

which describes the upper stability limit of H_2O. A plot of pH vs. pe will produce a straight line with a slope of -1. Furthermore, a number of lines can be produced for various partial pressures of $O_{2(g)}$ by introducing the partial pressure component into Equation 5.47 (Fig. 5.4):

$$pe = - pH + 20.88 + (1/4)\log (pO_{2[g]}) \qquad (5.50)$$

Similarly, for the lower stability limit of water described by the reaction,

$$2H^+ + 2e^- \Leftrightarrow H_{2(g)} \qquad (5.51)$$

and

$$K_{eq} = (^PH_{2[g]})/(H^+)^2(e^-)^2 \qquad (5.52)$$

where $pH_{2(g)}$ is the partial pressure of $H_{2(g)}$. Taking logs on both sides of Equation 5.52, assuming $pH_{2(g)} = 1$, and rearranging gives

$$pe = -pH + 1/2 \log K_{eq} - 1/2 \log pH_{2(g)} \qquad (5.53)$$

since $G^0 = 0$, $K_{eq} = 0$, and

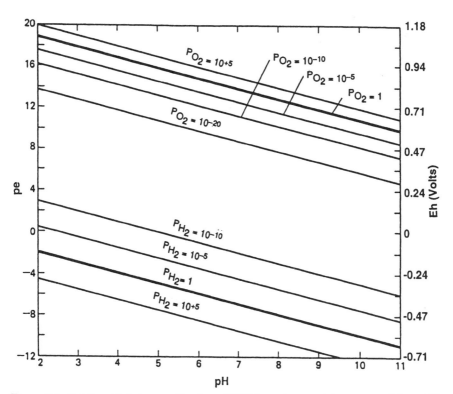

Figure 5.4. Relationship between pH and pe at 25°C for various partial pressures of O_2 and H_2 (adapted from Drever, 1982, with permission).

$$pe = -pH - 1/2 \log pH_{2(g)} \qquad (5.54)$$

Multiplying both sides of Equation 5.54 by 0.059 gives

$$Eh = -0.059 \, pH - 0.0295 \log pH_{2(g)} \qquad (5.55)$$

which describes the lower stability limit of water. A plot of pH versus pe will give a straight line with a slope of –1. Lines with constant $pH_{2(g)}$ will be parallel, as shown in Figure 5.4. When $pH_{2(g)} = 1$, pe = –pH.

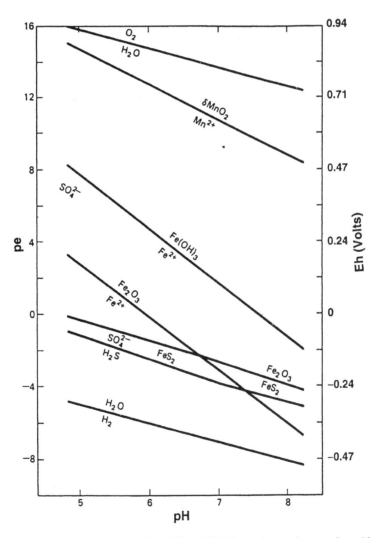

Figure 5.5. Relationship between pH and Eh at 25°C for various redox couples with solute activity = 10^{-6} M (adapted from Drever, 1982, with permission).

Using these approaches, pe–pH stability diagrams have been generated for a number of redox couples by various authors and are presented in Figures 5.5–5.7. For any particular redox couple, pe or Eh values above the stability line denote persistence of the oxidized species, whereas pe or Eh values below the stability line denote persistence of the reduced species. Furthermore, the relative positions, on the pe or Eh scale, of the redox couples signify the relative stability of the reduced or oxidized species. For example, in the pe range of 16 to 9 and corresponding pH range (4.5 to 8.5) (Fig. 5.5) δ–MnO_2 and Mn^{2+} will persist in the environment, but in the case of $Fe(OH)_3/Fe^{2+}$ or SO_4/H_2S in the same pe-pH range, only $Fe(OH)_3$ and SO_4 will persist in the environment. Furthermore, because the MnO_2 redox plateau is above that of NH_4^+ (compare Figs. 5.5 and 5.6 with respect to MnO_2/Mn^{2+} and NO_3^-/NH_4^+ pe range) and Fe^{2+} (Fig. 5.5), it reveals that MnO_2 has the potential to oxidize NH_4 to NO_3 and Fe^{2+} to Fe^{3+}. In the process, Mn^{2+} is produced.

In a similar manner, one may also derive equations describing complex pe–pH stability diagrams for Fe–O–H_2O at 25°C and 1 atm pressure involving gas, solution, and solid phases $H_{2(g)}$, $O_{2(g)}$, H_2O, Fe^{2+}, Fe^{3+}, Fe_2O_3, and Fe_3O_4 (Drever, 1982). In this case, stability lines can be drawn using the equations describing upper and lower boundary limits of H_2O (referred to as step 1) as demonstrated above (Fig. 5.8).

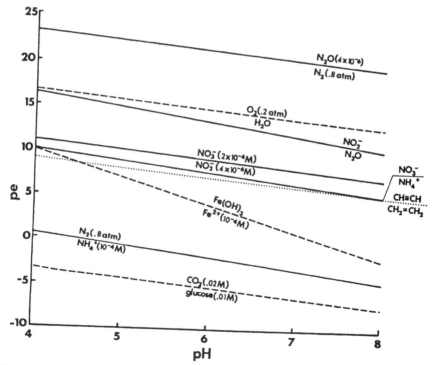

Figure 5.6. Relationship between pH and Eh at 25°C for various redox couples (from Bartlett, 1981, with permission).

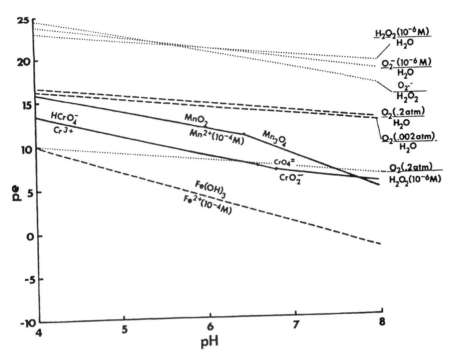

Figure 5.7. Relationship between pH and Eh at 25°C for various redox couples (from Bartlett, 1981, with permission).

Step 2 would be to derive the equation describing the boundary between Fe_2O_3 and Fe_3O_4:

$$3Fe_2O_3 + 2e^- + 2H^+ \Leftrightarrow 2Fe_3O_4 + H_2O \qquad (5.56)$$

and

$$pe = 1/2 \log K_{eq} - pH \qquad (5.57a)$$

Considering that K_{eq} for Reaction 5.56 is $10^{-5.41}$ (Drever, 1982), Equation 5.57 becomes

$$pe = 2.70 - pH \qquad (5.57b)$$

The boundary will be a straight line with a slope of -1 and a y intercept of $1/2 \log K_{eq} = 2.70$ (Fig. 5.8). Multiplying both sides of Equation 5.57 by 0.095 produces an equation in terms of Eh (V).

Step 3 would be to derive the equation describing the boundary between Fe^{3+} and Fe_2O_3:

$$Fe_2O_3 + 6H^+ \Leftrightarrow 2Fe^{3+} + 3H_2O \qquad (5.58)$$

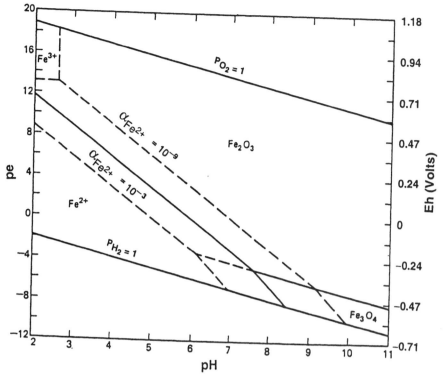

Figure 5.8. Relationship between pH and pe for $Fe-O-H_2O$ at 25°C. The solid lines are boundaries for Fe^{2+} activity set at 10^{-6}, whereas the $Fe_2O_3-Fe^{3+}$ boundary for Fe^{3+} activity is set at 10^{-6} below pH 2 (adapted from Drever, 1982, with permission).

and

$$\log K_{eq} = 2 \log(Fe^{3+}) + 6 \text{ pH} \tag{5.59}$$

where $\log K_{eq} = -2.23$ (Drever, 1982) and Fe^{3+} is assumed to be 10^{-3} or 10^{-9}. Since pe is not involved in this equation, the boundary plots as a vertical line on the pe–pH diagram. Multiplying both sides of Equation 5.59 by 0.095 produces an equation in terms of Eh (V).

Step 4 would be to derive the equation describing the boundary between Fe^{3+} and Fe^{2+}:

$$Fe^{3+} + e^- \Leftrightarrow Fe^{2+} \tag{5.60}$$

and

$$pe = 13.04 - \log[(Fe^{2+})/(Fe^{3+})] \tag{5.61}$$

In this case $(Fe^{2+})/(Fe^{3+})$ is set to 1 and, therefore, the boundary is a horizontal line (Fig. 5.8). Multiplying both sides of Equation 5.61 by 0.095 produces an equation in terms of Eh (V).

Step 5 would be to derive the equation describing the boundary between Fe_2O_3 and Fe^{2+}:

$$Fe_2O_3 + 2e^- + 6H^+ \Leftrightarrow 2Fe^{2+} + 3H_2O \tag{5.62}$$

and

$$pe = 1/2 \log K_{eq} - \log (Fe^{2+}) - 3 \text{ pH} \tag{5.63}$$

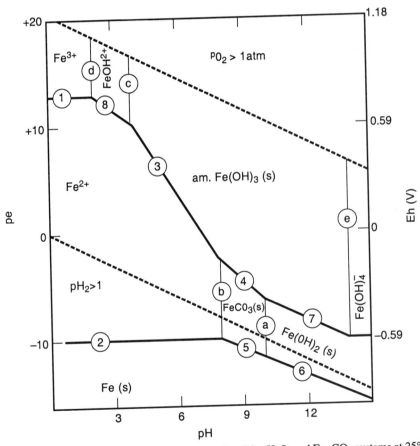

Figure 5.9. Relationship between pH and pe for Fe, CO_2, H_2O, and Fe–CO_2 systems at 25°C. Solid phases considered were: amorphous $Fe(OH)_3$, $FeCO_3$ (siderite), $Fe(OH)_2$, Fe, $C_t = 10^{-3}$ M, [Fe] = 10^{-5} M. The equations used to construct the diagram are given in Table 5.6 (from Stumm and Morgan, 1981, with permission).

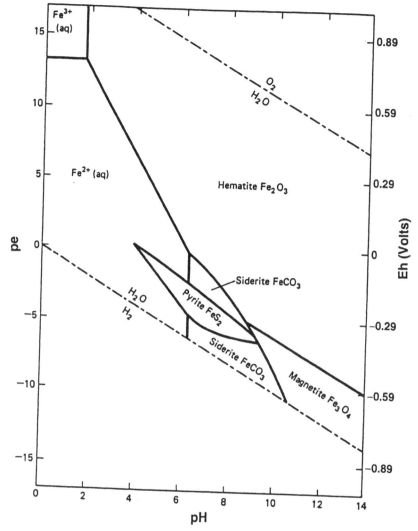

Figure 5.10. Relationship between pH and pe for Fe–O–H₂O–S–CO₂ at 25°C if $\Sigma S = 10^{-6}$ and $\Sigma CO_2 = 10°C$ (adapted from Drever, 1982, with permission).

Here Fe^{2+} is assumed to be 10^{-3} or 10^{-9} M. The slope of the boundary line will be –3 and when $Fe^{2+} = 10^{-9}$ M it passes through the intersection of the Fe_2O_3–Fe^{3+} and Fe^{3+}–Fe^{2+} boundaries; hence, there is no need to calculate K_{eq} (Fig. 5.8). Multiplying both sides of Equation 5.63 by 0.095 produces an equation in terms of Eh (V).

Step 6 would be to derive the equation describing the boundary between Fe_3O_4 and Fe^{2+}:

$$Fe_3O_4 + 2e^- + 8H^+ \Leftrightarrow 3Fe^{2+} + 4H_2O \qquad (5.64)$$

and

$$pe = 1/2 \log K_{eq} - 3/2 \log (Fe^{2+}) - 4\ pH \qquad (5.65)$$

The line is straight with a slope of -4 and passing through the intersection of the Fe_2O_3–Fe^{2+} and Fe_2O_3–Fe_3O_4 boundaries; hence, there is no need to calculate K_{eq} (Fig. 5.8.). Multiplying both sides of Equation 5.65 by 0.095 produces an equation in terms of Eh (V).

Similarly, one can add any number of additional solids, such as, $Fe(OH)_3$, $Fe(OH)_2$, and $FeCO_3$. The pe–pH stability diagram including these solid phases is shown in Figure 5.9 and the equations taken into consideration to produce the figure are shown in Table 5.6. Similarly, a stability diagram for a system containing iron as well as sulfur is shown in Figure 5.10. For additional information on these relationships see Garrels and Christ (1965), Lindsay (1979), Bartlett (1981), Blanchar and Marshall (1981), and Lowson (1982). In general, Eh–pH diagrams such as those shown in Figures 5.9 and 5.10 may be used to demonstrate whether a particular soil or geologic site is under $FeCO_3^-$, $Fe(OH)_2^-$, or $Fe(OH)_3$-forming conditions (Fig. 5.9) or under pyrite-forming conditions. For example, in Figure 5.10, Eh–pH values falling within the boundaries of $Fe(OH)_3$ represent acid mine drainage-producing conditions (Evangelou, 1995b). A particular site represented by such Eh–pH values would be under strong oxidative conditions and most iron released would likely precipitate as $Fe(OH)_3$. When Eh–pH

TABLE 5.6. Equations Needed for the Construction of a pe–pH Diagram for the Fe–CO$_2$–H$_2$ System (Fig. 5.9)

Equation	pe Functions	
$Fe^{3+} + e^- = Fe^{2+}$	$pe = 13 + \log\{Fe^{3+}\}/\{Fe^{2+}\}$	(1)
$Fe^{2+} + 2e^- = Fe(s)$	$pe = -6.9 + 1/2 \log\{Fe^{2+}\}$	(2)
$Fe(OH)_3(amorph,s) + 3H^+ + e^- = Fe^{2+} + 3H_2O$	$pe = 16 - \log\{Fe^{2+}\} - 3pH$	(3)
$Fe(OH)_3(amorph,s) + 2H^+ + HCO_3^- + e^- = FeCO_3(s) + 3H_2O$	$pe = 16 - 2\ pH + \log\{HCO_3^-\}$	(4)
$FeCO_3(s) + H^+ + 2e^- = Fe(s) + HCO_3^-$	$pe = -7.0 - 1/2\ pH - 1/2 \log\{HCO_3^-\}$	(5)
$Fe(OH)_2(s) + 2H^+ + 2e^- = Fe(s) + 2H_2O$	$pe = -1.1 - pH$	(6)
$Fe(OH)_3(s) + H^+ + e^- = Fe(OH)_2(s) + H_2O$	$pe = 4.3 - pH$	(7)
$FeOH^{2+} + H^+ + e^- = Fe^{2+} + H_2O$	$pe = 15.2 - pH - \log\{(Fe^{2+}\}/\{FeOH^{2+}\})$	(8)
$FeCO_3(s) + 2H_2O = Fe(OH)_2(s) + H^+ + HCO_3^-$	$pH = 11.9 + \log\{HCO_3^-\}$	(a)
$FeCO_3(s) + H^+ = Fe^{2+} + HCO_3^-$	$pH = 0.2 - \log\{Fe^{2+}\} - \log\{HCO_3^-\}$	(b)
$FeOH^{2+} + 2H_2O = Fe(OH)_3(s) + 2H^+$	$pH = 0.4 - 1/2 \log\{FeOH^{2+}\}$	(c)
$Fe^{3+} + 2H_2O = FeOH^{2+} + H^+$	$pH = 2.2 - \log(\{Fe^{3+}\}/\{FeOH^{2+}\})$	(d)
$Fe(OH)_3(s) + H_2O = Fe(OH)_4^- + H^+$	$pH = 19.2 + \log\{Fe(OH)_4^-\}$	(e)

Source: Reproduced from Stumm and Morgan (1981).

data fall within the boundaries of FeS_2 (Fig. 5.10), pyrite would be under stabilizing conditions.

5.5 HOW DO YOU MEASURE REDOX?

Redox potential (Eh) is commonly measured through the use of a reference electrode (calomel) plus a platinum electrode. Redox measurements are reported in volts or millivolts. Values greater than zero represent the oxidized state, whereas values less than zero represent the reduced state. The range of redox values (Eh) is determined by the points at which water is oxidized or reduced (it is pH dependent). Calibration of a potentiometer to measure the Eh of an unknown (e.g., soil sample) can be made by using reagents of known redox potential. First, attach a platinum (Pt) electrode to the plus terminal (in place of the glass electrode) of a pH meter with a millivolt scale and a saturated calomel electrode to the negative terminal. Second, lower the calomel and Pt electrodes into a pH 4 suspension of quinhydrone in 0.1 M K acid phthalate solution and adjust the potentiometer to read +213 mV. This reading is equivalent to an Eh of 463 mV or pe of 7.85 at 25°C (Bartlett, 1981).

A more extensive calibration procedure involves a multipoint approach using quinhydrone (Q_2H_2), which dissolves congruently, giving equimolar quantities of quinone (Q) and hydroquinone (Q_2H)

$$Q_2H_2 \Leftrightarrow Q + H_2Q \tag{5.66}$$

The half-cell reaction, as shown at the beginning of this chapter, is

$$Q + 2H^+ + 2e^- \Leftrightarrow QH_2$$

$$E^0_{Q,H_2Q} = 0.70 \text{ V} \tag{5.67}$$

The half-cell Nernst expression is

$$Eh = 0.70_{E^0_{Q,H_2Q}} - (0.059/2) \log \{(H_2Q)/(Q)(H^+)^2\} \tag{5.68}$$

Since $(H_2Q) = (Q)$

$$Eh = 0.70 - (0.059/2)\log\{(1)/(H^+)^2\} \tag{5.69}$$

or

$$Eh = 0.70 - 0.059 \text{ pH} \tag{5.70}$$

A plot of pH versus Eh will produce a straight line with a slope of −0.059 and a y intercept of $E^0_{Q,H_2Q} = 0.70$ V. The relationship between pH and Eh is given for a number of reagent-grade couples in Figure 5.11. However, because the Eh instrument also

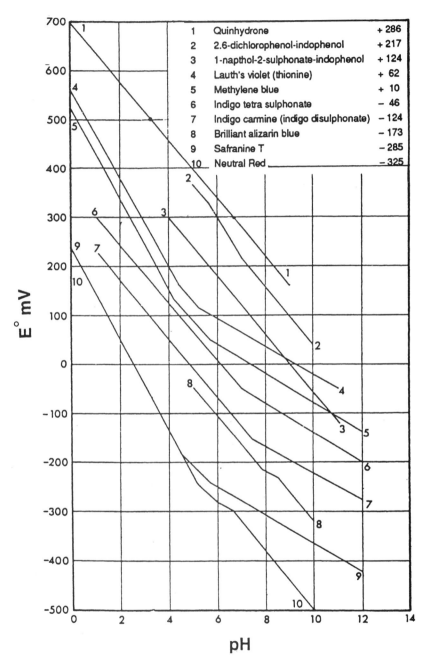

Figure 5.11. Relationship between pH and Eh for various redox couples (adapted from Kokholm, 1981).

contains a reference electrode (E_{ref}^0, calomel), it has to be included in the equation given above:

$$Eh_{meas} = E_{ref}^0 + 0.70 - 0.059 \text{ pH} \qquad (5.71)$$

and, therefore, Eh actual (Eh_{act}) is given by

$$Eh_{act} = Eh_{meas} - E_{ref}^0 \qquad (5.72)$$

where for the calomel electrode at 25°C, $E_{ref}^0 = 244.4$ mV. To generate the calibration equation, one may take quinhydrone (Q_2H_2) and dissolve it in buffers of 9–2 pH. The Eh measurements of the various pH-buffered, redox-coupled (Q,Q_2H) solutions should be in agreement with those of hydroquinone (Q_2H_2) in Figure 5.11. If they are not, potentiometer adjustments should be made to read appropriately.

5.5.1 Redox in Soils

Redox (Eh) measurements are based on sound scientific theory. Yet their accuracy is questionable because (1) the electrodes could react with gases such as O_2 or H_2S and form coatings of oxides or sulfides, (2) Eh measurements in soil most often represent mixed potentials owing to the heterogeneity of soil, (3) the degree and variability of soil moisture influences Eh, and (4) soil spatial variability is inherently very large. Redox reactions in soil could also be affected by specific localized electron-transfer effects and one may find it difficult to assign absolute meaning to a given Eh measurement. Nevertheless, field Eh measurements provide an excellent tool for detecting relative redox changes in the environment as a function of varying conditions (e.g., flooding).

Commonly, soils vary in Eh from approximately 800 mV under well-oxidized conditions, to –500 mV under strongly reducing conditions (Table 5.7; Figs. 5.12 and 5.13). The Eh values in a soil appear to be related to the redox reactions controlling it. A number of such reactions are shown in Table 5.8. It appears that a particular redox reaction with a given K_{eq} produces a given Eh plateau, which is referred to as poise (Fig. 5.14). Poise relates to Eh as buffer capacity relates to pH. It is defined as the potential of a soil to resist Eh changes during addition or removal of electrons. This occurs because Eh is related to ratios of oxidized species to reduced species.

In soils, two major sources/sinks for electrons are O_2 and plant organic residues. Generally speaking, O_2 acts as an electron acceptor in soils, while plant organic residues act as electron donors. In the case of O_2, the half-reaction is described by Equation 5.1, while the half-reaction of organic residue (e.g., sugar) is described by Equation 5.2. Neither of these reactions is reversible. For this reason, under well-aerated conditions, O_2 buffers the soil against reduction and plant organic residue buffers the soil against oxidation (Bartlett, 1981). Soil flooding is a mechanism by which O_2 is excluded from this soil, which allows the soil to be reduced, owing to the presence of organic residues, an electron source. During the process of reduction, however, the various inorganic

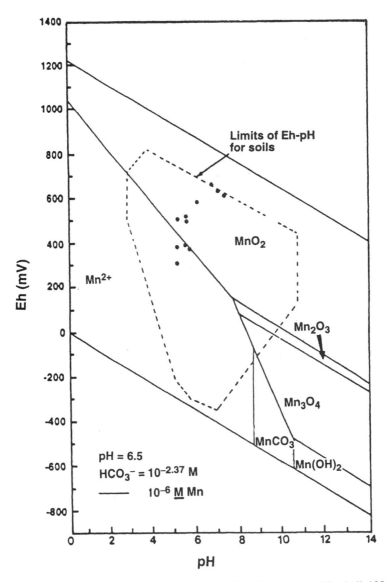

Figure 5.12. Relationship between pH and Eh in soils (from Blanchar and Marshall, 1981, with permission).

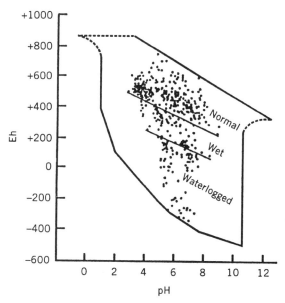

Figure 5.13. Relationship between pH and Eh for soils (from Baas Becking et al., 1960, with permission).

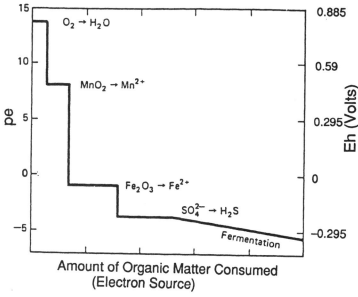

Figure 5.14. Change in pe for various redox couples as a function of organic matter (noted on a relative scale) decomposed (adapted from Drever, 1982, with permission).

TABLE 5.7. Range of Eh Measurements of Soil–Water Systems

Very well-oxidized soil	800 mV
Well-oxidized soil	500 mV
Poorly oxidized soil	100 mV
Much-reduced soil	−200 mV
Extremely reduced soil	−500 mV

mineral redox couples come into play, introducing poise (Fig. 5.14), which resists rapid Eh changes. The reverse is observed when Eh goes up during reoxygenation of soil.

Redox reactions in soils are affected by a number of parameters, including temperature, pH (see Chapter 7), and microbes. Microbes catalyze many redox reactions in soils and use a variety of compounds as electron acceptors or electron donors. For example, aerobic heterotrophic soil bacteria may metabolize readily available organic carbon using NO_3^-, NO_2^-, N_2O, Mn-oxides, Fe-oxides and compounds such as arsenate (AsO_4^{3-}) and selenate (SeO_4^{2-}) as electron acceptors. Similarly, microbes may use reduced compounds or ions as electron donors, for example, NH_4, Mn^{2+}, Fe^{2+}, arsenite (AsO_2^-), and selenite (SeO_3^{2-}).

Table 5.8. Principal Electron Acceptors in Soils, Eh of These Half–Reactions at pH 7, and Measured Potentials of These Reactions

Reaction	Eh pH 7 (V)	Measured Redox Potential in Soils (V)
O_2 Disappearance $1/2\ O_2 + 2e^- + 2H^+ = H_2O$	0.82	0.6 to 0.4
NO_3^- Disappearance $NO_3^- + 2e^- + 2H^+ = NO_2^- + H_2O$	0.54	0.5 to 0.2
Mn^{2+} Formation $MnO_2 + 2e^- + 4H^+ = Mn^{2+} + 2H_2O$	0.4	0.4 to 0.1
Fe^{2+} Formation $FeOOH + e^- + 3H^+ = Fe^{2+} + 2H_2O$	0.17	0.3 to 0.1
HS^- Formation $SO_4^- + 9H^+ + 6e^- = HS^- + 4H_2O$	−0.16	0 to −0.15
H_2 Formation $H^+ + e^- = 1/2H_2$	−0.41	−0.15 to −0.22
CH_4 Formation (example of fermentation) $(CH_2O)_n = n/2\ CO_2 + n/2\ CH_4$	—	−0.15 to −0.22

PROBLEMS AND QUESTIONS

1. Explain the role of frontier electron orbital configuration on environmental chemistry.

2. Explain, using equations, how redox chemistry affects mineral solubility.

3. The equation $Eh = 0.77_{Fe^{3+},Fe^{2+}} + 0.059\{\log[(Fe^{3+})/(Fe^{2+})]\}$ relates Eh (V) to the proportionality of Fe^{3+} to Fe^{2+}. Assuming that the total concentration of iron (Fe^{3+} plus Fe^{2+}) is 10^{-3} M, calculate the concentrations of Fe^{3+} and Fe^{2+}, when activities equal concentrations, for Eh of (a) 0.77 V, (b) 0.40 V, (c) 0.10 V, (d) 1.00 V, and (e) 1.20 V.

4. Consider all the parameters given in problem 3, with the only difference that activities are not equal concentrations (as expected). Calculate the concentrations of Fe^{3+} and Fe^{2+} assuming $\gamma_{Fe^{3+}} = 0.3$ and $\gamma_{Fe^{2+}} = 0.6$ at Eh of (a) 0.77 V, (b) 0.40 V, (c) 0.10 V, (d) 1.00 V, and (e) 1.20 V [hint: see Chapter 2 about the use of γ_i (i = any ion) in equilibria-type equations]. What can you conclude about the role of ion activity in redox systems?

5. Based on the equation $Eh = 0.77_{Fe^{3+},Fe^{2+}} + 0.059 \{\log[(Fe^{3+})/(Fe^{2+})]\}$, calculate Eh at the following Fe^{3+} to Fe^{2+} ratios: (a) 0.05, (b) 0.1, (c) 0.4, (d) 0.7, and (e) 0.95.

6. What does it mean when the stability redox line of a couple is beyond the upper or lower stability limits of water?

7. The redox equation describing the transformation of Na^+ to Na metal is as follows: $Na^+ + e^- \Leftrightarrow Na_{(metal)}$, $E^0 = -2.71$ or $\log K_{eq} = -47$. Produce the equation that relates Eh to $Na^+/Na_{(metal)}$ and explain if one could find $Na_{(metal)}$ in soil.

8. The redox equations describing the $H_2S-SO_4^{2-}$ couple and the MnO_2-Mn^{2+} couple are as follows: $SO_4^{2-} + 10H^+ + 8e^- = H_2S + 4H_2O$ and $MnO_2(s) + 4H^+ + 2e = Mn^{2+} + 2H_2O$. Produce the necessary redox constants by assuming that $Mn^{2+} = 10^{-3}$ M, $SO_4^{2-} = 10^{-5}$ M, and $H_2S = 10^{-2}$ M and plot the corresponding stability lines in terms of Eh–pH. Explain whether MnO_2 could oxidize H_2S to SO_4^{2-} in the absence of O_2.

9. A soil contains the mineral MnO_2 in rather large quantities; someone decides to dispose $FeCl_2$ in this soil. Based on redox reactions and some assumed concentrations of Mn^{2+}, Fe^{2+}, and Fe^{3+} one may find in a normal soil, explain what type of reactions would take place and what type of products these reactions would produce.

10. Propose a scheme using redox chemistry for the effective removal of NO_3 from soil.

6 Pyrite Oxidation Chemistry

6.1 INTRODUCTION

Pyrite is a mineral commonly associated with coal and various metal ores as well as marine deltas, wetlands, and rice fields. Often, pyrite becomes exposed to the atmosphere through various human activities including mining, land development, and construction of highways, tunnels, airports and dams. Pyrite exposure to the atmosphere leads to its oxidation and the production of extremely acidic drainages (as low as pH 2) enriched with Fe, Mn, Al, SO_4, and many other heavy metals (Table 6.1). Worldwide, large sums are spent to control or treat acid drainage (AD). Aside from this direct cost, however, there are environmental costs, for example, diminishing land and water quality. This chapter focuses on the oxidation chemistry and mechanisms of pyrite, an important component of the sulfur cycle. Additionally, much of the information presented in this chapter is covered in more detail by Evangelou (1995b).

TABLE 6.1. Concentrations of Environmentally Important Constituents (mg L^{-1}) in Acid Mine Drainages in the United States and Canada.

Substance	Coal Mine Drainage Throughout the United States	Acid Mine Drainage from Vancouver, Canada	Waste Rock Seepage from Saskatoon, Canada	Metal Mine Drainage from Colorado, U.S.A.	Drinking Water Standard in the United States
Fe	0.6–200	2,300	0.1–9.6	50	0.3
Mn	0.3–12	313	7–92	32	0.05
Cu	0.01–0.17	190	—	1.6	1.0
Zn	0.03–2.2	273	—	10	5.0
Cd	0.01–0.10	2.0	—	0.03	0.01
Pb	0.01–0.40	—	—	0.01	0.05
As	0.002–0.20	12	5.4–9.7	0.02	0.05
pH	3.2–7.9	—	3.94–5.20	2.6	6.5–8.5
SO_4^{2-}	—	20,000	86–1060	2100	250
Reference	1	2	3	4	5

Source: Data taken from Fyson et al., (1994); Rowley et al., (1994); and Wildeman (1991).

Figure 6.1. Electron microscope photograph of (**a**) framboidal pyrite sample and (**b**) a close-up of the framboidal pyrite crystals.

6.2 CHARACTERIZATION

Pyrite varies significantly in grain size and morphology, depending on environmental conditions and formation mechanisms (Evangelou, 1995b). A number of descriptions reported in the literature include pyrite with smooth crystal surfaces of octahedral, cubic, and pyritohedral atomic arrangement, conglomerates with irregular surfaces made up of many cemented particles, and framboids (strawberry-like) in which the cemented crystals form smooth spheres (Ainsworth and Blancher, 1984). An electron microscope photograph of framboid pyrite is shown in Figure 6.1. Framboid and polyframboid pyrites are more reactive than conglomerating pyrite owing to high specific surface and high porosity (Caruccio et al., 1977).

6.3 PYRITE OXIDATION MECHANISMS

Pyrite oxidation includes biological and electrochemical reactions, and varies with pH, pO_2, specific surface, morphology, presence or absence of bacteria and/or clay miner-

Figure 6.1. Continued

als, and hydrology. The chemical reactions governing pyrite oxidation are (Singer and
Stumm, 1970 and references therein):

$$FeS_2 + 7/2O_2 + H_2O \rightarrow Fe^{2+} + 2SO_4^{2-} + 2H^+ \qquad (6.1)$$

$$Fe^{2+} + 1/4O_2 + H^+ \rightarrow Fe^{3+} + 1/2H_2O \qquad (6.2)$$

$$Fe^{3+} + 3H_2O \rightarrow Fe(OH)_3(s) + 3H^+ \qquad (6.3)$$

$$FeS_2 + 7Fe_2(SO_4)_3 + 8H_2O \rightarrow 15FeSO_4 + 8H_2SO_4 \qquad (6.4)$$

Reactions 6.1 and 6.4 show that Fe^{3+} and O_2 are the major pyrite oxidants. Reaction
6.1 shows oxidation of pyrite with O_2 producing Fe^{2+}, which is then oxidized by O_2
to Fe^{3+} (Reaction 6.2). At low pH (< 4.5), Fe^{3+} oxidizes pyrite much more rapidly than
O_2 and more rapidly than O_2 oxidizes Fe^{2+} (Nordstrom, 1982a). For this reason,
Reaction 6.2 is known to be the rate-limiting step in pyrite oxidation (Singer and
Stumm, 1970). Iron-oxidizing bacteria, *Thiobacillus ferrooxidans* (an acidophilic
chemolithotrophic organism that is ubiquitous in geologic environments), can accel-
erate the rate of Fe^{2+} oxidation by a factor of 10^6 (Singer and Stumm, 1970). Reaction
6.3 is a readily reversible dissolution–precipitation reaction, taking place at pH values

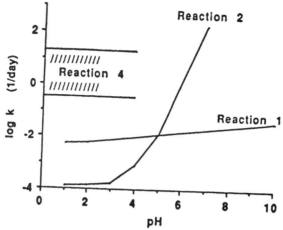

Figure 6.2. Comparison of rate constants as a function of pH for reaction 4, oxidation of pyrite by Fe^{3+}; reaction 2, oxidation of Fe^{2+} by O_2; and reaction 1, oxidation of pyrite by O_2. Reactions 1, 2, and 4 are given in the text as reactions 6.1, 6.2, and 6.4, respectively (from Nordstrom, 1982, with permission).

as low as 3, which serves as source or sink of solution Fe^{3+}, and is a major step in the release of acid to the environment. The data in Figure 6.2 show the relationship between reaction rate constants for the various reactions responsible for pyrite oxidation as a function of pH.

There is probably very little bacterial participation in pyrite oxidation at neutral to alkaline pH, and some researchers suggested that in such environments O_2 is a more important pyrite oxidant than Fe^{3+} (Goldhaber, 1983). Recent findings, however, showed that Fe^{3+} was the preferred pyrite oxidant at circumneutral pH and the major role played by O_2 was to oxidize Fe^{2+} (Moses et al., 1987; Moses and Herman, 1991).

6.4 BACTERIAL PYRITE OXIDATION

Thiobacillus ferrooxidans is an obligate chemoautotrophic and acidophilic organism and is able to oxidize Fe^{2+}, S^0, metal sulfides, and other reduced inorganic sulfur compounds. *Thiobacillus thiooxidans* has also been isolated from acid mine wastes and has been determined that can oxidize both elemental sulfur and sulfide to sulfuric acid ($S^0 + 1.5O_2 + H_2O \rightarrow H_2SO_4$ and $S^{2-} + 2O_2 + 2H^+ \rightarrow H_2SO_4$) (Brierley, 1982; Lundgren and Silver, 1980). However, *T. thiooxidans* cannot oxidize Fe^{2+} (Harrison, 1984).

The mechanisms of pyrite oxidation by bacteria are classified into (a) *direct metabolic reactions* and (b) *indirect metabolic reactions* (Evangelou, 1995b and references therein). Direct metabolic reactions require physical contact between bacteria and pyrite particles, while indirect metabolic reactions do not require physical contact between bacteria and pyrite particles. During indirect metabolic reactions, bacteria oxidize Fe^{2+}, thereby regenerating the Fe^{3+} required for the chemical oxidation of pyrite (Singer and Stumm, 1970).

6.5 ELECTROCHEMISTRY AND GALVANIC EFFECTS

Electrochemical pyrite oxidation is the sum of anodic (electron release) and cathodic (electron consumption) reactions occurring at the surface. The anodic process is a complex collection of oxidation reactions in which the pyrite reacts mainly with water to produce Fe^{3+}, sulfates, and protons,

$$FeS_2(s) + 8H_2O \rightarrow Fe^{3+} + 2SO_4^{2-} + 16H^+ + 15e^- \tag{6.5}$$

or to produce Fe^{2+} and S^0 when the acid strength increases,

$$FeS_2(s) \rightarrow Fe^{2+} + 2S^0 + 2e^- \tag{6.6}$$

The electrons are then transferred to a cathodic site where oxygen is reduced as shown by Reaction 6.7 (Lowson, 1982; Bailey and Peters, 1976):

$$O_2(aq) + 4H^+ + 4e^- \rightarrow 2H_2O \tag{6.7}$$

Elemental sulfur (S^0) can then be oxidized by the reaction

$$S^0 + 3Fe_2(SO_4)_3 4H_2O \rightarrow 6FeSO_4 + 4H_2SO_4 \tag{6.7a}$$

Note that the oxygen in the SO_4^{2-} is derived from water in an anodic reaction (Reaction 6.5).

Physical contact between two different metal–disulfide minerals in an acid–ferric sulfate solution creates a galvanic cell (Evangelou, 1995b and references therein). For example, when two metal–disulfides with different electrical rest potential make physical contact, only the mineral with lower electrical rest potential will be dissolved (Mehta and Murr, 1983). The galvanic reaction of an $CuFeS_2/FeS_2$ cell is described below by an anodic oxidation reaction on the $CuFeS_2$ surface and a cathodic oxygen reduction on the FeS_2 surface (Torma, 1988). The anodic oxidation reaction is

$$CuFeS_2 \rightarrow Cu^{2+} + Fe^{2+} + 2S^0 + 4e^- \tag{6.8}$$

while the cathodic oxygen reduction is

$$O_2 + 4H^+ + 4e^- \rightarrow 2H_2O \tag{6.9}$$

Summing Reactions 6.8 and 6.9,

$$CuFeS_2 + O_2 + 4H^+ \rightarrow Cu^{2+} + Fe^{2+} + 2S^0 + 2H_2O \tag{6.10}$$

Elemental sulfur (S^0) is then oxidized by Reaction 6.7a.

6.6 BACTERIAL OXIDATION OF Fe^{2+}

Abiotic oxidation of Fe^{2+} with O_2 is extremely pH-sensitive. The reaction is rapid above pH 5, and becomes extremely slow in very acidic solution. In the presence of *T. ferrooxidans*, however, Fe^{2+} oxidation is very rapid in acidic conditions. The conditions under which *T. ferrooxidans* activity is optimized is shown in Figures 6.3–6.5.

6.7 SURFACE MECHANISMS

The faster rate of pyrite oxidation by Fe^{3+} than O_2 occurs because Fe^{3+} can bind chemically to the pyrite surface whereas O_2 cannot (Evangelou, 1995b and references therein). The surface-exposed sulfur in the pyrite structure possesses an unshared pair of electrons, which produces a slightly negatively charged pyrite surface that can attract molecules or cations willing to share the pair of electrons. Compounds or ions that could accept this pair of electrons fall into three broad categories: (1) metallic transitional cations (e.g., Fe^{2+}, Fe^{3+}, and Cr^{2+}); (2) biomolecular halogens (e.g., F_2,

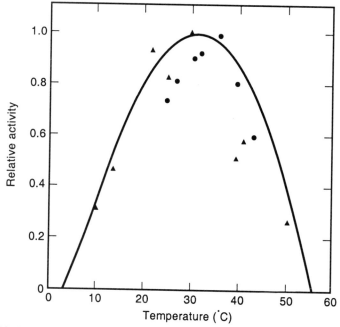

Figure 6.3. Influence of temperature on relative activity of *T. ferrooxidans* (from Jaynes et al., 1984 and references therein).

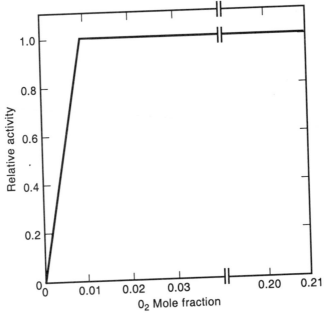

Figure 6.4. Influence of O_2 concentration on relative activity of *T. ferrooxidans* (from Jaynes et al., 1984 and references therein).

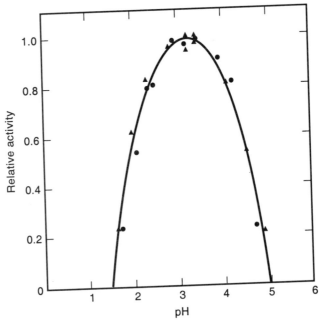

Figure 6.5. Influence of pH on relative activity of *T. ferrooxidans* (from Jaynes et al., 1984 and references therein).

Cl_2, I_2 and Br_2); and (3) singlet or light-activated oxygen ($^1O_2$) and H_2O_2 (Luther, 1990).

Luther (1987) described the oxidation of pyrite by Fe^{3+} in three major steps. In the first step, removal of a water ligand from $Fe(H_2O)_6^{3+}$ takes effect:

$$Fe(H_2O)_6^{3+} \rightarrow Fe(H_2O)_5^{3+} + H_2O \qquad (6.11)$$

In the second step, the $Fe(H_2O)_5^{3+}$ species can bind to the surface of pyrite by forming a persulfido bridge (two metals sharing a common ligand):

$$Fe(II) - \underline{\overline{S}}_A - \underline{\overline{S}}_B - Fe(H_2O)_5^{3+} \qquad (6.12)$$

In the third step, an electron transfer occurs and a radical is formed:

$$Fe(II) - \underline{\overline{S}}_A - \underline{\overline{S}}_B - Fe(H_2O)_5^{3+} \rightarrow$$

$$Fe(II) - \underline{\overline{S}}_A - \underline{\overset{+\bullet}{S}}_B - Fe(H_2O)_5^{2+} \qquad (6.13)$$

The product of Reaction 6.13 reacts with 5 mol $Fe(H_2O)_6^{3+}$ and 3 mol H_2O to transfer five electrons to $Fe(H_2O)_6^{3+}$ and three oxygens from H_2O to the pyrite surface (Taylor et al., 1984a,b). In the process, $6H^+$ and a $S_2O_3^{2-}$ are produced as follows:

$$Fe(II) - \underline{\overline{S}}_A - \underline{\overset{+\bullet}{S}}_B + 5Fe(H_2O)_6^{3+} + H_2O \rightarrow$$

$$Fe(II) \text{-} \overline{\underline{S_A}} \text{-} S_B \overset{\displaystyle |\overline{O}|}{\underset{\displaystyle |\overline{O}|}{|}} \text{-} \overline{O} \ + 5Fe(H_2O)_6^{2+} + 6H^+ \qquad (6.14)$$

Product detachment from the pyrite surface produces iron thiosulfate (FeS_2O_3) (for details, see Evangelou, 1995b). Summarizing Reactions 6.13 and 6.14 gives

$$FeS_2 + 6Fe(H_2O)_6^{3+} + 3H_2O \rightarrow Fe^{2+} + S_2O_3^- + 6Fe(H_2O)_6^{2+} + 6H^+ \qquad (6.15)$$

In the presence of excess Fe^{3+}, $S_2O_3^-$ is rapidly transformed to SO_4^{2-} according to Reaction 6.16:

$$5H_2O + S_2O_3^{2-} + 8Fe^{3+} \rightarrow 8Fe^{2+} + 10H^+ + 2SO_4^{2-} \qquad (6.16)$$

Moses et al. (1987) showed that SO_3^{2-}, $S_2O_3^{2-}$, and $S_nO_6^{2-}$ are produced when the oxidant is O_2. However, when the oxidant is Fe^{3+}, sulfoxy anions (SO_3^{2-}, $S_2O_3^{2-}$, and $S_nO_6^{2-}$) are rapidly oxidized by Fe^{3+} to SO_4.

6.8   CARBONATE ROLE ON PYRITE OXIDATION

Nicholson et al. (1988, 1990) reported that pyrite oxidation kinetics in a bicarbonate buffered system at pH values between 7.5 and 6.5 initially increased, reached a maximum at about 400 hrs, and then decreased to a final, relatively constant, low value (Fig. 6.6). They found that pyrite particles, when oxidized under alkaline conditions, were coated with a ferric oxide coating. They concluded that oxide accumulation on the pyrite surface resulted in a significant reduction in oxidation rate over time.

Hood (1991) reported that the rate of pyrite oxidation under abiotic conditions increased with increasing carbonate concentration. She concluded that the cause for the observed increase in pyrite oxidation was formation of a pyrite–surface Fe(II)–CO$_3$ complex which facilitated electron transfer to O$_2$ and consequently rapid oxidation of ferrous iron. Evangelou and Huang (1994) and Evangelou and Zhang (1995) showed that pyrite exposure to atmospheric air leads to formation of pyrite–surface Fe(II)–CO$_3$ complexes which may promote pyrite oxidation by promoting electron transfer, or may act as precursors to pyrite iron–oxide coating formation (Evangelou, 1995b). Millero and Izaguirre (1989) reported that Fe^{2+} oxidation is promoted in the presence of HCO$_3^-$/CO$_3^{2-}$ owing to Fe^{2+}–carbonate complex formation. Luther et al. (1992) pointed out that Fe^{2+} complexation by any ligands containing oxygen as the ligating atom promotes Fe^{2+} oxidation owing to the potential increase in frontier molecular-orbital electron density. The latter increases Fe^{2+} basicity and, consequently, Fe^{2+} behaves as a stronger electron donor.

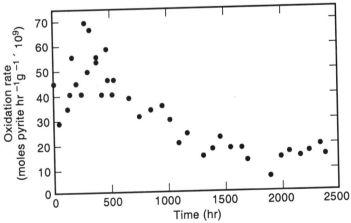

Figure 6.6. Variation in rate of pyrite oxidation in alkaline solutions with time (from Nicholson et al., 1990, with permission).

6.9   Mn– AND Fe–OXIDES

Ferric iron, as mentioned above, is the major oxidant of pyrite regardless of pH, which implies that any reactions producing Fe^{3+} will indirectly result in the oxidation of pyrite. Asghar and Kanehiron (1981) observed that Fe^{2+} can oxidize in the presence of manganese oxides:

$$MnO_2 + 4H^+ + 2Fe^{2+} \Leftrightarrow Mn^{2+} + 2H_2O + 2Fe^{3+} \qquad (6.17)$$

Luther et al. (1992) indicated that dissolution of ferric iron minerals such as goethite (FeOOH) leads to aqueous Fe^{3+} complexes:

$$FeOOH + 3H^+ + Ligand \rightarrow Fe^{3+} - Ligand + OH^- \qquad (6.18)$$

Reaction 6.18 represents dissolution of Fe(III) minerals by organic ligands. The ligands in Reaction 6.18 are weak-field organic acids produced by plant and bacteria and generally contain oxygen in the form of hydroxyl or carboxyl functional groups (Hider, 1984). Pyritic mine waste and coal residue are known to contain Mn– and Fe–oxides (Barnhisel and Massey, 1969). Ferric iron produced through the reactions shown above may oxidize pyrite (Luther et al., 1992).

6.10   PREDICTION OF ACID DRAINAGE

There are several methods or approaches for predicting the potential of pyritic material to produce acid drainage (AD). These approaches include: (1) determination of potential acidity in pyrite overburden, (2) acid–base accounting, and (3) simulated weathering. A brief discussion of each approach is given below (for details, see Evangelou, 1995b).

6.10.1   Potential Acidity

A direct determination of acid-producing potential is the rapid pyrite oxidation technique utilizing 30% H_2O_2. The actual acid produced during pyrite oxidation by H_2O_2 is termed potential acidity (Sobek et al., 1978). The technique determines the amount of acid produced during complete oxidation of Fe^{2+} and S_2^{2-} of pyrite as follows:

$$FeS_2 + 7.5H_2O_2 \rightarrow Fe(OH)_3(s) + 2H_2SO_4 + 4H_2O \qquad (6.19)$$

Reaction 6.19 shows that complete pyrite oxidation liberates 2 mol of H_2SO_4 for every mole of FeS_2 oxidized. This suggests that for each mole of pyrite, one needs 2 mol of $CaCO_3$ to neutralize the acid to be produced by oxidation.

6.10.2 Acid–Base Accounting

Acid–base accounting is the most widely used method for characterizing overburden geochemistry. The purpose of acid–base accounting on a geologic sample is to identify the acid-producing potential due to pyrite and the neutralization-producing potential due to alkaline material such as rock carbonates. The difference between the two potentials indicates whether there is enough base present to neutralize all acid produced from the oxidation of pyrite (Smith and Sobek, 1978). Table 6.2 presents an example of an acid–base account for a coal seam (Sobek et al., 1987). Note that a net deficiency indicates that the rock may lack sufficient neutralizer and will become an active acid producer when exposed to the atmosphere.

Acid–base accounting has been criticized because it does not consider differences between the rate of pyrite oxidation and the rate of carbonate dissolution. Furthermore, the technique considers that all pyrite present in the sample is oxidizable under natural oxidizing conditions.

6.10.3 Simulated Weathering

Simulated weathering involves leaching overburden or spoil in laboratory-scale experiments with the idea that results are applicable to the field. The effluent is collected and analyzed for pH, acidity, sulfate, iron and so on. The results from these analyses are used to evaluate acid drainage formation potential (Caruccio and Geidel, 1981;

TABLE 6.2. Acid–Base Account for an Eastern U.S. Coal Mine[a]

Sample No.	Thickness (ft)	Paste pH	Rock Type	Total Sulfur (%)	Maximum from Percent Total S^* (Acid Potential)	Amount Present (Neutralization Potential)	Excess (+) or Deficiency (−)
					T in $CaCO_3$ Equivalent/1000 T Material		
1	18.0	6.1	Siltstone	<0.01		2.6	+2.6
2	23.4	7.7	Sandstone	<0.01		4.8	+4.8
3	8.6	6.2	Shale	0.09	2.8	5.7	+2.9
4	16.5	7.3	Shale	0.26	8.1	9.3	+1.2
5	6.5	7.3	Shale	0.36	11.2	23.2	+12.0
6	6.9	7.8	Sandstone	0.15	4.7	24.7	+20.0
7	10.0	8.1	Shale	0.03	1.0	22.0	+21.0
8	25.2	7.6	Shale	0.53	16.6	20.0	+3.4
9	5.8	7.4	Shale	1.90	59.4	58.0	−1.4
10	5.3	7.5	Claystone	0.95	29.8	7.8	−22.0

Source: Sobek et al., 1987.

[a]Net for section = [(+713.14) + (−124.72)]/126 = +4.67 T $CaCO_3$ equivalent/1000 T material.

Sturey et al., 1982). One advantage of simulated weathering is that it considers relative weathering rates of pyrite and carbonate materials. However, it is difficult to say to what degree the technique resembles a profile, because the role of such factors as oxygen diffusion, water infiltration, bacteria effectiveness, and salt and clay mineral influences on pyrite oxidation remains elusive (Evangelou, 1995b).

Computer-simulation models are also used to predict acid drainage potential due to pyrite weathering. Such models incorporate a number of interactive variables controlling oxidation. They are based on reaction kinetics and are evaluated with acid drainage results from column leaching studies and/or field observations. By reviewing such models, Evangelou (1995b) concluded that those of Jaynes et al. (1984) and Bronswijk and Groenenberg (1993) represent some of the most up to date models for predicting acid drainage formation. However, more sophisticated computer models are needed for more accurate acid drainage predictions (Evangelou, 1995b).

PROBLEMS AND QUESTIONS

1. Write a balanced reaction between the products of Reaction 6.1 and $CaCO_3$.

2. Based on the balanced reaction (problem 1), calculate the amount of limestone that you would need to neutralize the total acidity produced by a geologic stratum containing 5% pyrite.

3. Explain how pyrite might oxidize in the absence of O_2.

4. Calculate the number of days that it would take to oxidize 1000 kg of pyrite. Assume that the oxidation process would be described by

$$\text{Pyrite}_{(remaining)} = (\text{original pyrite quantity})e^{-k(\text{days})}$$

   where $k = 0.01$ day^{-1}.

5. Name all the precautions that you would take to ensure that pyrite in nature would not oxidize.

6. Explain how organic ligands might enhance the pyrite oxidation rate.

7. Name the potential advantages and disadvantages of the use of limestone to control pyrite oxidation.

7 Reaction Kinetics in Soil-Water Systems

7.1 INTRODUCTION

There are many reactions in soil–water systems pertaining to nutrient availability, contaminant release, and nutrient or contaminant transformations. Two processes regulating these reactions are chemical equilibria (Chapter 2) and kinetics. The specific kinetic processes that environmental scientists are concerned with include mineral dissolution, exchange reactions, reductive or oxidative dissolution, reductive or oxidative precipitation, and enzymatic transformation. This chapter provides a quantitative description of reaction kinetics and outlines their importance in soil–water systems.

To understand reaction kinetics one needs to understand the difference between kinetics and equilibria. Generally, equilibria involves forward and reverse reactions and it is defined as the point at which the rate of the forward reaction equals the rate of the reverse reaction.

Consider the mineral AB (Reaction 7.1), where A denotes any cation (A^+) and B denotes any anion (B^-). Upon introducing H_2O, the mineral undergoes solubilization (forward reaction) until precipitation (reverse reaction) becomes significant enough so that the two rates (forward and reverse) are equal:

$$AB \underset{k_b}{\overset{k_f}{\Longleftrightarrow}} A^+ + B^- \tag{7.1}$$

The parameters k_f and k_b denote rate constants for the forward and reverse reactions, respectively. Reaction 7.1 demonstrates mineral equilibrium through two elementary reactions—one describes the forward reaction, while a second describes the reverse reaction. When the reverse reaction is inhibited, the forward reaction is termed dissolution (e.g., acid mineral dissolution).

Reaction 7.1 at the equilibrium point is described by

$$dA^+/dt = k_f(AB) - k_b(A^+)(B^-) = 0 \tag{7.2}$$

where dA^+/dt denotes the rate of the overall reaction, $k_f(AB)$ describes the rate of the forward reaction, and $k_b(A^+)(B^-)$ describes the rate of the reverse reaction. At equilibrium,

272

$$k_f(AB) = k_b(A^+)(B^-) \tag{7.3}$$

and

$$K_{eq} = k_f/k_b = (A^+)(B^-)/(AB) \tag{7.4}$$

where K_{eq} denotes the equilibrium product constant (note, in the example above K_{eq} = K_{sp}, see Chapter 2) and the parentheses denote activity. Equilibria constants (K_{sp}) are used to predict the concentration of chemical species in solution contributed by a given solid (assuming the solid's K_{sp} is known).

Equation 7.4 can also be derived using Gibb's free energy of formation (ΔG_f). Based on classical thermodynamics (Daniels and Alberty, 1975),

$$\Delta G_f(X) = \Delta G_f(X)^0 + RT \ln \alpha_x \tag{7.5}$$

where
$\Delta G_f^0(X)$ = Gibbs free energy of formation of ion X at the standard state, 25°C and 1 atm pressure
R = universal gas constant
T = temperature in degrees Kelvin
α_x = molar activity of ion X

At equilibrium, $\Delta G_r = 0$ and $\Delta G_r^0 = \Delta G_{f(products)}^0 - \Delta G_{f(reactants)}^0$, and the thermodynamic equilibrium constant (K_{eq}) is given by

$$K_{eq} = \exp - (\Delta G_r^0/RT) \tag{7.6}$$

where subscript r denotes reaction. By substituting each of the terms describing reactants and products in Equation 7.1 by Equation 7.5 and introducing the resulting equations into Equation 7.6,

$$K_{eq} = (A^+)(B^-)/(AB) = \exp - \{\Delta G_f(A^+)^0 + \Delta G_f(B^-)^0 - \Delta G_f(AB)^0\}RT \tag{7.7}$$

Based on the above, under standard pressure (1 atm) and temperature (25°C) (isobaric conditions) and under unit activity of reactants and products, a negative ΔG_r^0 denotes that the particular reaction will move spontaneously from left to right until an equilibrium state is met, whereas a positive ΔG_r^0, also under isobaric conditions and unit activity of reactants and products, denotes that the particular reaction will not move spontaneously from left to right. Finally, when ΔG_r equals zero, the particular reaction will be at equilibrium.

It follows then that the thermodynamic approach makes no reference to kinetics, while the kinetic approach is only concerned with the point at which the forward reaction equals the reverse reaction and gives no attention to the time needed to reach this equilibrium point. In nature, certain chemical events may take a few minutes to reach equilibrium, while others may take days to years to reach equilibrium; such phenomena are referred to as hystereses phenomena. For example, exchange reactions involving homovalent cations (forming outer-sphere complexes, e.g., Na^+–Li^+) may take only a few minutes to reach equilibrium, whereas exchange reactions involving heterovalent cations (e.g., Ca^{2+}–K^+ in a vermiculitic internal surface where Ca^{2+} forms an outer-sphere complex and K^+ forms an inner-sphere complex) may require a long period (e.g., days) to reach equilibrium.

The rate at which a particular reaction occurs is important because it could provide real-time prediction capabilities. In addition, it could identify a particular reaction in a given process as the rate-controlling reaction of the process. For example, chemical mobility in soils, during rain events, is controlled by the rate at which a particular species desorbs or solubilizes. Similarly, the rate at which a particular soil chemical biodegrades is controlled by the rate at which the soil chemical becomes available substrate.

7.2   RATE LAWS

Reaction rates are characterized by rate laws which describe rate dependence on concentration of reactants. For example, for the monodirectional reaction

$$A + B \rightarrow C \tag{7.8}$$

the reaction rate (dC/dt) can be described by the equation

$$dC/dt = k[A]^{n_1}[B]^{n_2} \tag{7.9}$$

where the brackets denote the concentration of the reacting species, k denotes the rate constant, and n denotes the order of the reaction. Assuming that $n_1 = 1$, the reaction is said to be first-order with respect to [A]. On the other hand, assuming that $n_2 = 2$, the reaction is second-order with respect to [B]. It is important to note that n_i are not the stoichiometric coefficients of the balanced equation; they are determined experimentally.

In soil–water systems, some of the most commonly encountered rate laws are first-, second-, and zero-order. A description of each order is given below.

7.2.1   First-Order Rate Law

Consider the monodirectional elementary reaction

$$A \rightarrow B \tag{7.10}$$

expressed by

$$dA/dt = -k[A] \tag{7.11}$$

rearranging

$$dA/[A] = -dt \tag{7.12}$$

Setting $[A] = A_0$ at $t = t_0$ and $[A] = A_i$ at $t = t_i$,

$$\int_{A_0}^{A_i} dA/[A] = -k \int_{t_0}^{t_i} dt \tag{7.13}$$

and integrating

$$\ln[A_i/A_0] = -k(t_i - t_0) \tag{7.14}$$

Assuming that $t_0 = 0$

$$\ln[A_i/A_0] = -kt_i \tag{7.15}$$

or

$$[A_i] = [A_0]e^{-kt_i} \tag{7.16}$$

A plot of A_i versus t_i would produce a curve with an exponential decay, approaching $[A_i] = 0$ asymptotically (Fig. 7.1). Taking logarithms to base 10 on both sides of Equation 7.16 gives

$$\log[A] = -kt_i/2.303 + \log[A_0] \tag{7.17}$$

A plot of $\log[A_i]$ versus t_i would produce a straight line with slope $-k/2.303$ (Fig. 7.2). In Equation 7.17, setting $[A_i/A_0] = 0.5$ at $t_i = t_{1/2}$ and rearranging gives

$$\log[0.5] = -kt_{1/2}/2.303 \tag{7.18}$$

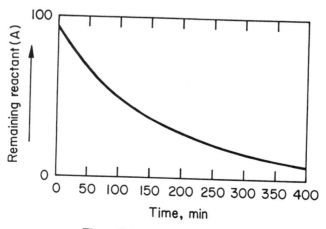

Figure 7.1. Ideal first-order plot.

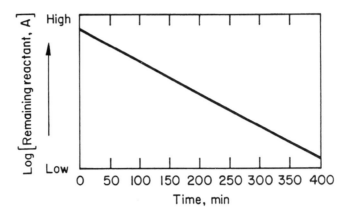

Figure 7.2. Linearized form of the first-order reaction.

and

$$t_{1/2} = (-\log[0.5])[2.303]/k = 0.693/k \tag{7.19}$$

where the term k is in units of t^{-1} (e.g., $\sec^{-1}$, $\min^{-1}$, hr^{-1}, or $days^{-1}$). The term $t_{1/2}$ represents the time needed for 50% of reactant A_0 to be consumed; it is also known as the half-life of compound A. In the case of a first-order reaction, its half-life is independent of the original quantity of A (A_0) in the system.

7.2.2 Second-Order Rate Law

Consider the monodirectional bimolecular reaction

$$A + B \rightarrow C \tag{7.20}$$

Assuming that A = B, its rate can be expressed by

$$dA/dt = -k[A]^2 \tag{7.21}$$

Rearranging

$$dA/[A]^2 = -dt \tag{7.22}$$

Setting $[A] = A_0$ at $t = t_0$ and $[A] = A_i$ at $t = t_i$

$$\int_{A_0}^{A_i} dA/[A]^2 = -k \int_{t_0}^{t_i} dt \tag{7.23}$$

and integrating

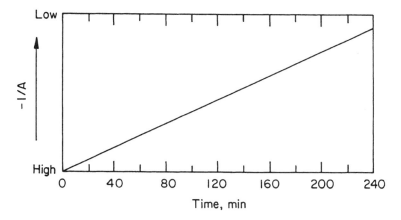

Figure 7.3. An ideal linear plot for a second-order reaction.

$$-[1/A_i] + [1/A_0] = -k(t_i - t_0) \tag{7.24}$$

Assuming that $t_0 = 0$,

$$-[1/A_i] = -[1/A_0] - k(t_i) \tag{7.25}$$

A plot of $-1/A_i$ versus t_i would produce a straight line with y intercept $-1/A_0$ and slope minus the second-order rate constant (Fig. 7.3). In Equation 7.25, setting $A_i = 0.5A_0$ at $t_i = t_{1/2}$ and rearranging gives

$$t_{1/2} = 1/k \cdot A_0 \tag{7.26}$$

where k is in units of mass$^{-1} \cdot t^{-1}$ (e.g., mol$^{-1} \cdot$ min^{-1}. The term $t_{1/2}$ represents the time needed for 50% of reactant A_0 to be consumed; it is also known as the half-life of compound A. In the case of a second-order reaction, its half-life is dependent on the original quantity of A (A_0) in the system.

7.2.3 Zero-Order Rate Law

Consider the monodirectional reaction

$$A \rightarrow B \tag{7.27}$$

expressed by

$$dA/dt = -k \tag{7.28}$$

Rearranging

$$dA = -k \, dt \tag{7.29}$$

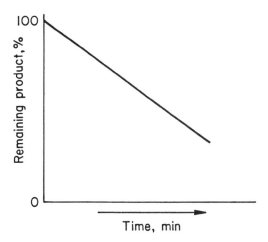

Figure 7.4. Zero-order reaction.

Setting $[A] = A_0$ at $t = t_0$ and $[A] = A_i$ at $t = t_i$,

$$\int_{A_0}^{A_i} dA = -k \int_{t_0}^{t_i} dt \qquad (7.30)$$

and integrating

$$A_i - A_0 = -k(t_i - t_0) \qquad (7.31)$$

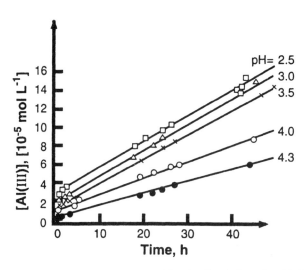

Figure 7.5. Linear dissolution kinetics observed for the dissolution of γ–Al_2O_3 (from Furrer and Stumm, 1986, with permission).

Assuming that $t_0 = 0$, then

$$A_i = A_0 - kt_i \qquad (7.32)$$

A plot of A_i versus t_i would produce a straight line with slope $-k$ in units of mass per unit time (e.g., mol min^{-1}) (Fig. 7.4). In the case of a zero-order reaction, its half-life is $1/2t_f$, where t_f represents the total time needed to decompose the original quantity of compound A (A_0). Another way to express such reactions is shown in Figure 7.5. The data show Al release from γ-Al$_2$O$_3$ at different pH values. The data clearly show that the reaction is zero-order with k dependent on pH.

7.3 APPLICATION OF RATE LAWS

Rate laws are employed to evaluate reaction mechanisms in soil–water systems. To accomplish this, kinetics are used to elucidate the various individual reaction steps or elementary reactions. Identifying and quantifying the elementary steps of a complex process allow one to understand the mechanism(s) of the process. For example, unimolecular reactions are generally described by first-order reactions; bimolecular reactions are described by second-order reactions.

When evaluating soil–water processes, one should distinguish the rate of an elementary chemical reaction from the rate of a process which is commonly the sum of a number of reactions. For example, the rate of an elementary chemical reaction depends on the energy needed to make or break a chemical bond. An instrument capable of measuring the formation or destruction of any given chemical bond could provide molecular data with mechanistic meanings. Instruments with such capabilities include nuclear magnetic resonance (NMR), Fourier transform infrared spectroscopy (FT–IR), and electron spin resonance (ESR). On the other hand, if the end product of a particular process represents several elementary events, data representing this end product may not have mechanistic meaning. For example, the rate of exchanging K$^+$ by Ca^{2+} in Al–hydroxy interlayered vermiculite may involve many processes. These processes may include partial loss of water by Ca^{2+}, cation diffusion, and cation exchange. Sorting out the reactions controlling the overall rate process is difficult. Researchers often overcome such limitations by evaluating kinetic processes using wet chemistry plus spectroscopic techniques, or through studying the kinetic processes by varying temperature, pressure, reactant(s), or concentration(s).

One important point to remember when using kinetics to study soil–water processes is that the apparatus chosen for the study is capable of removing or isolating the end product as fast as it is produced. A second point is that unimolecular reactions always produce first-order plots, but fit of kinetic data (representing a process not well understood) to a first-order plot is no proof that the process is unimolecular. Complementary data (e.g., spectroscopic data) are needed to support such a conclusion. On the other hand, rate-law differences between any two reaction systems suggest that the mechanisms involved may represent different elementary reactions.

7.3.1 Pseudo First-Order Reactions

Dissolution Kinetics. Pseudo first-order reactions are widely employed in the field of soil–water environmental science for evaluating physical, chemical, or biochemical events. A pseudo first-order dissolution example is given below to demonstrate the use of kinetics in identifying or quantifying minerals in simple or complex systems.

Consider a metal carbonate solid (MCO_{3s}) reacting with a strong acid (HCl):

$$MCO_{3s} + 2HCl \rightarrow CO_{2gas} + M^{2+} + 2Cl^- + H_2O \tag{7.33}$$

In the case where $HCl >>> MCO_{3s}$, so that the concentration of HCl does not change significantly when all MCO_3 is decomposed, the rate of Reaction 7.33 can be expressed by

$$dMCO_3/dt = -k[HCl][MCO_{3s}] \tag{7.34}$$

Assuming that during acid dissolution the newly exposed MCO_{3s} surface (S) remains proportional to the amount of unreacted MCO_3 (Turner, 1959; Turner and Skinner, 1959) such that

$$(S) = K[MCO_{3s}] \tag{7.35}$$

where K is an empirical constant. Rearranging Equation 7.34,

$$dMCO_3/MCO_3 = -k[HCl]dt \tag{7.36}$$

setting $[MCO_{3s}] = MCO_{3o}$ at $t = t_0$, and $[MCO_{3s}] = MCO_{3i}$ at $t = t_i$,

$$\int_{MCO_{3o}}^{MCO_{3i}} dMCO_3/[MCO_3] = -k[HCl]\int_{t_0}^{t_i} dt \tag{7.37}$$

and integrating,

$$\ln[MCO_{3i}/MCO_{3o}] = -k[HCl](t_i - t_0) \tag{7.38}$$

Assuming that $t_0 = 0$ and $k' = k[HCl]$, hence it is pseudo first-order, then

$$\ln[MCO_{3i}/MCO_{3o}] = -k't_i \tag{7.39}$$

or

$$[MCO_{3i}] = [MCO_{3o}]e^{-k't_i} \tag{7.40}$$

or

$$\log[MCO_{3i}] = -k't_i/2.303 + \log[MCO_{3o}] \tag{7.41}$$

A plot of $\log[MCO_{3i}]$ versus t_i would produce a straight line with slope $-k'/2.303$. The half-life ($t_{1/2}$) can be calculated by

$$t_{1/2} = 0.693/[(\text{slope})(2.303)] \tag{7.42}$$

The above theoretical analysis, however, does not reveal how MCO_{3s} can be accurately measured during acid dissolution. One approach would be to measure the carbon dioxide gas (CO_{2gas}) released during acid dissolution of MCO_{3s} (Reaction 7.33) in an air-tight vessel equipped with a stirring system and a transducer to convert pressure to a continuous electrical signal (Evangelou et al., 1982). A calibration plot between CO_{2gas} pressure and grams of MCO_{3s} (as shown in Fig. 7.6) can then be used to back-calculate remaining MCO_{3s} during acid dissolution. The data in Figure 7.7 describe dissolution kinetics of calcite ($CaCO_3$) and dolomite [$(CaMg(CO_3)_2$]. It is shown that calcite is sensitive to strong-acid attack, but dolomite is resistant to strong-acid attack; both minerals appear to obey pseudo first-order reaction kinetics. In the case where a sample contains calcite plus dolomite, the kinetic data reveal two consecutive pseudo-first-order reactions (Fig. 7.8). By extrapolating the second slope (representing dolomite) to the y axis, the quantity of calcite and dolomite in the sample could be estimated.

Additional information on metal–carbonate dissolution kinetics could be obtained by evaluating dissolution in relatively weak concentrations of HCl (Sajwan et al., 1991). A plot of pseudo first-order rate constants k' ($k' = k[HCl]$) versus HCl concentration would allow one to estimate first-order constants (k) as HCl $\rightarrow$ 0 by extrapolating the line representing k' to the y axis. Additional pseudo first-order dissolution examples are shown in Figure 7.9 where the linear form of the pseudo first-order acid dissolution of kaolinite in two different HCl concentrations is shown.

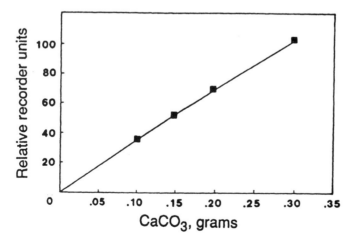

Figure 7.6. Pressure transducer electrical output in response to increases in grams of carbonate reacted with 5 mol L^{-1} HCl (from Evangelou et al., 1984a, with permission).

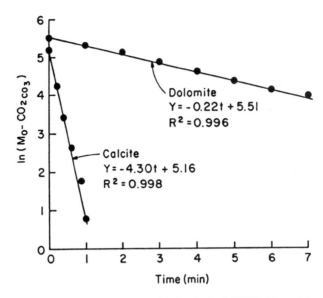

Figure 7.7. Rate of dissolution of calcite and dolomite in 5 N HCl. The calcite and dolomite analyzed were standard reference specimens (from Evangelou et al., 1984a, with permission).

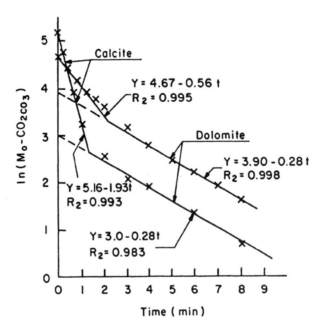

Figure 7.8. Plots illustrating differential rates of dissolution of calcite and dolomite in two samples of Mancos shale. Dissolution effected with 5 N HCl (from Evangelou et al., 1984a, with permission).

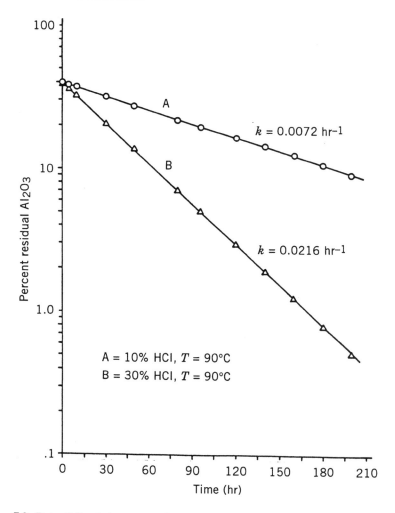

Figure 7.9. Rate of dissolution curves of aluminum from kaolinite in 10 and 30% HCl at 90°C (from Turner, 1966, with permission).

It is important to note that during pseudo first-order reactions, the concentration of the reactant responsible for the decomposition must remain constant throughout the reaction that is, it cannot decrease more than 10% of its original quantity. Thus, pseudo first-order dissolution reactions can be attained by using excess acids, bases, complexing agents, oxidizing agents, or reducing agents. Mineral dissolution kinetics in nature may require days or years to reach completion.

Exchange Reactions. Exchange reactions in soils commonly involve monovalent–monovalent (e.g., K^+–NH_4^+), monovalent–divalent (e.g., K^+–Ca, Na^+–Ca^{2+}), movovalent–trivalent (e.g., K^+–Al^{3+}), or divalent–trivalent (e.g., Ca^{2+}–Al^{3+}) cations. The notation $C_A^{n+} - C_B^{n+}$ denotes exchange between cations A and B where A represents the cation in the solution phase (displacing cation) and B represents the cation on the exchange phase (displaced cation). Consider the heterovalent exchange reaction

$$\text{ExCa}_{1/2} + \text{K}^+ \underset{k_b}{\overset{k_f}{\Leftrightarrow}} \text{ExK} + 1/2 \, \text{Ca}^{2+} \qquad (7.43)$$

where ExK and $\text{ExCa}_{1/2}$ denote exchangeable cations in units of $\text{cmol}_c \, \text{kg}^{-1}$ or meq/100 g, Ex denotes the soil exchanger with a charge of 1^-, and K^+ and Ca^{2+} denote solution cations in units of mmol L^{-1}. Reaction 7.43 is composed of at least two elementary reactions, a forward and a backward elementary reaction.

Reactions such as the one above are studied by first making the soil or clay mineral homoionic (ExCa) by repeatedly washing it with a solution of CaCl_2 (approximately $1 \, \text{mol}_c \, \text{L}^{-1}$) and then rinsing the sample with H_2O. The homoionic soil or clay material is then spread as a fine film on a filter in a filter holder. A solution of KCl, at a preset concentration, is pumped at a constant rate (e.g., $1 \, \text{mL min}^{-1}$) through the homoionic clay. Effluent is collected with respect to time using a fraction collector. The technique, known as miscible displacement, permits the study of any elementary forward or backward reactions using any appropriate homoionic soil or clay mineral with an appropriate displacing solution. A number of procedures and apparati are available to study kinetics of exchange reactions in soils (Sparks, 1995 and references therein).

The rate of the forward reaction (Reaction 7.43) can be expressed by

$$-d\text{ExCa}_{1/2}/dt = k[\text{K}^+][\text{ExCa}_{1/2}] \qquad (7.44)$$

where k is the rate constant in units of t^{-1}. Assuming that during cation exchange K^+ is kept constant, Equation 7.44 can be expressed as

$$-d\text{ExCa}_{1/2}/\text{ExCa}_{1/2} = k' \, dt \qquad (7.45)$$

where $k' = k[\text{K}^+]$. Setting the appropriate boundary conditions and integrating,

$$\ln \left[(1 - \text{ExCa}_{1/2t})/\text{ExCa}_{1/2\infty} \right] = -k't \qquad (7.46)$$

or

$$(1 - \text{ExCa}_{1/2t}) = (\text{ExCa}_{1/2\infty})e^{-k't} \qquad (7.47)$$

and

$$\log \left[(1 - \text{ExCa}_{1/2t})/\text{ExCa}_{1/2\infty} \right] = -k't/2.303 \qquad (7.48)$$

A plot of $\log[1 - \text{ExCa}_{1/2t}/\text{ExCa}_{1/2\infty}]$ versus t would produce a straight line with slope $-k/2.303$. Commonly, cation exchange reactions may take a few minutes to a few hours to reach completion depending on degree of diffusion.

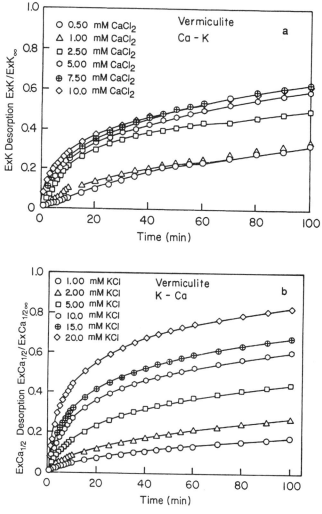

Figure 7.10. Influence of Ca and K concentration on the desorption of K and Ca, respectively, with respect to time - (vermiculite < 2 μm) (from Evangelou, 1997, unpublished data, with permission).

The data in Figure 7.10 describe the forward and reverse reactions of Ca^{2+}–K^+ exchange kinetics in vermiculite. These data clearly show that the exchange process, as expected, is dependent on the concentration of the exchanging cation. Linearized pseudo first-order plots of Ca^{2+}–K^+ and K^+–Ca^{2+} exchange in vermiculite are shown in Figure 7.11. These plots exhibit two slopes. An interpretation of the two-slope system in Figure 7.11 is that there are two consecutive pseudo first-order reactions. The justification for such a conclusion is that the forward and reverse exchange reactions in vermiculite are not simple elementary reactions. This is because the

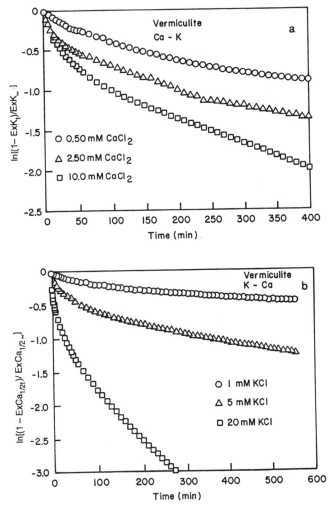

Figure 7.11. Pseudo first-order plots of Ca^{2+}-K^+ and K^+-Ca exchange under three different concentrations of $CaCl_2$ and KCl, respectively, in vermiculite (from Evangelou, 1996, unpublished data, with permission).

surface of vermiculite is rather complex and many elementary reactions are involved in the cation exchange process.

Vermiculite possesses internal and external surfaces and K^+ may diffuse faster than Ca^{2+} in the internal (interlayer) surface due to the ability of K^+ to decrease its hydration sphere by loosing some of its water molecules, which are held with less energy than the water molecules held by Ca^{2+} (Bohn et al., 1985). On the other hand, diffusion of Ca^{2+} and K^+ in the external surfaces does not appear to be limiting. The decision to conclude that the two slopes in Figure 7.11 represent two pseudo first-order reactions

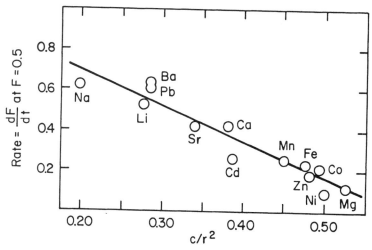

Figure 7.12. Relationship between rate of exchange and c/r^2 (from Keay and Wild, 1961, with permission).

is based on the fact that the complexities of the vermiculitic surface are well known. If this information was not known, the only conclusion one could have drawn was that pseudo first-order reactions do not describe heterovalent cation exchange reactions.

In 2:1 clay minerals (e.g., vermiculite) the rate of cation exchange depends on the ionic potential (c/r^2, c denotes charge of the cation and r denotes ionic radius) of the cations involved. Cation exchange data show that c/r^2 is inversely related to the rate of cation exchange (Fig. 7.12). The reason for this relationship is that the higher the ionic potential of an ion, the lower its entropy of activation, or the higher the energy by which hydration-sphere water is held (Bohn et al., 1985). Generally, cations that hold water tightly exhibit low diffusion potential in clay interlayer water.

7.3.2 Reductive and Oxidative Dissolution

Reductive Dissolution. Many substances in nature contain the same metal or metalloid, but under different oxidation states. For example, the metalloid arsenic may exist as arsenite (AsIII, AsO_3) or arsenate (AsIV, AsO_4) in the forms of ferrous-arsenite or ferric-arsenate, respectively. Ferrous-arsenite is more soluble than ferric-arsenate; for this reason, one may be interested in studying the kinetics of arsenate reduction to arsenite. Similar chemistry applies to all elements present in soil–water systems with more than one oxidation state (e.g., iron, manganese, selenium, and chromium).

Reductive dissolution kinetics of MnO_2 (MnIV) and MnOOH (MnIII) are presented below for demonstration purposes. The chemistry of manganese in nature is rather complex because three oxidation states are involved [Mn(II), Mn(III), and Mn(IV)] and form a large number of oxides and oxyhydroxides with various degrees of chemical stability (Bricker, 1965; Parc et al., 1989; Potter and Rossman, 1979a,b). One may characterize the chemical stability of the various manganese oxides through pseudo first-order dissolution. This can be done by reacting manganese oxides with concentrated H_2SO_4 plus hydrogen peroxide (H_2O_2). Under strong acid conditions, H_2O_2 acts as a manganese reductant (electron donor). In the case of MnO_2, reductive dissolution is given by

$$MnO_{2s} + H_2O_2 + 2H^+ \rightarrow Mn^{2+} + O_{2gas} + 2H_2O \qquad (7.49)$$

and in the case of MnOOH, reductive dissolution is given by

$$MnOOH_s + 1/2H_2O_2 + 2H^+ \rightarrow Mn^{2+} + 1/2O_{2gas} + 2H_2O \qquad (7.50)$$

The two reactions above could be quantified by either measuring the concentration of Mn^{2+} with respect to time or recording O_2 gas evolution with respect to time.

Reduction–dissolution kinetics of manganese-oxides in excess H_2O_2 plus H_2SO_4 (assuming complete Mn-oxide surface coverage by the reductant, irreversible electron transfer, and instantaneous product release) can be expressed by

$$-d[\text{Mn-oxide}]/[dt] = k'[\text{Mn-oxide}] \qquad (7.51)$$

where Mn-oxide denotes MnO_2 or MnOOH, $k' = k[H_2O_2][H_2SO_4]$ denotes the pseudo first-order rate constant, and brackets denote concentration. Rearranging and integrating Equation 7.51 using appropriate boundary conditions gives

$$\log([\text{Mn-oxide}]/[\text{Mn-oxide}]_0) = -(k'/2.303)t \qquad (7.52)$$

where $[\text{Mn-oxide}]_0$ represents the initial total Mn-oxide in the system and $[\text{Mn-oxide}]$ represents the quantity of Mn-oxide at any time t. A plot of $\log([\text{Mn-oxide}]/[\text{Mn-oxide}]_0)$ versus t provides a straight-line relationship with slope $-k'/2.303$.

The data in Figure 7.13 show reductive-dissolution kinetics of various Mn-oxide minerals as discussed above. These data obey pseudo first-order reaction kinetics and the various manganese–oxides exhibit different stability. Mechanistic interpretation of the pseudo first-order plots is difficult because reductive dissolution is a complex process. It involves many elementary reactions, including formation of a Mn–oxide–H_2O_2 complex, a surface electron-transfer process, and a dissolution process. Therefore, the fact that such reactions appear to obey pseudo first-order reaction kinetics reveals little about the mechanisms of the process. In nature, reductive dissolution of manganese is most likely catalyzed by microbes and may need a few minutes to hours to reach completion. The abiotic reductive–dissolution data presented in Figure 7.13 may have relative meaning with respect to nature, but this would need experimental verification.

Oxidative Dissolution. This is a process highly applicable to metal-sulfides. In general, under reducing conditions metal sulfides are insoluble solids. However, sulfide converts to sulfate (SO_4) under oxidative conditions and the metal–sulfate salts formed are relatively soluble (Singer and Stumm, 1970).

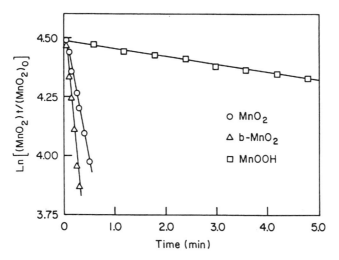

Figure 7.13. Pseudo first-order Mn–oxide reductive dissolution (from Sajwan et al., 1994, with permission).

Oxidative dissolution of metal-sulfides (e.g., pyrite, FeS_2) is a complex process involving surface adsorption of the oxidant (Fe^{3+}, O_2), surface electron transfer, and surface product formation and detachment. The overall oxidation process, without considering the detailed mechanisms, is demonstrated below using pyrite (FeS_2) (Evangelou, 1995b):

$$FeS_2 + 7/2O_2 + H_2O \rightarrow Fe^{2+} + 2SO_4 + 2H^+ \qquad (7.53)$$

or

$$FeS_2 + 14Fe^{3+} + 8H_2O \rightarrow 15Fe^{2+} + 2SO_4^{2-} + 16H^+ \qquad (7.54)$$

Both reactions (Reactions 7.53 and 7.54) oxidize pyrite and take place simultaneously, while the reaction responsible for regenerating Fe^{3+} is

$$Fe^{2+} + H^+ + 0.25O_2 \rightarrow Fe^{3+} + 1/2H_2O \qquad (7.55)$$

Based on Reactions 7.53 and 7.54, the rate of pyrite oxidation can be expressed by

$$-d[FeS_2]/dt = [(k_1(O_2)^{\nu_1} + k_2(Fe^{3+})^{\nu_2}](S) \qquad (7.56)$$

The parameters O_2 and Fe^{3+} refer to partial pressure and concentration, respectively, k_i refers to rate constants, and S denotes surface area. The exponents ν_i are experimentally determined (Daniels and Alberty, 1975). Considering that the rate of FeS_2 oxidation by O_2 is slow relative to that by Fe^{3+}, and Fe^{2+} oxidation by O_2 is slower than the rate of FeS_2 oxidation by Fe^{3+}, the latter (Fe^{2+} oxidation) is the pyrite oxidation rate-controlling process. This was demonstrated by Singer and Stumm (1970) and Moses et al. (1987).

According to Reaction 7.55, the rate of Fe^{2+} oxidation is given by

$$d[Fe^{3+}]/dt = k(H^+)^{v_1}(Fe^{2+})^{v_2}(O_2)^{v_3} \qquad (7.57)$$

Singer and Stumm (1970) demonstrated that at pH less than 3.5, Equation 7.57 takes the form of

$$d[Fe^{3+}]/dt = k'(Fe^{2+})(O_2) \qquad (7.58)$$

where k' denotes apparent rate constant. Equation 7.58 reveals that the rate of Fe^{2+} oxidation at pH less than 3.5 is independent of pH and first order with respect to Fe^{2+} and O_2 (Fig. 7.14). Thus, under the conditions stated above, the rate of Fe^{2+} oxidation is directly related to its concentration and partial pressure of bimolecular oxygen. However, at pH higher than 3.5, the rate expression for Fe^{2+} oxidation, according to Singer and Stumm (1970), is of the form

$$d[Fe^{3+}]/dt = k(Fe^{2+})(O_2)(OH)^2 \qquad (7.59)$$

Equation 7.59 reveals that Fe^{2+} oxidation is first order with respect to Fe^{2+} and O_2 and second order with respect to OH^- (Fig. 7.14).

Solution Fe^{3+} at pH higher than 3.5 would be controlled by the solubility of $Fe(OH)_{3s}$

$$Fe^{3+} + 3OH^- \rightarrow Fe(OH)_{3s} \qquad (7.60)$$

The solubility of $Fe(OH)_{3s}$ is described by

$$Fe(OH)_{3s} \Leftrightarrow Fe^{3+} + 3OH^- \qquad (7.61)$$

and

$$Fe^{3+} = K_{sp}/(OH^-)^3 \qquad (7.62)$$

where K_{sp} is the solubility product constant of $Fe(OH)_{3s}$. Introducing Equation 7.62 into Equation 7.56 gives

$$-d[FeS_2]/dt = \{k_1(O_2)^{v_1} + k_2[K_{sp}/(OH^-)^3]^{v_2}\}(S) \qquad (7.63)$$

According to Equation 7.63, abiotic FeS_2 oxidation is controlled by pH. As pH decreases (OH^- decreases), free Fe^{3+} in solution increases; consequently, pyrite oxidation increases. At low pH (pH < 4), FeS_2 oxidation is catalyzed by bacteria (Evangelou, 1995b) (see Chapter 6).

Experimental data show that no single model describes kinetics of pyrite oxidation because of the large number of variables controlling such process. These variables include crystallinity, particle size, mass to surface ratio, impurities, type and nature of

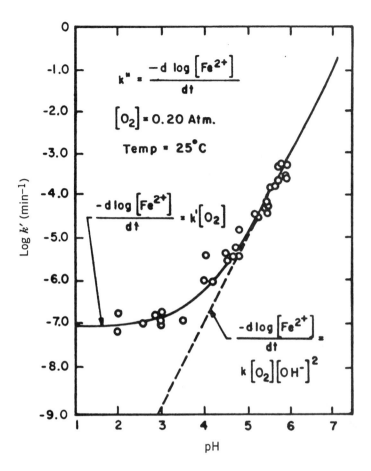

Figure 7.14. Relationship between Fe oxidation rate (k) and pH (from U.S. Government Publication, 1969).

impurities, crystal imperfections, presence/absence of other minerals or organics, ionic strength, pH, Fe^{3+}/Fe^{2+} ratio, type of oxidant, presence or absence of potentially determining ions, and nature of reaction products formed (Evangelou, 1995b).

7.3.3 Oxidative Precipitation or Reductive Precipitation

Oxidative Precipitation. This is a process that describes precipitation of metals, such as Fe^{2+} or Mn^{2+}, through oxidation. Oxidative precipitation is complex, involving various mechanisms. In general, however, it can be viewed as a two-step process and is demonstrated on Mn^{2+} below using unbalanced equations. The first step involves a slow reaction that generates a solid surface:

$$Mn^{2+} + O_2 \rightarrow MnO_{xs} \qquad (7.64)$$

The second step involves a faster surface-catalyzed reaction:

$$MnO_{xs} + Mn^{2+} \rightarrow 2MnO_{2s} \qquad (7.65)$$

The following rate law has been suggested (Stumm and Morgan, 1981):

$$-dMn^{2+}/dt = [k_1' (Mn^{2+}) + k_2'[MnO_{2s}] [Mn^{2+}] \qquad (7.66)$$

where

$$k_1' = k_1[O_{2aq}][OH^-]^2 \qquad (7.67)$$

and

$$k_2' = k_2[O_{2aq}][OH^-]^2 \qquad (7.68)$$

Quantifying Reactions 7.64 and 7.65 requires fixing the pH and partial pressure of O_2 (pO_2) at some predetermined value and providing OH^- upon demand. This is accomplished with a pH-stat technique. The technique utilizes a pH electrode as a sensor so that as OH^- is consumed, (during Mn^{2+} oxidation), the instrument measures the rate of OH^- consumption and activates the autoburete to replace the consumed OH^-. It is assumed that for each OH^- consumed, an equivalent amount of Mn^{2+} is oxidized.

The data in Figure 7.15 demonstrate kinetics of Mn^{2+} oxidation using the pH-stat technique. The data show at least two major slopes. The first slope (near the origin) represents Reaction 7.64, whereas the second slope represents the autocatalytic part of the reaction (Reaction 7.65). The data demonstrate that the reaction is pH-dependent. As pH increases, the autocatalytic part of the reaction represents the mechanism by which most Mn^{2+} oxidizes. Similar reactions for Fe^{2+} are shown in Figure 7.16. Note that Fe^{2+} oxidizes at a much lower pH than Mn^{2+}.

Reductive Precipitation. Reductive precipitation involves the production of reduced species with limited solubility. An example of reductive precipitation in the environment involves the reduction of SO_4 to H_2S and the precipitation of metals as metal-sulfides. In nature, the process of reductive precipitation is mostly microbiologically controlled. Production of H_2S is the rate-controlling reaction of metal–sulfide precipitate formation.

$$HS^- + M^{2+} \Leftrightarrow MSs + H^+ \qquad (7.69)$$

since Reaction 7.69 is known to be faster than SO_4 reduction. Such reductive precipitation reactions are known to reach completion within minutes to hours, depending on the degree of diffusion needed for the reactants to meet.

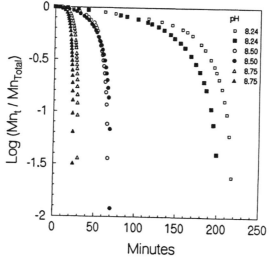

Figure 7.15. Pseudo first-order Mn^{2+} oxidation under various pH values at pO_2 of 0.2 in duplicate using a pH-stat technique (from Evangelou, unpublished data).

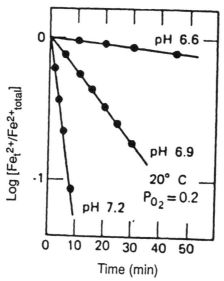

Figure 7.16. Pseudo first-order Mn^{2+} oxidation under various pH values at pO_2 of 0.2 in duplicate using a pH-stat technique (from Stumm and Morgan, 1970, with permission).

7.3.4 Effect of Ionic Strength on Kinetics

Reaction kinetics are known to be affected by ionic strength via two mechanisms. One mechanism is physical in nature and is related to the magnitude of ionic strength, whereas the second mechanism is considered chemical and is related to the charge of the ions. The two mechanisms affecting reaction rate can be explained by considering that

$$A^{z_A} + B^{z_B} \rightarrow AB^{z_A + z_A} \tag{7.70}$$

and

$$k_{exp} = k_{id}(\gamma_A \gamma_B / \gamma_{AB}) \tag{7.71}$$

where k_{exp} denotes experimental rate constant under a given ionic strength, k_{id} denotes rate constant at infinite dilution, and γ_A, γ_B denote activity coefficients of ions A and B, respectively. Taking logarithms on both sides of Equation 7.71 and using the Debuy–Huckle limiting law to express activity coefficients,

$$\log \gamma_A = -A z_i^2 (I)^{1/2} \tag{7.72}$$

then

$$\log (k_{exp}/k_{id}) = -A z_A^2 (I)^{1/2} - A z_B^2 (I)^{1/2} + A(z_{AB})^2 (I)^{1/2} \tag{7.73}$$

Considering that

$$z_{AB} = z_A + z_B \tag{7.74}$$

by substituting Equation 7.74 into Equation 7.73,

$$\log (k_{exp}/k_{id}) = A(I)^{1/2} \{ -z_A^2 - z_B^2 + (z_A + z_B)^2 \} \tag{7.75}$$

Collecting terms, and replacing A with 0.5, gives

$$\log(k_{exp}/k_{id}) = 1.0 z_A z_B (I)^{1/2} \tag{7.76}$$

A plot of $\log(k_{exp}/k_{id})$ versus $(I)^{1/2}$ would produce a straight line with slope $z_A z_B$. Benson (1982) pointed out that when one of the z_i values in Equation 7.76 is zero, the ionic strength would not have any influence on the reaction rate. When one of the z_i values is negative and the other is positive, the influence of ionic strength on the reaction rate should be negative, whereas when both z_i values are positive or negative, the influence of ionic strength on reaction rates should be positive. It follows that two factors (with respect to z_i) control the role of ionic strength on reaction rate constants. The first factor is the absolute magnitude of z_i, and the second factor is the sign of z_i. The statements above are demonstrated in Figure 7.17.

Millero and Izaguirre (1989) examined the effect various anions have on the abiotic oxidation rate of Fe^{2+} at constant ionic strength ($I = 1.0$) and found that this effect was on the order of $HCO_3^- > Br^- > NO_3^- > ClO_4^- > Cl^- > SO_4^{2-} > B(OH)_4^-$ (see also Fig. 7.18). Strong decrease in the rate of Fe^{2+} oxidation due to the addition of SO^{2-} and

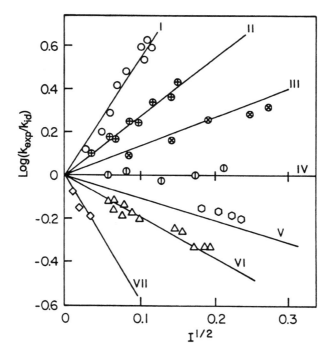

I: $[Co(NH_3)_5Br)]^{2+} + Hg^{2+} \rightarrow [Co(NH_3)_5(H_2O)]^{3+} + (HgBr)^+$
II: $S_2O_3^- + I^- \rightarrow ?[ISO_4 + SO_4^{2-}] \rightarrow I_3^- + 2SO_4^{2-}$ (not balanced)
III: $[O_2N\text{-}N\text{-}COOEt]^- + OH^- \rightarrow N_2O + CO_3^{2-} + EtOH$
IV: cane sugar + $H^+ \rightarrow$ invert sugar (hydrolysis reaction)
V: $H_2O_2 + Br^- \rightarrow H_2O + 1/2Br_2$ (not balanced)
VI: $[Co(NH_3)_5Br)]^{2+} + OH^- \rightarrow [Co(NH_3)_5(OH)]^{2+} + Br^-$
VII: $Fe^{2+} Co(C_2O_4)_3^{3-} \rightarrow Fe^{3+} + Co(C_2O_4)_3^{4-}$

Figure 7.17. The effect of ionic strength on the rates of some ionic reactions (from Benson, 1982, with permission).

B(OH)$_4$– pairs was attributed to the formation of FeSO$_4^0$ and Fe[B(O$_4$H)$_4$]$^+$ pairs which they assumed were difficult to oxidize. They also reported that the oxidation of Fe^{2+} is first-order with respect to HCO$_3^-$. This HCO$_3^-$ dependence of Fe^{2+} oxidation could be related to the formation of an FeHCO$_3^+$ pair which has a faster rate of oxidation than the Fe(OH)$_2^0$ pair.

7.3.5 Determining Reaction Rate Order

An approach to establish rate order for an experimental data set is as follows: Consider the generalized reaction

$$A \rightarrow products \tag{7.77}$$

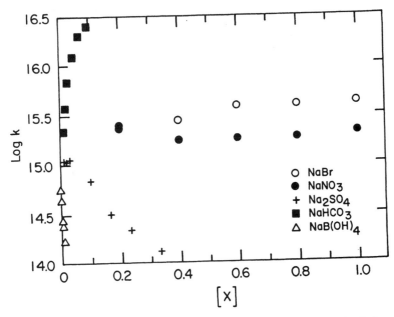

Figure 7.18. The effect of anions (X) on the oxidation of Fe(II) in NaCl–NaX solutions at $I = 1$ and 25°C (from Millero and Izaguirre, 1989, with permission).

The rate function is given by

$$-dA/dt = k[A]^n \tag{7.78}$$

Taking logarithms on both sides of the equation,

$$\log[-dA/dt] = \log k + \log[A]^n \tag{7.79}$$

or

$$\log[-dA/dt] = \log k + n\log[A] \tag{7.80}$$

If a plot of log $[-dA/dt]$ versus log[A] is a straight line, the slope of the line is the reaction order with respect to A. Figure 7.19 represents Mn^{2+} oxidation at pH 8.5 and pO_2 0.2 obtained by pH-stat technique. The technique utilizes a pH electrode as a sensor so that as OH^- is consumed (during Mn^{2+} oxidation), the instrument measures the rate of OH^- consumption and activates the autoburete to replace the consumed OH^-. It is assumed that for each OH^- consumed, an equivalent amount of Mn^{2+} is oxidized. These pH-stat data clearly show that the first part of the oxidation of manganese is zero-order, whereas the second part is first-order. Keep in mind, however, that these particular reaction orders are strictly empirical and without necessarily any mechanistic meaning.

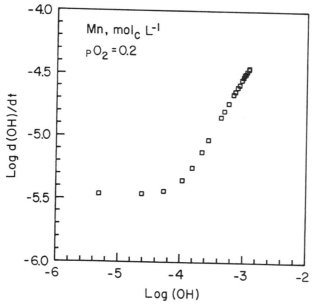

Figure 7.19. Relationship between log dOH/dt and log(OH) for Mn^{2+} using a pH–stat technique (note that OH represents dissolved manganese in mol L^{-1}; see explanation in the text).

7.4   OTHER KINETIC MODELS

A number of additional equations are often used to describe reaction kinetics in soil–water systems. These include the Elovich equation, the parabolic diffusion equation, and the fractional power equation. The Elovich equation was originally developed to describe the kinetics of gases on solid surfaces (Sparks, 1989, 1995 and references therein). More recently, the Elovich equation has been used to describe the kinetics of sorption and desorption of various inorganic materials in soils. According to Chien and Clayton (1980), the Elovich equation is given by

$$q_t = (1/\beta) \ln (\alpha/\beta + 1/\beta)\ln t \tag{7.81}$$

where q_t = amount of the substance that has been sorbed at any time t; α and β are constants. A plot of q_t versus $\ln t$ would give a linear relationship with slope $1/\beta$ and y intercept $1/\beta \ln (\alpha/\beta)$. One should be aware that fit of data to the Elovich equation does not necessarily provide any mechanistic meaning (Fig. 7.20).

The parabolic diffusion equation is used to describe or indicate diffusion control processes in soils. It is given by (Sparks, 1995 and references therein)

$$q_t/q_\infty = (4/\pi^{1/2})R_D t^{1/2} - C \tag{7.82}$$

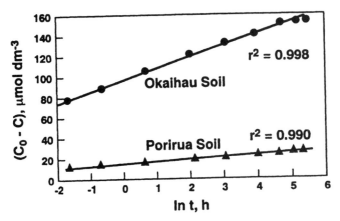

Figure 7.20. An Elovich equation plot representing phosphate sorption on two soils where C_0 is the initial phosphorus concentration added at time 0 and C is the phosphorus concentration in the soil solution at time t. The quantity (C_0-C) equals q_t, the amount sorbed at any time (from Chien and Clayton, 1980, with permission).

where q_t and q_∞ denote amount of the substance sorbed at time t and infinity (∞), respectively, R_D denotes the overall diffusion coefficient, and C is an experimental constant. The parabolic diffusion equation has been used to describe metal reactions (Sparks, 1995 and references therein).

The fractional power equation is

$$q = kt^v \tag{7.83}$$

where q is substance sorbed at any time t, k is empirical rate constant, and v is an experimental exponent. The equation is strictly an empirical one, without any mechanistic meaning.

Many reactions in actual soil–water systems are controlled by mass transfer or diffusion of reactants to the surface minerals or mass transfer of products away from the surface and to the bulk water. Such reactions are often described by the parabolic rate law (Stumm and Wollast, 1990). The reaction is given by

$$dc/dt = kt^{-1/2} \tag{7.84}$$

where c equals concentration, k is a rate constant, and t equals time. Integrating using the appropriate boundary conditions gives

$$C = C_0 + 2kt^{1/2} \tag{7.85}$$

where C_0 is original quantity of product and C is quantity of product at any time t. This equation has often been used to describe mineral dissolution. For more information, see Sparks (1995 and references therein).

7.5 ENZYME-CATALYZED REACTIONS (CONSECUTIVE REACTIONS)

Enzyme- or surface-catalyzed reactions involve any substrate reacting with a mineral surface or enzyme to form a complex. Upon formation of the complex, a product is formed, followed by product detachment and regeneration of the enzyme or surface (Epstein and Hagen, 1952 and references therein). These kinetic reactions are known as Michaelis–Menten reactions and the equation describing them is also known as the Michaelis–Menten equation.

The Michaelis–Menten equation is often employed in soil–water systems to describe kinetics of ion uptake by plant roots and microbial cells, as well as microbial degradation–transformation of organics (e.g., pesticides, industrial organics, nitrogen, sulfur, and natural organics) and oxidation or reduction of metals or metalloids. Derivation of the Michaelis–Menten equation(s) is demonstrated below.

7.5.1 Noncompetitive Inhibition, Michaelis–Menten Steady State

In the case of a steady state, the kinetic expression describing formation of product P_f can be written as follows (Segel, 1976):

$$\text{E–} + \text{S} \underset{k_{-1}}{\overset{k_1}{\Leftrightarrow}} \text{E–S} \overset{k_p}{\rightarrow} P_f \tag{7.86}$$

where

k_p = rate constant of product P_f formation
E– = enzyme surface
S = substrate
E–S = enzyme–substrate complex
k_1 = rate constant of the forward reaction
k_{-1} = rate constant of the reverse reaction

Equation 7.86 shows formation of complex E–S, product generation (P_f), and regeneration of E–. The equation describing the process is given by

$$V_{P_f} = V_{P_{fmax}} [\text{S}] / (K_m + [\text{S}]) \tag{7.87}$$

(see next boxed section for the derivation of the equation). Considering that when V_{P_f} equals 1/2 of the reaction rate at maximum (1/2 $V_{P_{fmax}}$), then

$$1/2\, V_{P_{fmax}} = V_{m_{fmax}} [\text{S}] / (K_m + [\text{S}]) \tag{7.88}$$

upon rearranging,

$$K_m = \text{S} \tag{7.89}$$

K_m denotes the concentration of the substrate at which one-half of the enzyme sites (surface-active sites) are saturated with the substrate (S) and for this reason the reaction rate is $1/2\ V_{P_{fmax}}$. Therefore, the parameter K_m denotes affinity of the substrate by the surface. The higher the value of K_m is, the lower the affinity of the substrate (S) by the surface (E–). This is demonstrated in Figure 7.21, which shows an ideal plot of V_{P_f} versus S producing a curvilinear line asymptotically approaching $V_{P_{fmax}}$.

Equation 7.87 can be linearized by taking the inverse. Thus, it transforms to

$$\frac{1}{V_{P_f}} = \frac{1}{V_{P_{fmax}}} + \frac{K_m}{V_{P_{fmax}}} \cdot \frac{1}{[S]} \tag{7.90}$$

A plot of $1/V_{P_f}$ versus $1/S$ produces a straight line with slope $K_m/V_{P_{fmax}}$ and y intercept $1/V_{P_{fmax}}$ (Fig. 7.22). Thus, the linear form of the Michaelis–Menten equation allows estimation of the so-called adjustable parameters (Segel, 1976). The adjustable parameters include K_m (in units of concentration) and $V_{P_{fmax}}$ (in units of product quantity per unit surface per unit time, or quantity of product per unit time), which denotes maximum rate of product formation.

The Michaelis–Menten equation thus provides the means for predicting rates of enzyme- or surface-catalyzed reactions. These predictions can be made because the concentration of enzyme- or surface-reactive centers is constant and small compared with the concentration of reactants with which they may combine. To generate data for a particular enzyme- or surface-catalyzed process, data representing product formation is plotted with respect to time, as in Figure 7.23. The rate of the reaction for a given substrate concentration C_{si} is determined by the slope of the curves as it approaches zero (see tangents in Figure 7.23). These slope data are then plotted against

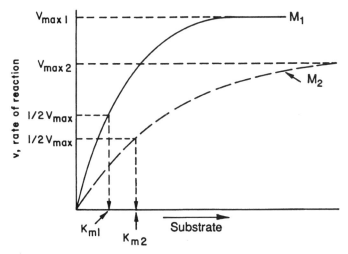

Figure 7.21. Ideal Michaelis–Menten plots showing the relationship between rate of NH_4 transformation, V, and substrate (S) concentration.

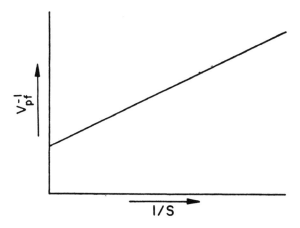

Figure 7.22. Ideal double reciprocal plot of the Michaelis–Menten equation.

substrate concentration, producing a Michaelis–Menten plot (Fig. 7.21). An example of actual experimental noncompetitive Michaelis–Menten data in the form of normal and linearized plots is shown in Figure 7.24, which demonstrates pyrite oxidation by Fe^{3+}, an electron acceptor.

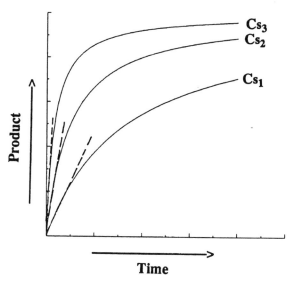

Figure 7.23. Ideal plot showing product formation (V_{P_f}) as a function of time and three different substrate concentrations [Cs_1 (lowest) to Cs_3 (highest)]. The tangent on the curve as $t \to 0$ denotes the experimentally defined steady state; its slope represents V_{P_f}.

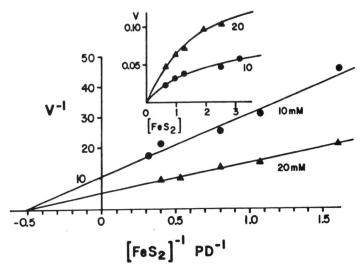

Figure 7.24. Effect of HP washed pyrite concentration on the indirect leaching of pyrite at two fixed Fe^{3+} concentrations (10 and 20 mM). FeS_2 % PD [PD = (g of pyrite per mL solution) $\times$ 100). HP = 0.1 g of K_2HPO_4, 0.4 g of $(NH_4)_2SO_4$), and 0.4 g of $MgSO_4 \cdot 7H_2O$ L^{-1}, adjusted to pH 2.3 with H_2SO_4 (from Lizama and Suzuki, 1989, with permission).

DERIVATION OF THE NONCOMPETITIVE EQUATION

The expression describing product (P_f) formation can be written as (Segel, 1976)

$$E- + S \underset{k_{-1}}{\overset{k_1}{\Longleftrightarrow}} E\text{-}S \overset{k_p}{\rightarrow} P_f \tag{A}$$

where
k_p = rate constant of product (P_f) formation
E– = enzyme surface
S = substrate
E–S = enzyme–substrate complex
k_1 = rate constant of the forward reaction and
k_{-1} = rate constant of the reverse reaction

Assuming that reaction $P_f \rightarrow$ E–S is rate limiting, for example, due to rapid removal of P_f, the reaction rate can be expressed as

$$d[\text{E–S}]/dt = k_1[\text{E–}][\text{S}] - k_{-1}[\text{E–S}] - k_p[\text{E–S}] \tag{B}$$

and under steady-state considerations,

$$d[\text{E–S}]/dt = k_1[\text{E–}][\text{S}] - k_{-1}[\text{E–S}] - k_p[\text{E–S}] = 0 \tag{C}$$

By rearranging Equation C,

$$k_1[\text{E–}][\text{S}] = \text{E–S}[k_{-1} + k_p] \tag{D}$$

Collecting terms and rearranging again,

$$K_m = [k_{-1} + k_p]/k_1 = [\text{E–}][\text{S}]/[\text{E–S}] \tag{E}$$

Considering that

$$E_{max} = [\text{E–}] + [\text{E–S}] \tag{F}$$

where E_{max} = maximum number of adsorption sites capable of forming E–S; substituting E– of Equation F with the expression

$$[\text{E–}] = [\text{E–S}][K_m]/[\text{S}] \tag{G}$$

which is obtained from Equation E, and rearranging,

$$[\text{E–S}] = [E_{max}][\text{S}]/(K_m + [\text{S}]) \tag{H}$$

The rate function describing product formation (P_f) is given by

$$d[P_f]/dt = V_{P_f} = k_p[\text{E – S}] \tag{I}$$

By substituting Equation H into Equation I and rearranging, rate of product formation (V_{P_f}) is given by

$$V_{P_f} = \frac{k_p[E_{max}][\text{S}]}{(K_m + [\text{S}])} \tag{J}$$

The rate of P_f formation (V_{P_f}) is at maximum ($V_{P_{fmax}}$) when S $\ggg K_m$. Thus,

$$V_{P_{fmax}} = k_p[E_{max}] \tag{K}$$

and Equation J takes the form

$$V_{P_f} = V_{P_{fmax}}[\text{S}]/(K_m + [\text{S}]) \tag{L}$$

A plot of V_{P_f} versus S produces a curvilinear line asymptotically approaching $V_{P_{fmax}}$. Equation L can be linearized by taking its inverse:

$$\frac{1}{V_{P_f}} = \frac{1}{V_{P_{fmax}}} + \frac{K_m}{V_{P_{fmax}}} \cdot \frac{1}{[\text{S}]} \tag{M}$$

7.5.2   Competitive Inhibition

The case of a competitive inhibition is described by (Segel, 1976):

$$
\begin{array}{c}
\text{E-} + \text{S} \quad \overset{K_S}{\Leftrightarrow} \quad \text{E-S} \quad \overset{k_p}{\to} \quad P_f \\
+ \\
\text{I} \\
\updownarrow K_i \\
\text{E-I}
\end{array}
\qquad (7.91)
$$

Reaction 7.91 shows that S and I (I = inhibitor) compete for the same E– sites. Under these conditions, the linearized form of the competitive equation is

$$
\frac{1}{V_{P_f}} = \frac{1}{V_{P_{fmax}}} + \frac{K_s}{V_{P_{fmax}}}\left(1 + \frac{[I]}{K_i}\right)\frac{1}{[S]} \qquad (7.92)
$$

The components K_s and K_i denote equilibrium constants for the E–S and E–I complexes, respectively. Equation 7.92, when plotted as $1/V_{P_f}$ versus $1/[S]$ for various quantities of inhibitor I, produces a straight line for each I with a common y intercept, but a different slope. The slope would differ from that of a system without I by the factor

$$
1 + ([I])/K_i \qquad (7.93)
$$

This is demonstrated ideally in Figure 7.25. The derivation of the competitive Michaelis–Menten equation is given in the next boxed section.

The experimental data in Figure 7.26 show normal and linearized competitive Michaelis–Menten plots of Rb^+ uptake by plant roots. The data clearly demonstrate

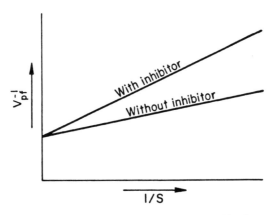

Figure 7.25. Ideal double reciprocal plots of competitive interactions.

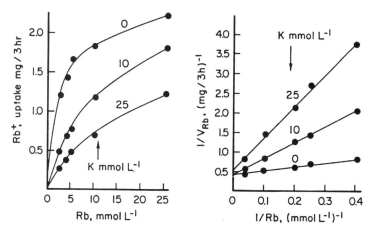

Figure 7.26. Normal and linearized Michaelis–Menten plots describing rubidium (Rb^+) uptake by plant roots under three different concentrations of K^+ (competitive process) (from Epstein and Hagen, 1952, with permission).

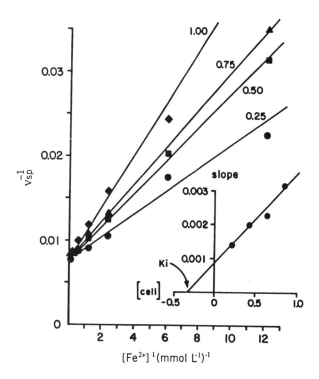

Figure 7.27. Competitive inhibition of Fe^{2+} oxidizing activity of SM-4 cells at concentrations of 0.25, 0.50, 0.75, and 1.0 mg mL^{-1}. The V_{sp}^{-1} was calculated by dividing rate V by cell concentration. The insert is a plot of the slope versus cell concentration to obtain the inhibition constant K_i or K'_{eq} in milligrams of cells per milliliter (from Suzuki et al., 1989, with permission).

that uptake of Rb^+ in the presence of K^+ is a competitive process. In addition, data in Figure 7.27 describe microbial oxidation of Fe^{2+}. These data show that pyrite oxidation under various microbial cell numbers is also a competitive process. In essence, as the number of microbial cells increases, they compete for available Fe^{2+} and the reaction rate appears to decrease. A plot of slope (obtained from the linearized competitive Michaelis–Menten plot) versus cell weight (mg L^{-1}) produces a linear relationship with slope $K_m/(K_i k_p)$, with x intercept $-K_i$ and y intercept K_m/k_p.

DERIVATION OF COMPETITIVE INHIBITION

The case of a competitive inhibition (Reaction 7.91) can be described as follows (Segel, 1976):
The total enzyme-surface reactive groups are given by

$$[E_{max}] = [E-] + [E-I] + [E-S] \tag{A}$$

and under rapid equilibrium

$$K_s = \frac{[E-][S]}{[E-S]} \tag{B}$$

and

$$K_i = \frac{[E-][I]}{[E-I]} \tag{C}$$

Letting V_{P_f} be the velocity of P_f production,

$$V_{P_f} = k_p[E-S] \tag{D}$$

and

$$V_{P_{fmax}} = k_p[E_{max}] \tag{E}$$

Dividing Equation D by Equation A and incorporating the quotient into Equation E gives

$$\frac{V_{P_f}}{V_{P_{fmax}}} = \frac{[E-S]}{[E-] + [E-I] + [E-S]} \tag{F}$$

By incorporating Equations B and C into Equation F, one obtains

$$\frac{V_{P_f}}{V_{P_{fmax}}} = \frac{[S]}{K_s\left(1 + \frac{[I]}{K_i}\right) + [S]} \tag{G}$$

and by inverting Equation G,

$$\frac{V_{P_{fmax}}}{V_{P_f}} = 1 + K_s\left(1 + \frac{[I]}{K_i}\right)\frac{1}{[S]} \tag{H}$$

Equation H can be linearized by taking its inverse with respect to reaction velocity (V_{P_f}) and concentration of S, giving

$$\frac{1}{V_{P_f}} = \frac{1}{V_{P_{fmax}}} + \frac{K_s}{V_{P_{fmax}}}\left(1 + \frac{[I]}{K_i}\right)\frac{1}{[S]} \tag{I}$$

7.5.3 Uncompetitive Inhibition

An uncompetitive Michaelis–Menten inhibition reaction is shown below (Segel, 1976)

$$E-+S \xrightarrow{K_S}\; E-S \xrightarrow{k_p} P_f \tag{7.94}$$
$$+$$
$$I$$
$$\updownarrow K_i$$
$$E-S\cdot I$$

where E–S·I denotes specific adsorption complex. In uncompetitive inhibition, inhibitor I competes for enzyme surface sites on E– occupied by S. The inhibitor I reacts specifically with E–S to form E–S·I. However, specific reaction of I with E–S alters the potential of S to form the product.

The linearized form of the uncompetitive Michaelis–Menten equation is given by taking its inverse with respect to reaction velocity (V_{P_f}) and concentration of S, giving

$$\frac{1}{V_{P_f}} = \frac{1}{V_{P_{fmax}}}\left(1 + \frac{[I]}{K_i}\right) + \frac{K_s}{V_{P_{fmax}}[S]} \tag{7.95}$$

When plotted as $1/V_{P_f}$ versus $1/[S]$ for various quantities of inhibitor I, Equation 7.95 produces a straight line for each quantity of I with a different y intercept for each chosen I, but the same slope (Segel, 1976). The y intercept differs from that of a system without inhibitor I by the factor

$$1 + ([I])/K_i \tag{7.96}$$

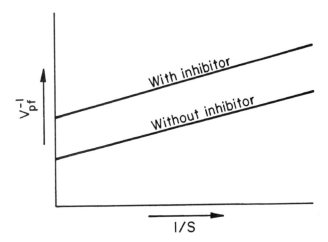

Figure 7.28. Ideal double reciprocal plots of uncompetitive interaction.

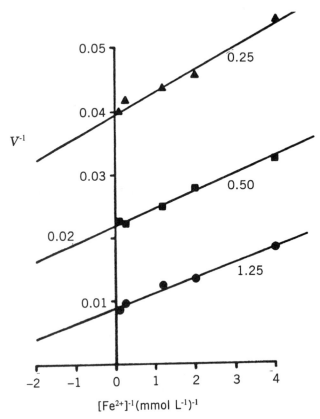

Figure 7.29. Effect of Fe^{2+} concentrations on the Fe^{2+}-oxidizing activity of SM-5 cells at concentrations 0.25, 0.50, and 0.25 mg mL^{-1} (from Suzuki et al., 1989, with permission).

The derivation of the uncompetitive equation is given in the next boxed section.

The ideal linearized plot representing Equation 7.95 is shown in Figure 7.28. Actual experimental data of the uncompetitive form is shown in Figure 7.29 which shows Fe^{2+} oxidation by *Thiobacillus*.

DERIVATION OF UNCOMPETITIVE INHIBITION

An uncompetitive Michaelis–Menten inhibition reaction is shown below (Segel, 1976)

$$E- + S \quad \overset{K_S}{\Leftrightarrow} \quad E\text{–}S \quad \overset{k_p}{\to} \quad P_f \tag{A}$$
$$+$$
$$I$$
$$\Updownarrow K_i$$
$$E\text{–}S\text{·}I$$

and

$$[E_{max}] = [E-] + [E\text{–}S] + [E\text{–}S\text{·}I] \tag{B}$$

Under rapid equilibrium,

$$K_s = \frac{[E-][S]}{[E\text{–}S]} \tag{C}$$

and

$$K_i = \frac{[E\text{–}S][I]}{[E\text{–}S\text{·}I]} \tag{D}$$

Letting V_{P_f} be the velocity of P_f production,

$$V_{P_f} = k_p[E\text{–}S] \tag{E}$$

and

$$V_{P_{fmax}} = k_p[E_{max}] \tag{F}$$

An expression relating V_{P_f}, $V_{P_{fmax}}$, [I], K_s, K_i, and [S] can be derived as in the case of competitive inhibition:

$$\frac{V_{P_f}}{[E_{max}]} = \frac{k_p[E\text{–}S]}{[E-] + [E\text{–}S] + [E\text{–}S\text{·}I]} \tag{G}$$

Upon rearranging,

$$\frac{V_{P_f}}{V_{P_{fmax}}} = \frac{[S]}{K_s + \left(1 + \dfrac{[I]}{K_i}\right)[S]}$$ (H)

Equation H can be linearized by taking its inverse with respect to reaction velocity and concentration of S, giving

$$\frac{1}{V_{P_f}} = \frac{1}{V_{P_{fmax}}}\left(1 + \frac{[I]}{K_i}\right) + \frac{K_s}{V_{P_{fmax}}[S]}$$ (I)

7.5.4 Competitive–Uncompetitive Inhibition

A competitive–uncompetitive inhibition is shown below (Segel, 1976):

$$\begin{array}{ccccc}
 & K_S & & k_p & \\
E-\ +\ S & \Leftrightarrow & E-S & \to & P_f \\
+ & & + & & \\
I & & I & & \\
\updownarrow K_i & K_S & \updownarrow K_i & & \\
E-I + S & \Leftrightarrow & E-S\cdot I & &
\end{array}$$ (7.97)

In this case, species S undergoes two types of surface reactions. In the first, species S may react specifically with the enzyme surface to produce the E–S·I complex,

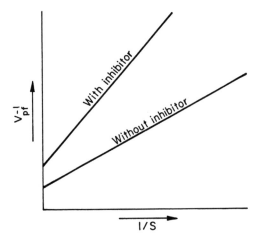

Figure 7.30. Ideal double reciprocal plots of competitive–uncompetitive interaction.

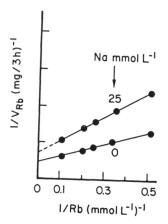

Figure 7.31. Kinetics of rubidium uptake by plant roots in the presence of sodium (Na^+) (adapted from Epstein and Hagen, 1952).

inhibiting product formation. In the second type of reaction, species S competes with I for sites on E−. The derived linear equation is

$$\frac{1}{V_{P_f}} = \frac{1}{V_{P_{fmax}}}\left(1 + \frac{[I]}{K_i}\right) + \frac{K_s}{V_{P_{fmax}}}\left(1 + \frac{[I]}{K_i}\right)\frac{1}{[S]} \qquad (7.98)$$

A plot of $1/V_{P_f}$ versus $1/[S]$ for the competitive–uncompetitive system would differ from a similar plot in the absence of inhibitor S in the slope and y intercept by

$$1 + [I]/K_i \qquad (7.99)$$

This is demonstrated in Figure 7.30. Actual competitive–uncompetitive data demonstrating uptake of Rb^+ in the presence of Na^+ are shown in Figure 7.31.

COMPETITIVE–UNCOMPETITIVE INHIBITION

A competitive–uncompetitive Michaelis–Menten inhibition reaction is shown below (Segel, 1976):

$$
\begin{array}{ccccc}
 & K_S & & k_p & \\
E- + S & \Leftrightarrow & E\text{-}S & \to & P_f \\
+ & & + & & \\
I & & I & & \\
\updownarrow K_i & K_S & \updownarrow K_i & & \\
E\text{-}I + S & \Leftrightarrow & E\text{-}S{\cdot}I & &
\end{array}
\qquad (A)
$$

In this case, the total number of reactive sites on the enzyme-surface is given by

$$[E_{max}] = [E-] + [E–I] + [E–S] + [E–S\cdot I] \tag{B}$$

Under rapid equilibrium

$$K_s = \frac{[E-][S]}{[E–S]} = \frac{[E–I][S]}{[E–S\cdot I]} \tag{C}$$

$$K_i = \frac{[E-][I]}{[E–I]} = \frac{[E–S][I]}{[E–S\cdot I]} \tag{D}$$

Letting V_{P_f} be the velocity of P_f production,

$$V_{P_f} = k_p[E–S] \tag{E}$$

and

$$V_{P_{fmax}} = k_p[E_{max}] \tag{F}$$

An expression relating V_{P_f}, $V_{P_{fmax}}$, [I], K_s, K_i, and [S] can be derived as in the case of competitive inhibition, producing

$$\frac{V_{P_f}}{[E_{max}]} = \frac{k_p[E–S]}{[E-] + [E–I] + [E–S] + [E–S\cdot I]} \tag{G}$$

and

$$\frac{V_{P_f}}{V_{P_{fmax}}} = \frac{[S]}{K_s + [S] + \left(\dfrac{K_s}{K_i}\right)[I] + \dfrac{[I]}{K_i}[S]} \tag{H}$$

Equation H can be linearized by taking its inverse with respect to reaction velocity and concentration of S, giving

$$\frac{1}{V_{P_f}} = \frac{1}{V_{P_{fmax}}}\left(1 + \frac{[I]}{K_i}\right) + \frac{K_s}{V_{P_{fmax}}}\left(1 + \frac{[I]}{K_i}\right)\frac{1}{[S]} \tag{I}$$

Double reciprocal plots have been extensively used in enzymatic studies to quantify the Michaelis–Menten adjustable parameters V_{max} and K_m. Such plots, however, provide excellent correlations but poor accuracy in predicting K_m and V_{max} (Dowd and Riggs, 1965; Kinniburgh, 1986; Evangelou and Coale, 1987). Kinniburgh (1986) reported that the double reciprocal plot produces erroneous Michaelis–Menten adjustable parameters because it alters the original error distribution in the data. He further concluded that the most acceptable linear data transformation is that which changes the original error distribution in the experimental data the least. The most acceptable linear transformations are those of V_P versus $V_P/[S]$ (Hofstee plot) or $[S]/V_P$ versus $[S]$ (Hanes–Woolf plot) (Segel, 1976; Dowd and Riggs, 1965; Evangelou and Coale, 1987). Kinniburgh (1986) concluded that the best approach to estimate adjustable parameters is through nonlinear regression.

7.6 FACTORS CONTROLLING REACTION RATES

Reaction rates are characterized by the number of successful collisions between reactants, which is represented by the product of the total number of collisions of the reactants, the number of collisions that have sufficient energy to cause a reaction event (energy factor), and the fraction of collisions that have the proper orientation (probability factor). The theoretical reaction rate (k) is given by

$$k = PZe^{-E_\alpha/RT} \tag{7.100}$$

where
P = probability factor
Z = collision frequency
E_α = activation energy
T = temperature
$e^{-E_\alpha/RT}$ = fraction of collisions with energy sufficiently large to cause a reaction event.

The product PZ is related to the preexponential factor (A) of the Arrhenius reaction and the entropy (S) of the reaction by the relationship

$$PZ = A\,e^{S/R} \tag{7.101}$$

Based on Reactions 7.100 and 7.101, the reaction rate constant is inversely related to the activation energy (E_α) of the reaction and directly related to the entropy of activation.

7.6.1 Temperature Influence

For heterogeneous reactions, the observed reaction rate is determined by the amount of surface covered by reacting molecules and by the specific velocity of the surface reaction. The influence of temperature on the rate, therefore, must include two factors, the effect on the surface area covered, and the effect on the surface reaction itself.

As for homogeneous reactions, the influence of temperature on the rate constant of heterogeneous reactions is given by the Arrhenius equation:

$$k = A \exp(-E_\alpha^*/RT) \tag{7.102}$$

where

k = specific rate constant (time^{-1})

A = frequency factor or preexponential factor (a constant)

E_α^* = activation energy (kJ mol^{-1})

R = universal gas constant, 8.31441 JK^{-1} mol^{-1} in degrees Kelvin

T = absolute temperature

The term E_α^* is often referred to as the "apparent energy of activation." This energy, evaluated from the observed rate constants, is a composite term and is not necessarily the energy required to activate the reactants on the surface, which is the true activation energy. The apparent energy of activation includes not only the true activation energy of the surface reaction, but also heats of adsorption of reactants, or reactants and products, to yield an apparent activation energy which may be quite different from the true.

If Equation 7.102 is integrated, and E_α is not itself temperature dependent, and hence a constant,

$$\ln k = (-E_\alpha^*/RT) + \ln A \tag{7.103}$$

where ln A is the constant of integration. In terms of logarithms to the base 10, Equation 7.103 may be written as

$$\log k = (-E_\alpha^*/2.303R)(1/T) + \log A \tag{7.104}$$

From Equation 7.104, it follows that a plot of the logarithm of the rate constant against the reciprocal of the absolute temperature should be a straight line with

$$\text{slope} = E_\alpha^*/2.303R = E_\alpha^*/19.15 \times 10^{-3} \tag{7.105}$$

and y intercept equal to $\log_{10} A$. By taking the slope of the line, E_α^* may be calculated readily from Equation 7.105:

$$E_\alpha^* = 19.15 \times 10^{-3}(\text{slope}) \text{ kJ mol}^{-1} \tag{7.106}$$

Activation energies less than 42 kJ mol^{-1} indicate diffusion-controlled reactions, whereas reactions with E_α values higher than 42 kJ mol^{-1} indicate chemical reactions or surface-controlled processes. The data in Figure 7.32 represent rate constants (k) for the acid dissolution of octahedral aluminum in kaolinite plotted against the reciprocals of the respective temperatures. From the slope of the line, the apparent energy of activation for dissolution of octahedral aluminum was found to be 101.7 kJ mol^{-1}.

Apparent activation energies of acid dissolution of calcite and dolomite and several samples of agricultural limestone indicate that the activation energy of dolomite varies

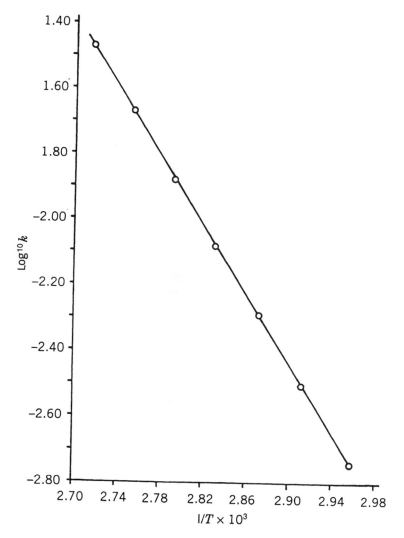

Figure 7.32. Graph of log k versus $1/T$ for the dissolution of aluminum in kaolinite (adapted from Turner 1966).

from 45.72 to 51.13 kJ mol^{-1}. The apparent activation energy of calcite is 12.47 kJ mol^{-1}. Apparent activation energies of the agricultural limestone samples varied from 7.48 to 51.13 kJ mol^{-1}. Samples with apparent activation energies ranging from approximately 40 to 50 kJ mol^{-1} represented mainly dolomitic samples, while limestone samples with apparent activation energies ranging from approximately 7 to 20 kJ mol^{-1} represented mainly calcite (Table 7.1). Apparent activation energies of the catalytic part of the Mn^{2+} oxidation reaction is in the range of 144 kJ mol^{-1} (Fig. 7.33).

TABLE 7.1. Specific Surface Area and Acid-Dissolution Kinetic Parameters of Calcite, Dolomite, and Limestone Samples

Sample Number	Sample Identification	Regression Equations	Specific Surface Area $(m^2\,g^{-1})$	Rate Constants (min^{-1})			Activation Energies (E_α)	
				14°C	21°C	30°C	$(kJ\,mol^{-1})$	r^2
1	Dolomite	$y = -6150x + 19.98$	0.35	0.21	0.41	0.72	51.13	0.9973
2	Dolomite	$y = -6100x + 19.12$	0.18	0.11	0.19	0.37	51.13	0.9967
3	Dolomite	$y = -5350x + 17.31$	0.19	0.24	0.43	0.70	44.48	0.9965
4	Calcite	$y = -1300x + 5.34$	0.18	2.24	2.41	2.91	10.81	0.9337
5	Calcite	$y = -900x + 3.66$	0.20	1.66	1.80	1.99	7.48	0.9959
6	Calcite	$y = -2300x + 7.85$	0.17	0.79	1.10	1.26	19.12	0.9407
7	Calcite	$y = -3150x + 11.57$	0.18	1.80	2.18	3.39	26.19	0.9501
	Ward's calcite	$y = -1500x + 5.96$	0.02	2.12	2.20	2.86	12.47	0.8479
	Ward's dolomite	$y = -5500x + 16.03$	0.02	0.04	0.07	0.12	45.72	0.9999
	Reagent grade $CaCO_3$ chips	$y = -2250x + 8.41$	0.02	1.68	2.18	2.64	18.71	0.9920

Source: From Sajwan et al., 1991.

316

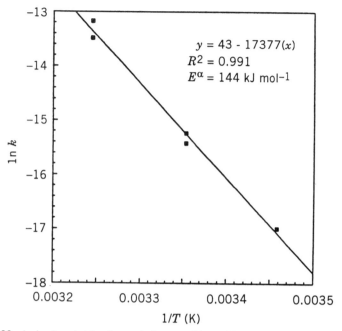

Figure 7.33. Arrhenius plot for the catalytic part of the oxidation reaction (Equation 7.103).

7.6.2   Relationships Between Kinetics and Thermodynamics of Exchange

An equilibrium state is defined as the point where the forward reaction equals the reverse reaction, thus

$$K_{eq} = k_f/k_b \qquad (7.107)$$

where K_{eq} is the thermodynamic exchange constant, k_f is the rate constant of the forward reaction, and k_b is the rate constant of the reverse reaction. The standard enthalpy of exchange (ΔH^0) can be calculated using the van't Hoff equation (see Chapter 4) or from the activation energies (E_a) of the forward (f) and reverse (b) reactions:

$$E_{af} - E_{ab} = \Delta H^0 \qquad (7.108)$$

From the classical thermodynamic relationship,

$$\Delta G^0 = \Delta H^0 - T\Delta S^0 \qquad (7.109)$$

and the standard entropy of exchange, ΔS^0, can now be calculated by rearranging Equation 7.109:

$$\Delta S^0 = (\Delta H^0 - \Delta G^0)/T \tag{7.110}$$

and

$$\Delta G^0 = -1.364 \log K_{eq} \tag{7.111}$$

When using activation energies to estimate ΔH^0, one must first demonstrate that the reactions, forward and reverse, are not limited by mass transfer or diffusion processes, which limits the application of this concept to external surfaces only under rapid solute–solid mixing.

PROBLEMS AND QUESTIONS

1. Explain the meaning of reaction kinetics with respect to soil–water systems concerning
   a. Solubility
   b. Exchange reactions
   c. Redox reactions

2. What is the fundamental difference between the half life of a first-order and a zero-order reaction? Why is the distinction important?

3. The reaction kinetics accounting for the solubility of O_{2g} in water are

$$O_{2g} \Leftrightarrow O_{2aq}$$

thus,

$$dO_{2aq}/dt = k_1(pO_{2g}) - k_{-1}(O_{2aq})$$

and at equilibrium

$$k_1/k_{-1} = (O_{2aq})/(pO_{2g}) = 1.3 \times 10^{-3} \text{ M atm}^{-1}$$

Knowing that atmospheric pO_2 is 0.20, calculate the solubility of O_2 in mg L^{-1}.

3. The first-order rate constant of Fe^{2+} oxidation is 0.14 min^{-1} at pH 7.2, 0.023 min^{-1} at pH 6.9, and 0.0046 min^{-1} at pH 6.6. Calculate the half life of Fe^{2+} and the time needed to convert 99% of the Fe^{2+} to Fe^{3+} at each pH. What can you conclude about the influence of pH on the oxidation rate of Fe^{2+}?

4. Calculate the number of bacteria that would be produced after 5, 10, 25, 50, and 300 h by one bacterial cell, assuming exponential growth with a generation time of 30 sec. Calculate the surface area of 1 g of spherical bacterial cells, 1 μm in diameter, having a density of 1.0 g cm^{-3}. Assuming that the surface charge density

of the spherical bacterial cells is 0.001 meq m^{-2}, calculate the CEC of the bacterial cells in meq/100 g.

5. Write a rate expression for the concentration of O_{2aq} with respect to time (dO_{2aq}/dt), following the introduction into the water body of an oxidizable substrate S:

$$O_{2g} \Leftrightarrow O_{2aq}$$

and

$$O_{2aq} + S \rightarrow products$$

6. Using the Michaelis–Menten equation, calculate the oxidation rate of NH_4 in a suspension assuming a V_{max} of 1 mmol L^{-1} day^{-1}, K_m of 2.5 mmol L^{-1}, and NH_4 of 10 mmol L^{-1}.

7. Demonstrate that the K_m represents the concentration of the substrate at which the rate of the reaction is half the maximum rate.

8. What is the meaning of a competitive inhibition? How is it distinguished from an uncompetitive inhibition?

9. How do colloid surfaces affect the rate of reactions? Why?

10. Why are some reactions in soil nonreversible? What is their significance?

11. What is the meaning of activation energy (E_α)? How is it related to rates of reactions and how may one decrease E_α?

12. Given the following data for rate of an enzyme reaction at different substrate concentrations, determine the Michaelis–Menten constant, K_m and the maximum reaction rate:

Substrate Concentration	Reaction Rate
1	16.7
5	50.0
10	66.7
50	90.9

APPLICATIONS

PART VI
Soil Dynamics and Agricultural–Organic Chemicals

8 Organic Matter, Nitrogen, Phosphorus and Synthetic Organics

8.1 INTRODUCTION

Soil is always in a dynamic state because of various biophysicochemical processes constantly taking place within it. One of the roles of soil–water environmental chemists is to understand and predict such processes. For example, in soil plant material, bacterial cells, manures, sludges, or synthetic organics decompose to satisfy microbial energy needs. In the process nitrogen, phosphorus, sulfur, and carbon are mineralized, various other inorganic compounds/minerals are released, and/or new organic compounds are synthesized. However, the buildup of certain organic and/or inorganic chemicals in soil could be undesirable and could bring negative results, for example, soil could become unsuitable for agricultural plants and/or pollute ground- and/or surface-water.

The purpose of this chapter is to introduce the student to the major biophysico-chemical processes taking place in soil and to demonstrate how soil–water chemistry affects such processes.

8.2 DECOMPOSITION OF ORGANIC WASTE

There are two major pathways responsible for decomposing organics (plant residue, manures, synthetic organics, sewage sludges, etc.) in the soil–water environment. One pathway involves *aerobic* processes, and a second involves *anaerobic* processes. During aerobic decomposition, carbohydrates are converted to carbon dioxide and water as follows:

$$CH_2O + O_2 \rightarrow CO_2 + H_2O \tag{8.1}$$

where CH_2O denotes carbohydrates, and therefore oxygen demand is relatively high, approximately 1.07 g of O_2 per gram of carbohydrate. Furthermore, since organic waste contains various elements, a number of products (e.g., NO_3, SO_4, and PO_4) are generated. During oxidative decomposition, assimilation processes by the microbial population also take effect:

$$5CH_2O + NH_3 \rightarrow C_5H_7O_2N + 3H_2O \tag{8.2}$$

where $C_5H_7O_2N$ denotes amino acid. Nitrogen assimilation by microbes requires 0.093 g NH_3 per gram of CH_2O. After rapid microbial cell synthesis subsides, NH_3 appears, if N is in excess, by deamination,

$$C_5H_7O_2N + 5O_2 \rightarrow 5CO_2 + 2H_2O + NH_3 \tag{8.3a}$$

followed by nitrification,

$$NH_3 + 2O_2 \Rightarrow HNO_3 + H_2O \tag{8.3b}$$

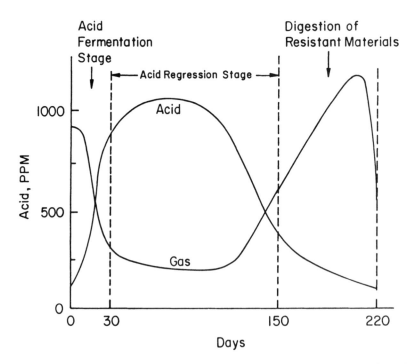

Figure 8.1. Stages of decomposition of organic material (unpublished class notes, Broadbent, U.C. Davis, 1978).

which further increases biological oxygen demand (BOD). Finally, biodegradation continues so slowly that BOD is extremely low and the organic products appear to be stable.

Anaerobic decomposition is a relatively slow process because of the slow rate of microbial growth. The process is composed of three stages (Fig. 8.1). The first stage is known as the *acid fermentation stage*, lasts approximately 1 month, and simple, relatively low molecular weight organic acids are produced, decreasing pH to 4. The main organic acids produced during this stage include acetic, lactic, propionic, and butyric. The second stage is referred to as the *acid regression stage* and it lasts 4–5 months. During this stage, organic acids are converted to methane, some metals or metalloids may be methylated and converted to gases (e.g., methylated mercury, arsenic, selenium), and pH rises to 7. In the third stage, often referred to as the *alkaline fermentation stage*, solids break down slowly and gas evolution decreases significantly.

In soil systems, both aerobic and anaerobic processes may be taking place simultaneously. In the case of aerobic processes, oxygen reaches organic-waste microbes by mass flow through macropores, whereas in the case of anaerobic decomposition, microbial demand for oxygen is not met because it diffuses slowly through micropores. Therefore, processes such as flooding actually favor anaerobic decomposition. Finally, during microbial decomposition, by-products such as CO_2, H_2O, heat, and stable humus-like organic material (organic matter) are also produced.

8.2.1 Some General Properties of Soil Organic Matter (SOM)

As noted in Chapter 3, organic matter is a high-complex heterogeneous substance with a number of properties beneficial to soils:

1. It contributes significantly to the soil's CEC
2. As pH increases, CEC increases and the preference for polyvalent cations also increases
3. It forms stable complexes with micronutrient elements such as zinc, iron, and copper
4. It exhibits low bulk density and low particle density
5. It possesses extensive surface area for adsorption and many other reactions
6. It exhibits low specific heat and low heat conductivity (its surface warms up easily)
7. It resists compaction
8. It improves water infiltration
9. It improves soil structure
10. Organic matter decomposes and thus elements are recycled

As noted above, one of the elements that derives from decomposing organic residues is nitrogen. Because of its importance as a plant nutrient and its potential role as a water pollutant when in large quantities, understanding the various processes responsible for its availability in nature is of great importance.

8.2.2 Nitrogen Mineralization–Immobilization

Nitrogen undergoes a complicated series of cyclic pathways in the ecosystem (Fig. 8.2). The atmospheric form of free nitrogen must be "fixed"—incorporated into chemical compounds (e.g., NH_3) which can be utilized by plants. This nitrogen fixation can be accomplished by bacterial action of both free-living soil bacteria such as azotobacter and chlostridium and symbiotic bacteria such as rhizobium. It can also be

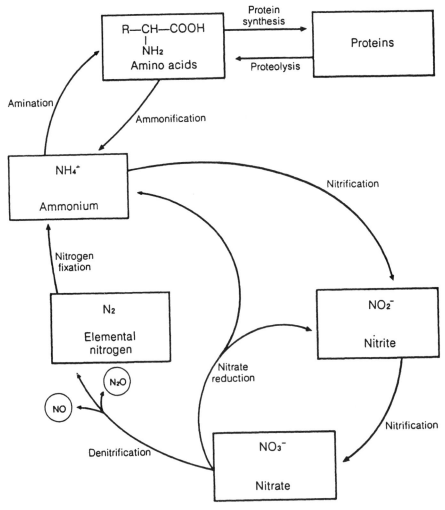

Figure 8.2. The nitrogen cycle (from Doefsch and Cook, 1974, with permission).

achieved as a physical process in the atmosphere by the ionizing effect of lightning and cosmic radiation.

In most soils, the bulk of the nitrogen is in the organic form, which is unavailable to plants and/or other soil microorganisms. However, a dynamic equilibrium exists between organic nitrogen and inorganic nitrogen or mineral nitrogen (N). This equilibrium can be expressed by

$$\text{Organic N} \underset{k_{-1}}{\overset{k_1}{\Leftrightarrow}} \text{Mineral N} \qquad (8.4)$$

The rate constants k_1 and k_{-1} refer to mineralization (m) and immobilization (i), respectively. Mineralization refers to production of NH_3 and/or NH_4 through microbial decomposition, whereas immobilization refers to incorporating N into bacterial cells (e.g., making up protein).

In soils under natural conditions, the source of the organic material is plants, and soil organisms are being offered a number of substances from such plant material. Cellulose forms the greatest portion, varying from 15 to 60% of the dry weight. Hemicellulose commonly makes up 10–30% of the dry weight, and lignin makes up 5–30%. The water-soluble fraction includes simple sugars, amino acids, and aliphatic acids, and contributes 5–30% of the tissue weight. Ether- and alcohol-soluble constituents include fats, oils, waxes, resins, and a number of pigments in smaller percentages. It is from these substances that the compost microorganisms derive all of their carbon for energy. Proteins have in their structure most of the plant nitrogen and sulfur. Organisms derive the nitrogen from the proteins, for their own nitrogenous substances, which are produced through the process of assimilation.

The factors that affect the ability of microorganisms to decompose organic material include type of organic material, temperature, pH, and redox potential (Eh). Fungi and actinomycetes are primarily responsible for the initial decomposition of organic waste. After that, bacteria are able to produce protease, a proteolytic enzyme which breaks protein down into simple compounds such as amino acids. The amino acids are absorbed by the microorganisms and ammonia is released by the following reactions:

Desaturative deamination:

$$RCH_2CHNH_2COOH \rightarrow RCH = CHCOOH + NH_3 \qquad (8.5)$$

Oxidative deamination:

$$RCHNH_2COOH + 1/2O_2 \rightarrow RCOCOOH + NH_3 \qquad (8.6)$$

Reductive deamination:

$$RCHNH_2COOH + 2H^+ \rightarrow RCH_2COOH + NH_3 \qquad (8.7)$$

Hydrolytic deamination:

$$RCHNH_2COOH \rightarrow RCHOHCOOH + NH_3 \qquad (8.8)$$

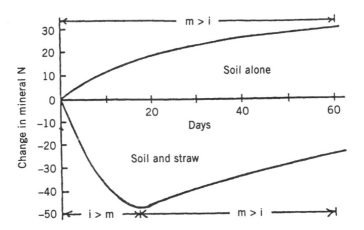

Figure 8.3. Mineral N transformations as a function of time with and without added straw (m = mineralization, i = immobilization) (unpublished class notes, Broadbent, U.C. Davis, 1978).

From this point on, NH_3 undergoes various reactions in soils, including assimilation, soil adsorption, nitrification, and volatilization.

The equilibrium between organic N and mineral N is highly dependent on environmental factors (e.g., temperature, moisture, and oxygen content) as well as the composition of organics with respect to the carbon–nitrogen (C/N) ratio. Generally, mineralization exceeds immobilization, and under such conditions, available nitrogen is supplied continuously for the use of growing plants. Immobilization or assimilation of nitrogen is due to the synthesis of large numbers of microbial cells which incorporate mineralized N (NH_4) into protein. The data in Figure 8.3 show that mineralization commonly exceeds immobilization in soils. However, when straw is added to soil, the immobilization rate exceeds the mineralization rate initially, and then mineralization exceeds immobilization. When crop residues containing less than approximately 1.5% N are incorporated in soil, the immobilization rate temporarily becomes greater than mineralization.

The factor that soil microbiologists employ to predict mineralization–immobilization in soil is the C/N ratio in organic residue. Generally, high C/N ratios in soil organic residue decrease with time because of loss of carbon as CO_2 gas and assimilation of available inorganic N. An equilibrium between the two pools of nitrogen (organic N–mineral N) is attained at C/N ratios of 8:20. When C/N ratios are less than 20, rapid net mineralization takes effect, whereas in systems with C/N ratios below

TABLE 8.1. Nitrogen Content and C/N Ratios of Mature Plant Materials

Material	N (%)	C/N
Wheat straw	0.5	90
Cornstalks	0.8	55
Sweet clover	1.7	26
Alfalfa	3.5	13

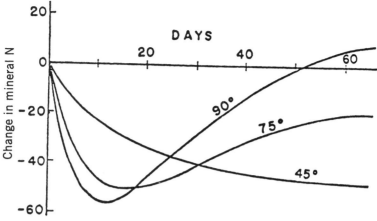

Figure 8.4. Immobilization of mineral N under various temperatures as a function of time (unpublished class notes, Broadbent, U.C. Davis, 1978).

30–35:1, net immobilization is rare. The data in Table 8.1 show nitrogen content and C/N ratios of some mature plant materials. The actual rates of mineralization–immobilization in soil depend on many factors such as temperature (Fig. 8.4), aeration (Eh), pH, and type of available N, for example $NH_4^+ - N$ is the preferred substrate over NO_3–N.

8.2.3   Ammonia Reactions in Soil–Water Systems

Ammonia in soil–water systems undergoes a number of reactions which are described below:

$$NH_3 + H^+ \Leftrightarrow NH_4^+ \tag{8.9}$$

or

$$NH_3 + H_2O \Leftrightarrow NH_4^+ + OH^- \tag{8.10}$$

Equation 8.9 shows that when NH_3 is introduced to an acid solution, it reacts directly with the acid and produces the ammonium ion (NH_4^+) (see Chapter 12). Concurrent with Equation 8.9, NH_3 may associate itself with several water molecules (NH_3nH_2O) without coordinating another H^+. This hydrated NH_3 is commonly referred to as unionized ammonia and is toxic to aquatic life forms at low concentrations. Because NH_3 is a volatile gas, some of it may be lost directly to the atmosphere (volatilization) without dissolving in solution. On the other hand, the ammonium ion may undergo various reactions in the soil water that may alter its availability to plants and/or other organisms. These reactions include formation of metal–ammine complexes, adsorption on to mineral surfaces, and chemical reactions with organic matter.

All metal ions in nature are surrounded by a shell of water molecules; an example is shown below:

$$
\begin{array}{ccc}
H_2O & & H_2O \\
 & \searrow M^{2+} \nearrow & \\
H_2O & \nearrow \qquad \searrow & H_2O
\end{array}
\qquad (8.11)
$$

where M^{2+} = any heavy metal. The water molecules of the metal hydration sphere can be replaced with NH_3 molecules as shown below:

$$[M(H_2O)_4]^{2+} + 4NH_3 \Leftrightarrow [M(NH_3)_4]^{2+} + 4H_2O \qquad (8.12)$$

and the reaction between M^{2+} and NH_3 involves the formation of a coordinated covalent bond where the element nitrogen (N) of NH_3 donates its single unshared electron pair to M^{2+}, forming a metal–ammine complex. For additional information on metal–ammine complexes, see Chapter 12.

Ammonium may be adsorbed by soil mineral surfaces. Generally, four types of clay "sites" are involved in NH_4^+ adsorption. One site may be planar, which justifies a relatively weak adsorption; a second site may be clay-edge or metal-oxide, forming a hydrogen bridge; a third site may involve interlayer sites where inner-sphere complexes are formed (vermiculite) or outer-sphere complexes are formed (montmorillonite). Finally, a fourth adsorption site may involve adsorption through –O–H bridging or polar covalent bonding with divalent heavy metals adsorbed weakly (electrostatically) on clay surfaces. Adsorption of NH_4^+ by H^+ bridging or polar covalent bonding depends on the pK_a of NH_4^+ and/or the strength of the polar covalent interactions between metal cation and NH_3. For example, hard divalent cations form –O–H-bridging (–O–H–NH_3), whereas electron-rich heavy metals tend to undergo polar-covalent interactions $S\equiv M–NH_3$, where $S\equiv$ denotes surface and M denotes adsorbed metal. For more details on these reactions, see Chapter 12. Because of these adsorption interactions, release or bioavailability of NH_3 or NH_4^+ in soil–water systems depends on pH and competitive cation interactions.

Other reactions leading to NH_4^+ removal from soil solution besides microbial nitrogen assimilation, metal–ammine formation, or adsorption onto mineral surfaces, involve NH_3 fixation by incorporating it as NH_2 in aromatic rings of humic acids (quinone) followed with aromatic ring condensation.

8.2.4 NH₃ Volatilization

An appreciable amount of NH_3 volatilizes from soil. In some fields, up to 50% of the added ammonium-N volatilizes. In addition to the economic importance of this nitrogen loss, there is also environmental importance. In industrialized regions of the world, NH_3 in the atmosphere approaches concentrations high enough to cause toxicity effects on forests.

Ammonium volatilizes as NH_3 through deprotonation of NH_4:

$$NH_4^+ \Leftrightarrow NH_3 + H^+ \qquad K_\alpha = 10^{-9.2} \qquad (8.13a)$$

(see Chapter 12). Based on Equation 8.13a, two factors determine potential NH_3 volatilization from soil. One factor is the dissociation constant of NH_4 and the second is the soil pH. Note that based on the pK_α of NH_4, at approximately pH 7, nearly 100% of the ammonium (e.g., NH_4Cl added to a solution) remains in the NH_4^+ form; at pH 9.2, approximately 50% of the ammonium is in the NH_3 form, and at approximately pH 11, nearly 100% of the NH_4 is in the NH_3 form. The latter species leaves the solution as gas because of the large NH_3 concentration gradient between the solution and the atmosphere. For all practical purposes, the partial pressure of NH_3 gas in the atmosphere is zero.

Soil pH buffering appears to play a major controlling role in NH_3 volatilization. The potential of soil to buffer pH depends on mineralogical composition, percent base saturation, type of exchangeable cations, and CEC. An empirical formula for soil pH buffering, noted as B, is given by (Avnimelech and Laher, 1977)

$$B = \Delta H^+/\Delta A \qquad (8.13b)$$

where ΔA is the amount of acid added as moles per liter or moles per kilogram soil–solution suspension and ΔH^+ is the resulting change in hydrogen activity (mol L^{-1}). When B approaches zero, the pH buffering capacity of the soil approaches infinity, and when B equals 1, the soil–solution system exhibits no buffering capacity. The amount of acid (moles) produced during NH_3 volatilization by the reaction $NH_4^+ \Leftrightarrow NH_3 + H^+$ is directly related to the moles of NH_3 that volatilized.

Generally, NH_3 volatilization is controlled by:

1. *Partial Pressure of NH_3 (pNH_3) in the Field's Atmosphere.* The factors controlling pNH_3 include nearby industries producing NH_3, the level of field fertilization with ammoniacal fertilizers, wind velocity, temperature, and moisture. Generally, the lower the pNH_3 is, the higher the volatilization rate.

2. *Soil pH.* This is only important when the pH buffering capacity of soil is very large (B approaches zero). Generally, when initial pH is high and soil B is very low, maximum NH_3 is lost (Avnimelech and Laher, 1977).

3. *Total Ammonium Present and Cation Exchange Capacity.* Generally, when large quantities of NH_4 are added to a field with high B (low pH buffering capacity), ammonium volatilization is rapidly minimized. Also, when the added NH_4 becomes adsorbed owing to high CEC, NH_3 volatilization declines (Avnimelech and Laher, 1977).

4. *Addition of Salts (e.g., $CaCl_2$, KCl, or KNO_3).* These salts have a tendency to minimize volatilization when basic sources of ammoniacal nitrogen are used (e.g., anhydrous NH_3 or urea). For example, calcium reacting with OH^- (released from basic fertilizer) and CO_2 produces $CaCO_3$ which buffers pH. In the case of K-salts, either the soil solution becomes low in pH because of a salt effect (K^+ displaces H^+ from the colloid surfaces) or K^+ displaces Ca^{2+} from the exchange phase, leading to formation of $CaCO_3$. The mechanisms limiting NH_3 losses become functional only with Ca-salts of high solubility, for example, $CaCl_2$ or $Ca(NO_3)_2$ versus $CaSO_4 \cdot 2H_2O$ (Evangelou, 1990 and references therein).

AN EQUILIBRIUM-BASED MODEL FOR PREDICTING POTENTIAL AMMONIA VOLATILIZATION FROM SOIL

As noted above, the potential for NH_3 volatilization depends on the soil's potential to buffer pH, B, which is given by (Avnimelech and Laher, 1977)

$$B = \Delta H^+ / \Delta A \tag{A}$$

where ΔA is the amount of acid added as moles per liter or moles per kilogram of soil–solution suspension and ΔH^+ is the resulting change in hydrogen activity (mol L^{-1}). Therefore,

$$\Delta A = [NH_4^+]_o - [NH_4^+]_f = \Delta[NH_4^+] \tag{B}$$

Equation B points out that the amount of acid produced during NH_3 volatilization (ΔA) equals the difference between original NH_4 ($[NH_4]_o$) minus final NH_4 ($[NH_4^+]_f$). By rearranging Equation 8.10,

$$[NH_4^+]_f = K_b[NH_3]_f/[OH^-]_f = [K_b/K_w][NH_3]_f[H^+]_f \tag{C}$$

where K_w is the dissociation constant of water and $K_b = 1.8 \times 10^{-5}$. The two independent variables in Equation C can be given by

$$[NH_3]_f = K(P_\alpha) \tag{D}$$

where K is a proportionality constant and P_α is the partial pressure of NH_3 in the air. If

$$[H^+]_f = [H^+]_o + \Delta H^+ \tag{E}$$

replacing ΔH^+ by substituting Equation B into Equation A gives

$$[H^+]_f = [H^+]_o + B ([NH_4^+]_o - [NH_4^+]_f) \tag{F}$$

Upon rearranging Equation C to solve for $[NH_3]_f$ and inserting Equation F

$$[NH_3]_t = [NH_4^+]_t[K_w/K_b]/\{[H^+]_o + B ([NH_4^+]_o - [NH_4^+]_t)\} \tag{G}$$

where NH_{4t}^+ denotes NH_4^+ concentration at any given time t. Equation G does not describe the rate of volatilization, but it does provide a potential for NH_3 volatilization. One may take the inverse of Equation G to produce a linear equation:

$$[1/NH_3]_t = -B/K_w/K_b + \{[H^+]_o + B [NH_4^+]_o/[K_w/K_b]\}(1/NH_4)_t \tag{H}$$

When plotted as $1/NH_3$ versus $1/NH_4$, Equation H would produce a straight line with slope $([H]_o + B\,[NH_4^+]_o)/(K_w/K_b)$ and y intercept $-B/[K_w/K_b]$. The equilibrium-dependent model described above was used by Avnimelech and Laher (1977) to demonstrate how pH and soil pH buffering capacity, B, control NH_3 volatilization (Fig. 8A). This can also be evaluated by carrying out a sensitivity analysis on Equation G, which shows that when B approaches zero (extremely high soil pH buffering capacity), or NH_4^+ in the system is very low, such that $H^o >> B([NH_4]_o - [NH_4]_t)$, Equation H reduces to

$$[NH_3]_t = [NH_4^+]_t[K_w/K_b]/[H^+]_o \qquad (I)$$

A plot of NH_3 versus $1/H^+$ would give a straight line under constant $[NH_4^+]$. Therefore, NH_3 volatilization would be inversely related to H^+ and directly related to NH_4^+ (Fig. 8B). When B is large (approaches 1) (soil pH buffering capacity is

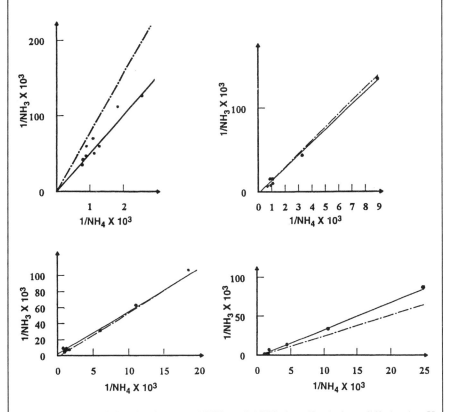

Figure 8A. The relationship between $1/NH_3$ and $1/NH_4$ in soil solutions differing in pH and buffering capacity. Experimental data are represented by the solid lines; broken lines represent calculated data (from Avnimelech and Laher, 1977, with permission).

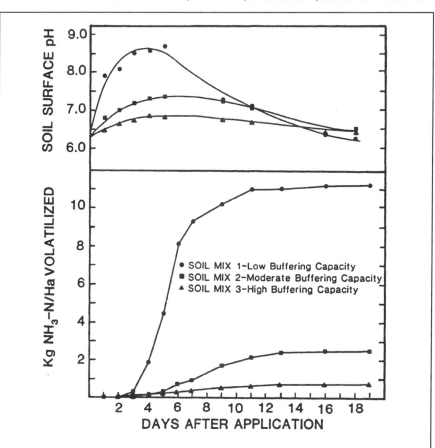

Figure 8B. Soil surface pH and cumulative NH_3 loss as influenced by pH buffering capacity. (from Ferguson et al., 1984, with permission).

very low) or added NH_4^+ is large, Equation G reveals that $B[NH_4^+]_o$ would be much greater than $[H^+]_o$ and, therefore, NH_3 volatilization would be independent of initial pH. In actuality, upon NH_4 addition, the pH would rapidly become too low to aid volatilization.

8.2.5 Nitrification

Nitrification is carried out by unique specialized bacteria in a two-step reaction. The first step, oxidation of ammonium (NH_4^+) to nitrite (NO_2^-), is catalyzed by a few species of bacteria that have names beginning with nitroso-, for example, nitrosomonas and nitrosospira. The reaction is as follows:

$$NH_4^+ + 1.5O_2 \rightarrow NO_2^- + H_2O + 2H^+ \qquad \Delta G^0 = -66.5 \text{ kcal} \qquad (8.14)$$

The second step, nitrite (NO_2^-) to nitrate (NO_3^-), is carried out by a different bacteria—nitrobacter:

$$NO_2 + 0.5O_2 \rightarrow NO_3^- \qquad \Delta G^0 = -17.5 \text{ kcal} \qquad (8.15)$$

Based on Reactions 8.14 and 8.15, nitrification is energetically favorable (overall $\Delta G^0 = -84$ kcal). The intermediate nitrogen form, NO_2^-, rarely accumulates in significant concentrations because nictrobacter normally acts as fast or faster than the NO_2-producing bacteria. However, nitrobacter is more sensitive to ammonia than nitrosomonas, and for this reason nitrite may accumulate under high concentrations of NH_4^+ (Fig. 8.5) but not under low concentrations of NH_4 (Fig. 8.6)

The rate of bacterial growth and hence the rate of nitrification is both temperature and pH dependent. Maximum bacterial activity occurs at about 28°C and a pH of about 8. Below a temperature of about 2°C, the reaction is very slow (Fig. 8.7). Below pH 5.5, the nitrifying bacteria decrease their activity, and below pH 4.5 the nitrification process is severely restricted; lack of oxygen also inhibits nitrification. As noted above, oxidation of NH_4^+ to NO_3^- is an enzyme-driven reaction and commonly the K_m (see Chapter 7) under optimum conditions is observed to be somewhere around 2.5 mM. Some K_m values below 2.5 mM are observed under high pH values when a large fraction of the ammonium is in the NH_3 form.

The overall nitrification rate which describes conversion of NH_4^+ to NO_3^- can be expressed by

$$dNO_3/dt = k[NH_4^+] \qquad (8.16)$$

which reveals that the rate of NH_4^+ oxidation is related to NH_4^+ in the soil solution and the magnitude of the first-order rate constant (k). In a ternary soil-exchange system

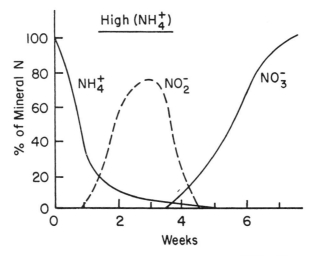

Figure 8.5. Nitrification as a function of time at high levels of NH_4 added to soil (unpublished class notes, Broadbent, U.C. Davis, 1978).

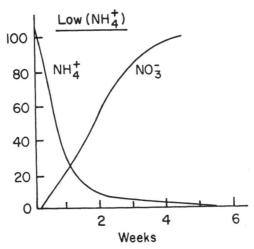

Figure 8.6. Nitrification as a function of time at low levels of NH_4 added to soil (unpublished class notes, Broadbent, U.C. Davis, 1978).

(e.g., K–NH_4–Ca) increasing K^+ concentration in the soil solution should influence the nitrification rate by increasing the concentration of NH_4^+ in the soil solution. Lumbanraja and Evangelou (1990) and Evangelou et al. (1994) reported that the potential of a ternary soil system to release NH_4^+ to the soil solution is dependent on the potential NH_4^+ buffering capacity of the soil, PBC_{NH_4} (Fig. 8.8). The PBC_{NH_4} of a ternary soil system at constant solution K^+ and Ca^{2+} and at concentration ratio (CR) values of NH_4^+ ($CR_{NH_4} = NH_4^+/[Ca^{2+}]^{1/2}$) commonly encountered in agricultural soils (Evangelou et al., 1994) is described by

$$PBC_{NH_4} = CECK_{V2}[4 + (K_{V1}CR_K)^2]^{-1/2} \qquad (8.17)$$

where K_{V1} and K_{V2} are the Vanselow exchange selectivity coefficients of K-Ca and NH_4-Ca, respectively, and CR_K equals $K^+/(Ca^{2+})^{1/2}$ (see Chapter 4). Equation 8.17

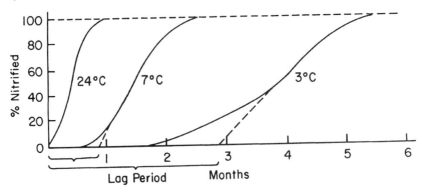

Figure 8.7. Temperature influence on nitrification in soil (unpublished class notes, Broadbent, U.C. Davis, 1978).

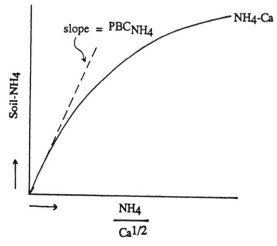

Figure 8.8. Schematic of NH_4 adsorption by a mineral surface in the binary mode.

demonstrates that under the soil-solution conditions stated above, PBC_{NH_4} depends on CEC, K_{V1}, K_{V2}, and on the magnitude of the soil-solution concentration of K^+ and Ca^{2+} or CR_K. Assuming that CEC remains constant under the various soil-solution changes imposed by fertilization practices, under a constant CR_K value, PBC_{NH_4} will depend directly on K_{V2} and inversely on K_{V1}. Furthermore, in accordance with Equation 8.17, an increase in CR_K would suppress PBC_{NH_4} Therefore, addition of K^+ to a soil system would increase solution NH_4^+ by decreasing PBC_{NH_4} (Fig. 8.9). Experimental data produced in the author's laboratory showed that for some soils PBC_{NH_4} is suppressed

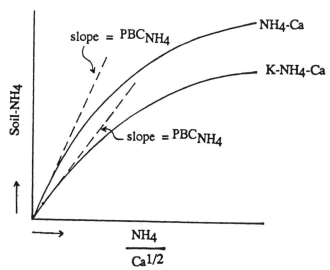

Figure 8.9. Schematic of NH_4^+ adsorption by a mineral surface in the binary (NH_4–Ca) and ternary modes (K–NH_4–Ca).

by the presence of K^+ as predicted, while in other soils PBC_{NH_4} increased when in the presence of K^+ (Fig. 8.10). The latter implied that the presence of K^+ increased the apparent affinity of the soil for NH_4^+. The above soil mineral surface reactions with NH_4^+ control release of NH_4^+ to the soil solution and consequently they may regulate

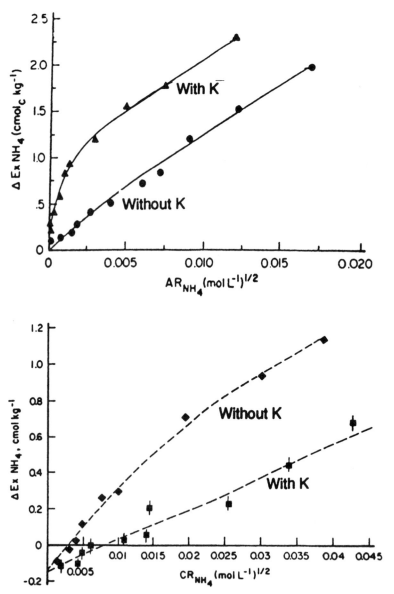

Figure 8.10. Experimental data on two soils demonstrating binary (NH_4–Ca) and ternary (K–NH_4–Ca) cation exchange behavior (from Lumbanraja and Evangelou, 1990, with permission).

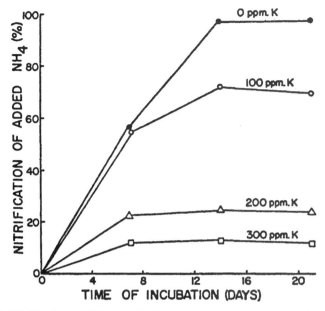

Figure 8.11. Nitrification of NH_4^+ adsorbed by vermiculite under various K^+ addition (from Welch and Scott, 1960, with permission).

volatilization, plant uptake, and biochemical transformation (e.g., nitrification) (see Fig. 8.11).

Production of NO_3 also leads to acidification of soil water systems. Based on the overall nitrification reaction,

$$NH_4^+ + 2O_2 \rightarrow NO_3^- + H_2O + 2H^+ \tag{8.18}$$

It is apparent that twice as much acid is generated by nitrification as is consumed in the initial reaction of NH_3 with water (Reaction 8.9). Because nitrification is pH controlled under a constant O_2 supply,

$$dNO_3/dt = k_1[NH_4]/H^+ \tag{8.19}$$

and its rate in soil systems would depend on the ability of soil to buffer pH. Based on Equation 8.13b, $B = \Delta H^+/\Delta A$ where B denotes soil pH buffering capacity, ΔH^+ is the resulting change in H^+ activity (mol L^{-1}), and ΔA is the amount of acid added as moles per liter. Since 1 mol of NH_4^+ produces 2 mol of H^+ (Equation 8.18),

$$\Delta A = 2([NH_4]_o - [NH_4^+]_f) = 2\Delta[NH_4^+] \tag{8.20}$$

Equation 8.20 points out that the amount of acid produced (ΔA) during nitrification is twice that of the NH_4^+ consumed. However, the actual impact of this acid on soil pH depends on the nature of the soil. For soils with initial pH near neutral and B approaching zero, nitrification would proceed rapidly. However, for soils with initial pH near neutral but B approaching 1 (low pH buffering), nitrification will rapidly decline. In soils with active plant growth, absorption of NO_3^- by roots also produces apparent pH buffering. For every equivalent of NO_3^- absorbed, an equivalent of OH^- is released to the soil solution by the root.

The nitrate produced may have adverse environmental effects in some situations. Some nitrate can be found in all natural waters, so low nitrate concentrations certainly do not constitute a problem. A large increase in nitrate concentration may have two results. The first, is eutrophication—in some waters, addition of nitrate may enhance the growth of algae and other aquatic organisms sufficiently to deplete atmospheric oxygen and foul the water. Second, consumption of water with very high nitrate concentrations may occasionally pose a health hazard. Cattle are generally more susceptible to nitrate toxicity than humans, and infants are more at risk than adults. Federal standards for nitrate in drinking water are 10 ppm (mg N L^{-1}).

8.2.6 Denitrification

Nitrogen is returned to its atmospheric form by the action of denitrifying bacteria such as *Pseudomonas thiobacillus* and *Micrococcus denitrificans*. The process is referred to as denitrification and represents the major mechanism of nitrogen loss in the overall nitrogen cycle whereby various forms of nitrogen in the soil revert to the N_2 form. The reactions and their energetics are given below:

$$4NO_3^- + 4H^+ \rightarrow 2N_2 + 5O_2 + 2H_2O \qquad \Delta G° = 31.48 \text{ kcal} \qquad (8.21)$$

Reaction 8.21 is not spontaneous because of its positive $\Delta G°$. Considering, however, that carbohydrates present in soil decompose spontaneously by the reaction

$$C_6H_{12}O_6 + 6O_2 \rightarrow 6CO_2 + 6H_2O \qquad \Delta G° = -686 \text{ kcal} \qquad (8.22)$$

In the absence of O_2 but in the presence of carbohydrates, the NO_3^- denitrifies through transferring electrons from the reduced organic carbon to nitrate N by the reaction

$$C_6H_{12}O_6 + 4NO_3^- \rightarrow 6CO_2 + 6H_2O + 2N_2 \qquad (8.23)$$

Summing the $\Delta G°$ values of Reactions 8.21 and 8.22 shows that Reaction 8.23 gives a negative $\Delta G°$; thus, it is spontaneous and justifies the fact that microorganisms catalyze Reaction 8.23 to obtain energy. The ratio of organic C oxidized per unit of N reduced is 1.28.

Nitrate in the soil environment, therefore, can undergo two distinct biochemical processes: (1) assimilatory, where NO_3 is used to produced protein, and (2) dissimilatory, where NO_3 is used to produce energy for microbes. The rate at which the latter process occurs depends indirectly on O_2 availability, or Eh, and directly on available moisture, organic carbon content, and pH. The optimum pH is around neutral and the

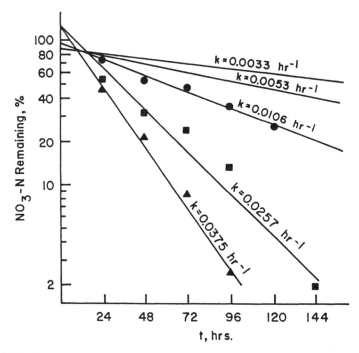

Figure 8.12. First-order reactions of denitrification data in soil (unpublished class notes, Broadbent, U.C. Davis, 1978).

critical Eh is around +225 mV. During denitrification, as long as NO_3^- is present, the soil system is poised at an Eh of around 200 mV.

Based on the many factors controlling denitrification, various soils are expected to exhibit varying rates of denitrification. This depends on clay content, organic C content, water holding capacity, texture, and structure. The data in Figure 8.12 show that denitrification in various soils exhibits first-order behavior. In addition to N losses through microbial denitrification, under very low pH, NO_3 may also react directly with amino acids and reduce to N_2 (Van Slyke reaction), and when in the HNO_2 form may also incorporate into lignin.

8.2.7 Eutrophication

The presence of ammoniacal nitrogen in ecosystems is a good measure of the balance between protein decomposition, bacterial action, and plant production. High levels of NH_4 in water (above 1 ppm) are generally indicative of some major source of decomposition within the system, in excess of that being utilized by bacterial action and plant growth. Field measurements of nitrates and nitrites are also good guides to the nutrient condition of lakes, streams, and estuaries.

Excessive nutrients, such as nitrates and phosphates, cause pollution primarily because they stimulate the growth of microorganisms, increasing the biological oxygen demand (BOD) of the water and reducing dissolved oxygen available for aquatic organisms. These nutrients stimulate algal growth, lead to plankton blooms which produce obnoxious tastes and odors in water, and disrupt aquatic ecology.

8.3 PHOSPHORUS IN SOILS

There are two forms of phosphorus in soil systems—organic and inorganic. The organic form of phosphorus in many soils represents more than half of the total phosphorus. It appears to be a component with three different sinks/sources. The first sink/source is the nucleic acids (e.g., quanine, adenine, thymine, and uracil). This source/sink represents a small part of total organic P, but nucleic acids readily undergo hydrolysis releasing all P. The second source/sink of organic P are phosphate esters of inositate (e.g., phytin) which represents the bulk of organic P in soil, but the hydrolysis of this sink/source is rather slow. Finally, the third source/sink of organic P are phospholipids such as lecithin. The amount of lecithin in soil is very small. A fraction of phosphorus in soils is shuttled between the organic and inorganic forms. Generally, if C/P < 200 mineralization occurs, whereas when C/P > 300, immobilization occurs.

The inorganic form of phosphate in soils (orthophosphate) exists in various forms: (1) oxide–phosphate complexes such as clay–mineral edge–phosphate complexes, and (2) minerals of apatite (Ca-phosphates), Mn-phosphates, Fe(III)-phosphates, Al-phosphates, and to a lesser degree pyrophosphate (e.g., $Ca_2P_2O_{7(s)}$). The concentration of P in soils from inorganic forms is very small (in the range of 0.03 mg L^{-1}). Therefore, the mobility of soluble inorganic phosphate in soil systems is for all practical purposes insignificant. However, most of the mobilized inorganic phosphate in soils is in the form of small mineral and/or colloidal particles. Often, the mobility of large amounts of P can be accounted for by the mobility of organic waste during intense rainstorms. Such organic waste may be carried by runoff as suspended

TABLE 8.2. Solubility Equilibria of Orthophoshates and Condensed Phosphates

Number	Equilibrium	Log Equilibrium Constant
1	$Ca_5OH(PO_4)_3(s) = 5Ca^{2+} + 3PO_4^{3-} + OH^-$	−55.6
2	$Ca_5OH(PO_4)_3(s) + 3H_2O = 2[Ca_2HPO_4(OH)_2]_{surface} + Ca^{2+} + HPO_4^{2-}$	−8.5
3	$[Ca_2HPO_4(OH)_2]_{surface} = 2Ca^{2+} + HPO_4^{2-} + 2OH^-$	−27
4	$CaHPO_4(s) = Ca^{2+} + HPO_4^{2-}$	−7
5	$FePO_4(s) = Fe^{3+} + PO_4^{3-}$	−23
6	$AlPO_4(s) = Al^{3+} + PO_4^{3-}$	−21
7	$Ca_2P_2O_7(s) = Ca^{2+} + CaP_2O_7^{2-}$	−7.9

Source: From Stumm and Morgan, 1970.

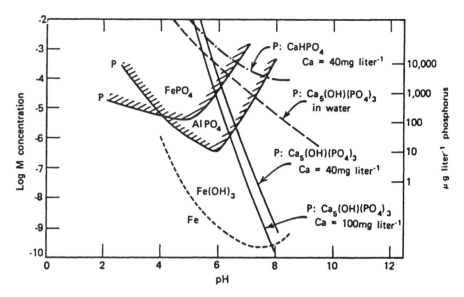

Figure 8.13. Influence of pH on the solubility of various phosphorus forms (from Stumm and Morgan, 1970, with permission).

particles, owing to this low density, as colloidal particles, owing to their high charge, and/or soluble organic components. The mobility of phosphorus to streams, rivers, and lakes is a major concern because of the contribution of P to eutrophication.

The solubility of metal–phosphates in soils is highly pH dependent because of the protonation potential of the phosphate species, and the various K_{sp} values of a number

TABLE 8.3. Complex Formation by Phosphates

Number	Equilibrium	Log Equilibrium Constant[a]
23	$Ca^{2+} + HPO_4^{2-} = CaHPO_4(aq)$	2.7
24	$Mg^{2+} + HPO_4^{2-} = MgHPO_4(aq)$	2.5
25	$Fe^{3+} + HPO_4^{2-} = FeHPO_4^+$	8.3
26	$Ca^{2+} + P_2O_7^{4-} = CaP_2O_7^{2-}$	5.6
27	$Ca^{2+} + HP_2O_7^{3-} = CaHP_2O_7^-$	2.0
28	$Mg^{2+} + P_2O_7^{4-} = MgP_2O_7^{2-}$	5.7
29	$Na+ + P_2O_7^{4-} = NaP_2O_7^{3-}$	2.2
30	$Fe^{3+} + 2HP_2O_7^{3-} = Fe(HP_2O_7)_2^{3-}$	22.0
31	$Ca^{2+} + P_3O_{10}^{5-} = CaP_3O_{10}^{3-}$	8.1
32	$Mg^{2+} + P_3O_{10}^{5-} = MgP_3O_{10}^{3-}$	8.6

Source: From Stumm and Morgan, 1970.

[a]Sugar phosphates and organic condensed phosphates form complexes with many cations. Stability constants for Ca^{2+} and Mg^{2+} are on the order of $10^3–10^4$.

of metal–phosphate species (Table 8.2; Fig. 8.13). Additionally, a significant portion of soluble phosphate encountered in soil solution could be attributed to the rather high stability constants of the various metal–phosphate pairs (Table 8.3). Phosphoric acid possess three pK_as—2, 7.2, and 12.2. Because of this wide range of all three pK_as, phosphate has a strong tendency to form mineral–surface complexes in a wide range of pH values (see Chapter 4). Such complexes are known to occur on $CaCO_3$ or $CaMg(CO_3)_2$ crystals, $Al(OH)_{3(s)}$, or $Fe(OH)_{3(s)}$, kaolinite, and 2:1 clays. For mineral surfaces exhibiting pH-dependent charge, phosphate adsorption generally increases as pH decreases. Generally, phosphate adsorption appears to involve two mechanisms, chemical bonding to positively charged edges, and substitution of phosphates for silicates in clay structure (see Stumm and Morgan, 1970 or 1981).

8.4 SULFUR IN SOILS

Generally, there are two forms of sulfur in soils—organic and inorganic. The organic form includes (1) the sulfhydryl group (–SH), as in cysteine, (2) disulfide (–S–S), as in cystine, (3) thioether (–S–CH$_3$), as in methionine, and (4) ester (O=S=O), as in taurine. During aerobic decomposition, sulfur oxidizes to form sulfate (SO_4). This requires the appropriate C/S ratio. Generally, C/S < 200 sulfate is produced, whereas C/S > 400 sulfate is incorporated into organic matter. Under anaerobic conditions, sulfur is reduced to sulfide. In the presence of divalent metals, sulfide precipitates as metal–sulfide, otherwise it is released as H_2S (see also Chapters 6 and 12). In the inorganic form, SO_4 exists mostly as $CaSO_4$ under oxidized conditions and in some geologic strata as $BaSO_4$. Under reduced conditions, sulfur exists as metal sulfide (MS) or pyrite (FeS_2) (Table 8.4).

TABLE 8.4. Oxidized and Reduced Forms of Certain Elements in Soils and the Redox Potentials (E_h) at which Change in Forms Commonly Occurs[a]

Oxidized Form	Reduced Form	E_h at Which Change of Form Occurs
O_2	H_2O	0.38 to 0.32
NO_3^-	N_2	0.28 to 0.22
Mn^{4+}	Mn^{2+}	0.28 to 0.22
Fe^{3+}	Fe^{2+}	0.18 to 0.15
SO_4^{2-}	S^{2-}	−0.12 to −0.18
CO_2	CH_4	−0.2 to −0.28

[a]Gaseous oxygen is depleted at Eh levels of 0.38–0.32 V. At lower Eh levels microorganisms utilize combined oxygen for their metabolism and thereby reduce the elements.

8.5   MICROBIAL ROLE IN SOIL REACTIONS

Microbiological processes in soil water systems are very important because they often control or drive certain soil reactions. For example, one of the parameters that dictates the type and extent of many reactions in soils is pH. The latter, however, is often controlled by biological processes because they either produce or consume acid. The microbial reactions that produce acid include:

1. Carbonic acid (H_2CO_3) formation due to CO_2 dioxide evolution by microbial respiration
2. Sulfide oxidation giving rise to sulfuric acid
3. Nitrification giving rise to NO_3 production from NH_4
4. Fe(II) oxidation

The microbial processes that increase pH include:

1. Fe(III) reduction
2. Sulfate reduction
3. Denitrification (see Chapter 5)

8.6   SYNTHETIC ORGANIC CHEMICALS

Synthetic organic chemicals are substances humans have invented and continue to invent and synthesize for beneficial purposes. Two broad groups of synthetic organic chemicals have found their way into soil–water systems. One group represents agricultural chemicals, the so-called pesticides, while the second group involves industrial organics. From the agricultural organic chemicals, the subgroup that soil environment chemists have given a great deal of attention are those used as herbicides. Table 8.5 lists the most widely used herbicides with some of their physicochemical properties.

The second broad group of synthetic organic chemicals involves various petroleum by-products, pharmaceuticals, plastics, resins, phonographical chemicals, and so on. The subgroup of these chemicals that is of major environmental concern is the so-called chlorinated hydrocarbons, because of their long persistence in the environment and potential carcinogenesis. Table 8.6 shows a number of these chemicals found in landfills. In addition to landfills, synthetic organic chemicals find their way to the environment by direct discharges, smokestack emissions, and emissions of internal combustion engines (e.g., cars).

8.6.1   Names of Organic Compounds—Brief Review

Organic compounds are made mainly by a small number of different elements (e.g., C, H, O, N, S, and P), but by combining them in various ways, a large number (actually

TABLE 8.5. Physical and Chemical Properties of Organics Used in Adsorption Studies

Family	Common Name	Chemical Name	Water Solubility g 100 mL or (ppm)³	pK^a	Analytical Wavelength, (mμ)[b]	Surface Area/Molecule (Å²)
Aniline	Aniline	Aniline	3,4 (20)	4.58(35)	230°, 280 H₂O	35.3
Anilide	Dicryl	3',4'-Dichloro-2-methylacrylanilide	8–9³		258 H₂O	58.7
	Propanil	3'4'-Dichloro-propionanlide	500³		248 H₂O	54.4
Amide	Sloan	3'-Chloro-2-methyl-p-valerotoluidide	8–9³		247 H₂O	66.4
Phenylcarbamate	CPIC (chloropropham)	m-Cholorolapropyl ester carbanilic acid	80³ 108³ (20)		237.5 H₂O	63.5
	IPC (propham)	Isopropyl ester carbanilic acid	20³–5 32³		234 H₂O	55.0
Benzoic acid	Amiben	3-Amino-2,5 dichlorobenzoic acid	700³		297 H₂O, 238 chl	52.4
Benzoic acid	Benzoic acid	Benzoic acid	0.27 (18)	4.12(50)	225[b], 270 H₂O	39.7
Picolinic acid	Picloram	4-Amino-3,5,6-tricloro-picolinic acid	430³ (25)		223 H₂O	49.5
α-Triazine	Atratone	2-Ethylamino-4-isopropylamino-6-methoxy-s-triazine	1800³ (20–22)		220 H₂O	67.9
	Atrazine	2-Chloro-4-ethylamino-6-isopropylamino-s-triazine	22³ (0) 70³ (27) 320³ (85)	1.68(22)	222, 263 H₂O	64.8
	Prometone	2,4-Bis (isopropylamino)-6-methoxy-s-triazine	750³		220 H₂O	71.8

						°C[a]
	Propazine	2-Chloro-4,6 bis (isopropylamino)-s-triazine	8.6^3 (20–2)		221 meal / 223 H_2O	68.7
	Simazine	2-Chloro-4,6 bis (ethylamino)-s-triazine	$2.0^3(0)$ / $5.0^3(20)$ / $84.0^3(85)$	1.65(18)	220 H_2O / 221 me. al.	55.3
	Simetone	2,4-Bis (ethylamino)-6-methoxy-s-triazine	3200^3 (20–2)	4.17	220 H_2O	58.4
	Trietazine	2-Chloro-4-diethylamino-6-ethylamino-s-triazine	20^3 (20–2)	1.88	226 me. al. / 228 H_2O	67.6
Substituted areas	Diuron	3-(3,4 Dichlorophenyl)-1,1-dimethylurea	42^3 (25)	–1 to –2	246 H_2O	64.5
	Fenuron	3-Phenyl-1, dimethylurea	2900^3 (24)		238 H_2O	55.3
	Monuron	3-(p-chlorophenyl)-1, 1-dimethylurea	230^3 (25)	–1 to –2	224 H_2O	60.2
	3-Phenylurea	3-Phenylurea	Solution		230 H_2O	46.2
Phenylalkanoic acid	2,4-D	2,4-Dichlorophenoxyacetic acid	400^3 / 725^3 / 900^3	2.64, 2.80 / 3.22(60) / 3.31	200^b, 230^b / 283 H_2O	56.1
	Phenoxyacetic acid	Phenoxyacetic acid	1.2 (10)		228^b, 269 H_2O	51.1
	2,4,5-T	2,4,5-Trichlorophenoxyacetic acid	200^3	3.14	$220,^b$ 289 H_2O	60.5
			238^3 (20) / 280^3 (25)	3.46 (60)	284 n-hex	

[a]The numbers in parentheses represent the temperature (°C).

[b]Analytical wavelength for low concentrations (generally < 10 ppm); chl = chloroform, meal = methylalcohol, n-hex = normal hexane.

Source: Barley et al., 1968.

TABLE 8.6. Most Frequently Detected Constituents in Groundwater of Disposal Sites

---

Dichloromethane (methylene chloride)
Trichloroethylene
Tetrachloroethylene
trans-1,2-Dichloromethane
Trichloromethane
1,1-Dichloroethane
1,1-Dichloroethylene
1,1,1-Trichloroethane
Toluene
1,2-Dichloroethane
Benzene
Ethyl benzene
Phenol
Chlorobenzene
Vinyl chloride

---

millions) of compounds can be made. These compounds are mainly characterized by carbon–carbon bonds in any number of ways, including (1) number of carbon–carbon bonds or number of carton atoms, (2) types of carbon bonds, (3) arrangement of carbon atoms or any other of the remaining atoms, and (4) combination of atoms. In the case of carbon–carbon bonds, the following molecules are identified: single carbon bond: alkanes; double carbon–carbon bonds: alkenes; triple carbon–carbon bonds: alkynes. The number of carbons associated with each other by either single, double, or triple bonds, up to 10-carbon atoms, are identified by the prefixes 1:meth-, 2:eth-, 3:prop-, 4:but-, 5:pent-, 6:hex-, 7:hep-, 8:oct-, 9:non-, 10:dec-.

In the case of arrangement of carbon atoms, organic compounds are separated into three major groups: (1) the chain or aliphatic group, (2) the cyclic group, commonly a molecule with six carbons forming a ring also known as aliphatic ring, and (3) the aromatic group, a molecule of six carbons also forming a ring. It differs from the aliphatic ring in that aromatic rings contain three double bonds and thus one less hydrogen per carbon atom. The number of hydrogens per carbon atom defines the degree of saturation which give unique properties to organic molecules. The structural formulas for a number of such organic compounds are given in Table 8.7.

Combinations of chains and aliphatic and aromatic rings produce various compounds with any number of properties. The presence and/or combination of atoms (e.g., carbon, nitrogen, or sulfur) gives rise to different compounds. For example, carbon atoms along with hydrogen and oxygen atoms give rise to organic acids while carbon, hydrogen, oxygen, and nitrogen give rise to amino acids (see Chapter 3), and a combination of various amino acids, through peptide bonds, give rise to various proteins.

In addition to these factors, the same atoms under the same arrangement could form the same compound but with different properties. These are known as chiral compounds. In essence, they represent identical compounds with different rotational momentum (e.g., D and L form, where D denotes dextroratory and L denotes leverotary). These two forms may seem to represent identical compounds, but they possess different reactivities. For example, one form of the compound may be deadly to humans, while the other form may have beneficial medicinal properties. Although chirality is slowly introduced to pharmaceuticals, their role on pesticides and various other products is not known mainly because of the expensive technology involved in separating these forms and the lack of knowledge about which form is beneficial.

TABLE 8.7. Structural Formulas of Various Organic Chemicals

Chlorinated Hydrocarbons

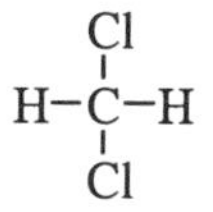

Dichloromethane[a]

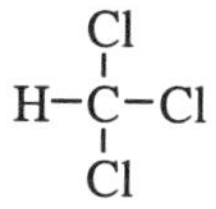

Trichloromethane[a]

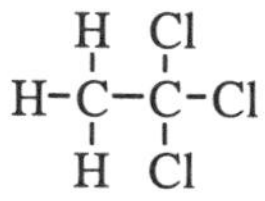

1,1,1-Trichloroethane[a]

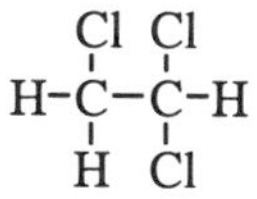

1,1,2-Trichloroethane[a]

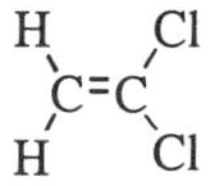

1,1-Dichloroethene[a]

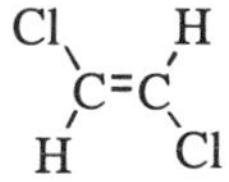

1,2-*trans*-Dichloroethene[a]

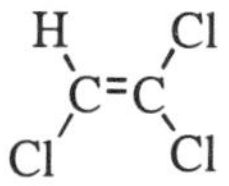

Trichloroethene[a]

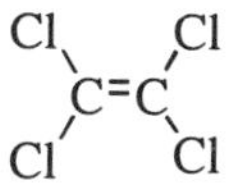

Tetrachloroethene[a]

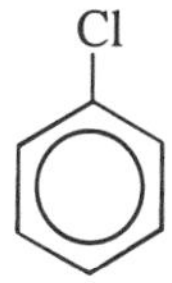

Chlorobenzene[a]

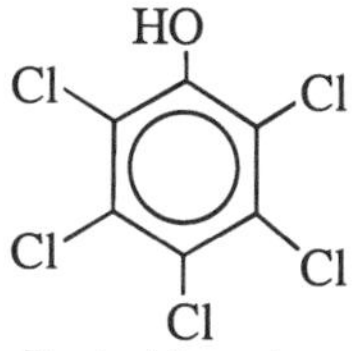

Pentachlorophenol[a]

(*continued*)

TABLE 8.7. Continued

Monocyclic Aromatics

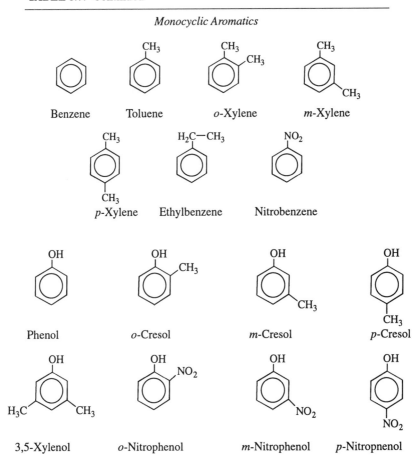

Benzene   Toluene   *o*-Xylene   *m*-Xylene

p-Xylene   Ethylbenzene   Nitrobenzene

Phenol   *o*-Cresol   *m*-Cresol   *p*-Cresol

3,5-Xylenol   *o*-Nitrophenol   *m*-Nitrophenol   *p*-Nitropnenol

Polycyclic Aromatic Hydrocarbons

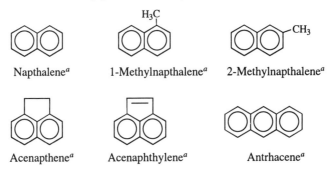

Napthalene[a]   1-Methylnapthalene[a]   2-Methylnapthalene[a]

Acenapthene[a]   Acenaphthylene[a]   Antrhacene[a]

(*continued*)

TABLE 8.7. Continued

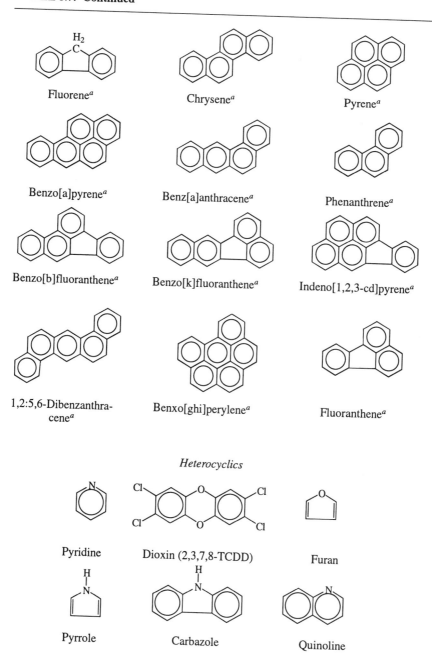

Fluorene[a]

Chrysene[a]

Pyrene[a]

Benzo[a]pyrene[a]

Benz[a]anthracene[a]

Phenanthrene[a]

Benzo[b]fluoranthene[a]

Benzo[k]fluoranthene[a]

Indeno[1,2,3-cd]pyrene[a]

1,2:5,6-Dibenzanthra-
cene[a]

Benxo[ghi]perylene[a]

Fluoranthene[a]

Heterocyclics

Pyridine

Dioxin (2,3,7,8-TCDD)

Furan

Pyrrole

Carbazole

Quinoline

(*continued*)

TABLE 8.7. Continued

Other Organics

1,2-Benzenedicarboxylic acid dibutylester (*n*-Butyl phthalate)

S=C=S   H—C≡N   Hydrazine

Carbon disulfide   Hydrogen cyanide   Hydrazine

*a*On the U.S. EPA-129 priority pollutants list.

8.6.2 Persistence of Organics in Soil–Water Systems

The persistence of organics in soils is determined by a number of processes taking place in it. These processes include: (1) photodecomposition, (2) microbial metabolism, (3) chemical reactions, (4) volatilization (5) adsorption by clay minerals and organic colloids and (6) leaching and plant uptake. A simple approach in modeling organics persistence is to consider that the overall decomposition reaction is a first-order process. Considering that any organic added to soil undergoes some form of decomposition, the following factors would need to be considered:

1. Organic concentration, O
2. Original amount of organic present in soil, O_o
3. Annual addition of organic, A
4. Decomposition rate of organic, r
5. Time, t

Based on the above,

$$dO/dt = A - rO \qquad (8.24)$$

Rearranging,

$$dO/A - rO = dt \qquad (8.25)$$

which can be integrated to

$$-1/r \ln (A - rO) = t + C \qquad (8.26)$$

Considering that at $t = 0$, $O = O_0$, and $C = -1/r \ln (A - rO_0)$, upon substituting C in Equation 8.26 with $-1/r \ln (A - rO_0)$ and rearranging,

$$\ln \{[A - rO]/A - rO_0]\} = -rt \tag{8.27}$$

or

$$\{[A - rO]/A - rO_0]\} = e^{-rt} \tag{8.28a}$$

and

$$O = A/r - [A/r - O_0] e^{-rt} \tag{8.29b}$$

Equation 8.29b points out that as t becomes large, the exponential function becomes small and O approaches the constant A/r, referred to as the equilibrium O value, O_e. Generally, for a large r such that $A < rO$, at a large t, O approaches asymptotically a minimum value. For a small r such that $A > rO$, at a large t, O approaches asymptotically a maximum value. The mathematical evaluation above applies to all organic chemicals added to soil, assuming that decomposition is proportional to the quantity of organic material present in the soil at any time t. In the case where when $t = 0$, $O_0 = 0$, the equation becomes

$$O = A/r(1 - e^{-rt}) \tag{8.29b}$$

Equation 8.29b reveals that O is related to the amount of organic added and its rate of decomposition.

The rate of decomposition of synthetic organic compounds in soil–water environments depends on the nature of the compound and the type of organisms present (e.g., algae, bacteria, or fungi), assuming all other factors (e.g., temperature, oxygen, moisture, level of organic matter in soils, and kinds and amounts of clay minerals) are optimum. Generally, organic gases (e.g., methane to pentane) are easily decomposed by organisms because they use such compounds as a sole source of energy. Light hydrocarbons (chains between 5 and 16 carbons, also known as aliphatics) are easily degraded and the relationship between degradation and chain length is inverse—the longer the chain, the faster the rate of decomposition. Decomposition is suppressed by branching of chains and increase in molecular weight. Nonaromatic cyclic compounds are easy to decompose via oxidation, ring cleavage, and carboxylation. On the other hand, aromatics such as benzene, toluene, napthalene, phenol, pyridine, and chlorobenzenes are more difficult to decompose than the ring aliphatics. The pathway involves hydroxylation (catechol formation), carboxylation, and ring cleavage. The compound is then decomposed as an aliphatic chain.

Additional factors controlling the rate of decomposition include carbon chain unsaturation, which increases rate of decomposition, number, and position of Cl atoms on an aromatic ring. For example, 2,4-dichlorophenoxyacetic acid is degraded readily, whereas 2,4,5-trichlorophenoxyacetic acid is more resistant. Finally, the position of attachment of a side chain alters the decomposition rate of aromatic compounds. For

TABLE 8.8. Hydrolysis Half-Lives for Several Groups of Organic Constituents

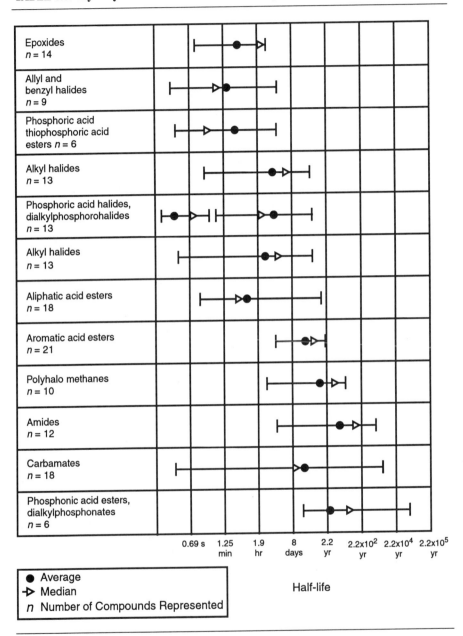

example, Ω-substituted phenoxyalkyl carboxylic acids are much more readily degraded than α-substituted phenoxyalkyl carboxylic acids. The half-lives of a number of organic compounds are reported in Table 8.8.

Pesticides represent a wide range of agricultural chemicals used to control various pests. The persistence of such chemicals in the soil is undesirable because of the potential of these chemicals to affect various crops differently, their bioaccumulation in various crops, and their leaching into ground and/or surface water. A brief description of the various classes of pesticides, with respect to their persistence behavior, is given below:

1. Chlorinated hydrocarbons—they represent the most highly persistent class, up to several years
2. Urea, triazine, and picloram—they represent a class of herbicides and exhibit a soil persistence in the range of 1–2 yr
3. Benzoic acid and amide—they represent a class of herbicides with a soil persistence of about 1 yr
4. Phenoxytoluidine and nitrile—they represent a class of herbicides with a persistence of about 6 months
5. Carbanate and aliphatic acid—a herbicide class with a soil persistence of approximately 3 months
6. Organophosphates—they represent a group of insecticides with persistence of few days to a few weeks

8.6.3 Adsorption–Sorption of Synthetic Organics

As previously discussed, synthetic organics are separated into two major groups—agricultural organic chemicals and industrial organics. Agricultural chemicals include pesticides and herbicides, organic chemicals specifically designed to kill weeds. In the case of industrial organics, the concern is mainly with chlorohydrocarbons. In general, pesticide behavior and fate in soils depends on (1) chemical decomposition, (2) photochemical decomposition, (3) microbial decomposition, (4) volatilization, (5) movement or leaching, (6) plant uptake, and (7) adsorption. It is generally agreed that adsorption–desorption of pesticides in soils directly and/or indirectly influence the magnitude of the effect of the seven factors listed above (Bailey et al., 1968). The two soil factors most important to adsorption–sorption or desorption are pH and organic matter content. In the case of acidic organic compounds, during the process of protonation–deprotonation, they become either noncharged or negatively charged. For example, herbicides containing carboxylic acid (COOH) groups or –OH groups exhibit a certain pK_a. The pK_a associated with the carboxylic acid is commonly less than 5 (Table 8.5) (e.g., benzoic and phenoxyacetic acid), whereas the –OH group, if present most likely exhibits a pK_a of 10 or higher. At pH values approximately two units below the pK_a, the carboxylate group would be fully protonated, therefore, the compound's charge as a result of its carboxylic group would be zero. When the pH equals its pK_a, 50% of the carboxylic acid groups would be protonated, thus contributing zero charge, while the other 50% would be deprotonated, thus contributing negative charge. When pH is approximately two pH units above the pK_a, approximately 100% of the functional groups would be fully deprotonated.

When the compound, representing a herbicide, contains an amino group, such a compound is referred to as basic. When fully deprotonated, its charge would be zero. When its pH equals its pK_a, half of its functional sites would be positively charged and the other half would carry no charge. On the other hand, when its pH is two units below its pK_a, all its functional sites would be positively charged because of 100% protonation.

Mineral surfaces appear to be more acid than the bulk solution. An equation that could be used to approximate clay surface pH (Stumm and Morgan, 1970) is

$$pH_s = pH_b + (1/2.3)(F\,\psi_0/RT) \qquad (8.30)$$

where pH_s denotes clay surface pH, pH_b denotes bulk solution pH, and the rest of the terms are as previously defined (Chapter 3). Substituting values for the constants in Equation 8.30 gives

$$pH_s = pH_b + 16.9\,\psi_0 \quad \text{at } 20°C \qquad (8.31)$$

Considering a mineral such as kaolinite at $I = 0.01$ and $\psi_0 = -0.1$ V, substituting ψ_0 into Equation 8.31 gives

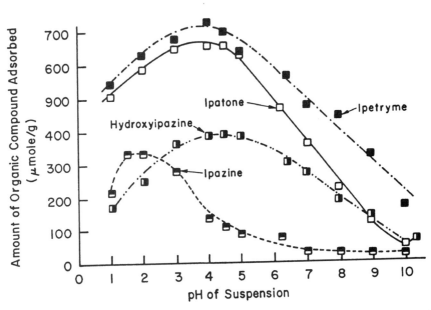

Figure 8.14. Effect of pH on the adsorption of four related s-triazines (4-isopropylamine-6-diethylaminme series) on montmorillonite clay (from Weber, 1966, with permission).

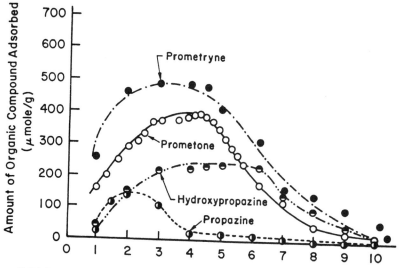

Figure 8.15. Effect of pH on the adsorption of four related s-triazines [4,6-bis(isopropyl-amino) series] on montmorillonite clay (from Weber, 1966, with permission).

TABLE 8.9. Chemical Properties of 13-s-Triazine Compounds

No.	Common Name	Groups on Triazine Ring			Solubility (%)	pK_a
		X	Y	Z		
1.	Desmetone	OCH_3	$NHCH_3$	NHC_3H_7	0.3500	4.15
2.	Simetone	OCH_3	NHC_2H_5	NHC_2H_5	0.3200	4.15
3.	Atratone	OCH_3	NHC_2H_5	NHC_3H_7	0.1800	4.20
4.	Prometone	OCH_3	NHC_3H_7	NHC_3H_7	0.0750	4.28
5.	Trietatone	OCH_3	NHC_2H_5	$N-(C_2H_5)_2$	0.0040	4.51
6.	Ipatone	OCH_3	$NHC_3H_7^5$	$N-(C_2H_5)_2$	0.0100	4.54
7.	Tetraetatone	OCH_3	$N-(C_2H_5)_2$	$N-(C_2H_5)_2$	—	4.76
8.	Prometryne	SCH_3	NHC_3H_7	NHC_3H_7	0.0048	4.05
9.	Ipatryne	SCH_3	NHC_3H_7	$N-(C_2H_5)_2$	—	4.43
10.	Propazine	Cl	NHC_3H_7	NHC_3H_7	0.0009	1.85
11.	Ipazine	Cl	NHC_3H_7	$N-(C_2H_5)_2$	0.0040	1.85
12.	Hydroxypropazine	OH	NHC_3H_7	NHC_3H_7	—	5.20 ~11.0
13.	Hydroxyipazine	OH	NHC_3H_7	$N-(C_2H_5)_2$	—	5.32 ~11.0

Source: From Weber, 1966.

$$pH_s = pH_b - 1.69 \tag{8.32}$$

Considering that a 2:1 clay mineral may exhibit a ψ_0 at least twice that of kaolinite, Equation 8.32 becomes

$$pH_s = pH - 3.38 \tag{8.33}$$

Therefore, the pH at the surface of clay minerals would be anywhere from 1.5 to 3.5 units lower than the bulk solution pH, depending on mineralogy. It has been determined experimentally that because of this lower pH at the surface, basic organic compounds would adsorb at the surface via hydrogen bonding at bulk pH values higher than expected. For example, the s-triazine group of herbicides exhibit pK_a in the range of 1.6 to approximately 4. One would expect that maximum adsorption for such compounds would take place on the surface of montmorillonite at bulk pH near their pK_a values (see Figs. 8.14 and 8.15 and Table 8.9). Experimental data demonstrated that maximum adsorption of basic compounds was independent of bulk pH and began to occur when clay surface acidity was 1–2 pH units lower than the lowest pK_a of the molecule.

Generally, in the case of acidic compounds (compounds containing carboxylic acid groups) (e.g., 2,4-D), experiments have shown that adsorption will begin when the pH of bulk solution is approximately 1–1.5 pH units above the pK_a of the compound and

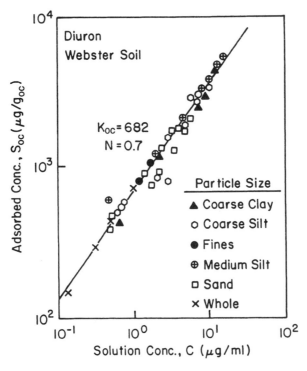

Figure 8.16. Equilibrium isotherms for adsorption of diuron on whole soil and particle-size separates of Webster soil (from Nkedi-Kizza et al., 1983, with permission).

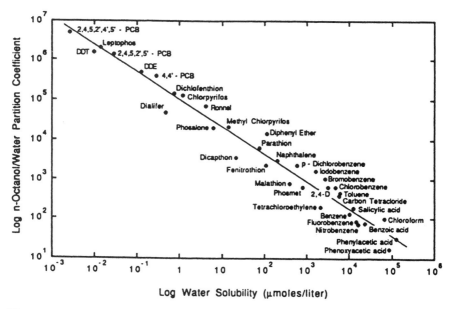

Figure 8.17. Relationship between octanol–water partition coefficients (K_{ow}) and solubility of several organic chemicals. Note the extensive range in solubilities of the organic chemicals (adapted from Chiou et al., 1977, with permission).

maximum adsorption would occur at bulk pH 1–2 pH units lower than the pK_a of the compound.

In the case of SOM, two mechanisms contribute to organic adsorption. One mechanism involves donation of H^+ by low-pK_a functional groups, thus forming a

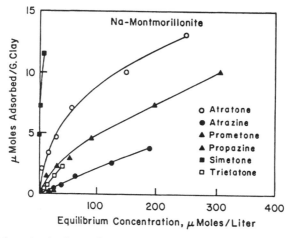

Figure 8.18. Adsorption isotherm of six s-triazines on Na–montmorillonite. Order of solubility: simetone > atratone > promatone > atrazine > trietazine > propazine (from Bailey et al., 1968, with permission).

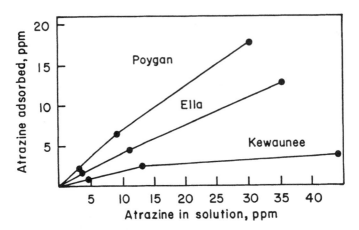

Figure 8.19. Atrazine adsorption by three different soils (from Armstrong, 1967, with permission).

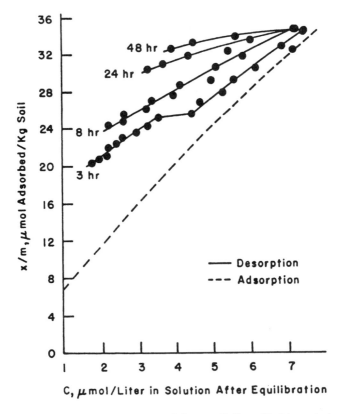

Figure 8.20. Desorption of 2.4.5-T from the Palouse soil (from Koskinen et al., 1979, with permission).

surface H$^-$ bridge with the adsorbing organic or producing excess positive charge on the adsorbing organic, hence surface cation adsorption occurs. The second mechanism involves hydrophobic–hydrophobic interactions between SOM and the sorbing organic. This is demonstrated in Fig. 8.16, where adsorption of diuron is shown to be related to organic carbon content, but not to the nature of the inorganic minerals making up the various sized particles. As expected, hydrophobic–hydrophobic interactions responsible for adsorption of organics by organic surfaces are related to the solubility of the compounds in question and are independent of pH. The data in Figure 8.17 show that the log-octanol partition coefficient is inversely related to the water solubility of the various organic compounds tested. The reverse is observed, however, in the adsorption on s-triazines by Na-montmorillonite. The data in Figure 8.18 show that, in general, the greater the solubility of a given s-triazine is, the greater its adsorption potential. Based on the above, because soils are made up of different clay minerals containing different quantities of SOM and may exhibit different pH, they will most likely exhibit different adsorption potential for herbicides (Fig. 8.19).

Other factors responsible for adsorption–sorption of hydrophobic organics by soil include the molecular size of adsorbing–sorbing organics. Generally, the greater the molecular size is, the greater the adsorption–sorption potential. For example, in the case of three chlorinated hydrocarbons—CHCl$_3$, CCl$_4$, and C$_2$Cl$_4$—adsorption in increasing order is CHCl$_3$ < CCl$_4$ < CCl$_4$ which follows the molecular size or weight order (McBride, 1994).

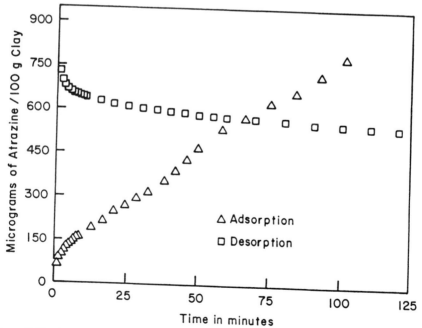

Figure 8.21. Atrazine adsorption and desorption at 0.5 mg L^{-1} by Maury silt loam soil under no-till management (from Evangelou, 1998, unpublished data, with permission).

In Chapter 4, data are presented which demonstrate that desorption of inorganics from clay mineral surfaces exhibits hysteresis. Similarly, in the case of desorption of organics, hysteresis appears to play a major role. For example, the data in Figure 8.20 show that the adsorption isotherm of 2.4.5-T is not similar to its adsorption isotherm and this difference increases as the soil sample incubation period increases. The data in Figure 8.21 also show that the kinetics of atrazine adsorption differs greatly from that of atrazine desorption. There are many potential causes for this difference, including particle diffusion effects and chemical thermodynamic effects.

In addition to adsorption of organics by surfaces owing to their low pH, the latter also promotes or catalyzes hydrolysis reactions of many organics (Evangelou, 1993 and references therein).

PROBLEMS AND QUESTIONS

1. List a number of soil processes outlined in this chapter and explain how the following principles might be used to quantify such processes:
   a. Acid–base chemistry
   b. Exchange reactions
   c. Reaction kinetics

2. Explain the role of pK_a in predicting adsorption of organics by clay minerals.

3. Explain how soil pH buffering potential controls NH_3 volatilization for
   a. Soil pH buffering less than 7
   b. Soil pH buffering higher than 8

4. Explain how soil CEC may regulate rate of nitrification at
   a. Low concentration of secondary cations (e.g., Ca^{2+})
   b. High concentration of secondary cations (e.g., Ca^{2+})

5. Consider the following degradation data of a particular organic in soil:

Time (wks)	Organic Decomposed (%)
1	16.7
5	50.0
10	66.7
50	90.9

Estimate the half-life of this particular organic (see Chapter 7)

6. Assuming that this particular organic (question 5) is added to soil once a year, continuously at the rate of 1 kg per acre, and the original quantity found in the soil was 0.05 kg per acre, calculate and plot the concentration behavior of the organic for the next 10 yr.

7. Assuming that for some unknown reason the same organic (question 5) in a different soil increases decomposition fourfold, calculate and plot the concentration behavior of the organic for the next 10 yr under the conditions stated in problem 6.

8. Assuming that for some unknown reason the same organic (question 5) in a third soil decreases decomposition fourfold, calculate and plot the concentration behavior of the organic for the next 10 yr under the conditions stated in problem 6.

9. What can you conclude about the influence of half-life on the accumulation of a particular organic in soil?

10. Explain the role of molecular size on soil sorption of nonionic organic compounds.

11. Two basic organic compounds that are fairly similar in size and structure exhibit two different pK_a values, 1.5 and 6.5. Which of the compounds would adsorb more at circumneutral pH by clay minerals?

12. Two nonionic compounds in the soil solution are in equilibrium with two different soils, one with low organic matter and the other with high organic matter. Which soil would contain a greater adsorbed quantity of this particular chemical?

13. Two inorganic weak acid compounds with pK_a values 2.5 and 6.5 are reacted with iron oxides. Which of the inorganic weak acid compounds would adsorb the most if the oxide is at pH 6?

14. Explain the general relationship between water solubility of organics and maximum sorption.

15. Explain the role of surface pH on the rate of hydrolysis of atrazine. Explain the practical significance of this phenomenon.

16. Explain the reversibility of adsorption of organics by soil surfaces. Explain the practical significance of your answer.

17. Using equilibrium equations demonstrate the solubility of phosphate in soil.

18. Describe various approaches that one may use to detoxify a given soil by an accidental overapplication of a particular pesticide. Explain the science of your approaches.

19. Name a number of factors that determine the half-life of synthetic organics in soil.

20. Give two mechanisms by which soil-added K^+ may inhibit nitrification.

PART V
Colloids and Transport Processes in Soils

9 Soil Colloids and Water-Suspended Solids

9.1 INTRODUCTION

Soil colloids (clay–organic colloids) are particles less than 2 μm (0.000008 in.) in diameter and commonly carry surface electrical charges. In the pH range of 6–9, most soil colloids carry a net negative charge. Elementary physics states that similarly charged particles repel each other, which brings about a condition known as colloid dispersion. Colloid dispersion has received little attention by soil scientists in comparison to the attention given to soil–water processes involving nutrient adsorption and/or release. Yet, clay–organic colloids are involved in many soil–water processes, including erosion, soil crusting, hydraulic conductivity, solute transport, pollutant transport, loss of nutrients, and presence of suspended solids in lakes, ponds, and even rivers.

Suspended solids have the potential to silt out stream channels, rivers, lakes, and reservoirs; they inhibit aquatic life and are expensive to remove from water. In some industries (e.g., mining) suspended solids, along with various other pollutants, are regulated by law, which requires that sediment ponds at the base of disturbed watersheds be built with sufficient detention time so that the water released meets certain sediment and water chemistry criteria (Tables 9.1 and 9.2).

Table 9.1. Effluent Limitations (mg L^{-1}) Except for pH

Effluent Characteristics	Maximum Allowable	Average of Daily Values for 30 Consecutive Discharge Days
Iron (total)	7.0	3.5
Manganese (total)	4.0	2.0
Total suspended solids	70.0	35.0
	pH—within range of 6.0 to 9.0	

TABLE 9.2. Water Content Composition of Selected Sedimentation Ponds in Kentucky Coal-Mine Fields

Sample	Location	Ca[a]	Mg[a]	Na[a]	K[a]	Al[a]	Cl[a]	SO$_4$[a]	Alkalinity HCO$_3$[a]	EC[b]	pH[b]	Ionic Strength	SAR[c]
1	Martin County	5.41	4.80	0.50	0.27	<0.0006	0.46	12.37	3.69	1.400	7.9	0.019	0.221
2	Martin County	0.46	0.42	0.26	0.10	<0.0006	0.47	1.31	0.75	0.160	6.6	0.002	0.392
3	Martin County	0.36	0.42	0.34	0.24	<0.0006	0.72	2.12	1.72	0.360	7.2	0.005	0.445
4	Martin County	2.04	1.02	0.29	0.05	<0.0006	0.28	2.62	1.17	0.320	7.1	0.005	0.235
5	Johnson County	2.04	1.60	0.62	0.07	1.3300	0.35	5.53	-1.67	0.760	3.6	0.011	0.457
6	Martin County	3.52	2.24	0.63	0.22	0.0006	0.45	6.95	1.63	0.830	7.5	0.012	0.371
7	Johnson County	3.09	4.30	0.75	0.21	—	0.42	7.42	1.11	0.870	7.6	0.012	0.390
8	Johnson County	2.65	3.66	0.63	0.19	—	0.31	5.98	1.15	0.790	7.6	0.011	0.355
9	Johnson County	0.32	0.36	0.38	0.08	—	0.58	0.61	0.53	0.070	7.3	0.001	0.652
10	Nolin River Lake	0.85	0.25	0.20	0.05	<0.0006	—	0.24	2.91	0.161	8.0	0.002	0.269
11	Lake Beshear	0.52	0.17	0.11	0.02	<0.0006	—	0.23	4.17	0.103	7.8	0.001	0.187
12	Rough River	0.85	0.21	0.12	0.05	<0.0006	—	0.27	2.85	0.158	8.1	0.002	0.165
13	Lake Malone	0.20	0.14	0.12	0.05	<0.0006	—	0.20	3.11	0.063	7.1	0.001	0.291
14	Hopkins County	3.28	3.24	0.30	0.08	<0.0006	—	8.30	0.12	0.985	6.8	0.014	0.166
15	Hopkins County	17.27	12.55	13.57	0.44	<0.0006	—	43.66	3.75	4.270	8.0	0.060	3.514
16	Hopkins County	2.22	2.83	0.95	0.06	<0.0006	—	7.30	0.12	0.780	6.9	0.011	0.598

Source: From Evangelou, 1989.

[a]Values given are in mmol$_c$ L^{-1}.

[b]Values given are in dS m^{-1}.

[c]Values given are in mmol L^{-1}.

The purpose of this chapter is to introduce the fundamental soil–water chemistry processes controlling behavior of colloids in soil–water environments.

9.2   FACTORS AFFECTING COLLOID BEHAVIOR AND IMPORTANCE

Colloid behavior in natural soil–water systems is controlled by dispersion–flocculation processes, which are multifaceted phenomena. They include surface electrical potential (El-Swaify, 1976; Stumm and Morgan, 1981), solution composition (Quirk and Schofield, 1955; Arora and Coleman, 1979; Oster et al., 1980), shape of particles, initial particle concentration in suspension (Oster et al., 1980), and type and relative proportion of clay minerals (Arora and Coleman, 1979). When suspended in water, soil colloids are classified according to their settling characteristics into *settleable* and *nonsettleable solids*.

By definition, *settleable solids* refer to the volume of particles that settle to the bottom of a 1-L volume *Imhoff cone* during a period of 1 h (Fig. 9.1). Federal regulations state that the critical volume of *settleable solids* in sedimentation reservoirs, as measured by the Imhoff Cone, should not exceed 0.5 mg L^{-1}. The assumption underlying settleable solids determination using the Imhoff cone method is that their volume is small relative to the total volume of the Imhoff cone. The practical meaning of this assumption is that the suspension tested is dilute enough so that particle settling is size and not solution viscosity or due to physical interactions between particles (i.e., larger particles forcing smaller particles to settle).

Nonsettleable solids are composed of solids that will not settle out within an hour. However, colloidal suspensions are very dynamic, and since the size of the particles distinguishes settleable from nonsettleable solids, the ratio between them depends on several variables (e.g., pH, type of cations in solution, and salt concentration). This is because colloids may unite (flocculate) via various mechanisms to form larger particles

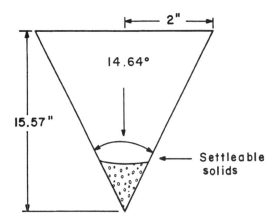

Figure 9.1. Schematic of Imhoff cone.

or separate (disperse) to form smaller colloidal particles. The increase in size of colloidal clusters takes place via a number of mechanisms involving physical and chemical forces. Commonly, two terms describe such colloidal processes—*flocculation* refers to the polymerization of colloidal particles and *coagulation* refers to aggregation or agglomeration of colloidal particles due to electrolytes. In this chapter, these terms are used interchangeably because natural soil colloids are structurally complex (include clays, oxides, and free or adsorbed organics) and both processes occur simultaneously.

The factors controlling the ratio between settleable and nonsettleable solids can be linked to various soil–water processes. Such processes include soil hydraulic conductivity, erosivity, crusting, structure form, and stability. For example, generally, when any two soils with the same textural composition (percent of sand, silt, and clay) are physically dispersed in water, the soil that maintains a higher concentration of nonsettleable solids would also most likely be subject to decreasing hydraulic conductivity or increasing erodibility.

9.2.1   Colloid Dispersion or Flocculation

Dispersion–flocculation processes are generally controlled by double layer swelling, adsorbed hydrolyzed Fe or Al, and chemical bridging (tactoid formation) (Stumm and O'Melia, 1968). Once dispersed, clay colloids are kept dispersed by repulsive double layers (Van Olphen, 1971). The force of repulsion is related to the thickness of the double layer (see Chapter 4). This dimension is represented by the ions concentrated near the oppositely charged colloid surface. Any colloid that has a net negative or a net positive charge repulses a like-charged colloid.

The classical theory of colloidal stability (DLVO theory) (Derjaguin and Landau, 1941; Verwey and Overbeek 1948) generally accounts for the influences of ion valence and concentration on suspended colloid interactions. According to the DLVO theory, the long-range repulsive potential, R_f, resulting from diffuse double layers (DDLs) (see Chapter 3) of like-charged colloids retards the coagulation or flocculation rate of soil colloids. This long-range repulsive potential, between like-charged colloids is given by

$$R_f = (64/k)\tanh(\nu F\psi_0/4RT)\ C_0 RT\exp(-kd) \tag{9.1}$$

and

$$k = (e\nu)\ (DKT/8C_0)^{-1/2} \tag{9.2}$$

where
  F = Faraday's constant
  C_0 = bulk solution concentration
  R = molar gas constant
  T = absolute temperature
  d = separation between planar surfaces

ψ_0 = colloid surface electrical potential
e = elementary charge of an electron
ν = valence of ions in solution
D = dielectric constant
K = Boltzman constant

Note that the inverse of k in Equation 9.2 represents the thickness of a single double layer.

The thickness of a double layer is controlled by the ionic strength of the solution (see Chapter 3) and ion valence (Fig. 9.2). Upon increasing ionic strength, colloids physically approach each other and van der Waals attraction approaches a maximum. The degree to which a given ionic strength will influence the double layer depends on the nature of a mineral's surface. Different mineral surfaces exhibit different electrical potentials, ψ_0 (Fig. 9.3). Similarly, upon increasing the valence of counterions in the system, the thickness of the double layers of two interacting colloid particles decreases (Fig. 9.2), and van der Waals attraction between particles increases hence, flocculation occurs.

The influence of counterion valence on the double layer thickness is described by the valency rule of Schulze and Hardy. It basically predicts that if a monovalent counterion is changed to a divalent counterion, the thickness of the double layer decreases by half; and if the divalent counterion is changed to a trivalent ion, the thickness of the double layer decreases by three-quarters (Fig. 9.2). The relative amounts of counterions required to induce flocculation are 100 for a monovalent, 2 for a divalent, and 0.04 for a trivalent ion.

At a constant ionic solution composition, the component ψ_0 is directly related to R_f. However, ψ_0 of variable charge colloid surfaces is also related to pH. This is

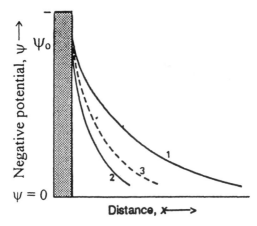

Figure 9.2. The influence of electrolyte concentration in the bulk solution on the thickness of the double layer: 1, monovalent ion at unit concentration; 2, ninefold increase in the concentration; 3, same concentration as the monovalent ion of curve 1 but for a divalent ion (adapted from Taylor and Ashroft, 1972).

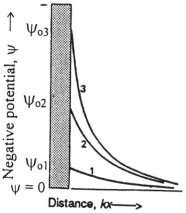

Figure 9.3. Schematic of the double layer: 1, low values of ψ_0; 2, intermediate ψ_0 values; 3, large ψ_0 values (adapted from Taylor and Ashroft, 1972).

demonstrated in Figure 9.4, which shows the maximum flocculation potential of iron oxide in the pH range where pH $\cong$ pH$_0$, otherwise known as the pH region of zero point of charge (PZC) (Table 9.3) (for details about changing the PZC of a surface, see Chapter 3). The component ψ_0 is controlled by the pH of the colloidal suspension, assuming that the colloids exhibit pH-dependent charge. The mathematical expression demonstrating the interrelationship between ψ_0 and pH is given by a Nernst-type equation (Uehara and Gillman, 1980):

$$\psi_0 = (KT/D)\,(pH_0 - pH) \tag{9.3}$$

where K, T, and D are as defined for Equations 9.1 and 9.2, pH$_0$ = pH of the colloidal suspension where $\psi_0 = 0$, and pH = pH of the colloidal suspension.

The influence of pH on ψ_0 of a colloid can be demonstrated indirectly by quantifying the increase in CEC as a function of pH (Marsi and Evangelou, 1991a). This relationship is demonstrated by the equation (Uehara and Gillman, 1980)

$$CEC_v = (2C_0\,DKT/\pi)^{1/2}\sinh \nu\,(pH_0 - pH) \tag{9.4}$$

where all terms are as defined for Equations 9.1–9.3 and CEC$_v$ = variable surface charge on a surface area basis. Generally, in soil colloids with variable charge, upon increasing pH, ψ_0 becomes more negative, CEC$_v$ increases, and R_f thus increases. Conversely, upon decreasing pH, ψ_0 becomes less negative and CEC$_v$ decreases. When ψ_0 or CEC$_v$ approach zero, R_f also approach zero, which leads to colloid coagulation or flocculation (Singh and Uehara, 1986; Emerson, 1964; Keren and Singer, 1988) (Fig. 9.3).

In addition to the components given above (ψ_0, ν, C_0) controlling colloidal flocculation or stability (stability is defined as the state or the conditions under which primary colloids are maintained) (Schofield and Samson, 1954; Shainberg and Letey,

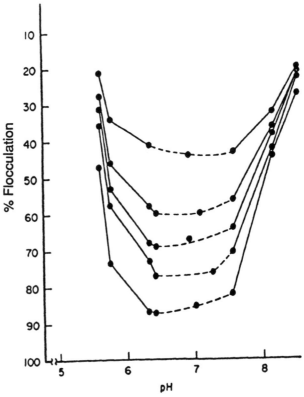

Figure 9.4. Settling characteristics of an iron oxide as a function of pH for every 3-hr time interval (from Evangelou, 1995b, with permission).

TABLE 9.3. Examples of Oxides and pH of Zero Net Charge

Oxide	pH_{PZNC}
Aluminum oxide	9.1
Aluminum trihydroxide	5.0
Iron oxide	6–8
Manganese oxide	2–4.5
Silicon oxide	2
Kaolinite	4.5
Montmorillonite	2.5

Source: From Stumm and Morgan, 1970.

1984; Evangelou and Sobek, 1989), in the case of soil colloids, additional components are also involved. These additional components include:

1. Relative proportion of monovalent to divalent cations in the bulk solution (Shainberg and Letey, 1984)
2. Type of cations and location of cations in the double layer (e.g., inner-sphere versus outer-sphere) (Hesterberg and Page, 1990)
3. Shape of particles and initial particle concentration in suspension (Oster et al., 1980)
4. Type of clay minerals present
5. Relative proportion of clay minerals (Arora and Coleman, 1979)

Based on Equations 9.1 and 9.2, colloidal stability (maximum dispersion) depends on maximum R_f, which describes the maximum repulsive energy between two planar colloidal surfaces. It also appears from these equations that R_f is controlled by C_0, ψ_0, or CEC_v. However, in addition to the repulsive force, there is an attraction force (A_f) between soil colloidal particles. The force of attraction (van der Waals force) between two particles separated by a distance of $2d$ (d = particle diameter) is described by

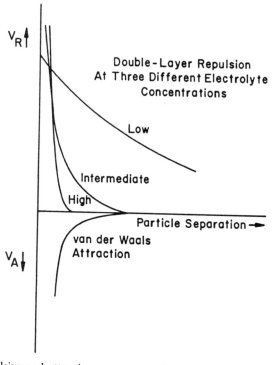

Figure 9.5. Repulsive and attractive energy as a function of particle separation at three electrolyte concentrations (from van Olphen, 1977, with permission).

$$A_f = A/48 \, \pi d^2 \qquad\qquad (9.5)$$

where A is an experimental constant. The force of attraction (A_f) appears to depend on the density and charge of the dispersed phase, but it is independent of the solution composition. This is demonstrated in Figure 9.5, which shows the inverse dependence of R_f on electrolyte and the independence of A_f from electrolyte. The difference between repulsive and attraction forces determines if a colloidal system would be in a dispersive or flocculative mode. When the force of attraction, A_f, is greater than the force of repulsion, R_f, the system would be flocculated. Conversely, when the force of attraction, A_f, is smaller than the force of repulsion, R_f, the system would be dispersed. This is demonstrated in Figure 9.6. These statements apply to charged colloids. However, many soil colloids possess pH-dependent charge; therefore, the force of repulsion, R_f, would also be pH dependent.

Clay colloids are irregular in shape and their charges are separated into *planar* charges and *edge* charges. Because of this charge distribution, two unique flocculation mechanisms are encountered. One is *phase to phase* flocculation, commonly encoun-

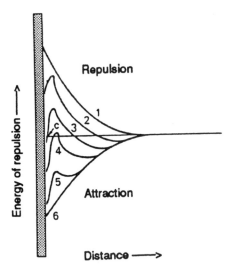

Figure 9.6. Schematic of repulsion or attraction forces (which vary with distance from the particle surface) between particles in suspension. Curves 1 and 6 are examples of repulsion and attraction curves, respectively, which vary with the colloid and the kinds and amounts of electrolytes. A summation of curves 1 and 6 for different conditions produces curves 2–5. In curve 2, the energy of repulsion predominates and a stable suspension is formed. Increasing electrolyte produces curves 3, 4, or 5 owing to suppression of the electric double layer. Curve 3 shows there is still an energy barrier to be overcome prior to flocculation. When the colloids surmount this energy barrier and approach closer than point C, flocculation occurs because the forces of attraction predominate. Curve 5 suggests spontaneous flocculation without redispersion unless there is a shift toward curve 2 by reexpanding the double layer through changing kinds and/or amounts of electrolytes (adapted from Kruyt, 1952).

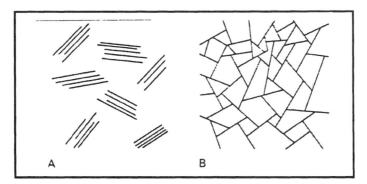

Figure 9.7. Schematic of modes of flocculation: A, phase to phase flocculation; B, edge to phase flocculation (adapted from Taylor and Ashroft, 1972, with permission).

tered in 2:1 clay minerals (e.g., smectites). A second mechanism is *edge to phase* flocculation, commonly encountered in 1:1 clay minerals (e.g., kaolin) (Fig. 9.7). The latter possesses planar negative charges and edge positive or negative charges, depending on pH, in significant quantities so that when positive charges encounter negative charges, edge to phase flocculation is induced. In phase to phase flocculation, increasing electrolyte concentration (e.g., $CaCl_2$) compresses the *electric double layer* and phase to phase flocculation is induced. The degree of hydration between phases dictates floccule stability. The less the amount of water, the more stable the floccules are. When complete dehydration takes effect, highly stable aggregates (*tactoids*) are formed.

9.2.2 Zeta Potential

The *zeta potential* is the electric potential of a double layer at the slipping point (Fig. 9.8). The slipping point is known to be in front of the colloid's surface at a distance of at least a single hydrated cationic layer (Stern layer) (the exact location of the slipping point is not clear). Based on what was said above with respect to pH dependence and/or ionic strength dependence of the electric double layer, it is also known that as the ionic strength increases, or pH approaches the PZC, the zeta potential decreases (Fig. 9.9).

The zeta potential is also related to a colloid's electrophoretic mobility, which describes the colloid's potential to move along an electric gradient. At the pH, or electrolyte concentration, where the zeta potential approaches zero, the electrophoretic mobility of the particle approaches zero. At this point, such particles would have a tendency to flocculate. When a high-valence cation, tightly adsorbed to the surface, is in excess of the negative charge of the colloid's surface, a phenomenon known as *zeta potential reversal* takes effect. This is demonstrated in Figures 9.9 and 9.10. Zeta potential reversal could induce colloid dispersion, depending on the type and concentration of electrolyte present.

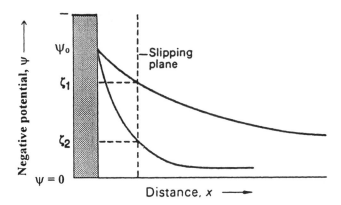

Figure 9.8. Schematic of the electric double layer under two different electrolyte concentrations. Colloid migration includes the ions within the slipping plane of the colloid; ζ_1 denotes the electric potential in dilute solution: ζ_2 denotes the electric potential in concentrated solution (adapted from Taylor and Ashroft, 1972)

Zeta potential can be measured by the use of a zeta meter, which is commonly employed in municipal water-treatment facilities to evaluate the flocculation potential of suspended biosolids.

9.2.3 Repulsive Index

The force of repulsion between like-charged particles manifests itself because of the hydration (osmotic potential) of the counter ions present in the double layer in relation

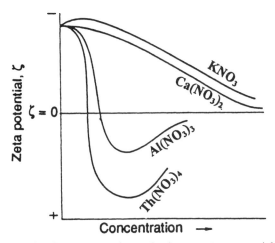

Figure 9.9. Influence of cation concentration and valence on zeta potential (from Taylor and Ashroft, 1972, with permission).

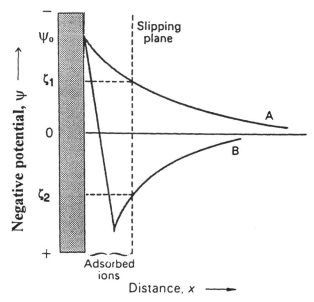

Figure 9.10. Schematic of zeta potential reversal: curve A is a normal curve with ζ_1; curve B is reversed owing to an inner-sphere adsorption (adsorption within the slipping plane) of multivalent ions producing ζ_2 (adapted from Taylor and Ashroft, 1972).

to the osmotic potential of the equilibrium solution (Shainberg et al., 1971; Bolt and Peech, 1953; Bolt and Miller, 1955; Singh and Uehara, 1987). This osmotic potential is directly related to the type and concentration of cations. For example, the presence of Na^+ in the double layer causes a relative increase in the osmotic pressure, whereas the presence of Ca^{2+} causes a relative decrease (Shainberg et al., 1971). The relative proportion of the various cations present in the double layer is related to factors such as proportion of cations in the bulk solution, ionic strength, surface potential, and type of clay minerals (e.g., variable vs. permanent-charge minerals) (Singh and Uehara, 1987).

Stumm and Morgan (1981) discussed the forces of repulsion and attraction between perfectly spherical or flat colloidal particles that have their charge uniformly distributed over their surface. Clay colloids, however, are not perfectly spherical or perfectly flat, nor is their charge uniformly distributed over the surface (Oster et al., 1980; Shainberg et al., 1980); for this reason, a less rigorous but more practical approach is used below to predict dispersion or flocculation phenomena. From the *Gouy–Chapman model* (Stumm and Morgan, 1981), which describes an ideal single, flat, double layer, the thickness of the double layer (R) in centimeters is given by

$$R = (DkT/8\pi e^2 NI)^{1/2} \tag{9.6}$$

where
e = charge of electron
D = dielectric constant

kT = Boltzman constant times absolute temperature
N = Avogadro's number
I = ionic strength

Only the ionic strength is variable if temperature remains constant. Therefore,

$$R = A(I)^{-1/2} \tag{9.7}$$

where A is a constant equal to $(DkT/8\pi e^2 N)^{1/2}$, hence,

$$R/A = (I)^{-1/2} \tag{9.8}$$

The value of R/A is referred to as the *repulsive index (RI)*.

From data representing the water chemistry of several lakes and sedimentation ponds, approximating water chemistry of many sedimentation ponds throughout the eastern U.S. humid region (Bucek, 1981), the relationship (Evangelou and Garyotis, 1985)

$$I = (EC)(0.014) \tag{9.9}$$

has been established where electrical conductivity (EC) is in units of millimhos per centimeter (dS m^{-1} or mmhos cm^{-1}) (Fig. 9.11). Equation 9.9 is in close agreement with that reported by Griffin and Jurinak (1973). The appropriate substitution into Equation 9.9 gives (Evangelou, 1990b)

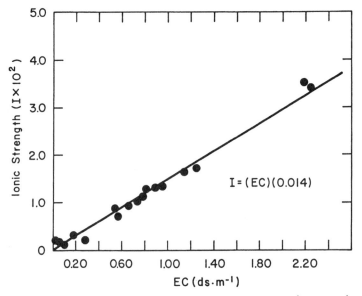

Figure 9.11. Relationship between electrical conductance (mmhos cm^{-1} or dS m^{-1} or mmhos cm^{-1}) and ionic strength (from Evangelou and Garyotis, 1985, with permission.)

$$RI = (0.014 \text{ EC})^{-1/2} \tag{9.10}$$

The term RI of Equation 9.10 can be related to the dispersive potential of clay colloids with net negative charge. However, because the force of repulsion is not directly controlled by RI, but rather by components controlling osmotic pressure in the double layer and in the bulk solution (Shainberg and Letey, 1984; Shainberg et al., 1971; El-Swaify, 1976; Arora and Coleman, 1979), the RI–suspended solid (SS) relationship is not expected to have universal application. It would be unique to each particular sedimentation system with its own mineralogy, surface chemistry, and water composition. This is demonstrated in Figure 9.12, which shows that the influence of electrolyte concentration (NaHCO$_3$) with respect to clay–colloid dispersion depends on clay mineralogy and soil type.

The data in Figure 9.13 relates RI to suspended solids after 90 min of settling time for four soil colloids by varying the EC. It shows a linear relationship between RI and suspended solids. It also shows that every colloid sample has two unique components, the extrapolated threshold value of RI corresponding to the highest flocculation point

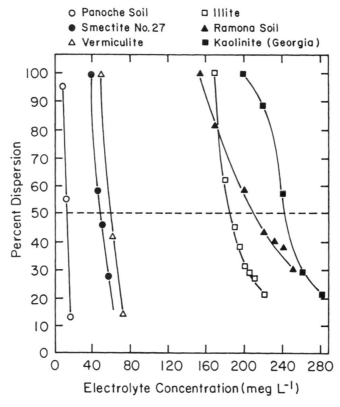

Figure 9.12. Effect of electrolyte concentration (NaHCO$_3$) on dispersibility and the determination of critical salt concentration (from Arora and Coleman, 1979, with permission).

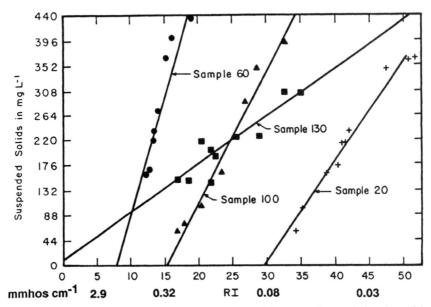

Figure 9.13. Relationship between flocculation after a 90-min settling period and repulsive index (RI) of four selected samples (from Evangelou, 1990, with permission).

and the responsiveness of the colloidal phase to the RI. For example, sample 20 has a threshold RI value for minimum suspended solids (100% flocculation) of near 30 (intercept of x axis), but sample 100 at RI 30 shows maximum concentration of colloidal suspended solids. Minimum sensitivity (smaller slope value) to the RI is exhibited by sample 130, while maximum sensitivity is exhibited by sample 60. At the practical level, these data demonstrate that a rainfall event with the same effects on water chemistry will have dramatically different effects on the level of colloidal suspended solids in ponds or shallow lakes, as represented by samples 60 and 130. Assuming that a rainfall event causes a dilution effect, the maximum amount of suspended solids (after a certain settling period) will be observed in the pond represented by sample 60.

Based on the above, a RI–SS relationship representing a particular colloidal system would be constant, assuming that pH is constant, because the RI–SS relationship is expected to be pH dependent. Furthermore, at a certain pH value, a colloidal system that exhibits both negative and positive surface potential would be expected to coflocculate when the net electrical potential is zero or the system is at its PZC (Schofield and Samson, 1953; Quirk and Schofield, 1955; Evangelou and Garyotis, 1985). Under these conditions, the settling characteristics of the suspended solids would be independent of RI (Fig. 9.14a and b).

Equation 9.10 does not make any distinction between different electrolytes. In other words, Equation 9.10 implies that two suspension systems with similar clay minerals and the same ionic strength should exhibit identical clay-settling behavior, even if one solution consists of NaCl and the other of $CaCl_2$. However, this is not valid because it

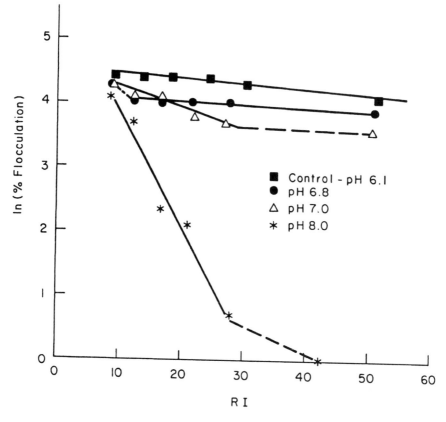

Figure 9.14a. Relationship between RI and percentage of flocculation immediately after pH adjustments were made (from Evangelou and Garyotis, 1985, with permission).

is known that Na^+ and Ca^{2+} have different clay-dispersive properties owing to their difference in selective preference by clay surfaces (Arora and Coleman, 1979; Oster et al., 1980; U.S. Salinity Lab Staff, 1954). The selective preference of the cations by the clay surface and its influence on clay-settling behavior can be demonstrated by the interrelationship between the sodium adsorption ratio (SAR) and exchangeable sodium percentage (ESP)

$$SAR = Na/(Ca)^{1/2} \tag{9.11}$$

and

$$ESP = \{ExNa/CEC\} \times 100 \tag{9.12}$$

where ExNa denotes exchangeable Na and CEC = ExNa + ExM = effective cation exchange capacity (M denotes Ca or Mg). At a certain SAR, a specific portion of the

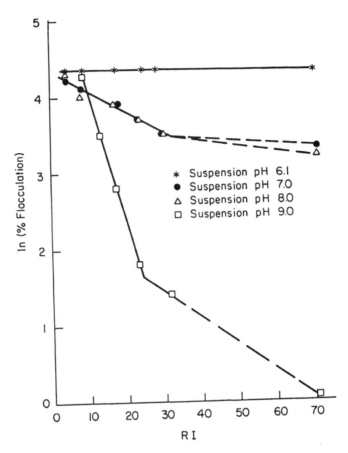

Figure 9.14b. Relationship between RI and percentage flocculation 48 h after pH adjustments were made (from Evangelou and Garyotis, 1985, with permission).

CEC, depending on the magnitude of preference of Na over M or vice versa, is occupied by sodium. A similar analysis can be conducted for both ammonium and potassium bases by simply using the ammonium adsorption ratio (AAR) or potassium adsorption ratio (PAR).

The U.S. Soil Salinity Laboratory Staff (1954) reported that SAR values of 10–15 $(mmol\ L^{-1})^{1/2}$ usually correspond to ESP values in the range of 10–15 at which values clays will undergo dispersion. This relationship may vary among colloids with different mineralogy (Oster et al., 1980) and/or mixtures of colloids with different mineralogy (Arora and Coleman, 1979). Consequently, the force by which given types of colloidal particles attract or repulse each other in a Na–Ca or Na–Mg solution is a function of the total concentration of the salt, the type of divalent cation (Ca or Mg), and SAR. Therefore, pH, salt concentration, type of divalent cation, and SAR are expected to play important roles on soil colloid flocculation.

The data in Figure 9.15 show the relationship between the critical salt concentration and SAR for various colloids. The relationship shows that under a given pH, two components need to be met for flocculation. One component is the salt solution concentration and the second is the SAR. Therefore, for any given critical salt concentration, SAR values above the one described by the linear relationship in Figure 9.15 predict colloid dispersion, while SAR values below that described by the linear relationship in Figure 9.15 predict colloid flocculation. The relationship in Figure 9.15 applies to various Ca^{2+} to Na^+ solution compositions. However, when SAR approaches infinity (solution contains only Na^+), the critical salt concentration appears to vary between soil colloid types and mixtures of different colloids (Tables 9.4 and 9.5). Arora and Coleman (1979) demonstrated that suspension mixtures of kaolinite and smectite at various proportions exhibit different critical $NaHCO_3$ concentrations (CSC) (Table 9.5). The latter is defined as the concentration of $NaHCO_3$ needed to settle out 50% of the colloidal clay particles within a 24-hr settling period. The CSC reaches a maximum when the mixture is 30% smectite and 70% kaolinite. The lowest CSC is that of the kaolinite suspension. There is a 40-fold difference between the lowest and highest CSC. The exact cause for this behavior is not well understood.

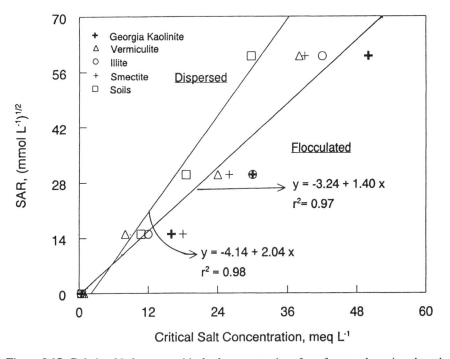

Figure 9.15. Relationship between critical salt concentrations for reference clay minerals and soil clays in mixed-ion systems (raw data were taken from Arora and Coleman, 1979).

TABLE 9.4. Influence of pH on the Critical Salt Concentration ($NaHCO_3$) of Reference Clay Minerals and Soil Clays

	CSC ($mmol_c\ L^{-1}$)[a]			
	pH 7	pH 8.3		pH 9.5
Sample	<2 µm	<2 µm	<0.4 µm	<2 µm
Reference clays				
Smectite No. 27	14	47	47	17
Smectite No. 23	20	48	—	68
Smectite No. 32	7	28	—	15
Nontronite No. 33A	10	60	—	60
Vermiculite	38	58	55	30
Illite No. 36	9	185	165	95
Kaolinite No. 9	[b]	8	—	30
Kaolinite (Georgia)	5	245	370	75
Soil clays				
Panoche	7	12	14	25
Hanford	9	45	—	90
Ramona	34	210	195	224
Grangeville	13	27	35	50

Source: Arora and Coleman, 1979.
[a]CSC = concentration of salt needed to settle out 50% of the colloidal clay particles within a 24-hr settling period.
[b]Flocculated.

TABLE 9.5. Critical Salt Concentrations (CSC) of Mixtures of Kaolinite and Smectite Determined in $NaHCO_4$ Solutions

Amount in Mixture (%)		CSC ($mmol_c\ L^{-1}$)[a]
Smectite	Kaolinite	
0	100	8
1	99	22
7	93	35
15	85	85
20	80	270
30	70	310
50	50	150
80	20	37
90	10	35
95	5	25
100	0	47

Source: Arora and Coleman, 1979.
[a]CSC = concentration of salt needed to settle out 50% of the colloidal clay particles within a 24-hr settling period.

9.3 FLOCCULATION AND SETTLING RATES

The rate of particle settling in a water column, representing a lake, pond, or any other water impoundment, can be described by Stokes' Law:

$$t = \{18hn/d^2g(P_s - P_f)\} \tag{9.13}$$

where

t = time (sec)
d = particle diameter (m)
h = distance for particle to fall, meter
n = viscosity of water at given temperature (kg ms^{-1}, 0.1 at standard temperature)
g = gravitational constant (9.8 kg ms^{-2})
P_s = density of the solid (kg m^{-3}, for clays, approximately 2650)
P_f = density of the fluid (kg m^{-3})

Equation 9.13 shows that nonphysically and nonchemically interacting particles of a given density and relatively large diameter settle faster than particles with similar composition but relatively smaller size. Additionally, particles with a given diameter and a relatively large density settle faster than particles with the same diameter but a relatively smaller density. The rate of settling (S_r) in meters per hour (m s^{-1}) can be described by

$$S_r = [d^2g(P_s - P_f)]/18n \tag{9.14}$$

where $g(P_s - P_f)/18n$ is a constant denoted as C^*. Therefore, the rate of colloid settling can be expressed by

$$S_r = d^2C^* \tag{9.15}$$

Equation 9.15 points out that a small increase in particle diameter due to colloid agglomeration has a large impact on the rate of particle settling (S_r).

Apparent particle size or particle diameter of colloids in natural bodies of water (e.g., lakes, ponds, and human-made water impoundments) varies depending on the type of interactions between the colloidal particles (physical versus chemical or outer-sphere, versus inner-sphere colloid to colloid interactions) and type of colloids participating in these interactions (e.g., inorganic versus organic colloids). Colloid to colloid interactions result in particle agglomeration or colloid flocculation. Therefore, the greater the degree of flocculation is, the greater the rate of settling. The rate of colloid flocculation or agglomeration depends on the frequency of successful colli-sions between colloids. Successful collisions depend on the force by which colloids collide with proper orientation. There are two types of colloid collision forces in a liquid medium (e.g., water). One force is due to Brownian (thermal) effects producing *perikinetic* agglomeration and a second force, which exceeds that of Brownian motion, is due to velocity gradients and is referred to as *orthokinetic* agglomeration (Stumm and Morgan, 1970).

Perikinetic agglomeration applies to a monodisperse suspension and can be represented by a second-order rate law:

$$-dp/dt = k_p p^2 \cdot \qquad (9.16)$$

or

$$(1/p) - (1/p_0) = k_p t \qquad (9.17)$$

where p = number of colloid particles at any time t, p_0 = initial number of particles, and k_p a conditional second-order rate constant. Orthokinetic agglomeration under a constant particle velocity applies to larger colloid particles and can be described by first-order kinetics:

$$-dp/dt = k_0 p \qquad (9.18)$$

or

$$\ln(p/p_0) = -k_0 t \qquad (9.19)$$

where p = number of colloid particles at any time t, p_0 = initial number of particles, and k_0 is a conditional first-order rate constant (Stumm and Morgan, 1970). In nature, it is difficult to distinguish *perikinetic* from orthokinetic agglomeration because both modes take place simultaneously. Thus, the overall agglomeration rate is the sum of orthokinetic and perikinetic agglomeration:

$$-dp/dt = k_p p^2 + k_0 p \qquad (9.20)$$

Agglomeration rate constants depend on many factors, including chemical makeup of colloids, size of colloids, surface charge of colloids, and solution concentration and ionic composition. For this reason, particle settling in a water column is difficult to predict. However, it is easy to produce experimental data describing the settling of a particular colloidal system under a given set of experimental conditions. Settling includes the overall rate of agglomeration as well as particle movement by gravity deeper in the water profile. Such systems are referred to as polydispersed systems. Figures 9.16a–c show that the settling behavior of suspended particles, as expected, is dependent upon ionic strength (EC). The settling rate at the highest ionic strength is initially very rapid because of high collision frequency (Oster et al., 1980) and then it declines. The majority of the suspended particles at the highest ionic strength settle out within a 60-min period. However, at the lowest EC values shown, an insignificant quantity has settled out, even after 7 hr. Also, the water EC in these systems is well within drinking standards (tap water may have an EC of up to 0.700 dS m^{-1} or mmhos cm^{-1}).

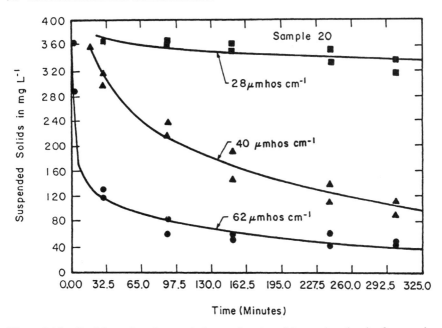

Figure 9.16a. Particle settling characteristics as a function of time at three levels of suspension electrical conductivity (EC) (from Evangelou, 1990, with permission).

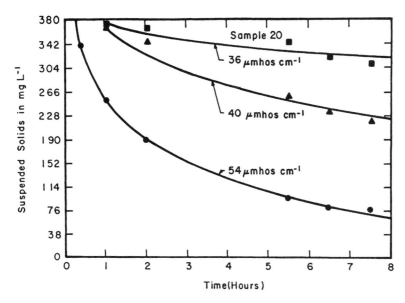

Figure 9.16b. Particle settling characteristics as a function of time at three levels of suspension electrical conductivity (EC) (from Evangelou, 1990, with permission).

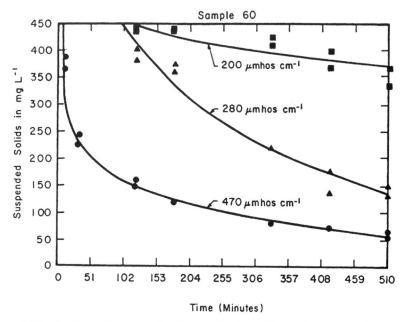

Figure 9.16c. Particle settling characteristics as a function of time at three levels of suspension electrical conductivity (EC) (from Evangelou, 1990, with permission).

The data in Figures 9.17 and 9.18 demonstrate the kinetics of settling characteristics of colloids with mixed mineralogy. By comparing data in Figures 9.17a and b and 9.18a and b, it can be seen that as SAR approaches zero with a total salt concentration of about 4 mmol$_c$ L^{-1}, Ca and Mg have approximately the same influence on the kinetics of settling of the clay colloids. Within a 2-hr settling period, the suspended solids for both systems approach 35 mg L^{-1} (Fig. 9.18a and b). However, at a SAR of approximately 24 (mmol L^{-1})$^{1/2}$ after an 8-hr settling period (Fig. 9.18a and b), the Na–Ca system maintains approximately 300 mg L^{-1} of suspended solids, while the Na–Mg maintains approximately 140 mg L^{-1} suspended solids.

These findings suggest that the particular colloids in the Na–Ca system are more dispersive than the Na–Mg system. This conflicts with some previous studies (Yadav and Girdhar, 1980) which showed that some soils have Na–Mg states that are more dispersive than the Na–Ca states. However, the soils employed by Yadav and Girdhar (1980) contained appreciable amounts of expanding minerals (smectites), whereas the material in Figures 9.17 and 9.18 is predominantly kaolinitic. Kaolin-type clay minerals exhibit stronger basic behavior than smectites and have a tendency to adsorb Mg^{2+} (a stronger acid) with higher affinity than Ca^{2+} (weak acid). This could explain the apparent low dispersivity of the Mg–clay system studied (Hesterberg and Page, 1990). Similar trends are shown in Figure 9.17c and d. A comparison of data shown in Figure 9.17a and c and 9.17b and d suggests that an increase in total salt concentration from 4 to 8 mmol$_c$ L^{-1} greatly increases the settling rate of the suspended solids. Note also that after 6 hr of settling time at SAR of 23 (mmol L^{-1})$^{1/2}$ and total salt

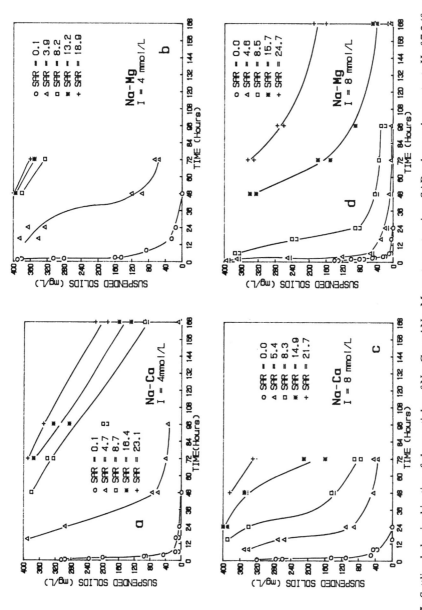

Figure 9.17. Settling behavior kinetics of clay particles of Na–Ca and Na–Mg systems at various SAR values and a constant pH of 7.2 (from Evangelou and Karathanasis, 1991, with permission).

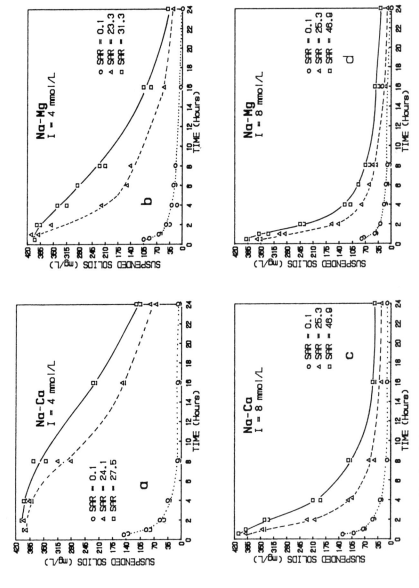

Figure 9.18. Settling behavior kinetics of clay particles of Na–Ca and Na–Mg systems at various SAR values and a constant pH of 5.2 (from Evangelou and Karathanasis, 1991, with permission).

concentration of 4 mmol$_c$ L^{-1} (Fig. 9.18b), the suspended solids are approximately 140 mg L^{-1}, as opposed to 70 mg L^{-1} for the same system at 8 mmol$_c$ L^{-1} (Fig. 9.18d).

The colloid suspensions systems shown in Figure 9.18a–d are analogous to the systems in Figure 9.17a–d with the only difference being the pH change from 5.2 to 7.2, respectively. It is apparent that the pH increase has caused a large decrease in the rate of settling of suspended solids (comparing corresponding graphs in Fig. 9.17 and 9.18). Furthermore, it is evident that the rate of clay-particle settling becomes highly dependent on the magnitude of SAR. For example, in sedimentation ponds with SAR values in the range of 23 to 25 and pH 7.2 (Fig. 9.17), it would take more than 168 hr of settling before suspended solids would approach the 200- to 350-mg L^{-1} range. In other words, a natural body of water with such sediment-settling behavior would be far from complying with the federal law requiring a maximum of 35 mg L^{-1} suspended solids. The trend effects of total salt concentration and type of divalent cation (Ca or Mg) are similar for both pH levels, as shown in Figure 9.17 (pH 7.2) and Figure 9.18 (pH 5.2), respectively.

9.4 FLOCCULANTS

Flocculants are substances that, when added to soil–water systems, induce colloid flocculation. Such substances may include simply an electrolyte (e.g., $CaCl_2$ or $CaSO_4$) or cations capable of hydrolyzing and inducing colloid to colloid attraction near circumneutral pH (e.g., $KAl_2SO_4 \cdot 12H_2O$ (Alum) or $FeCl_3$). Other chemicals that induce flocculation are synthetic long-chain organic polymers.

PROBLEMS AND QUESTIONS

1. Why are some colloidal particles considered nonsettleable?

2. Explain the repulsive index.

3. If you need to decrease the double-layer thickness by half in order to induce colloid flocculation, calculate the change in salt concentration you would need to introduce. Explain whether this is a realistic approach for cleaning up water.

4. Using Stokes' Law, calculate the amount of time needed to settle 1/1,000,000 m diameter particles 1 m depth. How much faster would the particles settle if their diameter was increased to 3/1,000,000 m? Finally, flocculation is found to increase the effective particle diameter to 1/100,000 m, calculate the time needed to settle these particles 1 m depth (consider the density of the particle to be 2500 kg m^{-3}).

5. How is question 4 related to pond water detention time? (Hint: detention time is the amount of time that water resides in a pond prior to exiting from the spillway to allow particle settling.)

6. Explain two approaches by which colloidal particles could be flocculated.

7. How is the zeta potential related to the two colloid flocculation approaches discussed in the answer to question 6?

8. How do potential determining ions affect colloid flocculation or dispersion?

9. Why is Na a dispersive ion while Ca is a flocculative ion?

10. Name two mechanisms of flocculation with respect to particle to particle interactions.

11. What is the practical importance of colloid flocculation–dispersion in nature?

12. Considering that the mechanism of flocculation or agglomeration of a pond is orthokinetic in nature (see Eq. 9.19), using the data given below, graphically estimate k_0 and the time needed for half of the particles to flocculate.

Time (sec)	Suspended Solids (mg L^{-1})
0	240
500	167
1000	125
2500	50
4000	20

13. Considering that the mechanism of flocculation or agglomeration of a pond is perikinetic in nature (see Eq. 9.17), using the data given below, graphically estimate k_p and the time needed for half of the particles to flocculate.

Time (sec)	Suspended Solids (mg L^{-1})
0	1400
400	450
1200	173
2000	100

10 Water and Solute Transport Processes

10.1 WATER MOBILITY

Water mobility in soil describes water transfer between any two points. Generally, this transfer involves horizontal and vertical flow. Both types of flow are important in soil because they describe water transfer as well as chemical transfer. For example, in the case of horizontal flow, water and its dissolved constituents may move horizontally through the soil and reach a stream or a lake. In the case of vertical movement, water and its dissolved constituents may move downward in the soil profile and reach the groundwater.

Soil solute movement is an extremely important phenomenon because it ensures the transfer of nutrients within the soil, thus making them available to plant roots and soil organisms. Unfortunately, solute movement may be undesirable at times because it may involve contaminants.

The purpose of this chapter is to introduce the various soil processes that influence water or solute movement and to demonstrate the role of soil–water chemistry in controlling these processes.

The transfer of water through a homogeneous medium under water-saturated conditions and constant temperature and pressure was first described by Darcy in the year 1856 by

$$q = -K(\Delta\phi/\Delta X) \tag{10.1}$$

where q denotes water flux in volume per unit cross-sectional area, ϕ = total potential ($\phi = h + z$; h = pressure head or height of the water column and z = gravitational head), X is the distance traveled by water, K is saturated hydraulic conductivity, and the minus sign denotes downward water movement. Originally, the Darcy equation was conceived for saturated water flow; it was later extended to include unsaturated flows. The term *saturated* describes the soil condition under which all soil pores are filled with water, while the term *unsaturated* describes the condition under which most of the large pores have drained and only a certain number of micropores may be saturated. It follows that during drainage the large pores drain first, followed by the smaller ones. Initially, water moves through soil because of a large hydraulic gradient, but as the soil desaturates, water movement is due mostly to the matrix gradient or the difference in relative force by which water is held by mineral surfaces.

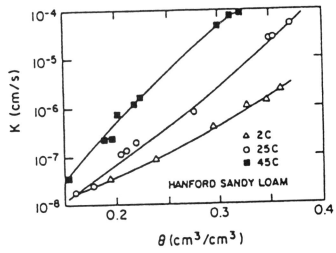

Figure 10.1. Unsaturated hydraulic conductivity as a function of water content (θ) at three different temperatures (from Constantz, 1982, with permission).

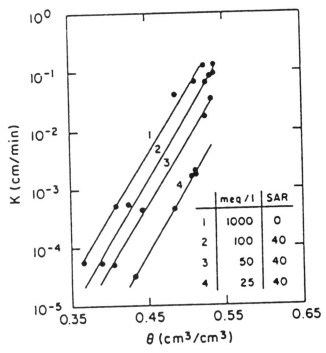

Figure 10.2. Hydraulic conductivity K as a function of water content of a Weld soil to several solutions of decreasing concentration and SAR (from Dane and Klute, 1977, with permission).

A great deal of research has been carried out for the purpose of modeling water flow through soil. It has now been generally concluded that certain soils, especially those under intense cultivation, appear to be homogeneous, and for this reason Darcian water flows are observed. However, for noncultivated soils or soils under no-tillage management (e.g., soils of the temperate regions of the United States and the world), non-Darcian flows are common. In such soils, often a large amount of water can be described as a bypass flow or water transfer through large cracks or channels in soil. Such cracks or channels can be made naturally or by animals, soil insects, worms, decaying roots, and so on. In general, however, Darcy's equation gives us a good conceptual understanding of water movement in soil.

In addition to soil factors such as texture and structure controlling water movement within a given soil, other factors include temperature, water content, and salt composition and concentration. For example, the data in Figure 10.1 show that as temperature increases, hydraulic conductivity also increases, perhaps because of increasing water fluidity. Furthermore, hydraulic conductivity increases as water content increases. The data in Figure 10.2 demonstrate that as salt concentration increases, hydraulic conductivity also increases, a phenomenon linked to the electric double layer (Chapters 3 and 9).

10.2 SOIL DISPERSION—SATURATED HYDRAULIC CONDUCTIVITY

The classical theory of colloidal stability (DLVO theory, Chapter 9) generally accounts for the influences of ion valence and concentration on colloid interactions. According to the DLVO theory, the long-range repulsive potential resulting from diffuse double layers (DDLs) of like-charged colloids retards the coagulation or flocculation rate of clay colloids. Colloid stability (potential maximum dispersion point) depends on the maximum repulsive energy between two colloidal surfaces and is controlled by the surface electric potential and the solution's ionic strength. The surface electrical potential is controlled by pH (assuming that the colloids involved exhibit pH-dependent charge). Generally, in clay colloids, upon increasing pH, the surface electrical potential increases (becomes more negative) and, therefore, soil colloid repulsion increases. Conversely, upon decreasing pH, the surface electrical potential decreases and soil colloid repulsion decreases. When the surface electrical potential approaches zero, colloid repulsion approaches zero; this leads to colloid flocculation. In the case of ionic strength, when it increases, electrical potential as a function of distance from the colloid decreases and colloid repulsion also decreases.

Many processes in soil are controlled by colloid flocculation or dispersion. One such process is hydraulic conductivity. The data in Figure 10.3 show that for a Mg^{2+}-saturated soil containing a solution of 3.16×10^{-2} M $MgCl_2$, its hydraulic conductivity decreased by 35% after 5 hr of leaching with distilled water (Quirk and Schofield, 1955). This demonstrates that as solution ionic strength approaches zero, soil hydraulic conductivity decreases significantly owing to soil dispersion induced by a decompressed electric double layer.

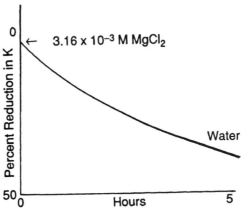

Figure 10.3. Influence of soil leaching by distilled water on saturated hydraulic conductivity (from Quirk and Schofield, 1955, with permission).

The presence of exchangeable Na^+ could also significantly decrease soil permeability. The mechanism(s) responsible for decreasing soil permeability in the presence of Na^+ can be demonstrated by looking into the components controlling water or soil solution movement potential under saturated conditions. Soil-saturated hydraulic conductivity is described by

$$K = kg/n \qquad (10.2)$$

where
  k = permeability of the soil (related to soil texture and structure)
  g = gravitational constant
  n = kinematic viscosity or the ratio of solution viscosity over the fluid density

For soil systems contaminated with Na^+, kinematic viscosity is not significantly affected, thus the components controlling water flow velocity are the hydraulic gradient ($\Delta\phi/\Delta X$) and soil permeability (k). The latter component (k) is influenced by clay dispersion, migration, and clay swelling. These processes may cause considerable alteration to such soil matrix characteristics as porosity, pore-size distribution, tortuosity, and void shape.

The deterioration of soil physical properties influencing k is accelerated directly or indirectly by the presence of high Na^+ on the soil's exchange complex and the electrolyte composition and concentration of the soil solution. To improve the physical properties of Na-affected soils, Ca^{2+} is usually added to replace Na^+ on the exchange sites. Calcium reduces clay swelling and enhances clay flocculation. The data in Figure 10.4 show that as salt concentration increases, saturated hydraulic conductivity increases and reaches a maximum which is independent of Na^+. However, as salt concentration decreases, the decrease in saturated hydraulic conductivity is related to the Na^+ in relationship to Ca^{2+} (SAR, see Chapter 11). The higher the SAR is, the lower

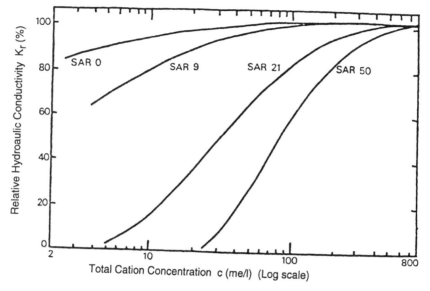

Figure 10.4. Influence of pH and SAR on saturated hydraulic conductivity.

the saturated hydraulic conductivity. Additional components influencing the effect of Na^+ on saturated hydraulic conductivity of soil include clay mineralogy, clay content, soil bulk density, Fe and Al oxide content, and organic matter content. The hydraulic properties of soils dominated by 1:1 type clay mineralogy (i.e., kaolinite) and Fe or Al oxides are relatively insensitive to variation in soil-solution composition and concentration, in contrast to those dominated by 2:1 type clay minerals (i.e., montmorillonite). Organic matter increases soil sensitivity to Na^+.

Generally, the same total quantity of Na^+ in a variably charged soil will reduce saturated hydraulic conductivity more effectively at a lower pH than at a higher pH. This is because in variably charged soils, as pH decreases, CEC decreases and soil-saturated hydraulic conductivity decreases because the same amount of Na^+ represents a greater ESP at a lower soil pH. Note also that, generally, for the same ESP or SAR value, the saturated hydraulic conductivity decreases as pH increases. The data in Table 10.1 show that as pH increases, a smaller SAR is needed to reduce saturated hydraulic conductivity by 20%. Furthermore, as expected, as total salt concentration increases, the SAR value needed to decrease saturated hydraulic conductivity by 20% increases.

The increase in soil pH could be implicated in increasing soil dispersion as well as in increasing clay-swelling potential. This is likely because of the removal of Al–OH polymers from the interlayer. The presence of Al–OH polymers at the lower pH values may limit interlayer swelling. Clays that have the basic 2:1 mineral structure may exhibit limited expansion because of the presence of Al–hydroxy islands which block their interlayer spaces. It is well known that these Al–hydroxy components are removed at low or high pH through dissolution mechanisms. This interlayer removal

TABLE 10.1. Sodium Adsorption Ratio (SAR) and Exchangeable Sodium Percentage (ESP) Values Associated with 20% Reduction in Saturated Hydraulic Conductivity (SHC) for Pembroke Soil (10- to 30-cm Incremental Depth) at Three pH Values

Cl (mmol L^{-1})	pH 4.3		pH 6.1		pH 7.5	
	SAR[a]	ESP	SAR	ESP	SAR	ESP
5	2.6	5.5	1.6	1.1	0.4	0.6
50	49.6	59.1	29.4	45.5	20.8	80.8
200	—[b]	—	—	—	90.4	80.5

Source: Marsi and Evangelou, 1991a

[a]SAR in (mmol L^{-1})$^{1/2}$.

[b]Threshold values are not reported because the reduction in SHC is less than 20%.

would be expected to increase the dispersion potential of the mineral by allowing free expansion. Similar phenomena of Al–hydroxy interlayer removal have been demonstrated to be the cause for failed septic systems. In addition to increased swelling, dispersion can also be enhanced in such systems. When removed from interlayer positions, these positively charged Al–hydroxy components would increase the effective surface charge available for sodium adsorption, thus increasing the probability for soil structural destabilization.

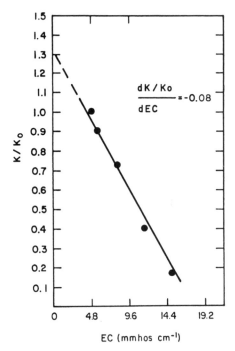

Figure 10.5. Influence of acid MgSO$_4$ solutions on saturated hydraulic conductivity (from Evangelou, 1997, unpublished data, with permission).

Saturated hydraulic conductivity may also be influenced by dramatic changes in solution viscosity as well as the soil's dispersive potential. The data in Figure 10.5 represent saturated hydraulic conductivity as a function of EC. It appears that as EC increases, hydraulic conductivity decreases. The soil material in this study represents a Kentucky mine spoil. The predominant salt in the solution was an acid, $MgSO_4$. Suspension data showed that as $MgSO_4$ concentration increased, colloid dispersion increased. This could be due to an increase in solution viscosity, which also has a suppressing effect on saturated hydraulic conductivity (see Eq. 10.2).

10.3 SOLUTE MOBILITY

Soils are composed of inorganic and organic minerals with surfaces possessing sites capable of producing chemical or physical bonds with compounds or minerals dissolved in water (see Chapter 3). These solute–mineral surface reactions regulate the potential of chemicals in the soil–water environment to become mobile. Such chemicals include plant nutrients, pesticides, and/or other synthetic organics making up soil–water pollutants. The potential of chemical species to move in the soil–water system depends on the potential of soil to conduct water and on the potential of solution minerals to react with soil minerals. In the case of a nonreactive chemical species (nonreactive solute), its mobility in the soil system will be equal to that of water. However, the mobility of a reactive solute would be less than that of water. The rate of downward movement of a chemical species (e.g., a monovalent cation X^+) can be predicted by the equation

$$\nu_{X^+} = \nu/[1 + (\rho_b/\tau)K_d] \tag{10.3}$$

where ν equals velocity of water, ρ_b is the bulk density of the soil, τ is the porosity of the soil, and K_d is the distribution coefficient (McBride, 1994). The distribution coefficient can be justified as follows:

$$SCa_{1/2} + X^+ \Leftrightarrow -S-X + 1/2\ Ca^{2+} \tag{10.4}$$

where $-S-$ denotes surface. Therefore,

$$K_{eq} = [-S-X/-S-Ca_{1/2}][(Ca^{2+})^{1/2}/(X^+)] \tag{10.5}$$

If we assume that $Ca^{2+}>>>X^+$ and $-S-Ca_{1/2}>>>-S-X$, we can rearrange Equation 10.5, producing

$$K_d = K_{eq}[-S-Ca_{1/2}/(Ca^{2+})^{1/2}] = [-S-X/(X^+)] \tag{10.6}$$

Therefore, the K_d of X^+ in a given soil equals the ratio of adsorbed X ($-S-X$) to that in solution (X^+). This can be established experimentally by meeting all of the conditions stated above. It appears, therefore, that this K_d is applicable when the solution composition considered is constant, the surface adsorption reaction is rapid and reversible, and the concentration of X^+ in the soil solution is low relative to any other competing cations.

Based on Equation 10.3, chemical mobility differs from water mobility by a factor of $1 + (\rho_b/\tau)K_d$. This factor is also known as the *retardation factor*. The larger the retardation factor, the smaller is the velocity of the chemical species in relationship to the velocity of water. Note, however, that the retardation factor contains a reactivity factor (K_d) and two soil physical parameters, bulk density (ρ_b) and porosity (τ). The two parameters affect retardation by producing a wide range of total porosity in soils as well as various pore sizes. Pore size regulates the nature of solute flow. For example, in very small pores, solute movement is controlled by diffusion, while in large pores, solute flow is controlled by mass flow.

Diffusion describes the number of molecules transferred across a boundary per unit time (e.g., seconds or minutes). It is given by Fick's law:

$$dc/dt = -D\alpha(dc/dx) \qquad (10.7)$$

where dc/dt describes the change in concentration with respect to time, D is the diffusion coefficient, α is the cross-sectional area of the boundary, and dc/dx is the concentration gradient ($x =$ distance). The diffusion coefficient (D) can be defined by

$$D = RT/6\pi r N_0 \eta \qquad (10.8)$$

where R is the universal gas constant, T is temperature, r is the molecular radium, N_0 is Avogadro's number, and η is the solvent viscosity. Equation 10.8 reveals that diffusion is controlled directly by temperature and inversely by the molecular radium and the viscosity of the solute. Therefore, in soil systems, diffusion can be the velocity-controlling factor of solutes by controlling the rate at which chemical species diffuse through the solution "film" that surrounds each soil particle (film diffusion) or by controlling the rate at which the chemical species diffuses through the particle (e.g., interlayer space). The role of mass flow and/or diffusion in solute transport is demonstrated through miscible displacement or breakthrough curves.

10.4 MISCIBLE DISPLACEMENT

When a solution containing a particular chemical species is displaced from a porous medium with the same solution but without the particular chemical species, this miscible displacement produces a chemical species distribution that is dependent on (1) microscope velocities, (2) chemical species diffusion rates, (3) physicochemical reactions of the chemical species with the porous medium, (e.g., soil), and (4) volume of water not readily displaced at saturation (this not-readily displaced water increases as desaturation increases (Nielsen and Biggar, 1961).

Experimentally, miscible displacement can be accomplished by putting together a column containing a porous material (e.g., sand or soil), wetting the column to establish a desirable moisture condition (level), and then pushing the solution with the desired chemical species through the column at a preset velocity. Such studies can be carried out under unsaturated or saturated conditions. Under saturated conditions, the column is set vertically and the solution is pumped from the bottom up. Under unsaturated conditions, the column is set horizontally or vertically and suction is applied (Nielsen and Biggar, 1961). The solution exiting the column is collected by a fraction collector and analyzed. Commonly, the original solute concentration is defined as C_0 and the concentration at any time t is defined as C. The volume of solution occupying the available column pore space is defined as V_0 (cm^3) and the rate of inflow or outflow of solute is defined as q (cm^3 hr^{-1}); therefore the ratio of q/V_0 is defined as pore volume.

A plot of C/C_0 versus pore volume (q/V_0) produces what is known as a breakthrough curve. Idealized breakthrough curves are given in Figure 10.6. These data show that the solute spreads owing to velocity distribution and molecular diffusion only. In other words, there is no interaction between the solute, solvent, and solid (Nielsen and Biggar, 1962). For this reason, the following equation is obeyed:

$$q/V_0 \int_0^{\infty} (1 - C/C_0) \, dt = 1 \qquad (10.9)$$

no matter what the shape of the curve is. Equation 10.9 points out that the quantity of the solute within the column will reach an equilibrium with that in the effluent and the influent such that the total quantity of the chemical species in the column is $C_0 \cdot V_0$. Furthermore, under the conditions stated above, the pore value below one pore volume equals the pore volume above one pore volume, regardless of the shape of the curve. This is a direct result of Equation 10.10:

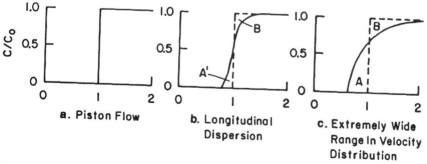

Figure 10.6. Types of breakthrough curves for miscible displacement. C/C_0 is the relative concentration of the chemical component measured in the effluent. Pore volume is the ratio of the volume of the effluent to the volume of solution in the column (from Nielsen and Biggar, 1962, with permission).

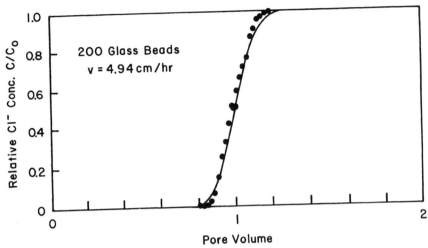

Figure 10.7. Chloride breakthrough curve measured for 200-μ glass beads (from Nielsen and Biggar, 1962, with permission).

$$q/V_0 \int\limits_{0}^{V_0/q} C/C_0 \, dt = q/V_0 \int\limits_{V_0/q}^{\infty} (1 - C/C_0) \, dt \qquad (10.10)$$

The data in Figure 10.6a describes ideal piston flow. It explains solute flow through a single capillary tube with a uniform radius. Experimental data involving a single-size glass bead approaches piston flow. On the other hand, Figure 10.6b shows that the

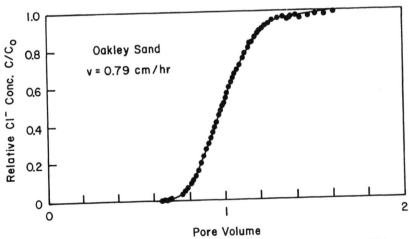

Figure 10.8. Chloride breakthrough curve measured for Oakley sand (from Nielsen and Biggar, 1962, with permission).

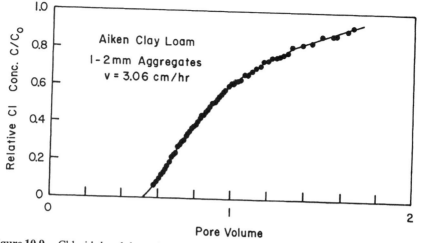

Figure 10.9. Chloride breakthrough curve for 1- to 2-mm aggregates of Aiken clay loam (from Nielsen and Biggar, 1962, with permission).

solute undergoes longitudinal dispersion and area A equals area B. Similar results were obtained when Cl$^-$ breakthrough curves were obtained using glass beads (Fig. 10.7). The latter figure shows that the area below one pore volume equals the area above one pore volume, hence no interaction between the solute and the solid phase has taken place. Finally, Figure 10.6c describes a breakthrough curve with a wide range in velocity distribution which is demonstrated in Figure 10.8. In Figure 10.9, where an Aiken clay loam sample ranging in particle size from 1 to 2 mm was used, the areas under the curve to the left and right of one pore volume are nearly equal or meeting the condition described by Equation 10.10, which is also demonstrated in Figure 10.10 where two different velocities were tested.

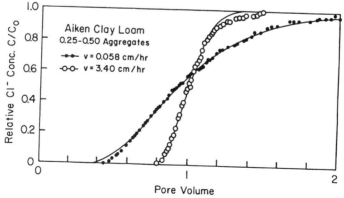

Figure 10.10. Chloride breakthrough curves measured for 0.25- to 0.50-mm aggregates of Aiken clay loam at two velocities (from Nielsen and Biggar, 1962, with permission).

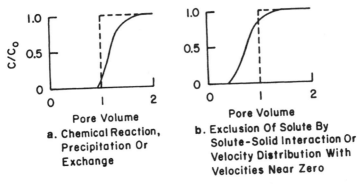

Figure 10.11. Types of breakthrough curves for miscible displacement. C/C_0 is the relative concentration of the chemical component measured in the effluent. Pore volume is the ratio of the volume of the effluent to the volume of solution in the column (from Nielsen and Biggar, 1962, with permission).

When the displacing solution or its solutes are retained by the column by any chemical or physical process, the breakthrough curve is shifted to the right of the one-pore volume border (Fig. 10.11a). When the solute is excluded from the surface, however, the breakthrough curve is shifted to the left (Fig. 10.11b). Experimental data demonstrating behavior analogous to Figure 10.11b is shown in Figure 10.12. The data in Figure 10.12 demonstrate miscible displacement of Cl⁻ in a variable-charge soil column. It is shown that at pH 7 and 9, where the soil sample exhibits negative charge (see Chapter 3), the breakthrough curve appears to the left of the one-pore volume boundary. However, at pH 4, the variably charged soil sample exhibits positive charge, thus adsorbing the negatively charged chloride; for this reason, the breakthrough curve has shifted significantly to the right of the one-pore volume boundary.

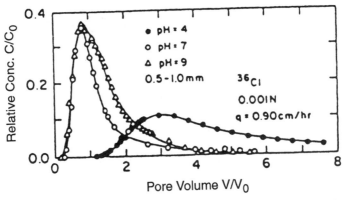

Figure 10.12. Experimental chloride breakthrough curves showing the effect of solution pH on the displacement process (from Nielsen et al., 1986, with permission).

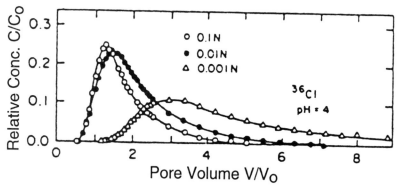

Figure 10.13. Experimental breakthrough curves showing the effect of total ionic concentrations of the displacement process (from Nielsen et al., 1986, with permission).

The data in Figure 10.13 (pH 4 sample) show that the breakthrough curve shifted to the left when Cl^- concentration increased. This was expected because the salt used was $CaCl_2$, a nonsymmetrical electrolyte. It is known that the surface potential of a variably charge soil depends on ions forming either inner- or outer-sphere complexes (e.g., Ca^{2+}), and it is completely independent of ions that mostly form diffuse layer complexes (e.g., Cl^-). Calcium (Ca^{2+}) has a tendency to form relatively strong outer-sphere or Stern layer complexes. This increases the negative surface potential and decreases the PZC of the surface; therefore, Cl^- is repulsed by the surface at the higher $CaCl_2$ concentration (see also Chapter 3).

Breakthrough curve shifts are also observed if the soil's surfaces are able to immobilize a proportionally large volume of water. The relative amount of water immobilized by mineral surfaces increases as undersaturation increases. The data in

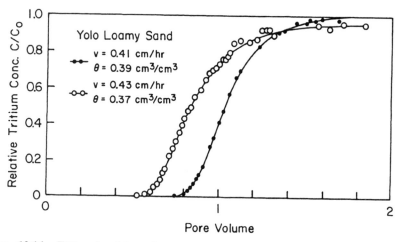

Figure 10.14. Tritium breakthrough curves from Aiken loamy sand at two water contents (from Nielsen and Biggar, 1962, with permission).

Figure 10.14 show breakthrough curves of tritiated water at approximately the same flow velocity, but at two different levels of water content. Note that at the lower water content, the breakthrough curve has been displaced to the left because of incomplete mixing of tritiated water with nontritiated water in the column. This water is also known as stagnant water of unsaturated soil. At the higher water content, the proportion of stagnant to unstagnant water decreases, and for this reason, the breakthrough curve is shifted to the right of the low-water content breakthrough curve. The data in Figure 10.14 also reveal that Equation 10.10 was not obeyed because tritium experienced isotopic exchange in the soil column.

The movement of chemicals undergoing any number of reactions with the soil and/or in the soil system (e.g., precipitation–dissolution or adsorption–desorption) can be described by considering that the system is in either the equilibrium or nonequilibrium state. Most often, however, nonequilibrium is assumed to control transport behavior of chemical species in soil. This nonequilibrium state is thought to be represented by two different adsorption or sorption sites. The first site probably reacts instantaneously, whereas the second may be time dependent. A possible explanation for these time-dependent reactions is high activation energy or, more likely, diffusion-controlled reaction. In essence, it is assumed that the pore–water velocity distribution is bimodal.

In a bimodal model, convective–dispersive transport is represented by only a fraction of the liquid-filled soil pores; the remainder represent stagnant water. This stagnant water is represented by micropore or interlayer water. An example of non-equilibrium bimodal transport is shown in Figure 10.15. The data clearly show an accelerated movement of boron through soil and an associated tailing phenomenon at higher pore volumes. The tailing phenomenon is believed to be due to adsorption and exchange sites present in dead-end pores or pores not located along the main liquid-

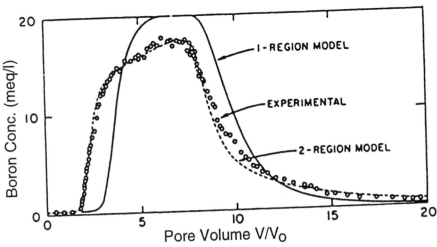

Figure 10.15. Experimental and predicted boron breakthrough curves from an aggregated clay loam (from Nielsen et al., 1986, with permission).

flow lines. Attempts to model the process using single and bimodal adsorption sites show that bimodal adsorption gives the best fit.

Solute movement through soil is a complex process. It depends on convective–dispersive properties as influenced by pore size, shape, continuity, and a number of physicochemical reactions such as sorption–desorption, diffusion, exclusion, stagnant and/or double-layer water, interlayer water, activation energies, kinetics, equilibrium constants, and dissolution–precipitation. Miscible displacement is one of the best approaches for determining the factors in a given soil responsible for the transport behavior of any given solute.

PROBLEMS AND QUESTIONS

1. Using Equation 10.2 ($K = kg/n$), answer the following questions:
   a. What will happen to K if the viscosity of the solution increases but density remains the same?
   b. What will happen to K if k decreases?
   c. What soil properties define k?
   d. How can one change k of a soil?
   e. What happens to the viscosity of a soil solution when it polymerizes in comparison to its density?
   f. Under what conditions would a soil solution polymerize?

2. Equation 10.3 describes the velocity of a chemical substance moving through soil relative to the velocity of water moving through soil. Any difference between these two velocities is due to the potential of soil to interact with the chemical substance physically and/or chemically. Answer the following questions using Equation 10.3.
   a. What will happen to the velocity of a chemical if $K_d = 0$?
   b. What will happen to the velocity of a chemical if $K_d = 1$?
   c. What soil parameters control chemical mobility when $K_d = 0$?
   d. What soil properties control τ of soil?
   e. What type of soil solution changes could alter soil τ?

3. Under what conditions can the velocity of a chemical in soil be larger than the velocity of water in that soil?

4. What is a breakthrough curve? Draw one for an adsorbing chemical species and explain it.

5. How is a breakthrough curve affected by large soil pores relative to very fine pores?

6. How is a breakthrough curve affected by a large K_d relative to a very small K_d?

7. Explain under what conditions C/C_0 would not reach one?

8. How would temperature affect the shape of a breakthrough curve? Why?

9. How would pH affect the shape of a breakthrough curve? Why?

10. When the breakthrough time of a chemical is delayed, what does it mean in practical terms?

11 The Chemistry and Management of Salt-Affected Soils and Brackish Waters

11.1 INTRODUCTION

The growth of most crops in salt-affected soils is adversely affected by soluble salts. Such soils include both saline and sodic soils. A saline soil is one that contains enough soluble salts to interfere with normal plant growth, while a sodic soil contains enough exchangeable sodium (ExNa) to also have an adverse effect on plant growth. A saline–sodic soil contains both soluble and exchangeable sodium at high levels.

Salt-affected soils are a common feature of arid and semiarid geographic regions. In humid regions, soils may also become salt-affected when they are irrigated with brackish water, intruded by sea water, or contaminated with oil-well brines. Some differences exist between salt-affected soils in arid and semiarid geographic regions and salt-affected soils in humid and tropical regions. Sodic soils in arid and semiarid regions are commonly associated with high pH and dominated by 2:1 type clay minerals. Salt-affected soils in humid or tropical regions generally have low pH, and they are often, but not always, dominated by 1:1 type clay minerals.

The presence of salinity in soil and water can affect plant growth in three ways: (1) it can increase the osmotic potential and hence decrease water availability; (2) it can induce specific-ion effects by increasing the concentration of ions with an inhibitory effect on biological metabolism; and (3) it can diminish soil–water permeability and soil aeration by adversely affecting soil structure. The adverse effects of soil salinity on plant growth and productivity varies with the type of plant being grown.

Managing salt-affected soils or brackish waters in natural environments (e.g., land, streams, rivers, and lakes) requires knowledge of the chemistry of soil and brine, how brines interact with soil–water systems, and how these systems are affected by such interactions. This chapter deals with the practical aspects of Na^+–Ca^{2+} exchange reactions and $CaCO_3$ solubility for the effective management of salt-affected soils and safe disposal of brines to soil–water environments.

11.1.1 Osmotic Effect

The osmotic effect due to salt in soil–water environments is related to water availability. Water availability is determined by the soil–water potential, ζ_w, which is composed of the osmotic potential, ζ_o, the matrix potential, ζ_m, and the gravitational potential, ζ_g:

$$\zeta_w = \zeta_m + \zeta_o + \zeta_g \tag{11.1}$$

At any given matrix potential, ζ_m, under a constant gravitational potential, ζ_g, as salinity increases, ζ_w decreases. This is because osmotic potential, ζ_o, is directly related to total dissolved solids (TDS). The relationship between osmotic potential and TDS can be expressed by

$$\zeta_o \text{ (bar)} = -5.6 \times 10^{-4} \times \text{TDS (ppm)} \tag{11.2}$$

Another way to express the relationship is through the EC of a soil's solution. Since EC is directly related to salt content, the U.S. Salinity Laboratory Staff (1954) gave the following relationship relating EC to osmotic potential:

$$\zeta_o \text{ (bar)} = -0.36 \times \text{EC (mmhos cm}^{-1}) \tag{11.3}$$

11.1.2 Specific Ion Effect

Excess Na ions in the soil solution can be inhibitory to certain plant processes. Plant sensitivity to various Na levels in soil is dependent on plant species and stage of plant development. Sodium toxicity to higher plants is characterized by leaf-tip burn, necrotic spots, and limited leaf expansion, thus reducing yield.

The specific effects of Na on plant physiological processes include antagonistic effects on Ca and Mg uptake and Ca deficiency. This is because Na^+ displaces Ca^{2+} from membranes, rendering them nonfunctional.

11.1.3 Physicochemical Effect

Excess exchangeable Na is harmful because it induces undesirable physical and chemical changes in soils. One such change includes soil dispersion, which is related to the highly hydrated nature of Na^+. Soils disperse when they are in equilibrium with a salt solution under the *flocculation value*. The flocculation value depends on solution composition (SAR), solution ionic strength, and clay mineralogy. For example, flocculation salt values for Na/Ca–montmorillonite are 3.0, 4.0, and 7.00 $mmol_c$ L^{-1}, and 6.0, 10.0 and 18.0 $mmol_c$ L^{-1} for Na/Ca–illite with exchangeable sodium percentage (ESP) values of 5, 10, and 20, respectively. Clay dispersion changes soil–pore distribution, which in turn influences soil hydraulic conductivity.

Exchangeable sodium percentage is related to the relative ratio of Na to $Ca^{1/2}$ in the solution phase, which is referred to as the SAR (see Chapter 4). An empirical relationship between SAR and ESP, representing soils of the arid west, was developed by the U.S. Salinity Laboratory Staff (1954):

$$ESP = \frac{100(-0.0126 + 0.01475\ SAR)}{1 + (-0.0126 + 0.01475\ SAR)} \tag{11.4}$$

where SAR is in millimoles per liter to the half power. When SAR is approximately in the range of 10 to 15, the ESP is also in the range of 10 to 15. In this ESP range, soils of the arid west will commonly undergo dispersion. However, this threshold ESP value represents soils of the arid west and may not be universally applied to all soils. There is a great deal of information on the behavior of sodium chloride in soils and in soil–solution suspensions. However, most of this research pertains to salt-affected soils of the arid west which are often alkaline and consist mostly of 2:1 clay minerals. In the humid regions, soils are often acid, their mineralogy is highly mixed (1:1 plus 2:1 clay minerals), and the 2:1 minerals are highly interlayered. It is also necessary to understand the sodicity and reclaimability of soils with mixed mineralogy, a condition more common in the humid regions of the United States.

DERIVATION OF THE EMPIRICAL SAR–ESP RELATIONSHIP

The equation most commonly used to describe heterovalent cation exchange, such as Na^+–Ca^{2+} exchange, is the Gapon exchange equation. For example,

$$ExCa_{1/2} + Na^+ \Leftrightarrow ExNa + 1/2\ Ca^{2+} \tag{A}$$

and

$$K_G = [ExNa/ExCa_{1/2}][Ca^{1/2}/Na] \tag{B}$$

Equation B can be rearranged to solve for ESR:

$$ESR = ExNa/ExCa = K_G(SAR) \tag{C}$$

where $SAR = Na/Ca^{1/2}$ $[(mmol\ L^{-1})^{1/2}]$, K_G is the Gapon exchange selectivity coefficient $[(mmol\ L^{-1})^{-1/2}]$, and Ex_i denotes exchangeable cation i (meq 100 g^{-1} soil).

Equation A can be used to solve for ExNa:

$$ExNa = K_G(CEC)(SAR)/[1 + SAR\ K_G] \tag{D}$$

and

$$ESP = ExNa/CEC \times 100 = 100 \cdot K_G(SAR)/[1 + SAR \cdot K_G] \tag{E}$$

Substituting SAR $\cdot K_G$ (Equation E) with ESR (Equation C) gives

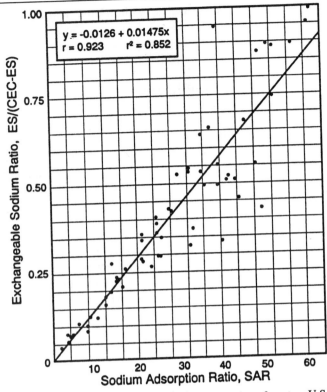

Figure 11A. Relationship between SAR and ESR for a number of western U.S. soils (from the U.S. Salinity Laboratory Staff, 1954).

$$ESP = (100)ESR/[1 + ESR] \qquad (F)$$

An experimental plot of SAR versus ESR, for only a fraction of the CEC of the soil or soils involved in the study, could produce a straight line with slope K_G (Fig. 11A). Substituting the regression equations from the experimental SAR versus ESR plot into Equation E produces Equation 11.4, which for convenience is given below:

$$ESP = \frac{100(-0.0126 + 0.01475 \, SAR)}{1 + (-0.0126 + 0.01475 \, SAR)} \qquad (G)$$

The y intercept (-0.0126) of the SAR–ESR relationship represents an empirical system constant and the slope [0.01474 $(mmol \, L^{-1})^{-1/2}$] represents the average K_G of a number of western U. S. soils (U. S. Salinity Laboratory Staff, 1954). Note that this empirical equation (Eq. G) does not necessarily have universal application. It appears that the K_G varies depending on soil type and the weather conditions under which a particular soil was formed. For additional information, see Chapter 4.

11.2 SALTS AND SOURCES

The sources of salts in arid environments include fertilizers and irrigation water. Under such environments, high evapotranspiration in relation to rainfall causes accumulation of salts in the upper soil horizons. In contrast, salts in the humid regions of the United States are introduced to soil as brine. Oil wells in humid U.S. regions are a major brine-source issue because environmental safeguards are often lacking. Such oil wells, also known as *stripper wells*, produce brine which is discharged onto agricultural lands and/or into natural water supplies. For example, in the state of Kentucky, more than 100,000 gal of brine per day, containing approximately 0.5 mol L^{-1} sodium chloride, are discharged onto land and surface waters. Similar brine problems exist in many southeastern and northeastern states.

Brine is a salty water trapped in rock formations and is often, but not always, associated with oil and gas deposits. It consists mostly of sodium chloride, but can also contain other constituents such as organics, bromide, some heavy metals, and boron. Releasing brine to the soil–water environment in the hope that dilution will minimize the problem is highly questionable because of the brine's toxicity potential. The causes and effects of salt in soil–water systems, or brine disposed into soil–water systems, are discussed below.

11.2.1 High Sodium

Sodium concentration in actual brines is greater than 1000 mg L^{-1} and varies widely. Concentrations in water greater than 69 mg L^{-1} can be toxic to crops. Sodium toxicity is closely related to the level of calcium (Ca) in the water or in the soil. If water of high sodium content is applied to a soil, it moves soil calcium to a greater depth. Under low root zone soil calcium levels, sodium can be highly toxic.

11.2.2 SAR and ESP Parameters

To assess the potential toxicity of sodium in the soil–water system, the SAR is used. The SAR is determined by obtaining soil solution from the soil after saturating it with water, removing the solution by vacuum, and analyzing it for Na, Ca and magnesium (Mg) in milligrams per liter. The following formula is then used to estimate the SAR:

$$SAR = (Na/23)/(Ca/40 + Mg/24)^{1/2} \qquad (11.5)$$

where 23, 40, and 24 are the atomic weights of Na$^+$, Ca^{2+}, and Mg^{2+}, respectively. Generally, SAR values greater than 15 are considered potentially toxic to plants.

The SAR magnitude reflects the quantity of sodium on the exchange sites of the soil. Most arid-region soils with SAR values of 15 have approximately 15% of their CEC loaded with sodium. This sodium load is known as the exchangeable sodium percentage or ESP. Soils with an ESP greater than 15 would be considered unproductive and, depending on the magnitude of ESP, such soils may also be classified as toxic. This information, however, comes from soils of the arid regions of the western United States, and one cannot be sure that the critical SAR–ESP threshold of these soils also applies to soils in humid regions.

11.2.3 SAR–ESP Relationships

The sodium adsorption potentials of two humid soils are presented in some detail and the SAR–ESP data are shown in Figures 11.1–11.4. For comparison purposes, these figures also include the SAR–ESP relationship of salt-affected soils found in arid-region soils (western United States). Figures 11.1–11.4 show that for any given SAR, the ESP for either one of the two humid soils is greater than the ESP of the western U.S. soils. This indicates that the two humid soils adsorb sodium on their exchange complex more effectively than the western U.S. soils.

Some differences in the SAR–ESP relationship between the two humid soils (Figs. 11.1–11.4) are also apparent. The data indicate that the SAR–ESP relationship of the Pembroke soil is independent of chloride (Cl^-) and to some degree pH, but this is not true for the Uniontown soil. It appears that as pH increases, the Uniontown soil shows a strong adsorption preference for Na^+, but as Cl^- concentration increases, it shows a

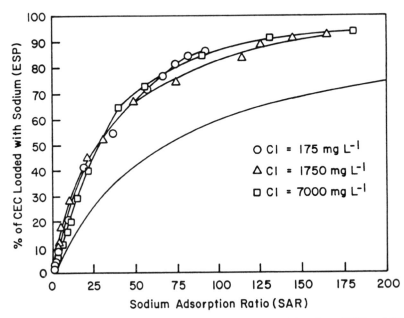

Figure 11.1. Relationship between percentage of CEC loaded with sodium (ESP) and SAR at three chloride concentrations of Pembroke soil at pH 4.3 (the solid line without data represents most salt-affected soils in the western United States; it was produced using Eq. 11.4) (from Marsi and Evangelou, 1991a, with permission).

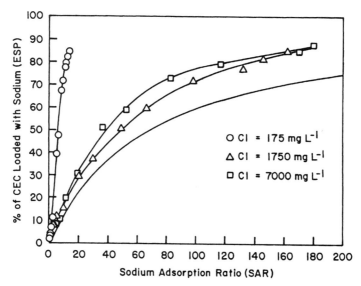

Figure 11.2. Relationship between percentage of CEC loaded with sodium (ESP) and SAR at three chloride concentrations of Uniontown soil at pH 4.3 (the solid line without data represents most salt-affected soils in the western United States; it was produced using Eq. 11.4) (from Marsi and Evangelou, 1991a, with permission).

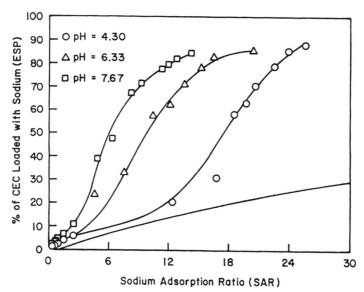

Figure 11.3. Relationship between percentage of CEC loaded with sodium (ESP) and SAR at a chloride concentration of 175 mg L^{-1} of Uniontown soil at three pH values (the solid line without data represents most salt-affected soils in the western United States; it was produced using Eq. 11.4) (from Marsi and Evangelou, 1991a, with permission).

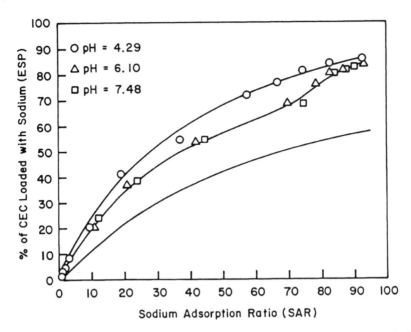

Figure 11.4. Relationship between percentage of CEC loaded with sodium (ESP) and SAR at a chloride concentration of 175 mg L^{-1} of Pembroke soil at three pH values (the solid line without data represents most salt-affected soils in the western United States (from Evangelou, 1998, unpublished data).

weak preference for Na$^+$. These observations play a very important role in decisions about managing brine discharges onto agricultural soils or into streams and lakes.

In summary, if one discharges brine onto a soil with a strong Na$^+$ adsorption potential, this soil will protect the groundwater from Na$^+$ contamination at the expense of its own potential Na$^+$ contamination. However, a soil with low Na$^+$ adsorption potential will protect itself from Na$^+$ contamination at the expense of potential groundwater contamination.

11.2.4 Adverse Effects of Na$^+$ in the Soil–Water Environment

Sodium adsorbs more water molecules per unit (mole) of charge than most other metal ions (K$^+$, Mg^{2+}, Ca^{2+}) commonly found in the soil–water environment. Hence, when brine (NaCl) is discharged in the soil–water environment, clay and organic particles tend to adsorb fully hydrated Na ions. This causes the particles to become waterborne, a process also known as *dispersion*. Under dispersion, soils become impermeable to water; lakes, streams, and rivers experience large increases in suspended solids (clays and organics). Most soil clays undergo dispersion at an ESP of around 15. At this ESP level, soils appear to be toxic because they lose the potential to function as porous media (water infiltration and gas exchange are restricted).

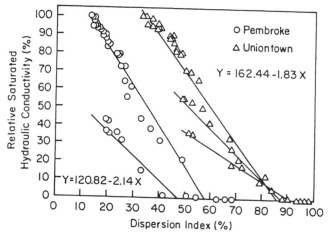

Figure 11.5. Relationships between relative saturated hydraulic conductivity and percent dispersion index of two Kentucky soils (dispersion index = percent of total clay remaining waterborne after 1 hr of settling in an Imhoff cone) (from Marsi and Evangelou, 1991c, with permission).

The dispersion phenomenon in the two humid soils (Pembroke and Uniontown) was evaluated through the use of an Imhoff cone test and a permeameter. The Imhoff cone is commonly used by engineers to determine settleable solids (see Chapter 9). The results of clay dispersion obtained by the Imhoff cone test are expressed as a dispersion index (percent of total clays in the soil sample dispersed), which is correlated with relative saturated hydraulic conductivity. This is shown in Figure 11.5. It demonstrates that each of the soils, depending on its clay content (Pembroke 59%; Uniontown 20%), exhibits unique saturated hydraulic conductivity behavior with respect to the dispersion index. Also, in each of the soils, various mechanisms (different line slopes) appear to control saturated hydraulic conductivity.

The phenomenon of soil dispersion with respect to Na^+ loads (magnitude of ESP or SAR) appears to be unique to all soils on at least one particular point. As the total salt or Cl^- concentration in the water increases, the dispersion index decreases and the saturated hydraulic conductivity increases (Fig. 11.6). When this occurs, the soil–water system becomes toxic to plants and organisms owing to high osmotic pressures. When chloride concentration in solution increases beyond 6000 mg L^{-1}, Na ions near clay surfaces begin to dehydrate because of high osmotic pressure in the surrounding solution. This causes clay particles to flocculate (flocculation is the reverse of dispersion) and, consequently, the saturated hydraulic conductivity of the soil increases.

11.2.5 Brine Chloride and Bromide

The chloride–bromide concentration in brine is greater than 1000 mg L^{-1} and varies widely among wells. At concentrations greater than 106 mg L^{-1} in water, it can be toxic to crops.

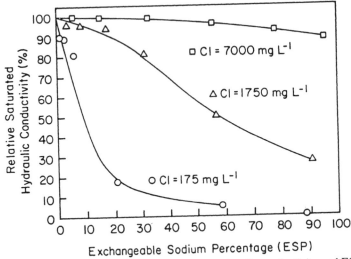

Figure 11.6. Relationship between relative saturated hydraulic conductivity and ESP at three levels of chloride (from Marsi and Evangelou, unpublished data).

Chloride plus bromide is also toxic to crops at elevated concentrations due to the salting-out effect (high osmotic pressure). This salting-out effect appears to become important at EC levels greater than 2 mmhos cm^{-1}. A level of 1 mmhos cm^{-1} is equal to approximately 640 mg L^{-1} dissolved solids or approximately 350 mg L^{-1} Cl^{-1}. Water normally used for human consumption has an EC value significantly less than 1 mmhos cm^{-1}. Because salting-out effects are generally independent of salt type, they can be caused by either sodium chloride plus bromide salt or calcium chloride plus bromide.

11.2.6   Heavy Metals

Although the concentration of heavy metals in brines is usually not high enough to cause alarm, iron can sometimes be quite high (10–100 mg L^{-1}), but it quickly oxidizes and precipitates out because of the high pH of the brine.

11.2.7   Boron

Boron concentration in brines can vary from 10 to 100 mg L^{-1}. In the soil solution, a boron concentration as low as 4 mg L^{-1} is toxic to some crops.

11.2.8   Alkalinity

As such, alkalinity does not cause toxicity. However, at concentrations greater than 90 mg L^{-1}, alkalinity can dramatically increase the toxicity of sodium by removing calcium from the water as calcium carbonate. If this occurs in an agricultural field owing to brine disposal, the soil–water environment could become highly toxic to crops.

Alkalinity in brines can be evaluated through the use of the pH_c. The mathematical expression for the pH_c (Langelier or saturation index) is as follows:

$$\text{Saturation index} = pH - pH_c \tag{11.6}$$

where pH denotes measured-solution pH, and pH_c denotes equilibrium pH for $CaCO_3$ under a given set of conditions (under a pCO_2 of 0.0003 and pure $CaCO_3$, $pH_c = 8.4$). When the saturation index >0, $CaCO_3$ precipitation is expected; when the saturation index <0, $CaCO_3$ dissolution is expected. The saturation index can be derived by estimating pH_c as follows:

$$K_{sp} = (Ca^{2+})(CO_3^{2-}) = 10.0^{-8.34} \tag{11.7}$$

and

$$K_2 = \frac{(H_c^+)(CO_3^{2-})}{(HCO_3^-)} = 10.0^{-10.33} \tag{11.8}$$

where K_{sp} is the solubility product constant of $CaCO_3$, K_2 is the second dissociation constant of H_2CO_3, and the parentheses denote solution ion activity. Rearranging and substituting Equation 11.7 into Equation 11.8 gives

$$K_2 = \frac{(H_c^+)(K_{sp}/Ca^{2+})}{(HCO_3^-)} = 10.0^{-10.33} \tag{11.9}$$

Taking logarithms on both sides of Equation 11.9 gives

$$-\log K_2 = -\log H_c^+ - \log K_{sp} + \log Ca^{2+} + \log HCO_3^- \tag{11.10}$$

Rearranging,

$$pH_c = pK_2 - pK_{sp} + pCa^{2+} + pHCO_3^- \tag{11.11}$$

A practical approach to estimating the pH_c of water moving through soil is as follows:

$$pH_c = (pK_2' - pK_c') + p(Ca + Mg) + pAl_k \tag{11.12}$$

where pK_2' and pK_c' represent pK_2 and pK_c ($pK_c' = pK_{sp}$) corrected for ionic strength (see Table 11.1). An estimated pH_c (using Eq. 11.12) of less than 8.4 suggests that Ca^{2+} will precipitate as limestone ($CaCO_3$). An estimated pH_c (using Equation 11.12) of greater than 8.4 suggests that $CaCO_3$, if present, will dissolve. The values of

$pK_2' - pK_c'$, $p(Ca + Mg)$, and pAlk are obtained from Table 11.1 after analyzing the water for Ca, Mg, Na, HCO_3, and CO_3. The concentration values in Table 11.1, columns 1, 3, and 5 are in milliequivalents per liter (meq L^{-1}). The values in columns 4 and 6 represent the negative logarithms of the corresponding values in columns 3 and 5, respectively. The values for $pK_2' - pK_c'$ are obtained from column 2 through the corresponding sum of Na, Ca, and Mg in column 1.

An example using Table 11.1 is demonstrated below. Assuming analysis of a water sample gives

TABLE 11.1. Tables for Calculating pH$_c$ Values of Waters

Concentration Ca + Mg + Na (1)	$pK_2'-pK_c'$ (2)	Concentration Ca + Mg (3)	p(Ca + Mg) (4)	Concentration CO_3 + HCO_3 (Alkalinity) (5)	pAlk (6)
0.5	2.11	0.05	4.60	0.05	4.30
0.7	2.12	0.10	4.30	0.10	4.00
0.9	2.13	0.15	4.12	0.15	3.82
1.2	2.14	0.2	4.00	0.20	3.70
1.6	2.15	0.25	3.90	0.25	3.60
1.9	2.16	0.32	3.80	0.31	3.51
2.4	2.17	0.39	3.70	0.40	3.40
2.8	2.18	0.50	3.60	0.50	3.30
3.3	2.19	0.63	3.50	0.63	3.20
3.9	2.20	0.79	3.40	0.79	3.10
4.5	2.21	1.00	3.30	0.99	3.00
5.1	2.22	1.25	3.20	1.25	2.90
5.8	2.23	1.58	3.10	1.57	2.80
6.6	2.24	1.98	3.00	1.98	2.70
7.4	2.25	2.49	2.90	2.49	2.60
8.3	2.26	3.14	2.80	3.13	2.50
9.2	2.27	3.90	2.70	4.0	2.40
11	2.28	4.97	2.60	5.0	2.30
13	2.30	6.30	2.50	6.3	2.20
15	2.32	7.90	2.40	7.9	2.10
18	2.34	10.00	2.30	9.9	2.00
22	2.36	12.50	2.20	12.5	1.90
25	2.38	15.80	2.10	15.7	1.80
29	2.40	19.80	2.00	19.8	1.70
34	2.42				
39	2.44				
45	2.46				
51	2.48				
59	2.50				
67	2.52				
76	2.54				

Source: From Ayers, 1977, with permission.

Na = 6.3 meq L^{-1}

Ca + Mg = 2.0 meq meq L^{-1}

Na + Ca + Mg = 8.3 meq L^{-1}

HCO$_3$ + CO$_3$ = 6 meq L^{-1}

it follows from Table 11.1 that

pK'_2 − pK'_c = 2.26 (column 2)

p(Ca + Mg) = 3.0 (column 4)

pAlk = 2.20 (column 6)

Therefore,

$$pH_c = 2.26 + 3.0 + 2.20 = 7.46 \qquad (11.13)$$

The example shows that since pH$_c$ < 8.4 (7.46 < 8.4), Ca^{2+} would precipitate as CaCO$_3$. From the analysis above, one can also calculate the SAR (Eq. 11.5):

$$SAR = 6.3/(1.0)^{1/2} = 6.3 \qquad (11.14)$$

This calculated SAR can be used to estimate the adjusted SAR (adj.SAR), which describes Na$^+$ potential in a CaCO$_3$-saturated solution to influence SAR. It is estimated as follows:

$$adj.SAR = (SAR) [1 + (8.4 - pH_c)] \qquad (11.15)$$

From the example above,

$$adj.SAR = (6.3) [1 + (8.4 - 7.46)] = 12.22 \qquad (11.16)$$

The adj.SAR is greater than the SAR (12.22 vs. 6.3). This suggests that Ca^{2+} would precipitate as CaCO$_3$ and the adverse effects of Na$^+$ on water quality would intensify.

11.3   MANAGEMENT OF BRINE DISPOSAL

Three factors determine how much brine can be disposed of in a field, assuming that the brine does not contain boron. The first factor is the type of crop crown. Different crops tolerate different levels of salt. For example, some clovers are extremely sensitive to salt, while some grasses, like tall fescue, are quite tolerant (Table 11.2). The second factor is the CEC of the soil. A soil with a CEC of 10 meq 100 g^{-1} can tolerate approximately 460 lb of sodium per acre (10% of CEC) before it reaches its critical toxicity threshold. However, a soil with a CEC of 20 meq 100 g^{-1} can tolerate up to 920 lb of sodium per acre before it reaches its critical threshold. The third factor is the texture of the soil. A sandy soil can take very little sodium chloride salt before it

TABLE 11.2. Crop Sensitivity to Salts, Based on the Saturation Extract Test

	Expected Yield Reduction			
	0%	10%	25%	50%
Crop	Electrical Conductivity (mmhos cm^{-1})			
Tall fescue	3.9	5.8	8.6	13.3
Vetch	3.0	3.9	5.3	7.6
Alfalfa	2.0	3.4	5.4	8.8
Clovers, (alsike, ladino, red)	1.5	2.3	3.6	5.7
Barley	8.0	10.0	13.0	18.0
Wheat	6.0	7.4	9.5	13.0
Soybean	5.0	5.5	6.2	7.5
Corn	1.7	2.5	3.8	5.9

Source: From the U.S. Salinity Laboratory Staff, 1954, with permission.

becomes toxic, but a clay soil can take a great deal more, owing to its higher CEC and water-holding capacity.

The data in Figure 11.7 show the relationship between EC and NaCl discharged onto a soil on an acre basis, 15 cm deep. For example, in a silt loam soil with a water saturation percentage of 50, assuming that the crop grown can withstand a solution composition of 3 mmhos cm^{-1}, the quantity of brine in NaCl equivalents that could be discharged should not exceed 2000 lb per acre (see dashed lines on Fig. 11.7). To determine whether the maximum possible quantity of brine has been disposed of in an agricultural soil, the adj.SAR and the EC of the soil solution at saturation must be found. The adj.SAR must be less than 5 and the EC less than the critical threshold of the crop grown.

11.3.1 Reclamation of Salt-Affected Soils

To reclaim a salt-affected soil, excess salt should be leached downward so that the EC of the soil solution becomes lower than the critical threshold of the crop grown, commonly less than 2 mmhos cm^{-1} (Table 11.2). The calcium lost because of leaching must be replenished so that the soil solution maintains a SAR somewhere around 5.

To leach the excess salt, it is necessary to pond water on the land or wait for natural rainfall to do it. However, because a lot of water is needed to leach the salt, relying on rainfall alone may require a wait of more than a year to accomplish the leaching process. Furthermore, as the salt in the soil is diluted by rainfall, the soil seals up and no water moves through it. To avoid the soil-sealing process, one has to supply calcium. This is generally done by applying calcium and incorporating it into the soil surface before the water is ponded on the treated area. The quantity of calcium to be applied depends on the quantity of exchangeable Na$^+$. Exchangeable Na$^+$ present in a soil can

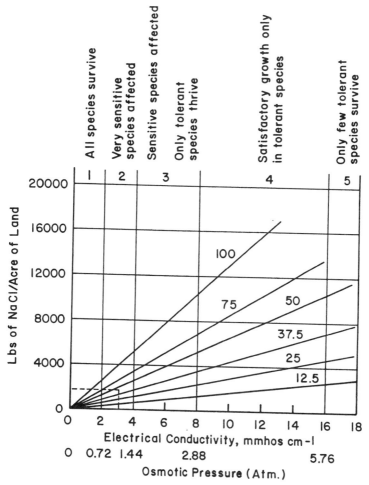

Figure 11.7. Relationship between electrical conductivity of soil solution and salt content (the numbers in the plot represent grams of water that are needed to saturate 100 g of soil (it takes 12.5 g of water to saturate 100 g of sand and 100 g of water to saturate 100 g of clay. Most Kentucky soils would require about 50 g of water to saturate 100 g of soil) (from U.S. Salinity Laboratory Staff, 1954).

be estimated from the concentrations of Na^+, Ca^{2+}, and Mg^{2+} (meq L^{-1}) in the soil solution by employing the nomogram in Figure 11.8 which gives the ESP. Figure 11.8 represents an average soil (Marsi and Evangelou, 1991a). The quantity of Na^+ can be estimated by multiplying the ESP/100 of the soil with the CEC of the soil. The CEC of the soil can be determined by a laboratory or an estimate can be obtained from the U.S. Soil Conservation Service.

One of the most commonly used calcium sources in the reclamation of brine-contaminated soil–water environments is gypsum ($CaSO_4 \cdot 2H_2O$). In highly calcare-

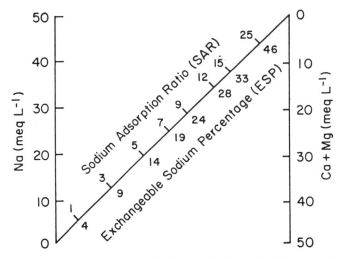

Figure 11.8. Nomogram for determining the SAR value of a soil solution and for estimating the corresponding ESP value of a soil that is at equilibrium with the water (adapted from the U.S. Salinity Laboratory Staff, 1954).

ous soils, elemental sulfur (S) can also be used because it reacts with calcium carbonate in the soil to produce gypsum. Road-deicing salt (calcium chloride) can also be used. Table 11.3 indicates the quantities of gypsum or sulfur that are needed to reclaim brine-contaminated soils based on the magnitude of exchangeable Na$^+$ present.

The quantity of water that must pass through a brine-contaminated soil to bring the SAR and EC within the critical thresholds depends on the hydraulic characteristics of

TABLE 11.3. Amounts of Gypsum and Sulfur Required to Replace Indicated Amounts of Exchangeable Sodium

Sodium (meq. 100 g^{-1} of soil)	Gypsum (CaSO$_4$·2H$_2$O)		Sulfur (ton/acre - ft)	Sulfur (ton/acre - 6 in.)
	(ton/acre - ft)	(ton/acre - 6 in.)		
1[a]	1.7	0.9	0.32	0.16
2	3.4	1.7	0.64	0.32
3	5.2	2.6	0.96	0.48
4	6.9	3.4	1.28	0.64
5	8.6	4.3	1.60	0.80
6	10.3	5.2	1.92	0.96
7	12.0	6.0	2.24	1.12
8	13.7	6.9	2.56	1.28
9	15.5	7.7	2.88	1.44
10	17.2	8.6	3.20	1.60

Source: From the U.S. Salinity Laboratory Staff, 1954, with permission.
[a]1 meq 100 g^{-1} equals 460 lb of Na per acre 6 in. deep.

the soil. In general, however, after one pore volume has passed through the soil, the SAR and the EC should be checked to determine the Cl^- level and the extent of exchange of Ca^{2+} for Na^+. A pore volume for a silt loam soil is approximately half the depth of the soil that one desires to reclaim from the brine. In other words, if one desires to reclaim a 1-f (30 cm) depth of soil, 6 in. (15 cm) of water should be applied. If the SAR and EC remain above the critical thresholds after 6 in. (15 cm) of water have passed through the soil profile, a second pore volume should be passed through the soil and the SAR and EC should be checked again. When the SAR and EC are within the critical thresholds at the desired depth, the soil-reclamation process is complete.

Large quantities of straw are often used in reclaiming brine-contaminated land, because incorporating it into the soil increases porespace, which makes the soil more permeable. This helps to move the salt out of the soil by increasing the leaching potential. However, to truly reclaim the soil, the SAR should be much less than 5 and the salinity level should be around 1–2 mmhos cm^{-1}. Large quantities of straw may improve these factors by increasing the CEC, supplying some calcium during decomposition, and increasing the water-holding capacity of the soil. Straw helps, but a supply of calcium may still be needed to restore soil productivity.

It is important to keep in mind that when one reclaims salt-contaminated land, groundwater contamination potential increases. Effective reclamation of soil means leaching salt deeper into the ground. However, before leaching salt downward, it is important to determine where it will go. For example, if it reaches the groundwater, one must ensure that if it empties into streams or lakes, the dilution would be high enough to bring the salt concentration to normal levels.

Finally, for a brine high in boron, toxicity levels can be determined by analyzing an extract of water from saturated soil. The critical threshold is around 2 mg L^{-1}. Fortunately, boron leaches quite easily from a soil at circumneutral pH, and if a soil is reclaimed from sodium, the chances are good that the boron would also be leached and its concentration would be less than 2 mg L^{-1}.

11.3.2   Brine Evaluation Prior to Disposal

Before a brine is disposed of onto a soil or into a water system, an evaluation should be carried out to determine the brine's potential to contaminate such systems. In carrying out such an evaluation, one must first determine the concentrations of heavy metals and toxic organics in the brine. Since the mechanism of detoxification is basically dilution of NaCl, certain procedures for brine disposition should be followed, depending on the content of heavy metals and boron. When the brine contains high concentrations of heavy metals and boron, the first consideration must be the potential of dilution to bring the concentration of heavy metals and boron to within an acceptable range. The next concern is to be sure that the magnitude of EC and SAR upon dilution will be within the critical thresholds (<2 and <5, respectively).

When NaCl is the dominant pollutant, the brine should be evaluated with respect to its chloride concentration (EC) and SAR. A classification scheme can be employed for this evaluation. In classifying brines an important consideration is the interaction

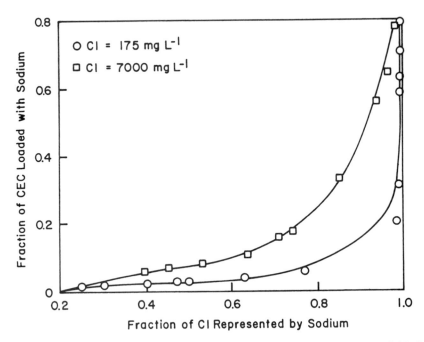

Figure 11.9. Relationship between fraction of CEC loaded with sodium and fraction of chloride (Cl⁻) represented by sodium at two levels of chloride concentration of the Uniontown soil (from Marsi and Evangelou, unpublished data).

of brine with a soil's exchange complex. This is demonstrated in Figure 11.9. It shows that for a brine composed of a mixture of NaCl and $CaCl_2$, the soil's ESP increases as the fraction of the chloride's negative charge represented by Na^+ increases. Furthermore, as Cl^- concentration in the brine increases, the soil's ESP also increases for any given Na^+/Cl^- ratio.

In summary, soils with low Cl^- (low salt concentration) appear to preferentially adsorb Ca^{2+}. At high Cl^- levels, soils appear to preferentially adsorb Na^+. The data in Figure 11.9 show that if soils or sediments are equilibrated with concentrated brines, the water becomes preferentially enriched with Na^+ relative to Ca^{2+} upon dilution because the sediments preferentially adsorb Ca^{2+} and desorb Na^+. The process removes Na^+ from the sediments and enriches the water with Na^+. The reverse occurs when brine becomes concentrated by evaporation of water or removal of water by plants. These observations are summarized in Figure 11.10.

The brine classification scheme shown in Figure 11.10 demonstrates the use of EC and SAR in evaluating a brine prior to disposal. The numerical order from left to right (1–4) shows increasing concentration of brine (NaCl + $CaCl_2$). The alphabetical notation A–D shows increments of SAR. The rating 1A refers to brines that can be disposed of in land and natural bodies of water without any special considerations. Rating 3B and 4B describe brine solutions for which dilution is necessary to bring the EC to the safe range (0.1–2 mmhos cm⁻¹). The same applies to brines 3A and 4A. The

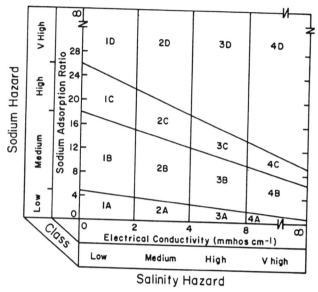

Figure 11.10. Diagram of brine classification (adapted from the U.S. Salinity Laboratory Staff, 1954).

only difference between 3A and 4A is that a higher dilution factor should be considered when discharge is taking place into a stream or lake. If brines 2A–3A or 4A are discharged onto land, the quantity to be discharged can be estimated from Figure 11.7. This estimation assumes that all the brine will percolate into the soil, reach the predetermined depth, and be evenly distributed.

When brine 1B (Fig. 11.10) is discharged into soil–water systems, it can create adverse effects because of its high SAR. Long-term disposal of brine 1B onto land could significantly increase the ESP of the soil and could induce soil dispersion. A dispersed soil would most likely be toxic to agricultural plants and to natural vegetation. Soils receiving brines with low Cl⁻ but high SAR should be checked periodically for SAR (which should be less than 5). If SAR is higher than 5, it can be reduced by the addition of high-quality gypsum. The quantity of gypsum to be added is based on the quantity of exchangeable Na⁺ (Table 11.3). Exchangeable Na⁺ can be estimated from Figure 11.8 and the CEC of the soil.

A similar approach can be taken for the disposal of brines 1C and 1D. The only difference between these brines and brine 1B is that a soil receiving brines 1C or 1D would require more gypsum than the soil receiving brine 1B. When brines 1B–1D are discharged into a stream, attention must be given to the long-term stability of the SAR of the receiving waters. If SAR is steadily rising with time, brine discharge should be stopped immediately. Gypsum should then be applied so that increases in suspended solids and bank instability are brought under control.

The most troublesome brines are the ones that have high SAR and high salt (e.g., brine 4D). Disposal of such brines onto soils should be carried out in accordance with

Figures 11.7 and 11.8. Ideally, at the end of every discharge event, EC and SAR should be checked. When either of these parameters is above the critical threshold, brine discharge should be stopped and the soil should be allowed to adjust to a lower ESP and lower salinity. Such changes may take place without treatment, but sometimes require amendments.

When brine 4D is discharged directly into rivers, streams, and lakes, brine dilution does not guarantee that the process can be continued without any adverse effects on water quality. Although no adverse effects would occur if SAR does not increase, brine discharges should be stopped even if Cl⁻ is well within the acceptable concentration limit if SAR exceeds 5. At this point, an ESP evaluation of the sediments should be carried out. If ESP is greater than 15, resumption of the brine discharge should not take place until it drops below 5. When disposing brine with high alkalinity levels into natural bodies of water, the adj.SAR of the mixture of the two types of water should always be considered.

PROBLEMS AND QUESTIONS

1. Calculate the osmotic potential of an irrigation water with an EC of 10 mmhos cm^{-1}. Explain whether the osmotic potential of this water would be too high for plants.

2. Calculate the osmotic potential of a water sample with a concentration of 3000 ppm dissolved solids. Convert ppm to (a) mg L^{-1}, (b) mmhos cm^{-1}, and (c) osmotic pressure in bars. Explain whether the salt content of this water would be too high for plants.

3. What additional analyses of the two water samples above would you carry out to determine their irrigation management?

4. A water sample contains the following cations:

   $Ca = 200$ mg L^{-1}

   $Mg = 50$ mg L^{-1}

   $Na = 1000$ mg L^{-1}

   a. Calculate the water's SAR.
   b. Using Equation G, calculate the soil's potential ESP.
   c. Explain whether the potential ESP is too high or too low.
   d. Based on the results of your calculation, explain if and what type of action you would take to avoid making the soil saline and/or sodic.

5. The ionic composition of a water sample is

   $Na = 10.3$ meq L^{-1}

   $Ca + Mg = 4.6.0$ meq meq L^{-1}

   $HCO_3 + CO_3 = 12$ meq L^{-1}

   a. Calculate the water's SAR.
   b. Calculate the water's adj.SAR.
   c. Estimate the difference between SAR and adj.SAR and explain its meaning.

6. A farmer wants to reclaim a soil contaminated with 200 mmol L^{-1} NaCl. He/she needs advice on how to do it. List the type of tests you would recommend and the reasons for each of the tests and outline the salt-leaching procedure.

7. Explain how you would reclaim a soil that has been contaminated with 200 mmol L^{-1} $CaCl_2$.

PART VI
Land-Disturbance Pollution and Its Control

12 Acid Drainage Prevention and Heavy Metal Removal Technologies

12.1 INTRODUCTION

Acid drainage (AD) has various anthropogenic and natural sources, but the most extensive and widely known AD source is the one related to mining coal and various metal ores including copper, gold, lead, and silver. Other human activities related to AD production include various forms of land disturbance such as industrial or residential development and farming (e.g., rice). Acid drainage also emanates from lands disturbed in the past (e.g., old surface or underground gold, silver, and coal mines) or lands which, because of tectonic processes, are continuously exposing acid-forming minerals. Generally, strong acid-forming processes in nature involve exposure of metal–sulfides enriched with heavy metals or metalloids (e.g., lead and arsenic) to atmospheric air, which leads to oxidation and the production of acid and/or heavy-metal-rich waters (Evangelou, 1995b, see also Chapter 6).

In addition to AD produced through oxidation of metal–sulfide minerals, AD and/or heavy-metal sources include human activities such as mineral processing; manufacturing or recycling of batteries; electronics; wood pulp, paper, and heavy steel industries such as the manufacturing of cars or heavy equipment; tanneries; textile manufacturing; food processing; and waste-disposal or waste-management industries.

The purpose of this chapter is to introduce the various technologies and mechanisms used to treat AD or heavy-metal-rich solutions and to demonstrate the use of soil–water chemistry principles for generating and/or improving contaminant treatment technologies.

428

12.2 MECHANISMS OF ACID DRAINAGE CONTROL

One may employ a number of mechanisms to control the release of acid, heavy metals, or metalloids to the soil–water environment. These mechanisms are listed below:

pH Control
Chemical Reactions
  Metal–hydroxide precipitation
  Metal–carbonate precipitation
  Metal–sulfide precipitation
  Metal–silicate precipitation
  Metal–organic complexation
Redox Potential

Sorption
Ion Exchange
Encapsulation
  Microencapsulation
  Macroencapsulation
  Embedment
Alteration of Waste Properties
Bioremediation

Some of these mechanisms are discussed in detail throughout this book; others are summarized in this chapter.

12.2.1 Precipitation

Metal–Hydroxides. Most heavy metals may precipitate via strong bases (e.g., NaOH and KOH) as metal–hydroxides [$M(OH)_n$]. These precipitation reactions are described in Chapter 2. As noted, metal–hydroxide solubility exhibits U-shape behavior and ideally its lowest solubility point in the pH range allowed by law (e.g., pH 6–9) should be lower than the maximum contaminant level (MCL). However, not all heavy metal–hydroxides meet this condition. The data in Figure 12.1 show the various metal–hydroxide species in solution when in equilibrium with metal–hydroxide solid(s). In the case of Pb^{2+}, its MCL is met in the pH range of 7.4–12, whereas the MCL of cadmium (Cd) the MCL is not met at any pH. Similar information is given by the solubility diagrams of Cu^{2+}, Ni^{2+}, Fe^{3+} and Al^{3+}.

Data showing the total solubility of the various metals listed in Figure 12.1 are given in Figure 12.2. It appears that almost each metal exhibits its lowest solubility point at a unique pH. This suggests that it is difficult to remove two or more heavy metals as metal–hydroxides simultaneously from solution by adjusting pH. One heavy metal may be precipitating at a given pH while another may be redissolving at the same pH. Under such conditions, a series of treatment systems may be a more effective heavy metal removal approach.

The solubility of metal–hydroxide precipitates in water varies depending on ionic strength and number of pairs and/or complexes (Chapter 2). A practical approach to determining the pH of minimum metal–hydroxide solubility, in simple or complex solutions, is potentiometric titration, as demonstrated in Figure 12.3. The data show that potentiometric titration of a solution with a given heavy metal is represented by a sigmoidal plot. The long pH plateau represents pH values at which metals precipitate; the equivalence point, or titration end point, indicates the pH at the lowest metal–

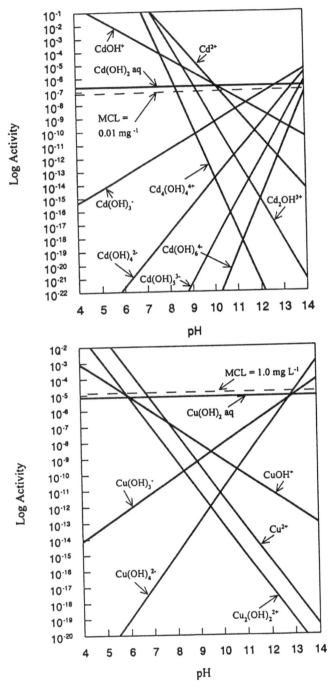

Figure 12.1. Stability lines for various metal–hydroxide species as a function of pH. The data were generated employing MINTEQA2 by assuming that pCO$_2$ = 0 and is in equilibrium with metal–hydroxide solids (from Evangelou, 1997, unpublished data, with permission).

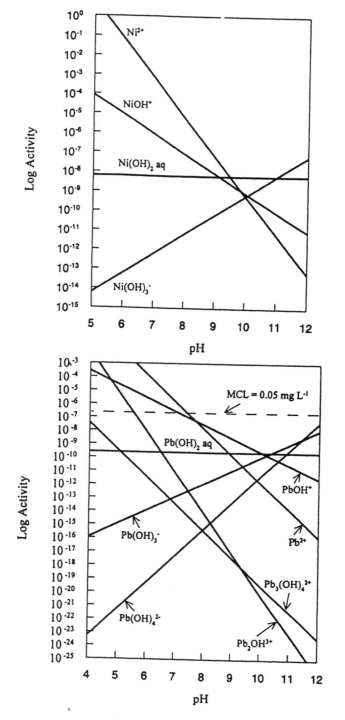

Figure 12.1. Continued.

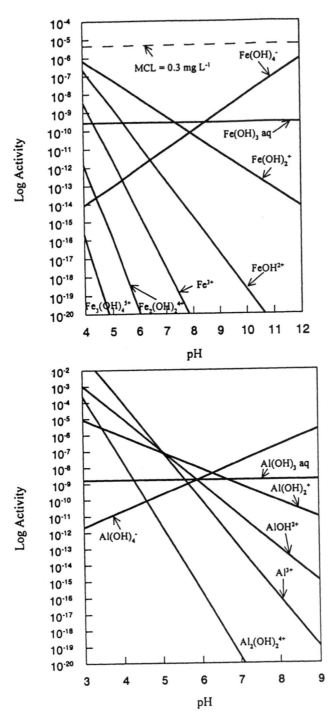

Figure 12.1. Continued.

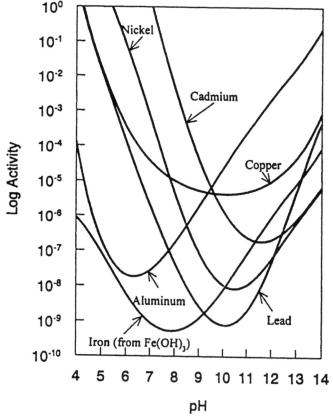

Figure 12.2. Relationship between total metal dissolved from various metal–hydroxides in an equilibrium state as a function of pH. The data were generated employing MINTEQA2 by assuming $pCO_2 = 0$ (from Evangelou, 1997, unpublished data, with permission).

hydroxide solubility potential. The titration equivalence point suggests that most, if not all, metal ions are hydroxylated, but this does not mean that all metal ions are necessarily part of a metal–hydroxide solid. This requires verification by metal analysis, because maximum metal precipitation depends on the overall solution composition. Potentiometric titrations of simple or complex solution systems do show, however, the pH range at which maximum metal–hydroxide precipitation is expected to take effect.

Metal Carbonates. In certain cases, metal carbonates are less soluble than their corresponding hydroxides. Natural formation of carbonates from carbon dioxide in the air, referred to as carbonation, depends on pCO_2 and pH. The carbonation process in the alkaline pH range is

$$M(OH)_2s + H_2CO_3 \rightarrow MCO_3s + 2H_2O \qquad (12.1)$$

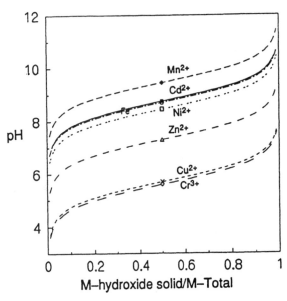

Figure 12.3. Potentiometric titration of various metal–chloride solutions (from Evangelou, 1997, unpublished data, with permission).

The pH at which carbonation occurs depends on the solubility products of the carbonate and hydroxide species and on pCO_2. Patterson et al. (1977) reported that hydroxide precipitates controlled the solubility of zinc and nickel over a wide range of pH values, but cadmium and lead solubilities were controlled by carbonate precipitates. The data in Figure 12.4 show the various metal species in solution in equilibrium with metal carbonates as a function of pCO_2. These data were produced using MINTEQA2 without allowing formation of metal–hydroxides. They show, as expected, that metal solubility depends on pCO_2. For any given pH, the concentration of the total metal–carbonate dissolved can be predicted using Figure 12.5. For example, at pH 6 ($pCO_2 = 10^{-3}$), total Cd in solution would be approximately 10^{-5}. However, at the same pH (pH = 6.0), the total Cd dissolved from $Cd(OH)_2$ (Fig. 12.2) would be higher than one mole per liter. This comparison suggests that cadmium carbonate is less soluble than $Cd(OH)_2$ and the former ($CdCO_3$) would be controlling the amount of Cd that one may find dissolved in water at the specified equilibrium. For this reason, a more effective way of removing Cd^{2+} from solution is through precipitation as $CdCO_3$ instead of $Cd(OH)_2$. A similar evaluation can be carried out for all other metals in Figures 12.2–12.5.

Metal–Phosphates. Another way of removing metals from water is through precipitation by phosphate (e.g., $FePO_4$, $AlPO_4$, or $PbHPO_4$), which is demonstrated in Figure 12.6. It shows that metal–phosphates exhibit U-shaped solubility behavior. At low pH, metal–phosphates dissolve because of M^{2+}–H^+ competitive interactions. At high pH, metal–phosphates undergo incongruent interactions, allowing the heavy metal to precipitate as metal–hydroxide or metal–carbonate (depending on the type of heavy metal) and the phosphate to precipitate as calcium–phosphate ($Ca_5(PO_4)_3OH$).

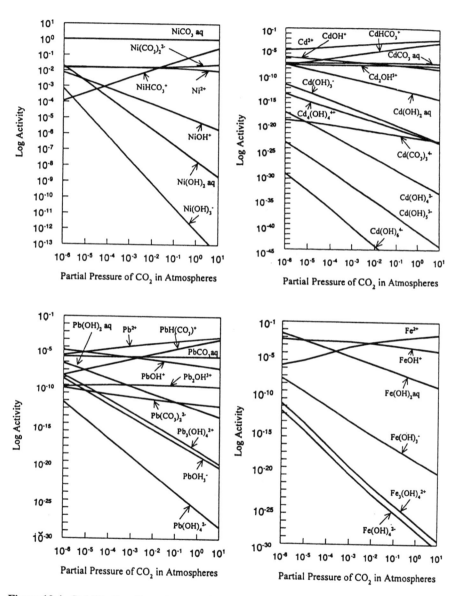

Figure 12.4. Stability lines for various metal–carbonate or metal–hydroxide species in solution as a function of pCO_2. The data were generated employing MINTEQ2A by assuming equilibrium with metal–carbonate solids without allowing formation of metal–hydroxide solids (from Evangelou, 1997, unpublished data, with permission).

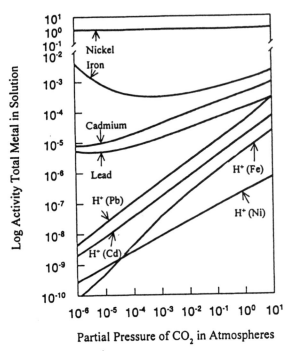

Figure 12.5. Relationship between total metal dissolved from various metal–carbonates as a function of pCO$_2$. The data were generated by MINTEQA2 without allowing formation of metal–hydroxide solids (from Evangelou, 1997, unpublished data, with permission).

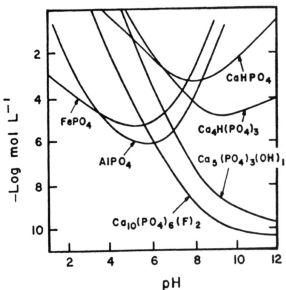

Figure 12.6. Solubility of various metal–phosphates as a function of pH (from Stumm and Morgan, 1981, with permission).

Metal–Sulfides. Sulfide precipitation has been one of the most widely used methods to precipitate many of the heavy metals. The low MCLs required for some of the highly toxic metals (e.g., mercury), are often achievable only by precipitation as sulfides, since they generally have solubilities several orders of magnitude lower than the hydroxides, carbonates, or phosphates throughout the pH range of 6–9. However, metal sulfides can resolubilize in an oxidizing environment.

Sulfide produces an undesirable "rotten-egg" odor and is toxic when in the H_2S gas form. Since the first pK_a of H_2S is 7.24, it is necessary to maintain pH 9 or above to completely prevent evolution of H_2S gas (Fig. 12.7). Although excess H_2S is necessary for the precipitation reaction, the excess must be kept to a minimum. Furthermore, although metal–sulfide solubility with respect to pH exhibits U-shaped behavior (Fig. 12.8), its solubility within the desirable pH range is extremely small (MCLs are met) (Fig. 12.9). Precipitation of metal–sulfides is normally carried out using Na_2S or NaHS. However, not all metals precipitate effectively by sulfide. For example, chromium (Cr^{3+}) precipitates effectively as a hydroxide rather than sulfide.

Metal–Silicates. Another important method of metal removal from solution is by silicate precipitation using soluble silicates. The insoluble precipitates formed by interacting soluble silicate ions with metals are not well characterized. Metal silicates

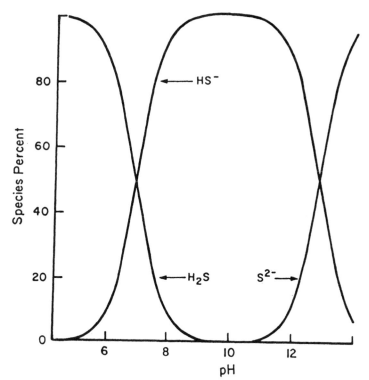

Figure 12.7. Speciation of sulfide as a function of pH (from Evangelou, 1997, unpublished data, with permission).

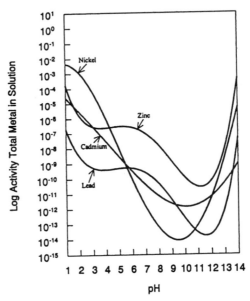

Figure 12.8. Solubility of metal–sulfide minerals as a function of HS$^-$ activity (mol L^{-1}). The data were generated by MINTEQA2 using an initial concentration of metal 10^{-2} mol L^{-1} and pH fixed at 6 (from Evangelou, 1997, unpublished data, with permission).

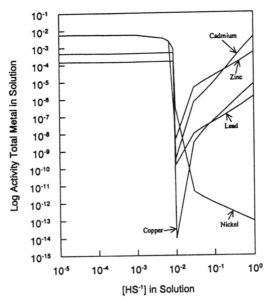

Figure 12.9. Relationship between total metal in solution in equilibrium with metal–sulfides as a function of pH. The data were generated by MINTEQA2 using an initial concentration of metal 10^{-2} mol L^{-1}. The data were produced employing an initial equimolar concentration of dissolved metal and dissolved HS$^-$ (from Evangelou, 1997, unpublished data, with permission).

are nonstoichiometric compounds in which the metal is coordinated to silanol groups (SiOH), in an amorphous silica matrix.

Metal–Organics. Many organic materials also form low-solubility species with certain metals. Among these are humic acids. The most widely publicized insoluble substrate for heavy-metal immobilization has been insoluble starch xanthate (ISX). In contact with metal ions, the metal links to the sulfur group much as it would with the S^{-2} in inorganic sulfides:

$$\text{starch}-\text{O}-\underset{S}{\overset{S}{C}} \;\; + \; Cu^{2+} \;\; \Leftrightarrow \;\; \text{starch}-\text{O}-\underset{S}{\overset{S}{C}}\,Cu \qquad (12.2)$$

Cellulose xanthates are reported to operate in much the same way.

12.2.2 Redox Potential

The redox potential (Eh) controls the ratio of oxidants and reductants within the soil–water system (see Chapter 5). The presence of strong oxidants or reductants can change the valence state of a number of metals or metalloids, affecting their chemical speciation and, therefore, their solubilities and mobilities. For some metalloids, such as arsenic and selenium, both valence and speciation can change easily with redox potential. The metals and metalloids of interest, with more than one possible valence state in aqueous systems, include As, Cr, Fe, Hg, Mn, Ni, and Se. Also, nitrogen and sulfur have multiple valence states which affect the speciation of the metals in a given system. The behavior of metals such as Cu, Cd, and Zn can be strongly but indirectly influenced by redox processes even though they persist in natural aqueous systems in only one valence state. The change in metal behavior involves reductive dissolution of metal–oxides and, thus, loss of surfaces where metal sorption takes effect.

Iron and Manganese Chemistry. The Surface Mining Control and Reclamation Act of 1977 requires that sediment ponds be used to improve water quality with respect to pH, iron (Fe), and manganese (Mn). Iron and manganese are classified as water contaminants, but manganese is also used as an indicator of heavy metal removal from water. The argument is that removal of Mn^{2+} from water, a particularly hard task because of high $Mn(OH)_2$ solubility and extremely slow oxidation kinetics at circum-neutral pH, ensures removal of most heavy metals from water, thus making monitoring of heavy metals unnecessary. Low pH and high concentrations of iron and manganese are common water-quality problems found in eastern U.S. coal fields. The law specifies that certain minimum chemical standards are to be met for water released from surfaced-mined areas (Table 12.1), and water treatment is necessary in many ponds.

To understand the release of Mn and/or Fe to water one needs to understand the redox chemistry of the two elements as well as the solubility of the solids formed under the various redox potentials present in a natural or disturbed environments. The data in Figure 12.10 show the stability of various Fe species as a function of pe and pH. The diagram shows that between approximately pH 4 and 12, and in the presence of

TABLE 12.1. Effluent Limitations (mg L^{-1}) Except for pH

Effluent Characteristics	Maximum Allowable	Average of Daily Values for 30 Consecutive Discharge Days
Iron (total)	7.0	3.5
Manganese (total)	4.0	2.0
Total suspended solids	70.0	35.0
	pH—within range of 6.0–9.0	

atmospheric CO_2, three solids control Fe in solution. These solids are a number of iron–oxides, not shown in the diagram, as well as $Fe(OH)_3$, $FeCO_3$, and $Fe(OH)_2$. The first mineral, $Fe(OH)_3$, controls the release of Fe^{3+}, while the other two minerals, $FeCO_3$ and $Fe(OH)_2$, control the release of Fe^{2+}. Based on Figure 12.10, the stability of these solids is controlled by pH and Eh or pe (Eh = [pe][59]) and Fe^{3+}, present mostly under oxidizing conditions, is the easiest to remove from water using a base (e.g., NaOH), while Fe^{2+}, stable under reducing conditions, is most stable in the pH range of 7 to approximately 10 as carbonate solid rather than hydroxide solid.

The data in Figure 12.11 show the stability of various Mn species as a function of pe and pH. It appears that between approximately pH 4 and 12, and in the presence of atmospheric CO_2, various solids control Mn in solution. These solids are MnO_2, MnOOH, Mn_3O_4, $MnCO_3$, and $Mn(OH)_2$. Under reducing conditions, the solids controlling the release of Mn^{2+} are $MnCO_3$ and $Mn(OH)_2$. Manganese–carbonate is most stable in the pH range of 7.5 to approximately 11.2. At circumneutral pH, the most stable form of Mn, in the form of manganese–oxides [e.g., manganese–dioxide, $Mn(III)O_2$, or manganese oxyhydroxide, $Mn(IV)OOH$] is controlled by pH and Eh or pe. In general, removal of the soluble Mn^{2+} from solution can effectively be attained by high pH and strongly oxidizing conditions.

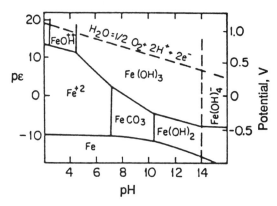

Figure 12.10. Iron Eh–pH (or pe–pH) speciation diagram with $C_T = 2 \times 10^{-3}$ and soluble metal 10^{-5} mol L^{-1} (adapted from Stumm and Morgan, 1981, with permission).

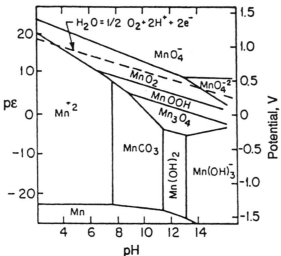

Figure 12.11. Manganese Eh–pH (or pe–pH) speciation diagram with $C_T = 2 \times 10^{-3}$ and soluble metal 10^{-5} mol L^{-1} (from Stumm and Morgan, 1981, with permission).

It is important to understand the terms *dissolved* concentration and *total* concentration when dealing with metals in water (e.g., Mn and Fe in mine sedimentation ponds). Federal law regulates total metal concentration. Dissolved concentration refers to actual iron or manganese dissolved. Dissolved metals in water are either in some kind of ionic or complexed form [e.g., Mn^{2+}, $MnOH^+$, $MnSO_4^0$, Fe^{2+}, $FeOH^+$, $FeSO_4^0$, $R-(COO-)_2 \cdot Mn^{2+}$, and $R-(COO-)_2 \cdot Fe^{2+}$, where R denotes any organic molecule]. Total concentration of iron and manganese refers to the dissolved plus that which is solid. The solid particles of iron and manganese could be maintained in suspension because of their potential colloidal nature. Generally, these solids can be filtered out by conventional laboratory filtering procedures employing the appropriate size filter. Analysis of the filtered and nonfiltered sediment pond water can be used to determine the contribution of dissolved and suspended solid iron and manganese to the total concentration (Table 12.2). This information can be used to appropriately treat such ponds to decrease the high concentrations of total iron and manganese. For example, when most of the metal(s) is in the dissolved form, oxidation should be carried out first. When most of the metal is in the colloidal form, flocculants should be used instead.

For the stability diagrams shown in Figures 12.10 and 12.11, it is assumed that the kinetics of metal transformation between the various species (e.g., reduced to oxidized) are too rapid or that the detention time of treated water is long enough so that the oxidation time needed is much shorter than detention time. However, this is not always the case because the oxidation kinetics of iron and manganese are pH dependent. The data in Figures 12.12 and 12.13 show that the kinetics of Fe^{2+} oxidation above pH 7 are rapid, while Mn^{2+} oxidation becomes rapid above pH 8.7. This suggests that kinetics would be the controlling factor in regulating Mn^{2+} in solution in ponds with

TABLE 12.2. Total and Dissolved Concentrations of Iron and Manganese of Selected Coal Mine Sediment Ponds

	Total (ppm)		Dissolved (ppm)		Total Dissolved Solids (ppm)	
Pond	Iron	Manganese	Iron	Manganese	Iron	Manganese
West Virginia	6.56	3.30	6.75	3.09	0.41	0.21
Kentucky[a]	350.00	26.00	303.00	23.00	47.00	3.00
Illinois	61.00	0.51	3.98	0.05	57.02	0.46
Kansas	18.30	0.19	1.64	0.02	16.66	0.17
North Dakota	965.00	17.83	0.18	0.11	964.82	17.22
Other sites	231.00	10.90	4.60	10.17	226.40	0.73

Source: From Bucek, 1981.
[a]From Nicholas and Foree. 1979.

short detention times. Furthermore, Fe^{2+} removal from solution through oxidation would be easier than Mn^{2+} removal. This is also the general experience of field environmental practitioners.

Controlling Dissolved Iron. Two factors regulate concentrations of dissolved iron (Fe^{2+}) in sedimentation ponds: (1) dissolved gaseous oxygen (O_2) and (2) pH. Generally, in the absence of dissolved gaseous oxygen, there is a large potential for high concentrations of dissolved Fe^{2+}. The potential for maintaining a high concentration of dissolved iron increases with decreasing pH. Two steps should be taken to control dissolved Fe^{2+}: (1) raise the pH and (2) increase the availability of gaseous oxygen.

The following reaction explains the mechanism by which dissolved oxygen and pH control the concentrations of free dissolved Fe^{2+}:

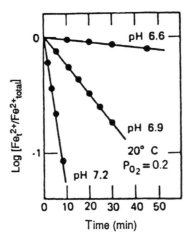

Figure 12.12. Kinetics of Fe^{2+} oxidation under various pH values (from Stumm and Morgan, 1981, with permission).

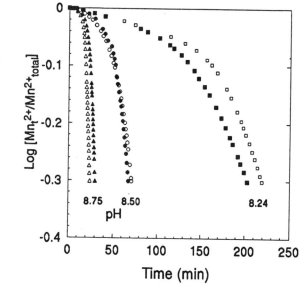

Figure 12.13. Kinetics of Mn^{2+} oxidation under various pH values (from Evangelou, 1997, unpublished data, with permission).

$$Fe^{2+} + 1/4O_2 + 2OH^- + 1/2H_2O \rightarrow Fe(OH)_3s \qquad (12.3)$$

Reaction 12.3 shows that dissolved iron (Fe^{2+}) in the presence of dissolved gaseous oxygen and lime forms ferric (Fe^{3+}) hydroxide [$Fe(OH)_3$], which is nearly insoluble at circumneutral pH. Gaseous oxygen can be increased through the use of pumps to force air into the water, or to spray the water into the air so that it becomes aerated. These methods of increasing dissolved oxygen are rarely used in the field because they are costly and inconvenient. A practical way to increase dissolved oxygen in the water and effectively remove dissolved iron (Fe^{2+}) is to create a series of small waterfalls in the creeks leading into the ponds and then heavily lime these waterfall structures with calcitic granulated limestone. As shown in Equation 12.3, the role of dissolved gaseous oxygen and high pH is to convert Fe^{2+} to Fe^{3+}. This conversion takes place rapidly at about pH 7 (Fig. 12.12). Therefore, to effectively remove dissolved iron (Fe^{2+}) from water, it is only necessary to raise the pH to 7 and make available gaseous oxygen.

It is often thought that when raising pH up to 9 and above, Fe^{2+} will precipitate. A practice commonly used in the field is to raise pH to 9 or above using sodium hydroxide (NaOH) or hydrated lime [$Ca(OH)_2$]. This causes Fe^{2+} to precipitate as ferrous hydroxide, $Fe(OH)_2$, and in the presence of CO_3^{2-} to form $FeCO_3s$. However, both of these precipitates are soluble enough that, given the right water chemistry, they may contribute dissolved iron at levels higher than those permitted by law. The most effective way to remove dissolved iron is by precipitating it as $Fe(OH)_3$.

Controlling Dissolved Manganese. Manganese (Mn^{2+}) removal from water presents a similar problem to that of iron (Fe^{2+}). As with iron, the most effective way to remove manganese from water is to raise the pH and allow Mn^{2+} to oxidize to Mn^{3+} or Mn^{4+}. The only important difference between manganese and iron in this regard is that manganese oxidation requires pH values greater than 9 for near-complete removal.

To facilitate the oxidation of manganese at pH values near neutral or slightly alkaline, a stronger oxidizer than gaseous oxygen is needed. Sodium hypochlorite and calcium hypochlorite are effective oxidizers. Calcium hypochlorite is preferable to sodium hypochlorite because sodium is a clay dispersant (i.e., it causes high amounts of suspended solids). There are various other chemical oxidizing agents available that one may use in the natural environment. These include ozone (O_3), hydrogen peroxide (H_2O_2), and calcium dioxide (CaO_2). Permanganate, a very effective oxidizer, is not appropriate for the natural environment because it increases the total manganese in the system and, under reducing conditions, would most likely convert to soluble manganese (Mn^{2+}).

The reaction below shows how calcium hypochlorite oxidizes dissolved manganese and converts it to manganese oxide, an insoluble solid:

$$1/2Ca(OCl)_2 + H_2O + Mn^{2+} \rightarrow MnO_2 + Cl^- + 2H^+ + 1/2Ca^{2+} \qquad (12.4)$$

The oxidation of manganese (Reaction 12.4) shows that the process produces excess hydrogen ions. The best way to alleviate this acid-producing problem is to lime the ponds with excess calcium carbonate. The pH will be maintained at about 7–8 and there will always be enough hydroxyls to neutralize the acid generated by the oxidation of manganese.

The Relationship Between Manganese and Iron in Sediment Ponds. To under-stand the behavior in and removal of iron and manganese from water, it is important to know the interactions of these two metals. A common occurrence in sediment ponds is the sudden development of a dissolved manganese problem. The cause may be ferrous iron from the incoming water due to the disturbance of a new site. The ferrous iron can react with insoluble manganese oxide (MnO_2) in the sediments at the bottom of the pond according to Equations 12.5 and 12.6:

$$MnO_2 + 4H^+ + 2Fe^{2+} \rightarrow Mn^{2+} + 2Fe^{3+} + 2H_2O \qquad (12.5)$$

$$Fe^{3+} + 3H_2O \rightarrow Fe(OH)_3 + 3H^+ \qquad (12.6)$$

The reactions show that soluble iron coming into a pond with runoff may be oxidized to form insoluble $Fe(OH)_3$s. Iron(II) oxidation however, reduces manganese(III) or manganese(IV) of manganese oxides to soluble manganese (Mn^{2+}). Therefore, dis-solved manganese becomes a problem in the pond.

There are some secondary relationships between iron and manganese that should be considered. As iron and manganese precipitate as oxides and hydroxides, they act as sinks for the soluble manganese. However, these sinks are pH dependent. For example, Mn^{2+} adsorption by oxide–hydroxide surfaces takes place at pH values 8–8.5. Such pH-dependent manganese adsorption sinks are efficient in removing much, but not all, of the soluble manganese from water. To comply with regulations, removal of the last few milligrams of soluble manganese will have to be accomplished by creating oxidative conditions—either by aeration or addition of an oxidation agent such as calcium hypochlorite or sodium hypochlorite.

Controlling pH alone cannot control the solubility of metals. Often, as the pH increases, it appears that metals become more soluble. This is true only because, at the higher pH values, a certain fraction of the organic matter becomes soluble and tends to complex metals, which means that the metals in solution become associated with the soluble organic fraction. This can be avoided by building ponds in areas where tree leaves would not accumulate and where plants would not likely grow in or near the ponds. Organic matter is a source of energy for microorganisms, but this energy-acquiring process needs an oxidant (e.g., O_2). Since the bottom of a pond may be devoid of O_2, microorganisms use manganese oxides as electron sinks, reducing the manganese to a soluble form. The reaction is

$$MnO_2 + microorganisms \rightarrow Mn^{2+} + CO_2 + H_2O \qquad (12.7)$$

In summary, the two most important factors in removing iron and manganese from water are high pH and the presence of an oxidizer. One of the most effective available oxidizers for the oxidation of manganese is calcium hypochlorite.

Calculation of Chemicals for the Control of Mn^{2+} and/or Fe^{2+}. To estimate the quantity of a chemical oxidant needed to oxidize a given quantity of Mn^{2+}, the stoichiometry of the reaction is needed. In the case of hydrogen peroxide (H_2O_2),

$$Mn^{2+} + H_2O_2 \rightarrow MnO_2 + 2H^+ \qquad (12.8)$$

Therefore, for each mole of Mn^{2+}, 1 mol of H_2O_2 is needed. By dividing the atomic weight of Mn^{2+} by the molecular weight of H_2O_2 we get $54.9/34 = 1.61$, or for every one part of manganese, 0.62 parts of H_2O_2 are needed [concentration of manganese in parts per million or milligrams per liter) × (volume of water in liters) × $(0.62/1000) =$ grams of H_2O_2]. One may also need to consider the purity of the peroxide (typically 30%) and its efficiency.

In the case of calcium hypochlorite [$Ca(OCl)_2$], the reaction with Mn^{2+} is

$$1/2Ca(OCl)_2 + H_2O + Mn^{2+} \rightarrow MnO_2 + Cl^- + 2H^+ + 1/2Ca^{2+} \qquad (12.9)$$

and for each mole of Mn^{2+}, half a mole of $Ca(OCl)_2$ is needed. By dividing the atomic weight of Mn^{2+} by one half of the molecular weight of $Ca(OCl)_2$, we get $54.9/71.4 = 0.77$, or for every one part of manganese, 1.3 parts of $Ca(OCl)_2$ are needed [concentration of manganese in parts per million or milligrams per liter) × (volume of water in liters) × $(1.3/1000) =$ grams of $Ca(OCl)_2$]. One may also need to consider the purity of the hypochlorite sample and its efficiency, which is highly variable depending on its environmental stability. However, other factors also affect the actual amount of an oxidizing agent needed: (1) size of the pond, (2) temperature of the water, (3) water turnover time, (4) depth of the pond, (5) rate of water mixing, and (6) method of treatment application.

Arsenic. Arsenic (As) has strong affinity for oxygen and forms various species in the environment, depending on Eh and pH. It can be found as As^0, As_3^0 gas, $As(III)O_2^-$, $As(III)O_3^{2-}$, and $As(V)O_4^{3-}$. The solubility of each of these species varies depending on the presence of adsorbing surfaces, soluble cation type, and concentration. Commonly, arsenic is present in geologic strata as arsenides (e.g., Cu_3As), or

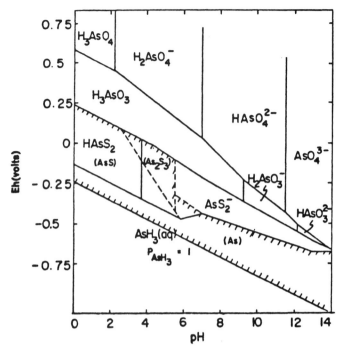

Figure 12.14. The Eh diagram for As at 25°C and 1 atm with total arsenic 10^{-5} mol L^{-1} and total sulfur 10^{-3} mol L^{-1}. Solid species are enclosed in parentheses in cross-hatched area, which indicates solubility less than $10^{-5.3}$ mol L^{-1} (from Ferguson and Gavis, 1972, with permission).

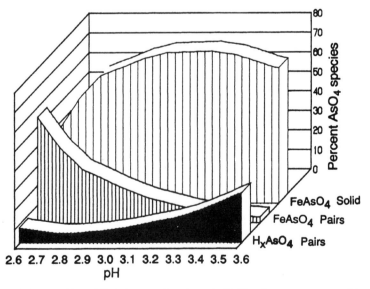

Figure 12.15. Solubility of $FeAsO_4$ as a function of pH. The data were generated by GEO-CHEM–PC with initial Fe^{3+} to arsenate ratio of 1:1 (from Evangelou, 1997, unpublished data, with permission).

sulfides (e.g., AsS or arsenopyrite, FeAsS). These minerals are more stable under reducing conditions. Upon exposure to the atmosphere, oxidation follows and the species of arsenite $[As(III)O_3]$ or arsenate $[As(V)O_4]$ persist, depending on Eh and pH (Fig. 12.14). Under these conditions, iron and iron–oxides appear to control the solubility of arsenic. For example, at low pH, Fe^{3+} reacts with H_3AsO_4 to form insoluble scorodite $(FeAsO_4 \cdot 2H_2O)$ (Fig. 12.15).

Commonly, arsenite exhibits low adsorption potential for oxides or clay edges because the former exhibits high pK_as (9.22 and 13.52). Recall that maximum adsorption of an oxyanion by an oxide takes place at pH closest to its pK_a. Arsenate adsorption is nearly at maximum in the pH range of 3 to 11 because it encompasses the range of its three pK_a values, 2.22, 6.98, and 11.52 (see Chapter 4).

Iron(III)–arsenate compounds are stable under oxidizing conditions (Fig. 12.15). Assuming that redox conditions in the stratum become reductive, iron(III) converts to iron(II) and arsenate becomes arsenite $(AsO_3^{3-}$ or $AsO_2^{-})$. As conditions reduce further, arsenic solubility is regulated by sulfides and pH (arsenic MCL is set at 0.05 mg L^{-1}, arsenosulfides exhibit a solubility near 1 mg L^{-1}). Arsenic redox reactions can be carried

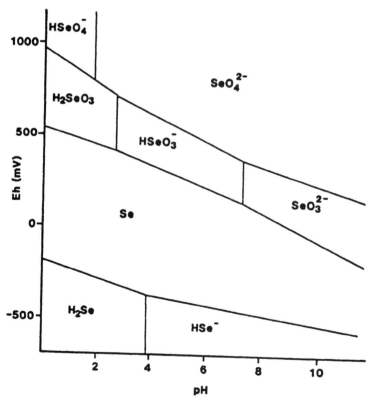

Figure 12.16. Speciation diagram of selenium (selenium = 1 µmol L^{-1}) (from Neal et al., 1987, with permission).

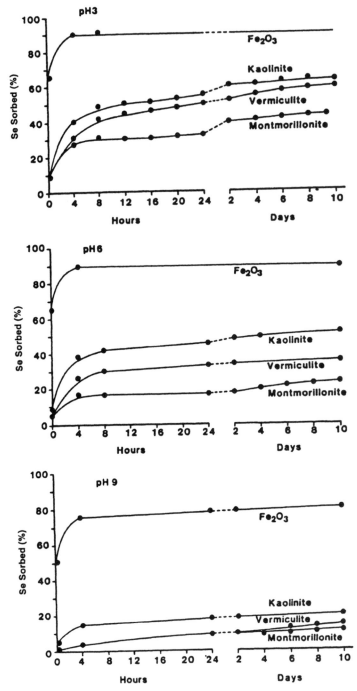

Figure 12.17. Relationship between selenium (SeO$_3$) sorbed and in solution for various pH values (time effects) (from Hamdy and Gissel–Nielsen, 1977, with permission).

out microbiologically as well as inorganically. Arsenic in the arsenite form is toxic and relatively water soluble. Two additional highly toxic arsenic products are arsines and methylarsines, which are produced under highly reduced environments in the form of gases.

Selenium. Selenium (Se) at low concentrations serves as an essential element for animals, but at high concentrations induces toxicity. Its role in plants as an essential element has not been substantiated. In soils, Se may be found as hydrogen selenide (H_2S_2), elemental Se (Se^0), selenite (SeO_3^{2-}), and selenate (SeO_4^{2-}). A stability diagram of selenium under various pH and Eh values is shown in Figure 12.16. These data show that at the lowest redox potential, selenium exists as selenide, and in this state it is present mostly as relatively insoluble metal–selenide. Its transformation in soil–water systems to H_2Se gas is doubtful because of its relatively low pK_a (3.9) and its strong affinity for metal cations. As Eh increases, the most stable species is elemental selenium, followed by selenite [Se(IV)] and selenate [Se(VI)]. Selenite is commonly found in soils and is more toxic than selenate.

Since selenite has two pK_a values, 2.3 and 7.9, as opposed to selenate which exhibits a pK_a of approximately 1.7, it follows that selenite adsorption is more likely to be pH dependent. The data in Figure 12.17 show adsorption of selenite by various minerals. As expected, iron–oxide is more effective in adsorbing selenite than vermiculite or montmorillonite.

12.3 ACID DRAINAGE PREVENTION TECHNOLOGIES

Acid drainage prevention technologies refer to approaches used to limit production of acidic drainages from sulfide-rich geologic strata. These technologies are generally based on the mechanisms discussed in Section 12.2.

12.3.1 Alkaline Materials

Alkaline products such as limestone or strong bases (e.g., sodium hydroxide) are usually applied or pumped into sites (e.g., surface or groundwater) with AD problems (Evangelou, 1995b). Alkalinity derived from limestone and/or strong bases acts as pH buffer, AD neutralizer, and precipitator of heavy metals such as hydroxides or carbonates. Materials such as alkaline fly-ash and topsoil, or their mixtures with lime, significantly reduce iron in the drainage as well as manganese and sulfate (Jackson et al., 1993).

Another approach to controlling AD production by buried geologic strata is through the use of alkaline recharge trenches (Ladwig et al., 1985; Caruccio et al., 1985). Neutralizers such as $CaCO_3$ or Na_2CO_3 can be moved by dissolution with percolating water deep in the strata to sites where AD is produced. However, effectiveness lasts only as long as there is alkaline material in the recharge trenches (Evangelou, 1995b).

Limestone is the most widely used material in treating AD because of its cost advantage over other alkaline materials. However, because of limestone's relatively

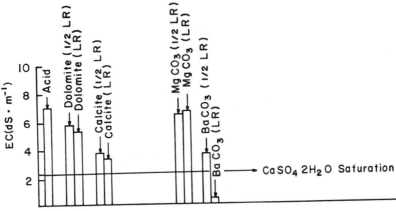

Figure 12.18. Influence on salt production by liming with various alkaline earth carbonates at half the needed quantity and all the needed quantity (from Evangelou, 1997, unpublished data, with permission).

limited solubility at near-neutral pH and its tendency to armor with ferric hydroxide, it is not as effective in controlling AD as one might expect (Wentzler and Aphan, 1972). When waters enriched with Fe^{2+} contact limestone in an oxidizing environment, the limestone is rapidly coated with ferric hydroxide precipitates and the rate of alkalinity production by coated limestone is significantly diminished (Evangelou, 1995b).

The addition of calcium bases to acid–pyritic waste (e.g., mine waste) causes acid neutralization and produces certain metal–salts. For example, when $CaCO_3$ is intro-

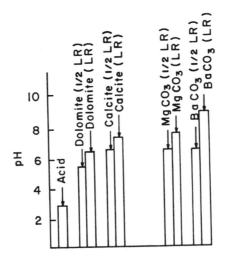

Figure 12.19. Influence on pH by liming with various alkaline earth carbonates at half the needed liming rate and at the full rate (from Evangelou, 1997, unpublished data, with permission).

duced to acid–sulfate systems, gypsum ($CaSO_4 \cdot 2H_2O$) is the by-product. In the case of Ca–Mg bases, for example, dolomite [$CaMg(CO_3)_2$], magnesium sulfate, a relatively high-solubility salt, is the by-product as well as $CaSO_4 \cdot 2H_2O$ (Evangelou, 1985b). Finally, assuming that one used $BaCO_3$ then $BaSO_4$, a relatively highly insoluble mineral would be the by-product. When neutralizing acid–pyritic waste, the goal is to maximize pH and reduce soluble salts. Some examples of the effects on soluble salts and pH resulting from types of bases used are shown in Figures 12.18 and 12.19. These two figures show that the base, in this case $BaCO_3$, which forms the most insoluble sulfate salt ($BaSO_4$), is also the most effective in raising pH. Note however, that $BaCO_3$ is not recommended as an alkaline source owing to the toxic nature of barium.

12.3.2   Phosphate

The potential of Fe^{3+} to act as a pyrite oxidant (see Chapter 6) can be reduced by the addition of phosphate. Phosphate can precipitate Fe^{3+} in an insoluble form as $FePO_4$ or $FePO_4 \cdot 2H_2O$ (strengite) (Baker, 1983; Hood, 1991; Huang and Evangelou, 1994; Evangelou, 1995a, 1996; Evangelou and Huang, 1992). Apatite and phosphate by-products control AD production by inhibiting metal–sulfide oxidation (Spotts and Dollhopf, 1992). However, the materials are effective only temporarily for the control of pyrite oxidation because of their potential for iron armoring (Evangelou, 1995b).

12.3.3   Anoxic Limestone Drains

An anoxic limestone drain (ALD) is an excavation filled with limestone and then covered by plastic and clay to inhibit oxygen penetration and loss of carbon dioxide gas. Under these conditions, limestone produces higher rates of alkalinity by lowering pH caused by the higher partial pressure of carbon dioxide (see Chapter 2). Iron-armoring of limestone is diminished owing to the inhibition of iron oxidation (Evangelou, 1995b).

The water discharged from ALD contains a significant concentration of HCO_3^- and, in some cases, relatively high concentrations of iron and manganese. This strongly buffered alkaline water, when oxygenated, causes metal oxidation, hydrolysis, and precipitation to occur in a settling pond or constructed wetland. Anoxic limestone drains are currently widely used for treating AD (Turner and McCoy, 1990; Nairn et al., 1991 and 1992; Watzlaf and Hedin, 1993; Brodie et al., 1991).

12.3.4   Hydrology

Generally, solutes in micropores tend to move as a pulse and, therefore, the solute concentration in this pulse tends to increase with depth (Evangelou et al., 1982; Evangelou and Phillips, 1984; Evangelou, 1995b). It is known that for any given disturbed land, soil, or geologic waste material, macropore flow gives different leachate chemistries than micropore flow. Based on such results, to improve water quality in geologic waste, one should artificially introduce macropore flow. Field data, however, are needed to verify the effectiveness of this approach (Evangelou, 1995b).

12.3.5 Microencapsulation Technologies

Recently, Evangelou (1995a, 1996a and b) developed two laboratory microencapsulation (coating) methodologies for preventing pyrite oxidation and acid production in coal pyritic waste. The first coating methodology involves leaching coal waste with a solution composed of low but critical concentrations of H_2O_2, KH_2PO_4, and a pH buffer. During leaching, H_2O_2 oxidizes pyrite and produces Fe^{3+} so that iron phosphate precipitates as a coating on pyrite surfaces, inhibiting further oxidation.

A second coating methodology is the use of an iron–oxide–silica coating. Oxidation of pyrite by H_2O_2 in the presence of Si and a pH buffer leads to formation of an iron–oxide silicate coating. These two pyrite-inhibition methodologies are still in the experimental stage.

12.3.6 Organic Waste

A number of organic compounds, from simple aliphatic acids, amino acids, sugars, and alcohols to complex materials like peptone, are inhibitory to *T. ferrooxidans* and *T. thiooxidans* (Evangelou, 1995b, and references therein). Additionally, formation of Fe^{3+}–organic complexes limits oxidation of pyrite by Fe^{3+}, and specific adsorption of organic materials on the pyrite surface prevents either Fe^{3+}, dissolved oxygen, or oxidizing microbes from reaching the pyrite surface (Pichtel and Dick, 1991). Furthermore, organics may combine with Fe–oxide to form stable colloids (Hiltunen et al., 1981). Pichtel and Dick (1991) tested various amendments, including composted sewage sludge, composted paper-mill sludge, water-soluble extract from composted sewage sludge, and pyruvic acid on pyrite oxidation. They found that the pH of the amended coal waste increased and sulfate-S and total soluble Fe decreased (Table 12.3), and concluded that the organic material reduced acid production from pyrite by preventing Fe^{2+} oxidation and removing soluble Fe from the solution.

Organic waste, however, may also promote pyrite oxidation under certain conditions by solubilization of $Fe(OH)_3$ through formation of Fe^{3+}–carboxylate complexes. Such complexes, especially if positively charged, could adsorb onto the pyrite surface and act as electron acceptors in an outer-sphere mode (Luther et al., 1992).

12.3.7 Bactericides

Anionic surfactants (common cleaning detergents) have been used as pyrite oxidation inhibitors by controlling bacterial growth (Erickson and Ladwig, 1985; Kleinmann, 1981; Dugan, 1987). In the presence of such compounds, hydrogen ions cause bacteria cell–membrane deterioration (Evangelou, 1995b).

TABLE 12.3. pH and Concentrations of Sulfate-S and Total Soluble Fe in Spoil Suspensions Incubated with Various Organic Amendments

Amendment	pH Time of Incubation (days)					Sulfate-S[a] Time of Incubation (days)					Total Soluble Fe[a] Time of Incubation (days)				
	0	7	14	21	28	0	7	14	21	28	0	7	14	21	28
Nonamended	5.90	4.96	4.35	4.06	3.80	5.2	16.7	22.9	67.7	81.3	5.5	3.4	21.5	87.7	91.3
Composted sewage sludge	5.90	5.75	5.22	5.15	5.00	5.2	29.1	31.2	26.6	37.5	1.2	0.4	0.5	0.5	0.5
Composted papermill sludge	5.90	8.05	8.07	7.90	7.50	10.4	33.9	49.5	44.3	78.7	1.2	0.7	0.0	0.8	0.5
Water-soluble extract (composted sewage sludge)	5.90	6.30	4.90	4.80	4.70	5.2	16.1	27.1	31.8	66.2	6.1	6.3	9.8	2.7	8.9
Pyruvic acid	5.90	5.75	5.60	5.75	5.70	5.2	10.9	18.2	27.1	45.3	6.4	7.0	11.5	13.6	13.4
LSD$_{0.05}$		—	0.17	0.10	0.27	0.23	0.5	13.1	10.9	38.9	25.1	0.7	1.3	2.5	49.6

Source: Pichtel and Dick, 1991.

[a]Concentrations are given in mmol kg^{-1} spoil.

However, the use of anionic surfactants to control sulfide oxidation is limited because (1) they are very soluble and move with water, (2) they may be adsorbed on the surfaces of other minerals and may not reach the pyrite–bacteria interface (Erickson and Ladwig, 1985; Shellhorn and Rastogi, 1985), and (3) bactericides do not have much effect on acid-metallic drainages produced prior to treatment.

12.3.8 Wetlands

Wetlands have the potential to remove metals from AD by metal adsorption on ferric oxyhydroxides, metal uptake by plant and algae, metal complexation by organic materials, and metal precipitation as oxides, oxyhydroxides, or sulfides. However, only metal precipitation as either oxides or sulfides has long-term metal-removal potential (Evangelou, 1995b).

The data in Table 12.4 provide a summary of chemical and physical characteristics of influent and effluent water from constructed wetlands at Tennessee Valley Authority (TVA) facilities. Water quality generally improved in all cases, and most met the state effluent guidelines for total Fe < 3.0 mg L^{-1}, total Mn < 2.0 mg L^{-1}, pH 6.0–9.0, and nonfilterable residues (NFR) < 35.0 mg L^{-1}. These data also indicate that wetlands are more effective in the removal of Fe^{2+} than Mn^{2+}.

Wetlands, at times, may be a poor environment for the formation of metal oxides and/or oxyhydroxides because of the typically low redox potential (Eh). Optimizing the activity of sulfate-reducing bacteria (e.g., *desulfovibrio*) in the anaerobic zone would be a more effective way of removing metals and sulfates from AD (Kleinmann, 1989). These sulfate-reducing bacteria consume acidity and most of the hydrogen sulfide they produce reacts with heavy metals to create insoluble precipitates. The reactions are shown below:

$$\text{Carbohydrates } (2CH_2O) + SO_4^{2-} \rightarrow H_2S_{gas} + 2HCO_3^- \tag{12.10}$$

$$\text{Metal}(M^{2+}) + H_2S \Leftrightarrow MS_s + 2H^+ \tag{12.11}$$

and

$$2H^+ + 2HCO_3^- \Leftrightarrow CO_{2gas} + H_2O \tag{12.12}$$

Laboratory studies have shown that trace metals such as Co, Cu, Cd, Ni, Pb, and Zn can be removed as sulfides (Staub and Cohen, 1992; Eger, 1992; Hammack and Edenborn, 1991).

12.3.9 Inundation

Underwater disposal of pyritic materials has been used with some success (Ritcey, 1991). Oxygen diffusion is greatly reduced upon inundation because the diffusion

TABLE 12.4. TVA Acid Drainage Wetlands Treatment Summary

Wetlands System	Date Initiated Operation	Area (m²)	Number of Cells	Influent Water Parameters (mg L⁻¹)				Effluent Water Parameters (mg L⁻¹)				Flow (L min⁻¹)		Treatment Area (m²/mg/min)	
				pH	Fe	Mn	NFR[a]	pH	Fe	Mn	NFR[a]	Ave	Max	Fe	Mn
WC 018	6-86	4,800	3	5.6	150.0	6.8		3.9	6.4	6.2		70	1495	0.2	4.2
King 006	10-87	9,300	3	4.2	153.0	4.9	40.0					379	2271	0.2	5.0
Imp 4	11-85	2,000	3	4.9	135.0	24.0	42.0	4.6	3.0	4.0	6.0	42	49	0.4	2.0
950 NE	9-87	2,500	2	6.0	11.0	9.0	19.0	6.6	0.5	0.2	49.0[b]	348	1673	0.7	0.8
RT-2	9-87	7,300	3	5.7	45.2	13.4		6.7	0.8	0.2	2.0	238	681	0.7	2.3
Imp 2	6-86	11,000	5	3.1	40.0	13.0	9.0	3.1	3.4	14.0	0.8[c]	400	2200	0.7	2.1
Imp 3	10-86	1,200	3	6.3	13.0	5.0	28.0	6.8	0.8	1.9	4.7	87	379	1.1	2.8
WC 019	6-86	25,000	3	5.6	17.9	6.9		4.3	3.3	5.9		492	6360	2.8	7.4
950-1&2	1976	3,400	3	5.7	12.0	8.0	20.0	6.5	1.1	1.6	5.4	83	341	3.4	5.1
Imp 1	5-85	5,700	4	6.3	30.0	9.1	57.0[d]	6.5	0.9	2.1	2.8	53	227	3.6	11.8
Col 013	10-87	9,200	5	5.7	0.7	5.3		6.7	0.7	13.5		288	408	45.6	6.0

Source: Modified from Brodie et al., 1988.

[a]Nonfilterable residues.

[b]One effluent sample to date.

[c]One sample, July 1987.

[d]From preconstruction in-stream sample.

455

coefficient of O_2 through the covering water table is only 1/10,000 of that through air. A shallow water cover (0.5–1.0 m in depth) of acid-generating waste (through oxidation) is commonly effective in controlling acid production. However, complete inhibition of mineral oxidation (e.g., metal–sulfides) by flooding may never be possible because of the potential availability of Fe^{3+} (from an external source) as an alternate oxidant. Various studies (Foreman, 1972; Watzlaf, 1992; Pionke et al., 1980) have shown that when pyritic waste was flooded, there was usually a significant, albeit incomplete, reduction of acidity. Additional concerns with underwater disposal include the potential to maintain complete and continuous water saturation. It is known, for example, that biotic oxidation of pyrite is not limited until pore gas oxygen is reduced to less than 1% (Carpenter, 1977; Hammack and Watzlaf, 1990; Evangelou, 1995b).

12.4   NEUTRALIZATION TECHNOLOGIES

Land disturbance and exposure of buried geologic strata to the open environment leads to sulfide oxidation (if present) and, as a consequence, water-quality degradation of runoff. For water-quality-control purposes, sedimentation ponds required by law are used as water treatment basins. Often, the pH of such basin waters is below 6, and the concentration of heavy metals is above acceptable levels. Water treatments include neutralization and removal of heavy metals as precipitates. Similar water-quality problems arise from other industrial sources, including heavy steel industries, electronics, food processing, mineral processing, and waste-disposal leachates. This portion of the chapter deals with some of the chemical agents used for neutralization purposes and some of their limitations.

Several bases increase the pH of acidic water. Bases, depending on solubility and reactivity mechanisms, are separated into three classes: calcium bases, sodium or potassium bases, and gaseous bases such as ammonia. Examples of the three classes of bases are listed in Table 12.5. There are numerous advantages and disadvantages to using either class of bases.

12.4.1   Calcium Bases

Calcium bases are relatively low in solubility compared to sodium or potassium. When added to sulfate-rich AD, low-solubility bases such as calcium carbonate ($CaCO_3$–lime), calcium hydroxide [$Ca(OH)_2$–hydrated lime], and calcium oxide (CaO–burned

TABLE 12.5. Bases for Increasing pH

Calcium	Sodium and Potassium	Ammonia
Ground limestone ($CaCO_3$)	Caustic soda (NaOH)	NH_3 (anhydrous)
Hydrate lime ($Ca(OH)_2$)	Potassium hydroxide (KOH)	NH_4OH (aqueous)
Burned lime (CaO)	Sodium bicarbonate ($NaHCO_3$)	

TABLE 12.6. Properties of Liming Agents

Type	Solubility	pH Maximum	Sludges
Burned lime (CaO)	Low ($40\ meq\ L^{-1}$)	12.2	High
Hydrated lime ($Ca(OH)_2$)	Low ($40\ meq\ L^{-1}$)	12.2	High
$CaCO_3$ (Calcite)	Very low ($1\ meq\ L^{-1}$)	$9.2^a(CO_2 \approx 10^{-6})$	High
$CaMg(CO_3)_2$	Very low	$9.3^a(CO_2 \approx 10^{-6})$	High

[a]Values were estimated by MINTEQA2.

lime) convert to carbon dioxide (gas), water, and calcium sulfate. Under certain conditions, addition of calcium bases will precipitate gypsum ($CaSO_4 \cdot 2H_2O$) (see Chapter 2). In general, if the EC of sulfate-rich AD is greater than 2.2 mmhos cm^{-1} and a calcium base is added, gypsum will probably precipitate. The higher the EC is, the more gypsum will precipitate. Note that gypsum is a sludge, and as such would need proper disposal.

Limestone ($CaCO_3$) addition to sulfate-rich AD will raise pH above 9 (maximum 9.3) only when dissolved salts are relatively low and pCO_2 is below that of the atmosphere (e.g., 10^{-6}). Since AD reacting with $CaCO_3$ generates dissolved salts and microbial action produces CO_2, $CaCO_3$ will not increase the pH above 9; in fact, it may not even go much above 8 (see Chapter 2, Table 12.6). Hydrated lime and burned lime (CaO) respond similarly to $CaCO_3$ with respect to solubility and gypsum precipitation. However, excess $Ca(OH)_2$ or CaO can temporarily raise the pH above 9 (Table 12.6). This high pH will not persist very long because $Ca(OH)_2$ is unstable in CO_2-rich water and will convert spontaneously to $CaCO_3$.

12.4.2 Sodium and Potassium Bases

Sodium and potassium bases are in the high-solubility category of neutralizing agents (Table 12.7). Their major advantage is ease of application. The most common high-

TABLE 12.7. Properties of Liming Agents

Type	Solubility	pH Maximum	Sludges
Sodium hydroxide (NaOH)	Very high	>12	High-suspension solids
Sodium carbonate (Na_2CO_3)	Very high	≈12	High-suspension solids
Sodium bicarbonate ($NaHCO_3$)	Very high	8.3	High-suspension solids
Potassium hydroxide	Very high	>12	High-suspension solids (Na > K)

TABLE 12.8. Additional Neutralizers

Type	Solubility	pH	Sludges
Na_3PO_4	Very high	>12	High-suspension solid
K_3PO_4	Very high	>12	High-suspension solid
Ca-phosphates[a]	Extremely low	≈8	No particular influence

[a]Not to be used in sediment ponds.

solubility bases employed to neutralize AD are sodium hydroxide (NaOH), potassium hydroxide (KOH), sodium bicarbonate ($NaHCO_3$), and potassium bicarbonate ($KHCO_3$). The pollution potential of using NaOH or $NaHCO_3$ is greater than that of KOH or $KHCO_3$. Generally, sodium (Na^+) accumulates in natural water systems while potassium is utilized as a nutrient by plants and aquatic life and/or becomes fixed by 2:1 clay minerals (see Chapters 3 and 4).

Because of the high solubility of these bases, over-application can raise the pH above 9, which is undesirable. Owing to the monovalent nature of the associated cations (Na^+ or K^+), such bases act as colloid dispersants (see Chapter 9). Thus, they increase suspended solids. Additional acid ameliorates are shown in Table 12.8. These phosphate compounds are highly effective in precipitating heavy metals as well as manganese and iron. However, the use of such compounds may increase suspended solids by increasing the surface electrical potential of clays or metal–oxides (see Chapters 3 and 9).

Another high-solubility base is NH_3 or NH_4OH; its chemistry is presented in the next section.

12.4.3 Ammonia

Ammonia, a highly water-soluble gas base (Table 12.9) is being used by various industries to neutralize acidity. Injection of ammonia into water results in a rapid pH increase. The basic reactions of ammonia in water are shown below:

$$NH_{3gas} + H^+ \Leftrightarrow NH_4^+ \tag{12.13}$$

or

$$NH_{3gas} + H_2O \Leftrightarrow NH_4^+ + OH^- \tag{12.14}$$

and

$$M^{2+} + 2OH^- \Leftrightarrow M(OH)_2s \tag{12.15}$$

where M^{2+} denotes any heavy metal. Equation 12.13 shows that when NH_3 is introduced into an acid solution, some of the ammonia reacts directly with the acid,

TABLE 12.9. Properties of Liquid Ammonia at Various Temperatures

Temperature (°F)	Liquid Density		Specific Gravity of Liquid Compared to Water (4°C)
	lb/ft^3	lb/U.S. gal	
−28	42.57	5.69	0.682
−20	42.22	5.64	0.675
−10	41.78	5.59	0.669
0	41.34	5.53	0.663
10	40.89	5.47	0.656
20	40.43	5.41	0.648
30	39.96	5.34	0.641
40	39.49	5.28	0.633
50	39.00	5.21	0.625
60	38.50	5.14	0.617
65	38.25	5.11	0.613
70	38.00	5.08	0.609
75	37.74	5.04	0.605
80	37.48	5.01	0.600
85	37.21	4.97	0.596
90	36.95	4.94	0.592
95	36.67	4.90	0.588
100	36.40	4.87	0.583
110	35.84	4.79	0.573
115	35.55	4.75	0.570
120	35.26	4.71	0.565
125	34.96	4.67	0.560
130	34.66	4.63	0.555
135	34.04	4.55	0.545

consuming acidity, raising pH, and producing the ammonium ion NH_4^+. A fraction of the introduced NH_3 associates itself with several water molecules ($NH_3 \cdot nH_2O$) without becoming NH_4. This hydrated NH_3 is commonly referred to as unionized NH_3, and it is toxic to aquatic life forms (Table 12.10). Reaction 12.14 shows that in addition to Reaction 12.13, NH_{3gas} may react directly with water, producing NH_4^+ and OH^-. The quantity of unionized NH_3 produced is small, but may be significant because of its toxic effects. Finally, Reaction 12.15 shows that OH^- reacts with dissolved metals (e.g., divalent or trivalent) to form insoluble hydroxides. Reactions 12.13–12.15 occur rapidly in water.

The discussion above points out how NH_3 behaves as a strong base. Two reasons that NH_3 is popular for neutralizing AD are that it is inexpensive relative to other bases and it is convenient to use. However, using NH_3 in this manner does create some problems which are outlined below.

TABLE 12.10. Critical NH_3 Concentrations for Various Aquatic Organisms

Organism	Concentration (mg L^{-1})	Effects
Invertebrates	0.530–22.8	Acutely toxic
Fish	0.083–4.60	Acutely toxic
Trout	0.083–1.09	96 hr LC-50
Non-trout	0.140–4.60	96 hr LC-50
Invertebrates	0.304–1.2	Chronic effects
Fish	0.0017–0.612	Chronic effects

Metal–Ammine Complexes. All metal ions in water are surrounded by a shell of water molecules (see Chapter 1):

$$\begin{array}{ccc} H_2O & & H_2O \\ & \diagdown \quad \diagup & \\ & Cd^{2+} & \\ & \diagup \quad \diagdown & \\ H_2O & & H_2O \end{array} \qquad (12.16)$$

These water molecules of metal hydration can be replaced with other molecules. For example, in the case of NH_3 injection in water containing Cd^{2+}, the reaction between Cd^{2+} and NH_3 is

$$[Cd(H_2O)_4]^{2+} + 4NH_3 \Leftrightarrow [Cd(NH_3)_4]^{2+} + 4H_2O \qquad (12.17)$$

This reaction (between Cd^{2+} and NH_3) involves the formation of a coordinated covalent bond where the element nitrogen of NH_3 shares its single unshared electron pair with Cd^{2+} (see Chapter 1). The number of water molecules that could be displaced from the cation's hydration sphere depends on the concentration of NH_3 and the strength by which it associates with the metal ion. In these complex ions, otherwise known as metal–ammine complexes, the metal is called the *central atom* and the associated molecule or ion is called the *ligand*.

The behavior of metal–ammine complexes in water is different from that of the noncomplexed metal ion. For example, if sodium hydroxide is added to a solution containing heavy metals, they would precipitate as metal–hydroxide [$M(OH)_2$]. However, if sodium hydroxide is added to a solution containing heavy metals and excess ammonium (NH_4), no metal precipitation takes place because metal–ammine complexes are soluble in alkaline solutions. Consider the reaction

$$M^{2+} + NH_3 \Leftrightarrow MNH_3^{2+} \qquad (12.18)$$

and

$$K_{st_1} = (MNH_3^{2+})/(M^{2+})(NH_3) \qquad (12.19)$$

Furthermore

$$MNH_3^{2+} + NH_3 \Leftrightarrow M(NH_3)_2^{2+} \tag{12.20}$$

and

$$K_{st_2} = [M(NH_3)_2^{2+}]/[M(NH_3)^{2+}](NH_3) \tag{12.21}$$

The larger the value of the constant is, the more stable the complex (Table 12.11). The metal–ammine formation constants K_{st_1} and K_{st_2} are known as stepwise formation constants. Stepwise formation constants could be used to estimate overall formation constants. For example,

$$M^{2+} + 2NH_3 \Leftrightarrow M(NH_3)_2^{2+} \tag{12.22}$$

and

$$K_{st_1} \cdot K_{st_2} = K_{0_2} = [M(NH_3)_2^{2+}]/(M^{2+})(NH_3)^2 \tag{12.23}$$

Rearranging

TABLE 12.11. Stepwise Formation Constants

Ligand	Cation	$\log K_{eq1}$	$\log K_{eq2}$	$\log K_{eq3}$	$\log K_{eq4}$	$\log K_{eq5}$	$\log K_{eq6}$
CH_3COO^-	Ag^+	0.4	−0.2				
	Cd^{2+}	1.3	1.0	0.1	−0.4		
	Cu^{2+}	2.2	1.1				
	Hg^{2+}	$\log K_{eq1}K_{eq2} = 8.4$					
	Pb^{2+}	2.7	1.5				
NH_3	Ag^+	3.3	3.8				
	Cd^{2+}	3.6	2.1	1.4	0.9	−0.3	−1.7
	Co^{2+}	2.1	1.6	1.0	0.8	0.2	−0.6
	Cu^{2+}	4.3	3.7	3.0	2.3	−0.5	
	Ni^{2+}	2.8	2.2	1.7	1.2	0.8	0.0
	Zn^{2+}	2.4	2.4	2.5	2.1		
SCN^-	Ag^+	$AgSCN(s) + SCN^- \Leftrightarrow Ag(SCN)_2^-$ $\log K_{s2} = -7.2$					
	Cd^{2+}	1.0	0.7	0.6	1.0		
	Co^{2+}	2.3	0.7	−0.7	0.0		
	Cu^{2+}	$CuSCN(s) + SCN^- \Leftrightarrow Cu(SCN)_2^-$ $\log K_{s2} = -3.4$					
	Fe^{3+}	2.1	1.3				
	Hg^{2+}	$\log K_{eq1}K_{s2} = 17.3$	2.7	1.8			
	Ni^{2+}	1.2	0.5	0.2			

Source: Meites, 1963.

$$[M(NH_3)_2^{2+}] = K_{o_2} (M^{2+})(NH_3)^2 \qquad (12.24)$$

where K_{o_2} denotes overall formation constant. Note that $K_{st_1} = K_{o_1}$. In a similar manner, one may proceed to describe all possible stepwise and overall formation constants.

Based on the above, total M dissolved (M_T) would be described by the sum of all metal–ammine complexes in solution:

$$M_T = MNH_3^{2+} + M(NH_3)_2^{2+} + M(NH_3)_3^{2+} + \ldots \qquad (12.25)$$

Taking the inverse of the above equation,

$$1/M_T = 1/\{M^{2+} + MNH_3^{2+} + M(NH_3)_2^{2+} + M(NH_3)_3^{2+} + \ldots\} \qquad (12.26)$$

Replacing the metal–ammine complexes in the denominator as a function of NH_3, M^{2+}, and overall formation constants, as demonstrated in Equation 12.24, gives

$$1/M_T = 1/M^{2+}\{1 + K_{o_1}NH_3 + K_{o_2} (NH_3)^2 + K_{o_3} (NH_3)^3 + \ldots\} \qquad (12.27)$$

For simplicity, Equation 12.27 is written as

$$1/M_T = 1/M^{2+}\left\{1 + \sum_{i,j=1}^{i,j=6} K_{o_i} (NH_3)^j\right\} \qquad (12.28)$$

and rearranging,

$$\alpha_0 = M^{2+}/M_T = 1/\{1 + \sum_{i,j=1}^{i,j=6} K_{o_i} (NH_3)^j\} \qquad (12.29)$$

$$\alpha_1 = M(NH_3)^{2+}/M_T = K_{o_1} NH_3/\{1 + \sum_{i,j=1}^{i,j=6} K_{o_i} (NH_3)^j\} \qquad (12.30)$$

$$\alpha_2 = M(NH_3)^{2+}/M_T = K_{o_2} (NH_3)^2/\{1 + \sum_{i,j=1}^{i,j=6} K_{o_i} (NH_3)^j\} \qquad (12.31)$$

and

$$\alpha_3 = M(NH_3)^{2+}/M_T = K_{o_3} (NH_3)^3/\{1 + \sum_{i,j=1}^{i,j=6} K_{o_i} (NH_3)^j\} \qquad (12.32)$$

In the same manner we may solve for any α_i. Using the equations above, the percent of metal–ammine complexes may be estimated as a function of NH_3. The data in Figures 12.20–12.22 represent stability diagrams for Cu^{2+}–ammine, Zn^{2+}–ammine, and Cd–ammine complexes.

Based on the data presented in Figures 12.20–12.22, it is clear that the potential of NH_3 to solubilize heavy metals depends on metal softness and on the concentration of NH_3. Soft metals (see Chapter 1) are metals that are electron rich with high polarizability (e.g., Cd^{2+}, Ni^{2+}, Hg^{2+}, Co^{2+}, Cu^{2+}, Zn^{2+}, and Ag^+. Hard or intermediate metals such as Fe^{2+}, Mn^{2+}, Al^{3+}, Fe^{3+}, Ca^{2+}, and Mg^{2+} do not solubilize in ammoniated waters because of their inability to form metal–ammine complexes.

To predict the potential concentration of metal–ammine complexes in solution, one needs to understand the relationship between pH and NH_3 formation. Consider the equation

$$N_T = NH_3 + NH_4^+ \qquad (12.33)$$

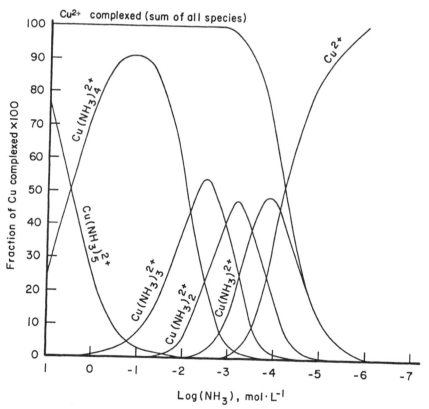

Figure 12.20. Percent copper–ammine complexes as a function of NH_3 (from Evangelou, 1997, unpublished data, with permission).

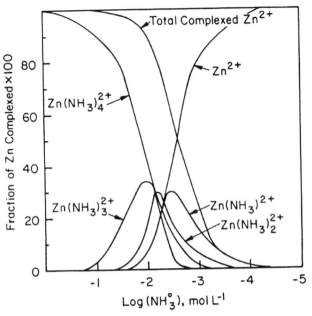

Figure 12.21. Percent zinc–ammine complexes as a function of NH_3 concentration (from Evangelou, 1997, unpublished data, with permission).

Taking the inverse of Equation 12.33,

$$1/N_T = 1/\{NH_3 + NH_4^+\} \tag{12.34}$$

Replacing the metal–ammine complexes in the denominator as a function of NH_4 and H^+,

$$1/N_T = 1/\{NH_3 + K_f(NH_3)(H^+)\} \tag{12.35}$$

where $K_f = (NH_4^+)/(NH_3)(H^+) = 10^{9.2}$ (K_f is the inverse of K_α of NH_4^+, which is $10^{-9.2}$), and rearranging,

$$\alpha_1 = (NH_3)/N_T = 1/\{1 + K_f(H^+)\} \tag{12.36}$$

and

$$\alpha_2 = (NH_4^+)/N_T = K_f(H^+)/\{1 + K_f(H^+)\} \tag{12.37}$$

Using Equations 12.36 and 12.37, one may produce a plot of pH versus percent species (NH_3 and NH_4^+). This plot is shown in Figure 12.23. It shows that for a given amount of NH_4^+ added to a solution, the amount that would convert to NH_3 depends on the equilibrium pH. Generally, 2 pH units below the pK_a of NH_4, or at approximately pH

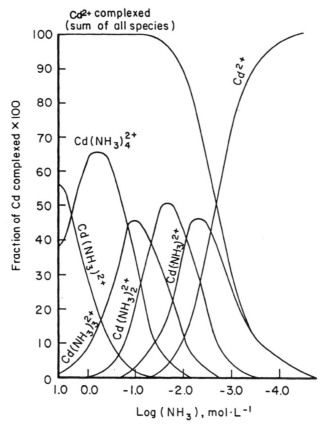

Figure 12.22. Percent cadmium–ammine complexes as a function of NH_3 concentration (from Evangelou, 1997, unpublished data, with permission).

7, all added NH_4^+ would remain in the NH_4^+ form. At the point where pH equals pK_a, NH_4^+ would equal NH_3; at 2 pH units above the pK_a, all of the added NH_4 would convert to NH_3. Therefore, for a given fixed concentration of total N (N_T) ($NH_3 + NH_4$), the higher the pH is, the higher the heavy-metal solubilization potential due to the formation of metal–ammine complexes.

In the case of hard or intermediate metals whose potential for forming metal–ammine complexes is very low, the precipitates forming because of NH_3 addition are those of metal–hydroxides and/or metal–oxyhydroxides.

Surface Adsorption Behavior of Metal–Ammine Complexes. Metal–hydroxides or oxyhydroxides possess variably charged surfaces. Since pH is expected to be around the PZC, addition of NH_3 to the newly formed metal–oxyhydroxide leads to surface adsorption of NH_3 by protonation. This is demonstrated below (see also Chapters 3 and 4):

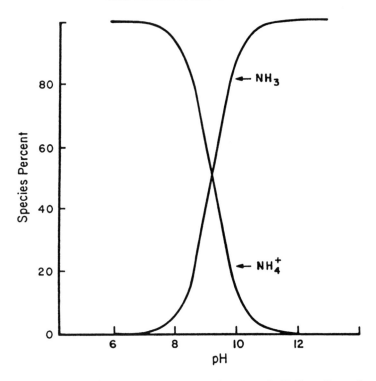

Figure 12.23. Percent NH_4^+ and NH_3 species as a function of pH (from Evangelou, 1997, unpublished data, with permission).

$$\begin{array}{ccc} | & & | \\ -\text{S}-\text{OH} + \text{NH}_3 & \Leftrightarrow & -\text{S}-\text{OHNH}_3 \\ | & & | \\ \text{Metal or} & & \\ \text{clay surface} & & \end{array} \qquad (12.38)$$

For this reason, NH_3-treated waters containing certain transitional metals may not possess much NH_4^+ in solution. Note that the NH_4^+ is part of the so-called metal–hydroxy sludges and care should be given when disposing such sludges (e.g., disposing on land versus underground burial).

When NH_3-treated water contains dissolved hard metals, heavy metals, and clay colloids, a number of reactions may take effect. Many of the metals would be precipitated as metal–hydroxides or oxyhydroxides. Any heavy metals capable of forming metal–ammine complexes would react differently when in the presence of clay colloids or any other charged surfaces. Metal–ammine complexes could form in solution as well as on the colloidal surfaces. The mechanism is shown below:

$$\begin{array}{c|} \\ \hline \end{array}\!\!-S\text{–}O\text{–}M_h + NH_3 \Leftrightarrow \begin{array}{c|} \\ \hline \end{array}\!\!-S\text{–}OM_h\text{:}NH_3 \qquad\qquad (12.39)$$

Metal or
clay surface

where M_h denotes adsorbed heavy metal. The reaction above shows the formation of a metal–ammine complex on a charged surface. At least two conditions must be met in order for such a complex to form. One condition is that the metal (M_h) would have the potential to form a surface outer-sphere complex, and a second condition is that the metal should have the potential to form a relatively strong metal–ammine complex. Experimental data supporting the above are shown in Figures 12.24–12.31.

The data in Figure 12.24 show the potentiometric titrations of a metal-exchange synthetic resin (Dowex 50W-X8) known to produce relatively weak metal–surface complexes (outer-sphere complexes). These data show that when the resin was saturated with Ca^{2+}, no titration plateau (region of Na^+, NH_4^+ or NH_3 adsorption) was exhibited. When the resin was saturated with Cu^{2+}, up to two titration plateaus were exhibited, depending on the type of titrant used. When NaOH was the titrant, one apparent titration plateau was exhibited, whereas when NH_4OH was the titrant, two

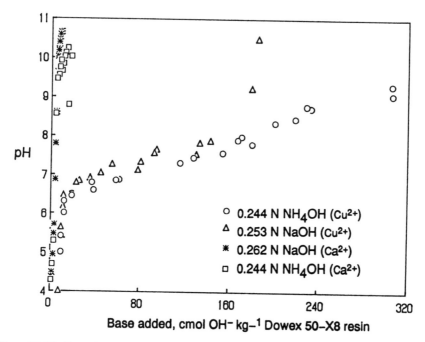

Figure 12.24. Potentiometric titration of Dowex 50W-X8 cation-exchange resin saturated with Cu^{2+} or Ca^{2+} and titrated with NaOH or NH_4OH (from Evangelou, 1997, unpublished data, with permission).

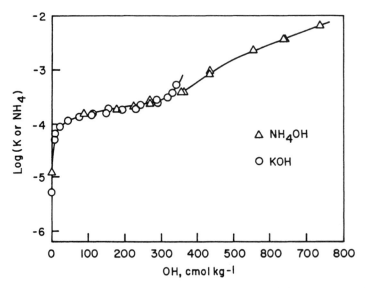

Figure 12.25. Relationship between K or NH$_4$ in solution and surface coverage of Cu^{2+}–Dowex 50W-X8 cation-exchange resin (from Evangelou, 1997, unpublished data, with permission).

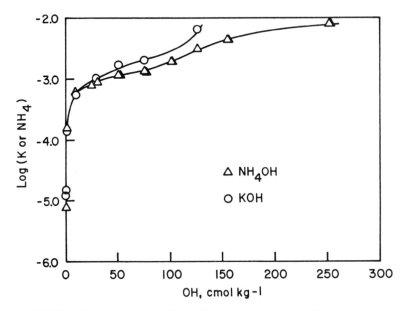

Figure 12.26. Relationship between K or NH$_4$ in solution and with surface coverage of Cu^{2+}–bentonite (from Evangelou, 1997, unpublished data, with permission).

apparent titration plateaus were exhibited. The first titration plateau represented hydroxylation of the adsorbed metal and generation of surface-negative charge by dissociation of the resin-surface Cu–OH complex. Adsorption of Na^+ or NH_4^+ took effect by displacing the H^+, whereas the second plateau (higher pH range) took effect by formation of a Cu–ammine complex on the surface of the resin. Further support for this interpretation is given in Figures 12.25 and 12.26.

The data in Figures 12.25 and 12.26 show that NH_4^+ adsorption by Cu^{2+}–resin and Cu^{2+}–bentonite (a clay mineral known to form metal outer-sphere complexes) exhibits two major adsorption plateaus, whereas K^+ adsorption exhibits a single titration plateau. The data in Figures 12.27–12.29 also show that adsorption of metals by surfaces in the presence of ammoniated solutions was greater than adsorption of metals at similar pH values but in the absence of NH_3. These data are also exhibited in the form of Freudlich plots (Fig. 12.30 and 12.31) and the adjustable parameters are summarized in Table 12.12.

The data above were presented to demonstrate that in treating acid heavy-metal-rich colloidal suspensions with NH_3, the latter introduces some undesirable complexities owing to the potential of the heavy metals to form metal–ammine complexes in solution and/or the exchange complex. Disposal of such waters and/or such sludges requires prior knowledge. For example, the formation of metal–ammine complexes in solution would not permit precipitation of heavy metals in the treated water. On the other hand, formation of precipitate–NH_4 complexes and/or metal–ammine complexes on the surface of colloidal particles causes disposal problems for such sludges. Nitrification of this NH_4 would release NO_3 and heavy metals owing to acidification.

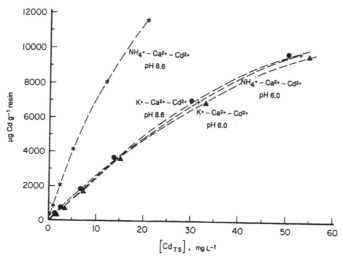

Figure 12.27. Influence of pH and background electrolyte of cadmium adsorption by DowexZ 50W-X8 cation-exchange resin (background electrolyte included 20 mmol L^{-1} KCl or NH_4Cl plus 5 $mmol_c$ L^{-1} $CaCl_2$). The exchanger was first repeatedly washed with the background electrolyte (from Evangelou, 1997, unpublished data).

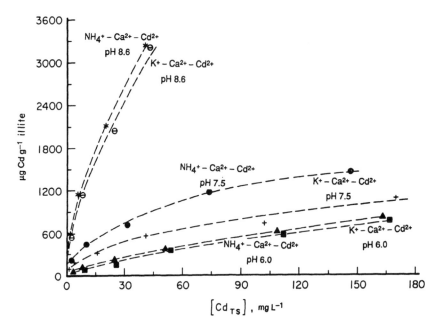

Figure 12.28. Influence of pH and background electrolyte on cadmium adsorption by illite (background electrolyte included 20 mmol L^{-1} KCl or NH$_4$Cl plus 5 mmol$_c$ L^{-1} CaCl$_2$). The exchanger was first repeatedly washed with the background electrolyte (from Evangelou, 1997, unpublished data).

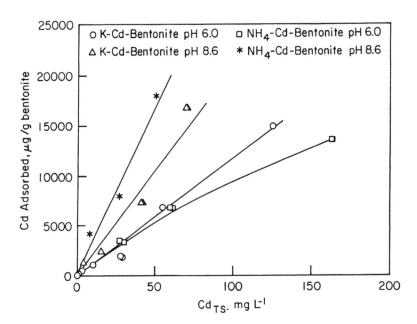

Figure 12.29. Influence of pH and background electrolyte on cadmium adsorption by bentonite (background electrolyte included 20 mmol L^{-1} KCl or NH$_4$Cl plus 5 mmol$_c$ L^{-1} CaCl$_2$). The exchanger was first repeatedly washed with the background electrolyte (from Evangelou, 1997, unpublished data).

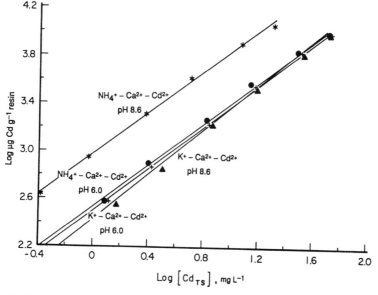

Figure 12.30. Freundlich plots of Cd adsorption by Dowex 50W-X8 cation-exchange resin with a background electrolyte of 20 mmol L^{-1} KCl or NH_4Cl plus 5 $mmol_c$ L^{-1} $CaCl_2$. The exchanger was first repeatedly washed with background electrolyte (from Evangelou, 1997, unpublished data).

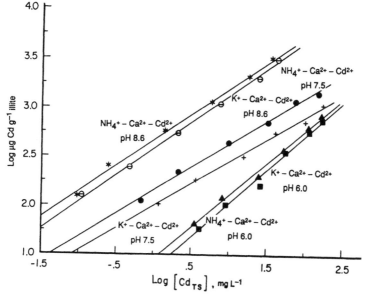

Figure 12.31. Freundlich plots of Cd adsorption by illite with a background electrolyte of 20 mmol L^{-1} KCl or NH_4Cl plus 5 $mmol_c$ L^{-1} $CaCl_2$. The exchanger was first repeatedly washed with background electrolyte (from Evangelou, 1997, unpublished data).

TABLE 12.12. Freundlich Cadmium Adsorption Constants (k_f) for Various Exchangers

pH	Cation	Resin k_f	$1/n$	r
		Dowex-50W-X8		
6.1	K	3.16×10^2	0.87	.994
6.1	NH$_4$	3.16×10^2	0.87	.998
8.7	K	3.16×10^2	0.93	.995
8.7	NH$_4$	10.0×10^2	0.86	.998
		Illite		
6.0	K	0.31×10^2	0.66	.995
6.0	NH$_4$	0.19×10^2	0.67	.995
8.5	K	3.98×10^2	0.51	.993
8.5	NH$_4$	5.01×10^2	0.52	.996
		Bentonite		
6.0	K	1.32×10^2	0.89	.973
6.0	NH$_4$	1.44×10^2	0.93	.979
8.6	K	6.28×10^2	0.58	.985
8.6	NH$_4$	6.76×10^2	0.75	.973

Nitrification. The physicochemical reactions of NH$_3$ described in the previous section occur rapidly, within seconds to minutes. These are followed by a slower biological transformation of ammonium (NH$_4^+$) to nitrate (NO$_3^-$). This reaction is nitrification. Nitrification is carried out by unique, specialized bacteria called, for example, *Nitrosomonas, Nitrospora,* and *Nitrobacter.* The nitrifying microbes use the oxidation of ammonium (NH$_4^+$) as a source of energy. The carbon dioxide (CO$_2$) in the atmosphere supplies the carbon required to build their cells. Thus, unlike almost all other microbes, they do not need to decompose organic matter to grow. In this way, they resemble the bacteria that grow by oxidizing the iron and sulfur in pyrite, and so create many of the acidity problems in reclamation. Nitrifying bacteria are particularly sensitive to pH, being most active in neutral or slightly alkaline conditions. They are severely inhibited by pH below about 5.5, and the reaction hardly goes on at all below about 4.5. Addition of ammonium to water or soil can tremendously increase the potential nitrification rate because this causes the nitrifying population to grow.

The nitrification reaction can be written as

$$NH_4^+ + 2O_2 \rightarrow NO_3^{2-} + H_2O + 2H^+ \qquad (12.40)$$

Comparing Reaction 12.13 with Reaction 12.40, it is apparent that twice as much acid is generated by nitrification as is consumed in the initial reaction of ammonia with water. Therefore, the final result is the production, not the consumption, of one equivalent of acidity for every equivalent of ammonia used. The only reason that ammonia can increase the pH in a sediment pond is the much greater rate of the neutralization reaction (Eq. 12.13), compared to the nitrification reaction (Eq. 12.40).

The nitrate produced from the added ammonia may also have adverse environmental effects in some situations. Some nitrate can be found in all natural waters, so low nitrate concentrations certainly do not constitute a problem. A large increase in nitrate concentration may have two results. The first is eutrophication; in some waters, addition of nitrate may enhance the growth of algae and other aquatic organisms sufficiently to deplete atmospheric oxygen and foul the water. Second, consumption of water with very high nitrate concentrations may occasionally pose a health hazard. Cattle are generally more susceptible to nitrate toxicity than humans, and infants are more at risk than adults. Federal standards for nitrate in drinking water are 10 ppm (mg N L^{-1}). If 10 meq of acidity per liter are neutralized with ammonia, this will eventually produce 10 meq of nitrate. This is equal to 140 ppm, or 14 times greater than the standard.

Denitrification. Nitrogen is returned to its atmospheric form by the action of denitrifying bacteria such as *Pseudomonas thiobacillus* and *Micrococcus denitrificans*. The process is known as denitrification and the reaction is as follows:

$$H^+ + NO_3^- \rightarrow 1/2N_{2gas} + 5/4O_{2gas} + 1/2H_2O \qquad (12.41)$$

This reaction, however, is not spontaneous (ΔG = 7870 cal mol^{-1}). Denitrification however, becomes spontaneous when in the absence of O_2, bacteria are using NO_3 instead of O_2. The reaction is as follows:

$$Sugar + 4.8H^+ + 4.8NO_3^- \rightarrow 2.4N_{2gas} + 6CO_{2gas} + 2.4H_2O \qquad (12.42)$$

Reaction 12.42 shows that for each mole of NO_3 consumed, 1 mol of H^+ is also consumed. The net effect with respect to H^+ of Reactions 12.13, 12.40, and 12.42 is zero. In other words, as long as the applied nitrogen in the form of NH_3 undergoes its complete cycle (NH_3–NO_3–N_2), there is no pH influence in the aquatic system. If, for whatever reason, this cycle is interrupted, the net effect would be more acid and lower pH.

PROBLEMS AND QUESTIONS

1. Describe three mechanisms by which heavy metals may be precipitated.

2. Calculate the solubility of Fe(OH)$_3$ (K_{sp} = 2.0 × 10^{-43}) at pH 2, 3, and 4. What can you conclude about the relationship between Fe(OH)$_3$ and pH?

3. Describe a setup by which one may remove two different heavy metals having a minimum solubility at two different pH values. What two metals exhibit different pH at their minimum solubility?

4. Demonstrate the potential minimum concentration of Cd in milligrams per liter that one may attain in water by treating it with NaOH [$Cd(OH)_2$ $K_{sp} = 6.0 \times 10^{-15}$ and K_{eq} for $Cd^{2+} + OH^- \Leftrightarrow Cd(OH)_2^0$ equals $10^{7.7}$].

5. Explain the impact of CO_2 on the release of metals from freshly disposed metal–hydroxides in the open environment.

6. Name two approaches by which one may remove Mn^{2+} from solution. Which approach is most effective? Why? Why does the U.S. government regulate Mn levels in mining drainages?

7. Describe a potential electrochemical interaction between MnO_2 and Fe^{2+}. Explain why the interaction is environmentally important.

8. Using the gravimetric formula, calculate the amount of sodium hypochlorite one needs to oxidize all the Mn^{2+} in one acre–foot pond containing 100 mg L^{-1} Mn.

9. Discuss some of the advantages and some of the disadvantages of using hydrogen sulfide to precipitate heavy metals in natural waters.

10. Using Equation 12.40, calculate the pH of an unbuffered natural body of water containing 100 mg L^{-1} NH_4 when the latter is converted to NO_3.

11. You have been asked to design a cation-exchange resin column to remove 10 meq L^{-1} Cu^{2+} from a given water. Calculate the size of the column, assuming that the CEC of the resin is 250 meq/100 g, its density 1.5 g cm^{-3}, and the amount of water to be treated is 10,000 L per month per single column.

12. Assuming that the Cu^{2+} in problem 11 is a copper–ammine [$Cu - (NH_3)_2^{2+}$] and is adsorbed as such by the resin, calculate the total amount of NH_4 in milliequivalents per column per month or killigrams of N per column per month to be removed.

13. Three metals, Zn^{2+}, Cu^{2+}, and Cd^{2+}, were discussed in some detail with respect to their potential to form metal–ammine complexes. Which metal would dissolve the most in an ammoniated solution? If this represented a real problem, how would you solve it?

14. A lake receiving sulfate-rich waters from metal-ore processing plants exhibits a pH near 2. After the industry closes down, the pH slowly rises and within several years reaches 7. Explain the reaction mechanism(s) responsible for the pH rise to 7.

15. Using the denitrification data given below, graphically estimate the first-order rate constant k and the time needed for half of the NO_3 to become N_2. Derive the half-life algebraically.

Time (days)	NO_3-N (mg L^{-1})
0	240
5	167
10	125
25	50
40	20

PART VII
Soil and Water: Quality and Treatment Technologies

13 Water Quality

13.1 INTRODUCTION

Living organisms, including humans, depend on water because a significant portion of their bodies is water. Water is the medium in which the reactions necessary for living functions take place. It is an active participant in many biological reactions; water acts as the carrier of nutrients in the bodies of living organisms and it serves as the temperature regulator. One may find water in nature in a relatively pure state in various forms: (1) vapor, as in fog and clouds; (2) liquid, as in rain; and (3) solid, as in ice. However, in the current state of the world's industrialization, it has become increasingly difficult to find water in its pure state (Table 13.1).

Anthropogenic and/or natural chemical substances come in contact with water in its vapor or liquid phases, and because of the potential of these substances to dissolve, water loses its purity; at times it becomes contaminated. These dissolved substances

TABLE 13.1. Industrial Sources of Metal Contamination

Industry	Common Environmental Metal Contaminants
Battery recycling	Cd, Cu, Ni, Pb, Zn
Chemical/pharmaceutical	As, Cd, Cr, Cu, Hg, Pb
Fossil fuels/energy	As, Be, B, Cd, Hg, Ni, Pb, Se, V
Metal finishing/plating electronics	As, Cd, Cr, Cu, Fe, Ni, Pb, Zn
Mining/ore processing/smelting	Ag, As, Cd, Cr, Cu, Fe, Hg, Mn, Pb, Zn
Oil and solvent recycling	As, Cr, Pb, Zn
Paint	Cd, Co, Cr, Hg, Pb
Wood treatment	As, Cr, Cu

Source: Table reproduced from Zuiderveen, 1994. Adapted from Manahan, 1991 and 1992; Wilmoth et al., 1991; and Chang and Cockerham, 1994.

476

include inorganics (e.g., metals, metalloids, and stable and/or radioactive isotopes), and organics (e.g., pesticides, other industrial organics such as by-products of the plastics industry, and solvents). Included in these organic–inorganic substances are decaying animal and vegetable matter, sediments, road salts, algae, bacteria, and viruses.

The impact of the various substances listed above may result in low-quality water with bad taste, odor, and turbidity or toxicity from the high concentration of heavy metals, chlorinated hydrocarbons, and/or pathogenic bacteria and viruses. Water purity, however, is not a prerequisite to good water quality with respect to human consumption or agricultural and industrial uses. When water contacts soil, the latter contributes dissolved minerals (e.g., Ca, Mg, K) which may increase the potential water quality for biological uses because these minerals serve as nutrients.

Global industrialization has made it difficult to maintain good natural water quality, and substances other than those with nutrient value are introduced to natural water by human activities. Therefore, it has become necessary for government to intervene by introducing appropriate legislation. For example, the federal Safe Drinking Water Act (SDWA) (P. L. 93-523), which was signed into law in 1974 and amended in 1986, was designed to protect drinking water from contamination by human activities. The federal Safe Drinking Water Act paved the way for setting drinking water quality standards through the introduction of maximum contaminant levels (MCLs). An MCL is defined as the highest amount of a specific contaminant allowed in the water delivered to any customer of a public water system (Shelton, 1989). The MCLs are

TABLE 13.2. Characteristics of Metals

Metal	Valence No.	Cation Classification[a]	Relative Toxicity[b]	Reference for Relative Toxicity
Aluminum	3+	A	M	Biesenger and Christensen, 1972
Cadmium	2+	B	E	Biesenger and Christensen, 1972
Chromium	3+	T	M	Biesenger and Christensen, 1972
Chromium	6+	GA	H	U.S. EPA, 1985b
Copper	2+	T	H	Biesenger and Christensen, 1972
Iron	3+	T	N	Biesenger and Christensen, 1972; Birge et al. 1985
Lead	2+	B	H	Biesenger and Christensen, 1972
Nickel	2+	T	H	Biesenger and Christensen, 1972
Silver	1+	B	E-H	U.S. EPA, 1980d
Zinc	2+	B	H	Biesenger and Christensen, 1972

Source: Table reproduced from Zuiderveen, 1994.

[a]From Morgan and Stumm, 1991. A = type A metal cation, nonheavy metals; T = transition-metal cation (various oxidation states); B = type B metal cation, heavy metals; GA = generally anionic (U.S. EPA, 1980a).

[b]Based on chronic values (CV) for *Daphnia magna* in waters with E= extremely toxic (CV < 5 mg L^{-1}); H = highly toxic (5 < CV < 100 mg L^{-1}); M = moderately toxic (100 < CV < 1000 mg L^{-1}); N = relatively nontoxic (CV > 1000 mg L^{-1}).

TABLE 13.3. U.S. Environmental Protection Agency Classification of Compounds Inducing Carcinogenicity

Group A—Human carcinogen (sufficient evidence from epidemiological studies)

Group B—Probable human carcinogen

Group B_1—At least limited evidence of carcinogenicity in humans

Group B_2—Usually a combination of sufficient evidence in animals and inadequate data in humans

Group C—Possible human carcinogen (limited evidence in carcinogenicity in the absence of human data

Group D—Not classifiable (inadequate human and animal evidence of carcinogenicity)

Group E—Evidence of noncarcinogenicity for humans (no evidence of carcinogenicity in at least two adequate animal tests in different species or in both epidemiological and animal studies)

Source: From Shelton, 1989.

expressed in milligrams per liter (mg L^{-1} or parts per million, ppm) or micrograms per liter (μg L^{-1} or parts per billion, ppb). A brief summary of metals and their toxicity potential is presented in Table 13.2.

Commonly, contaminants affect human health in two ways; by producing acute or chronic health effects. Acute health effects include immediate body reaction to the contaminant, inducing vomiting, nausea, lung irritation, skin rash, dizziness, and in extreme situations death. Chronic health effects may include cancer, birth defects, organ damage, nervous system disorders, and potential damage to the immune system. The chronic health effects are most commonly associated with long-term consumption of contaminated water. An example of the U.S. EPA contaminant classification with respect to the most serious human health effect, carcinogenicity, is given in Table 13.3. Note, however, that any classification of the so-called water pollutants with respect to their carcinogenicity is expected to be controversial because, in the view of many critics, approaches taken to establish pollutant carcinogenicity are less than perfect.

Two types of water quality standards, primary and secondary, are presently enforced by the federal government. Primary standards involve contaminants believed to induce acute and chronic health effects (Table 13.4). The primary standards (MCLs) for drinking water contaminants are based on the following factors: (1) the contaminant causes adverse health effects; (2) instruments are available to detect the particular contaminant in the drinking water; and (3) the contaminant is known to occur in the drinking water (Shelton, 1989). Secondary standards deal with what is believed to be the aesthetic quality of drinking water (e.g., taste, odor, color, and appearance) (Table 13.4). Detailed drinking water standards are given in Table 13.5a and 13.5b. In addition, the U.S government regulates industrial effluents (e.g., mining, water treatment plants, and nuclear facilities, see Table 13.6) and provides water quality guidelines, but does not regulate agricultural irrigation water quality (Table 13.7).

This chapter presents a brief summary on water quality with respect to the various pollutants one may encounter in the natural environment. The purpose is to make the

TABLE 13.4. Water Drinking Standards as Described by a Kentucky Water Company (Kentucky American)

Primary Standards[a]		Secondary Standards[a]	
Heavy Metals[b]		Heavy Metals	
Arsenic	0.05	Copper	1.0
Barium	1.0	Iron	0.3
Cadmium	0.010	Manganese	0.05
Chromium	0.05	Zinc	5.0
Fluoride	4.0		
Lead	0.05	Other	
Mercury	0.002	Total dissolved solids	500.0
Nitrate	10.0	Chloride	250.0
Selenium	0.01	Color[g]	15.0 PCU
Silver	0.05	Odor[h]	3.0 TON
Clarity		Sulfate	250.0
Turbidity[c]	1.0 N.T.U.	Hydrogen sulfide	0.05
Microbiological		Phenols	0.001
Coliform bacteria[d]	4.0 in 5% samples or 1.0 in total samples		
Organics			
Endrin	0.0002		
Lindane	0.004		
Methoxychlor	0.10		
2, 4, 5-TP(Silvex)	0.01		
Toxaphene	0.005		
2, 4-D	0.1		
THM[e]	0.10		
Radionuclides[f]			
Gross alphas	15.0 pCi L^{-1}		
Gross beta	50.0 pCi L^{-1}		
Tritium	20,000.0 pCi L^{-1}		
Strontium-90	8.0 pCi L^{-1}		

[a]All numbers are expressed in milligrams per liter unless otherwise indicated.

[b]The analyses of heavy metals are done by standard atomic absorption spectrophotometry.

[c]Turbidity is the measure of suspended material in the water. Drinking water turbidity is measured by nephelometric turbidity units (NTU). The standard is based on a monthly average, not to exceed 1.0 mg L^{-1}. Also, the level must not exceed 5.0 mg L^{-1} on two consecutive days.

[d]The coliform bacteria standard is expressed as the number of coliform colonies in 100 mL of water. The level cannot exceed 4 colonies per 100 mL in 5% of samples, or 1 coliform colony per 100 mL of the average of all monthy samples. The standard is based on a monthly reporting period.

[e]The trihalomethanes (THM) standard is a measurement of total trihalomethanes for a 4-quarter running annual average, based on average values throughout the distribution system.

[f]Radionuclides are measured by the picocurie, the official unit of one million-millionth of a curie per liter of water.

[g]Color is measured in color units and based on platinum cobalt units (PCU) for comparative purposes.

[h]Odor is measured by the threshold odor number (TON), based on the number of dilutions with odor-free water necessary to cause the odor to be nondiscernible in warmed water under prescribed conditions.

TABLE 13.5a. Proposed Maximum Contaminant Levels for Public Drinking Water[a]

Chemical	Maximum Contaminant Levels
Inorganic	
Arsenic[b]	0.03 mg L^{-1}
Asbestos	7 million fibers L^{-1} (longer than 10 μm)
Barium[b]	5 mg L^{-1}
Cadmium[b]	0.005 mg L^{-1}
Chromium[b]	0.1 mg L^{-1}
Copper[c]	1.3 mg L^{-1}
Lead[b]	0.005 mg L^{-1}
Mercury[b]	0.002 mg L^{-1}
Nitrate[b]	10.0 mg L^{-1} (as N)
Nitrite	1.0 mg L^{-1} (as N)
Nitrate + nitrite (total)	10 mg L^{-1} (as N)
Selenium[b]	0.05 mg L^{-1}
Organic	
Alachlor	2 μg L^{-1}
Aldicarb	10 μg L^{-1}
Aldicarb sulfoxide	10 μg L^{-1}
Aldicarb sulfone	40 μg L^{-1}
Atrazine	3 μg L^{-1}
Carbofuran	40 μg L^{-1}
Chlordane[b]	2 μg L^{-1}
Chlorobenzene[b]	100 μg L^{-1}
Dibromochloropropane	0.2 μg L^{-1}
o-Dichlorobenzene[b]	600 μg L^{-1}
cis-1,2-Dichloroethylene[b]	70 μg L^{-1}
trans-1,2-Dichloroethylene[b]	70 μg L^{-1}
2,4-D (2,4-dichlorophenoxyacetic acid)[b]	70 μg L^{-1}
1,2-Dichloropropane	5 μg L^{-1}
Ethylbenzene	700 μg L^{-1}
Ethylene dibromide	0.05 μg L^{-1}
Heptachlor	0.4 μg L^{-1}
Heptachlor expoxide	0.2 μg L^{-1}
Lindane[b]	0.2 μg L^{-1}
Methoxychlor[b]	400 μg L^{-1}
Pentachlorophenol	200 μg L^{-1}
Polychlorinated biphenyls	0.5 μg L^{-1}
Styrene[d]	5 μg L^{-1}
Tetrachloroethylene[b]	5 μg L^{-1}
Toluene	2000 μg L^{-1}
Toxaphene[b]	5 μg L^{-1}
2,4,5-TP (silvex)[b]	50 μg L^{-1}
Xylenes[c]	10,000 μg L^{-1}
Microbiological	
Total coliforms	0/100 mL
Turbidity	1 TU

Source: From Shelton, 1989.
[a]These MCLs are pending final adoption by the EPA.
[b]An MCL currently exists for this contaminant.
[c]An SMCL currently exists for this contaminant (SMCL = secondary MCL).
[d]Pending final carcinogenicity classification by EPA.

TABLE 13.5b. Maximum Contaminant Levels for Public Drinking Water in New Jersey

Inorganic Chemicals	MCL (in ppm or mg L^{-1})	Organic Chemicals	MCL (in ppb or µg L^{-1})
Arsenic	0.05	1,1,1-Trichloroethane	26
Barium	1	Trichloroethylene	1
Cadmium	0.010	Vinyl Chloride	2
Chromium	0.05	Xylene(s)	44
Fluoride	4	**Pesticides**	
Lead	0.05		
Mercury	0.002	Endrin	0.2
Nitrate	10	Lindane	4
Selenium	0.01	Methoxychlor	100
Silver	0.05	Toxaphene	5
		2,4-D (2,4-Dichlorophe-noxyacetic acid)	100
Organic Chemicals	**MCL (in ppb or µg L^{-1})**	2,4,5-TP (Silvex)	10
		Trihalomethanes (Total)	100
Benzene	1	**Microbiological Contaminants**	
Carbon tetrachloride	2		
Chlordane	0.5	Coliform bacteria Membrane filter technique	One coliform bacteria/100 ml
Chlorobenzene	4		
meta-Dichlorobenzene	600		
ortho-Dichlorobenzene	600	Multiple tube technique	No more than 10% portions (sample tubes) positive
para-Dichlorobenzene	75		
1,2-Dichlorobenzene	2		
1,1-Dichloroethane	2	**Turbidity**	
1,2-Dichloroethylene (cis and trans)	10	Turbidity	1 Turbidity Unit
Methylene chloride	2	**Radiological Contaminants**	
Polychlorinated biphenyis	0.5		
Tetrachloroethylene	1	Gross alpha activity	15 picoCuries L^{-1}
Trichlorobenzene(s)	8	Radium 226/228	5 picoCuries L^{-1}

Source: From Shelton, 1989.

TABLE 13.6. Summary of Current EPA Effluent Limitations for the Coal-Mining Point Source Category

Type of Operation and Category	Total Fe (mg L^{-1})	Total Mn (mg L^{-1})	Total Suspended Solids TSS (mg L^{-1})	Settleable Solids SS (ml L^{-1})	pH
Coal preparation plant and associated areas					
NSPS[a]	6.0/3.0	4.0/2.0	70/35	—	6–9
BAT	7.0/3.5[b]	4.0/2.0[c]	70/35	—	6–9
Active Mining					
Surface disturbance and underground workings					
NSPS	6.0/3.0	4.0/2.0[d]	70/35	—	6–9
BAT	7.0/3.5	4.0/2.0[d]	70/35	—	6–9
Post-Mining (Reclamation)					
Surface disturbance					
NSPS, BAT	—	—	—	0.5	6–9
Underground workings					
NSPS	6.0/3.0	4.0/2.0[d]	70/35	—	6–9
BCT	7.0/3.5	4.0/2.0[d]	70/35	—	6–9
Applicable Time Period					
Discharge resulting from precipitation > 10 yr, 24-hr storm					
All operations except underground workings (NSPS, BAT)	—	—	—	0.5	6–9

[a]NSPS: new source performance standards; BPT: best practicable control technology currently available; BAT: best available technology economically achievable; BCT: best conventional pollutant control technology. Effluent Limitations are not contained in this table.
[b]7.0—Maximum concentration for one day; 3.5—average concentration for 30 consecutive days.
[c]Manganese applicable only if the pH is normally less than 6.0 in untreated discharge.
[d]Manganese applicable only if the pH is normally less than 6.0 or iron is normally equal to or greater than 10 mg L^{-1}.

TABLE 13.7. Guidelines for Interpretation of Water Quality for Irrigation[a]

Irrigation Problem	No Problem	Increasing Problem	Severe Problem
		Degree of Problem	
Salinity (affects water availability to crop)			
EC_w (mmhos cm^{-1})	< 0.75	0.75–3.0	> 3.00
Permeability (affects infiltration rate of water into soil)			
EC_w (mmhos cm^{-1})	> 0.5	0.5–0.2	< 0.2
Adj.SAR	< 6	6.9	> 9
Specific toxicity (affects only sensitive crops)			
Sodium (Adj.SAR)	< 3	3.9	> 9
Chloride (meq L^{-1})	< 4	4.10	> 10
Boron (mg L^{-1})	< 0.5	0.5	2.0–10.0
Miscellaneous effects (affects only susceptible crops)			
$NO_3 \cdot N$ (or) $NH_4 \cdot N$ (mg L^{-1})	< 5	5.30	> 30
HCO_3 (meq L^{-1}) overhead sprinkling	< 1.5	1.5–8.5	> 8.5
pH		Normal range 6.5–8.4	

[a]<: less than; >: more than; EC_w: electrical conductivity, a measure of salinity (see Chapter 2); Adj.SAR: adjusted sodium adsorption ratio (see Chapter 14); $NO_3 \cdot N$: nitrogen in the water in the form of nitrate; $NH_4 \cdot N$: nitrogen in the water in the form of ammonia.

student (reader) aware of the types and concentrations of pollutants found in water environments and their possible impact on the various uses of such water.

13.2 AQUATIC CONTAMINANTS

Water contaminants include a large number of chemicals ranging from aromatic hydrocarbons, organic solvents, and pesticides to metals. Metals occur naturally and are commonly found in areas where industrial and municipal effluents are being discharged. Metals discharged into freshwater environments can have adverse effects on bioecosystems.

The first method for detecting water pollution is to carry out comprehensive chemical monitoring and look for concentrations that exceed water quality criteria. However, since over 1500 substances have been listed as freshwater pollutants (Mason, 1981), extensive monitoring can be costly. Additionally, acute and chronic toxicity data are limited for commonly tested organisms and are almost nonexistent for many native species. However, chemical monitoring alone may not detect water pollution because it is assumed that pollutants do not act additively, synergistically, or antagonistically toward each other.

A second method for detecting water pollution is to conduct toxicity identification evaluations (TIEs). According to Zuiderveen (1994), the U.S. EPA has compiled manuals to aid in determining the cause of acute toxicity using TIEs. Three phases are involved in this process. Phase I is used to determine the general class(es) of the toxicants (U.S. EPA, 1991a). Phase II involves identification of the toxicant. Phase III confirms toxicant identification by demonstrating its potential toxicological effects.

13.3 TOXICITY INDICATORS

Ceriodaphnia, commonly known as "water fleas," is used as a test organism for studying toxicity in freshwater aquatic systems. The U.S. EPA recommends using *Ceriodaphnia* in all three phases of acute TIEs (Zuiderveen, 1994).

13.4 METALS

Metals in natural waters differ with respect to their potential to affect water quality (Table 13.2). Metal aquatic toxicity is based on (1) valence, (2) relative hardness or softness, referred to as Type A, transitional, and Type B metal cations, and (3) differences in relative toxicity (i.e., extremely toxic to relatively nontoxic). Type A metal cations, with electron configurations resembling inert gases, are referred to as hard metals (see Chapter 1) and have low reactivity owing to their relatively low electron-acceptance potential. Anions such as fluoride or oxyanions (ligands having oxygen as a donor, e.g., OH^-, CO_3^{2-}) react with these cations to form insoluble precipitates. Less-reactive molecules, such as halides (e.g., Cl^- and I^-) and sulfur or nitrogen donors, rarely complex with hard cations. Type B metal cations, or soft metals (see Chapter 1), are readily reactive owing to their relatively high electron-accepting potential (Zuiderveen, 1994).

13.5 PRIMARY CONTAMINANTS

13.5.1 Arsenic (MCL 0.05 mg L^{-1})

Arsenic is a by-product of the smelting of copper, lead, and zinc ores. It has been shown to produce acute and chronic toxic effects, with the trivalent (3+) form as the most toxic. Arsenic has been classified in the EPA's Group A (human carcinogen), and it is regulated by the U.S. government.

13.5.2 Barium (MCL 1.0 mg L^{-1})

Barium is naturally occurring in many types of rock. Deposits of fossil fuels and peat may also contain high levels of barium, often in the form of barium sulfate. It has been classified in the EPA's Group D (not classifiable). Because of its presence in drinking water and some adverse health effects on humans, it is regulated by the U.S. government.

13.5.3 Aluminum (MCL not regulated)

Aluminum is one of the most abundant elements in the earth's crust. Acid rain and acid mine drainage are two major causes of increased aluminum in freshwater systems. As acid water goes through soil, pH decreases and aluminum dissolves. The process may increase aluminum concentrations to toxic levels (>2 mg L^{-1}). Aluminum is toxic to both humans and aquatic organisms, especially to humans undergoing dialysis.

According to Zuiderveen (1994), the phytotoxic responses to aluminum are believed to involve activation of root-growth control mechanisms, nutrient disorders, aluminum binding to membranes and biologically important molecules, and disturbances of key regulatory processes (e.g., inhibition of cell division in the root) (Bennett and Breen, 1991). Aqueous calcium has been found to decrease aluminum toxicity (Birge et al., 1987). Lead, zinc, iron, and phosphate also act antagonistically to aluminum toxicity (Sayer et al., 1991; Murungi and Robinson, 1992).

13.5.4 Cadmium (MCL 0.01 mg L^{-1})

Cadmium is found in low concentrations in most soils and waters. It is produced as a by-product of zinc and lead mining and smeltering. Industrial use of cadmium has led to a dramatic increase in environmental problems caused by this element. Cadmium is used in semiconductors, nickel–cadmium batteries, electroplating, polyvinyl chloride (PVC) manufacturing, and control rods for nuclear reactors. The most important sources for aquatic contamination are active and inactive lead–zinc mines, land application of sewage sludge, zinc–cadmium smelters, effluents from plastic and steel production, and wastewaters from the production of nickel–cadmium batteries and electroplating (Zuiderveen, 1994).

Chronic exposure to cadmium has several different toxic effects in humans, targeting the lungs, kidneys, bone, blood, liver, and testes (Zuiderveen, 1994). Chronic inhalation exposure, which can come from smoking, can cause obstructive lung disease, resulting in dyspnea (Klassen, 1985). Cadmium problems of the skeletal and renal systems are the major symptoms of Itai–Itai disease, first diagnosed in Japanese individuals who consumed rice and water contaminated with high cadmium levels (Zuiderveen, 1994; Kjellstroem, 1986). Cadmium has been classified in the EPA's Group B$_1$ (probable human carcinogen).

13.5.5   Chromium (MCL 0.05 mg L^{-1})

Chromium is used for metal plating, in stainless steel, wear-resistant, and cutting-tool alloys, and as an anticorrosive additive to cooling water (Manahan, 1991). Other products containing chromium include pigments, primer paints, fungicides, and wood preservatives. The leather-tanning industry also makes use of chromium. The major sources of chromium to the aquatic environment are electroplating and metal-finishing industrial effluents, sewage and wastewater treatment plant discharge, and chromates from cooling water (Manahan, 1991; Zuiderveen, 1994).

Chromium exists in several oxidation states (e.g., di-, tri-, penta-, and hexa-), but only Cr^{3+} and Cr^{6+} are biologically important (Goyer, 1986). The trivalent form is less toxic. However, long-term exposure to trivalent chromium can cause allergic skin reactions and cancer (Eisler, 1986). Chromium in the aquatic environment tends to speciate into Cr^{3+} and Cr^{6+}, with the trivalent ion precipitating out of solution or oxidizing into the hexavalent form (Gendusa and Beitinger, 1992). Aquatic chromium toxicity is increased with decreasing pH, alkalinity, and hardness. It has been classified in the EPA's Group A (human carcinogen) (Shelton, 1989; Zuiderveen, 1994).

13.5.6   Fluoride (MCL 4.0 mg L^{-1})

Federal regulations require that fluoride not exceed a concentration of 4.0 mg L^{-1} in drinking water. Chronic exposure to levels above 4.0 mg L^{-1} may result in some cases of crippling skeletal fluorosis, a serious bone disorder. Fluoride in children's drinking water at levels of approximately 1 mg L^{-1} reduces the number of dental cavities. Federal law also requires that notification take place when monitoring indicates that the fluoride exceeds 2.0 mg L^{-1}.

13.5.7   Lead (MCL 0.05 mg L^{-1})

Lead is used in storage batteries, gasoline additives, pigments, and ammunition (Manahan, 1991). It is also used for bearings, cable cover, caulk, glazes, varnishes, plastics, electronic devices, flint glass, metal alloys, and insecticides (Sittig, 1979; Manahan, 1992; Venugopal and Luckey, 1978). The major sources of lead into the aquatic environment (U.S. EPA, 1985b) include effluents from industry and mining, coal use, plumbing, and deposition of gasoline exhausts (Manahan, 1991). Lead has been classified in the EPA's Group B$_2$ (probable human carcinogen).

13.5.8   Mercury (MCL 0.002 mg L^{-1})

Mercury exists as inorganic salt and as organic mercury (methyl mercury). Mercury levels in coal range from 10 to 46,000 ppb. Mercury enters the environment through its industrial uses (e.g., batteries) as well as from mining, smelting, and fossil fuel combustion. Inorganic mercury is the form detected in drinking water, and because of its potential adverse health effects, it is regulated.

13.5.9 Nitrate (MCL 10 mg L^{-1})

Nitrates occur in mineral deposits, soils, seawater, freshwater systems, the atmosphere, and biota. Lakes and other natural bodies of water usually have less than 1.0 mg L^{-1}. The sources of nitrates in drinking water include fertilizer, sewage, and feedlots. The toxicity of nitrate in infants is due to the reduction of nitrate to nitrite. Methemoglobinemia, or "blue-baby disease," is an effect in which hemoglobin is oxidized to methemoglobin, resulting in asphyxia (death due to lack of oxygen). Infants up to 3 months of age are the most susceptible subpopulation. The effect of methemoglobinemia is rapidly reversible.

Nitrate and nitrite have been classified in the EPA's Group D (not classifiable). Owing to its potential toxicity and occurrence in drinking water, it is regulated.

13.5.10 Selenium (MCL 0.01 mg L^{-1})

Selenium is mostly found in arid environments under irrigated agriculture. Selenium has toxic effects at high dose levels, but at low dose levels is an essential nutrient. It is classified in the EPA's Group D (not classifiable), and because of its potential chronic health effects, it is regulated by the U.S. government.

13.5.11 Nickel (MCL not given)

Nickel is used to make stainless steel, cast iron, permanent magnets, and storage batteries (Duke, 1980b). It is also used as a catalyst and for electroplating (Duffus, 1980; Manahan, 1991). Because of its use, nickel can be found as a contaminant in industrial water discharges (U.S. EPA, 1986a).

Nickel causes reductions in the growth and/or photosynthesis of aquatic plants (U.S. EPA, 1980c; Wang, 1987). It is also teratogenic to rainbow trout, channel catfish, goldfish (Birge and Black, 1980), and carp (Blaylock and Frank, 1979).

13.5.12 Silver (MCL 0.05 mg L^{-1})

Silver is used in the manufacture of silverware, coins, jewelry, and storage batteries (Sittig, 1979). Silver is also utilized for photographic materials and mirrors and as a bactericide (Petering and McClain, 1991; Venugopal and Luckey, 1978). The contamination of natural bodies of water occurs from mining, electroplating, and film processing (Manahan, 1991). Although silver is moderately toxic to man, acute poisoning via ingestion can cause violent abdominal pain, vomiting, diarrhea, convulsions, severe shock, paralysis, and death (U.S. EPA, 1980d; Venugopal and Luckey, 1978).

Silver has been classified in the EPA's Group D (not classifiable); because of its potential toxic effects on humans, it is regulated by the U.S. government.

13.6 SECONDARY CONTAMINANTS

13.6.1 Copper (MCL 1.0 mg L^{-1})

Copper is used throughout the electrical industry for wire, armature windings, water pipes, cooking utensils, stills, roofing materials, pigments, and chemical and pharmaceutical equipment (Scheinberg, 1991). Copper salts are used as algicides and fungicides.

Anthropogenic sources of copper in the aquatic environment include mining, metal plating, and domestic and industrial wastes (Manahan, 1991). Copper is extremely toxic to aquatic biota. Algae are especially sensitive to copper, with both marine and freshwater species being adversely impacted at concentrations as low as 1–5 mg L^{-1} (U.S. EPA, 1985a, 1980b).

13.6.2 Iron (MCL 0.3 mg L^{-1})

Iron is used mainly for steel and steel alloys, dyes, and abrasives (Manahan, 1991; Duffus, 1980). The contamination of aquatic habitats by iron is often the result of acid mine drainage (Evangelou, 1995b). Iron pollution from acid mine drainage is considered one of the main causes of fish kills in fresh waters (Duffus, 1980).

13.6.3 Zinc (MCL 5.0 mg L^{-1})

Zinc is used for the production of galvanized iron and steel, brass alloys, and paint pigments. Other products that use zinc include wood preservatives, dry-cell batteries, dyes, deodorants, cosmetics, and pharmaceuticals (Zuiderveen, 1994). The major sources of zinc contamination in the aquatic environment are industrial wastes, metal plating, plumbing, and acid mine drainage (Manahan, 1991; Diamond et al., 1993; Evangelou, 1995b). Presently, zinc is not considered mutagenic, carcinogenic, or teratogenic to humans.

13.6.4 Foaming Agents (MCL 0.5 mg L^{-1})

Foaming agents or detergents are used as indicators of undesirable pollutants such as sewage. They are measured by the methylene blue test (Table 13.8).

13.6.5 Chloride (MCL 250 mg L^{-1})

High concentrations of chloride in water often imply high sodium. The latter may cause adverse health effects on humans if such water is used for drinking purposes (Table 13.8).

TABLE 13.8. Chemical Concentrations That Cause Odor or Taste

Chemical	Concentration (mg L^{-1})
Chlorides	100–250
Total dissolved solids	500–1000
Copper	1
Hydrogen sulfide	0.1–0.2
Iron	1.0–2.0
Zinc	5
ABS (detergent)	0.5
Phenols	0.001

13.6.6 Color (SMCL 10 CU)

In drinking water, color may be used as an index of large quantities of organic chemicals from plants and soil organic matter. Metals such as copper, iron, and manganese may also introduce color (Table 13.8).

13.6.7 Corrosivity

The aggressive index (AI) was established as a criterion for determining the quality of the water that can be transported through asbestos cement pipe without adverse effects. It is calculated from the pH, calcium hardness [mg L^{-1} as $CaCO_3$–(H)], and total alkalinity [mg L^{-1} as $CaCO_3$–(A)] by the formula AI = pH + log [(A)(H)] (Shelton, 1989). Values of AI between 10.0 and 12.0 indicate moderately aggressive water, and values greater than 12.0 indicate nonaggressive water.

13.6.8 Hardness

Water hardness is related to polyvalent metallic ions in water, and it is reported as an equivalent concentration of calcium carbonate ($CaCO_3$). A commonly used hardness classification is given in Table 13.9. Hardness is associated with the hard scale (metal–carbonate precipitates) forming in cooking utensils, pipes, hot water tanks, and boilers. This scale reduces the capacity of pipes to carry water and transmit heat well.

13.6.9 Manganese (MCL 0.05 mg L^{-1})

Manganese produces a brownish color in clothes washed in manganese-rich waters. Since the Reclamation Act of 1977, manganese is used as an index metal in coal-mining operations because it was assumed by U.S. government regulatory agencies that manganese removal from industrial waters would ensure that all other heavy metals also be removed.

TABLE 13.9. Water Hardness Classification

| Quality | Hardness as $CaCO_3$ | |
	Milligrams per liter	Grains per gallon[a]
Soft	0–60	$0-3\frac{1}{2}$
Moderate	61–120	$3\frac{1}{2}-7$
Hard	121–180	$7-10\frac{1}{2}$
Very hard	More than 180	More than $10\frac{1}{2}$

[a]Grains per gallon = (mg L^{-1}/17.1).

13.6.10 Odor (MCL 3 TON)

Odor affects the drinkability of water. It is measured by the threshold odor number (TON). This is the dilution factor necessary before the odor is perceptible. A TON of 1 indicates that the water has characteristics comparable to odor-free water.

13.6.11 pH

A pH range of 6.5–8.5 was adapted for best environmental and aesthetic results.

13.6.12 Sodium (MCL 50 mg L^{-1})

Sodium is present in soils and water as NaCl. In arid environments, NaCl accumulates in the surface and groundwater owing to irrigation and high evapotranspiration. Other activities such as road salting and water softening may also contribute NaCl to natural waters. For additional information on human health effects and drinking water levels, see Shelton (1989).

13.6.13 Sulfate (SMCL 250 mg L^{-1})

High levels of sulfate in water tends to form hard scales in pipes and other equipment under high temperature, gives water a bad taste, and may induce diarrhea. Sulfate-rich waters are commonly found in areas where the geologic strata is rich in sulfides (e.g., pyrite). For additional information, see Shelton (1989).

13.6.14 Taste

The taste of drinking water is a qualitative factor determined by selected individuals testing water samples. For details, see American Public Health Association (1981).

13.6.15   Total Dissolved Solids (MCL 500 mg L^{-1})

Total dissolved solids (TDS) pertains mostly to dissolved salts in water which may have adverse effects on health and/or the durability of household appliances.

13.7   MICROBIOLOGICAL MCLs

The maximum contaminant levels for coliform bacteria are applicable to community and noncommunity water systems (Shelton, 1989). Two methods are used to measure coliform bacteria in water. One is the membrane filter technique and the other is the fermentation tube method. For details, see Shelton (1989).

13.8   MAXIMUM CONTAMINANT LEVELS FOR TURBIDITY

Turbidity reflects the amount of suspended matter in water (e.g., clay, silt, organic and inorganic matter, and plankton). The standard measure of turbidity is the turbidity unity (TU), which is based on the optical property of a water sample, causing transmitted light to be limited. As the number of particles increases, turbidity increases. The measuring instrument is called a nephelometer, and the readings are expressed as nephelometric turbidity units (NTU).

The maximum turbidity permitted in drinking water is 1 TU. In industrial waters, turbidity is measured in milligrams per liter of suspended solids. For example, in the case of mining sedimentation ponds, the law permits a monthly average concentration of 35 mg L^{-1}, with a maximum of 70 mg L^{-1}. High concentrations of suspended solids are known to inhibit aquatic life. Also, colloidal matter is known to carry various inorganic and organic pollutants.

13.9   RADIOACTIVITY (RADIONUCLIDES)

The purpose of the MCLs for radioactivity (radionuclides) is to limit human exposure. Some waters in contact with radioactive geologic strata (e.g., certain shales) are known to possess radioactivity. Since radioactivity in organisms is cumulative, monitoring should be carried out.

A. Gross alpha particle activity, Radium-226 and Radium-288
   1. Combined Radium-226 and Radium-228, 5 pCi L^{-1}.
   2. Gross alpha particle activity (including Radium-226 but excluding radon and uranium), 15 pCi L^{-1}.
B. Man-made radioactivity

1. The average annual concentration of beta particle and photon radioactivity from man-made radionuclides in drinking water shall not produce an annual dose equivalent to the total body or any internal organ greater than 4 millirem/yr. Tritium must be less than 20,000 pCi L^{-1}, and strontium must be less than 8 pCi L^{-1}.

2. When gross beta particle activity exceeds 50 pCi L^{-1}, radioactive constituents must be identified; total body doses must be calculated.

13.10 AMMONIA

Anhydrous ammonia (NH_3) is known to exert toxic effects on aquatic life. Some other concerns include:

1. Potential release or formation of toxic concentrations of ammonia into streams (ammonia is highly toxic to trout)
2. Stream acidification by its conversion to nitrate
3. Nitrate buildup inducing eutrophication in the receiving bodies of water and/or methemoglobinemia (lack of oxygen in infants)

Table 13.10 gives some critical threshold values of NH_3 in the environment. In-stream standards in many states allow a maximum concentration of 0.02 mg L^{-1} unionized ammonia in trout waters and 0.50 mg L^{-1} in all other waters. No limits are set in drinking water standards. Ammonia is often used to neutralize acid mine drainages (AMD) as well as other industrial acid- or metal-rich drainages (e.g., woodpulp and electronics). Many states have banned or discourage the use of ammonia for such purposes unless extensive monitoring is carried out.

TABLE 13.10. Some Critical Thresholds of Ammonia (NH_3)

Least perceptible odor	5 ppm
Rapidly detectable odor	20–25 ppm
USDOL Regs. 29 CFR, 1910.1000 limit (OSHA)	25 ppm
No impairment of health for prolonged exposure	50–100 ppm
General discomfort, eye tearing, no lasting effect on short exposure	150–200 ppm
Severe irritation to eyes, nose, throat	400–700 ppm
Coughing, bronchial spasms	1,700 ppm
Dangerous; exposure of 1/2 hr may be fatal	2,000–3,000 ppm
Serious edema, asphyxia, rapidly fatal	5,000–10,000 ppm
Immediately fatal	5,000–10,000 ppm

13.11   INDUSTRIAL ORGANICS

13.11.1   Benzene (MCL 1 μg L^{-1})

Benzene is a natural component of crude oil and natural gas. It is listed as a human carcinogen (EPA Group A).

13.11.2   Carbon Tetrachloride (MCL 2 μg L^{-1})

Carbon tetrachloride is used in the manufacture of chlorofluorocarbons. It is classified as a probable human carcinogen (EPA Group B$_2$).

13.11.3   Chlordane (MCL 0.5 μg L^{-1})

Chlordane is a wide-spectrum insecticide, and it is being phased out of use. It is classified as a probable human carcinogen (EPA Group B$_2$).

13.11.4   Chlorobenzene (MCL 4 μg L^{-1})

Chlorobenzene is used in chemical and pesticide production. It is classified as a probable human carcinogen (EPA Group B$_2$) or possible human carcinogen (EPA Group C).

13.11.5 m-Dichlorobenzene (MCL 600 μg L^{-1}), o-Dichlorobenzene (MCL 600 μg L^{-1}), and p-Dichlorobenzene (MCL 75 μg L^{-1})

Chlorinated benzenes, otherwise known as DCBs, are used in the production of organic chemicals (e.g., herbicides, pesticides, and fungicides). Chronic exposures can result in liver injury.

13.11.6   1,2-Dichloroethane (MCL 2 μg L^{-1})

1,2-Dichloroethane (ethylene dichloride) is used in the production of vinyl chloride. The odor threshold in water is 20 mg L^{-1} and has been shown to cause cancer in rats and mice.

13.11.7   1,1-Dichloroethylene (MCL 2 μg L^{-1}) and 1,2-Dichloroethylene (cis and trans)

1,1-Dichloroethylene (1,1-DCE) is used by the packaging industry. 1,2-Dichloroethylenes are used as solvents and preservatives. Direct evidence of carcinogenicity in humans is lacking.

13.11.8 Methylene Chloride (MCL 2 μg L^{-1})

Methylene chloride, a volatile chlorinated hydrocarbon, is used as a paint remover, metal degreaser, and aerosol propellant. The odor threshold for methylene chloride in the air is 100 ppm. According to Shelton (1989), it has been shown to induce tumors in mice.

13.11.9 Polychlorinated Biphenyls (MCL 0.5 μg L^{-1})

Polychlorinated biphenyls (PCBs) have been used in the manufacturing of electrical transformers. They are highly persistent in the environment. Rodent tests have suggested that PCBs are carcinogenic and the U.S. government strongly regulates their production (Shelton, 1989).

13.11.10 Tetrachloroethylene (MCL 1 μg L^{-1})

Tetrachloroethylene is used as a solvent. It is highly volatile and, for this reason, has been found only in groundwaters. Its odor threshold in water is 300 μg L^{-1}, and it is classified by the EPA as a probable human carcinogen (Group B$_2$).

13.11.11 Trichlorobenzene(s) (MCL 8 μg L^{-1})

1,2,4-Trichlorobenzene (1,2,4-TCB) is used as a solvent. According to Shelton (1989), it is formed in small quantities during the chlorination of drinking water. Chronic exposure can adversely affect various organs.

13.11.12 1,1,1-Trichlorethane (MCL 26 μg L^{-1})

1,1,-Trichloroethane is a solvent also with an odor threshold of 50 mg L^{-1}. Chronic exposure leads to liver damage.

13.11.13 Trichloroethylene (MCL μg L^{-1})

Trichloroethylene is a solvent with an odor threshold of 0.5 mg L^{-1}. It is classified by the EPA as a probable human carcinogen (Group B$_2$).

13.11.14 Vinyl Chloride (MCL 2 μg L^{-1})

Vinyl chloride is used in the production of polymer (polyvinyl chloride), for use in plastics. It is classified as a human carcinogen (EPA Group A).

13.11.15 Xylene(s) (MCL 44 μg L^{-1})

The xylenes are widely used as solvents and various other industrial products. In general, xylenes are acutely toxic to animals and humans only at higher concentrations and through chronic exposure.

13.12 PESTICIDES

13.12.1 Endrin (MCL 0.2 μg L^{-1})

Endrin is a commercially used insecticide and rodenticide and is persistent through the aquatic food chains. Production of this chemical in the United States has been stopped. It has been classified in EPA Group E.

13.12.2 Lindane (MCL 4 μg L^{-1})

Lindane (1,2,3,4,5,6-hexachlorocyclohexane) is an insecticide registered for commercial and home use, and it is also used in some shampoos. Lindane is slightly soluble in water and volatilizes readily. Lindane is classified as "B$_2$–C" (i.e., between the lower half of the B category of "probable" and the C category of "possible" carcinogen classifications).

13.12.3 Methoxychlor (MCL 100 μg L^{-1})

Methoxychlor, a chemical closely related to DDT, has been used as an insecticide. The half-life for methoxychlor in water is rather short (46 days) and it is not considered to be persistent. Methoxychlor has been classified in the EPA's Group D (not classifiable). At high doses, methoxychlor has been shown to exhibit chronic toxic effects and has been detected in drinking water.

13.12.4 Toxaphene (MCL 5 μg L^{-1})

Toxaphene (approximate overall empirical formula of $C_{10}H_{10}Cl_5$) is a persistent, broad-spectrum insecticide with some occurrence in drinking water. Its registered uses are currently limited. The EPA reported that toxaphene is highly persistent and accumulates in the environment. Acute exposure to toxaphene affects various organs in humans. Toxaphene has been classified in the EPA's Group B$_2$ (probable human carcinogen).

13.12.5 2,4-D (2,4-Dichlorophenoxyacetic Acid) (MCL 100 µg L^{-1})

2,4-D (2,4-dichlorophenoxyacetic acid) is a systemic herbicide used to control broad-leaf weeds. It undergoes both chemical and biological degradation in the environment and usually does not accumulate. According to the EPA, 2,4,-D has been detected in surface and groundwaters. It has been classified in the EPA's Group D (not classifiable).

13.12.6 2,4,5-TP (Silvex) (MCL 10 µg L^{-1})

2,4,5-TP [2-(2,4,5-trichlorophenoxy propionic acid], or silvex, is an herbicide that has been used for weed and brush control on rangeland and rights of way. It is soluble in water and its environmental resistance is expected to be relatively short. 2,4,5-TP is contaminated to varying extents with 2,3,7,8-TCDD, a toxic polychlorinated dibenzo-p-dioxin. It has been classified in the EPA's Group D (not classifiable).

13.12.7 Trihalomethanes (100 µg L^{-1})

Trihalomethanes (TTHMs) are organic chemicals that contain one carbon atom, one hydrogen atom, and three halogen atoms. The most common trihalomethanes found in water are: trichloromethane (chloroform), bromodichloromethane, dibromochloromethane, and tribromoethane (bromoform). Chloroform is found in the highest concentrations. It is formed by the reaction of free chlorine with certain natural organic compounds in the water.

The primary drinking water regulations provide an MCL of 0.10 mg L^{-1} for total trihalomethanes. Recently, public water suppliers have switched from the use of chlorine as a water disinfectant to the use of chloroammines. The latter limits the formation of trihalomethanes, but it could be very toxic to aquatic organisms.

13.13 CHELATORS

Chelators are synthetic or natural organic compounds capable of forming metal complexes with more than one donor atom. The resulting metal–chelate has a structure with one or more rings, making it quite stable (Martell and Calvin, 1952). Chelators are used to reduce the toxicity of metals in the bloodstream by binding them, and in soil extraction for determining the mobility and/or bioavailability of metals. Because of their uses, chelators find their way into natural bodies of water through hospital and other industrial waste (Zuiderveen, 1994).

13.13.1 EDTA

Ethylenediaminetetraacetic acid (EDTA) is an industrial and analytical reagent because of its ability to complex with many divalent and trivalent metals up to a hexadentate mode (six bonds). It is currently recommended by the EPA for acute toxicity identification evaluations (U.S. EPA, 1991a). This chelator has been used with mammals and aquatic organisms to treat poisoning due to various heavy metals (Zuiderveen, 1994; U.S. EPA, 1991a).

13.13.2 NTA

Nitrilotriacetic acid (NTA) is similar to EDTA (N and COOH functional groups) and forms a maximum of four bonds (tetradentate) with metal ions. It is used for reducing metal toxicity in humans and in aquatic and microbial life.

13.13.3 DTPA

Diethylenetriaminepentaacetic acid (DTPA) possesses amine and carboxylic acid functional groups and could form octadentate complexes with heavy metals. It is used as a chelator to determine metal bioavailability in soil to plants and mobility to the aquatic environment.

13.13.4 DMPS

The chelator, 2,3-dimercapto-1-propanesulfonic acid (DMPS) binds with metals via a sulfate and two sulhydryl groups. It is used for the removal of inorganic and methyl mercury and may reduce the toxicities of copper, nickel, and cadmium (Zuiderveen, 1994).

13.13.5 Citrate

Citrate has three oxygen-containing ligands (tridentate). It can complex heavy metals, and certain yeasts produce excess citric acid to reduce metal toxicity. Citrate and citric acid have been used to extract metals from soil particles to determine their bioavailability and fate (Zuiderveen, 1994).

13.14 SUMMARY

Most of the pollutants listed in this chapter find their way into water directly by human or natural discharges (from diffuse or point sources) or indirectly through discharges into soil. It is therefore important to realize that in order to manage or control such direct or indirect pollution, the chemistry of such chemicals must be understood and their physicochemical behavior in soil–water systems predicted. To accomplish these tasks, and to minimize adverse effects on the environment, one needs to have a background in water chemistry, soil mineralogy, soil surface chemistry, and the chemistry of anthropogenic or industrial pollutants and an understanding of soil–water treatment or management technologies and their principles.

PROBLEMS AND QUESTIONS

1. Water contaminants affect human health in two possible ways. Name and briefly explain these two ways.

2. Define primary and secondary water standards.

3. Name the three conditions needed to be met by a primary standard.

4. Explain how metal softness may induce toxicities.

5. Name three heavy metals that are classified as carcinogens or probable human carcinogens.

6. What beneficial effects does fluoride have for humans?

7. What human disease is fluoride associated with?

8. Name some industrial organic contaminants and briefly discuss the sources of these contaminants.

9. Explain what chelators are and why they are considered potential water contaminants.

14 Soil and Water Decontamination Technologies

14.1 INTRODUCTION

Water contamination by organics and/or inorganics in nature occurs because of human activities or natural processes. In the case of soil contamination by organics and inorganics, it is caused by a number of industries, including the chemical, electronics, pharmaceutical, plastics, and automobile industries. Treating water and/or soils contaminated by organics and/or inorganics generally involves two technologies. One is called "pump and treat" and uses an external energy source to remove the pollutant from the soil and then treat it using various approaches such as heat, incineration, ultraviolet radiation, or supercritical water oxidation. A second technology involves removing contaminated soil, treating it by any of the approaches listed above, and then returning it back to its place. In the case of drinking water, various treatments are used, including filtration, air stripping, disinfection, distillation, ion exchange, and reverse osmosis.

The purpose of this chapter is to briefly introduce the student to the various technologies used for cleaning and remediating soil and/or water from various contaminants. The scientific basis for these technologies is discussed throughout the chapters of this book. Such technologies are based on the principal of acid–base chemistry, solubility–precipitation, ion exchange, redox, kinetics and catalysis, complexation, surface sorption, and phase tranformation. For additional information on these technologies, see Just and Stockvell (1993) and Tester et al. (1993).

14.2 METHODS OF SOIL TREATMENT

Two approaches are commonly employed to treat contaminated soil. The first involves a *phase transfer*. In this approach, the contaminant(s) is moved from one phase (either solid, liquid, or gas) into another. The second approach involves *destruction* or *transformation* of the contaminant(s). This approach is more advantageous because the contaminants are broken down into harmless products as opposed to being simply removed through phase transfer (e.g., from liquid phase to gaseous phase).

The two soil treatment approaches can be utilized by incorporating a number of technologies, including: (1) high–low temperature thermal treatments, (2) radio frequency heating, (3) steam stripping and vacuum extraction, (4) aeration, (5) in situ bioremediation, and (6) soil flushing/washing. The treatments can either be applied to soil in situ or to soil that has been removed from the contamination site.

14.2.1 High–Low Temperature Treatment

Generally, high-temperature systems operate at temperatures above 1000°F (500–600°C), whereas low-temperature systems operate below 1000°F. High-temperature processes include (1) incineration, (2) electric pyrolysis, and (3) in situ vitrification. Low-temperature treatment systems include (1) soil roasting, (2) low-temperature incineration, (3) low-temperature thermal aeration, (4) infrared furnace treatment, and (5) low-temperature thermal stripping.

High-temperature treatment systems involve destruction of contaminant(s) through complete oxidation, whereas low-temperature systems increase the rate of phase transfer (e.g., liquid phase to gaseous phase), and thus encourage contaminant partitioning from soil. Some of the disadvantages of heat treatment include its high cost and its ineffectiveness with some contaminants (e.g., low volatilization potential or incineration actually produces more toxic substances).

14.2.2 Radio Frequency Heating

Radio frequency (RF) heating is used for in situ thermal decontamination of soil. This process was originally developed in the 1970s for use in recovering hydrocarbons from materials such as oil shales and tar sands. The treatment is effective for volatile and semivolatile organics only.

14.2.3 Steam Stripping

Steam stripping is an in situ process where air and steam or hot water are injected into the ground, resulting in increased volatilization of the contaminants (Fig. 14.1). A vacuum is then applied to bring the air, hot water, and contaminants to the surface for further treatment. It is effective for volatile and semivolatile organic compounds and its effectiveness can be increased by the the use of chemical agents capable of increasing the volatility of the contaminants. Such agents include chemicals to change pH, redox potential, and contaminant destruction, and/or to enhance bioremediation.

14.2.4 Vacuum Extraction

Vacuum extraction involves aeration followed by vacuum. It represents one of the most commonly used in situ treatment technologies. The technigue is effective when employed under buildings; it is relatively cost effective but lengthy, and is not effective in water-saturated soils.

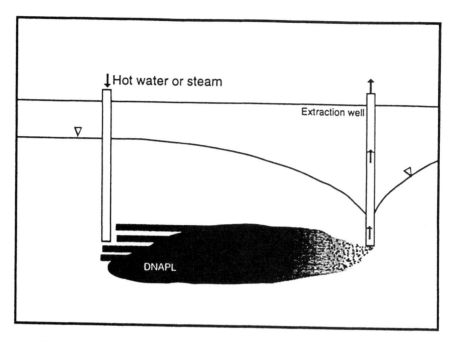

Figure 14.1. Schematic of hot water or steam extraction (from U.S. EPA, 1995e).

14.2.5   Aeration

Aeration involves soil contaminant removal through volatilization and/or transformation. There are various ways of applying this type of treatment. It includes soil vapor extraction or soil disturbance. The latter is rather simple to use in shallow soils. It is accomplished by tilling the soil to expose its surfaces to the air, thus enhancing volatilization and/or transformation. Henry's constant (K_h = gas phase/aqueous phase; see Chapter 1) is commonly used as an index of volatilization potential. Organic chemicals with K_h values greater than 10^{-4} exhibit high volatilization potential, whereas organic chemicals with K_h values less than 10^{-4} exhibit low volatilization potential.

14.2.6   Bioremediation

Bioremediation enhances natural transformation by optimizing the conditions necessary for the process. The process is carried out by organisms capable of degrading the compounds. Both anaerobic and aerobic conditions may be involved in enhancing transformation. Commonly, four possible approaches for bioremediation of hazardous contaminants are employed:

1. Improving the conditions for enhancing the biochemical mechanisms

2. Introducing specialized microorganisms

3. Introducing cell-free enzymes

4. Improving vegetative growth

14.2.7 Soil Flushing or Washing

Soil flushing and soil washing involves the removal of organic or inorganic contaminants through the use of water. In this technique, fluid is used to wash contaminants from the saturated zone by injection and recovery systems. A solvent, surfactant solution, or water with or without additives is applied to soil to enhance contaminant release and mobility, resulting in increased recovery and decreased soil contaminant levels. This technology can be applied to volatile organics such as halogenated solvents. However, it has more often been applied to heavier organic compounds such as oils or wood-treating compounds.

The K_{ow} (K_{ow} = n-octanol phase/water phase; see Chapter 4) is used as an index of washing potential for various organics. Three classes of K_{ow} are recognized: low sorption potential or easily washed, log K_{ow} < 2.5; moderate sorption potential, 2.5 < log K_{ow} < 4; and high sorption potential or not easily washed, log K_{ow} > 4.

14.3 IN SITU TECHNOLOGIES

14.3.1 Surfactant Enhancements

Surfactant enhancement is a technology used to remove contaminants from soils and water at hazardous waste sites. The application of surfactants enhances remediation by (a) increasing contaminant mobility and solubility, (b) decreasing the mobility of contaminants, and (c) increasing the rate of biodegradion of contaminants in soil.

Surfactants increase the apparent solubility of the contaminant in water and thus water becomes more effective in the removal of nonaqueous phase liquids (NAPLS). Surfactants also reduce the interfacial tension between the water and the NAPL, thus NAPL mobility increases. However, cationic surfactants increase the capacity of soil to sorb hydrophobic organic chemicals such as polyaromatic hydrocarbons (PAHs).

The application of this technology involves the construction of two wells. One well serves to introduce the surfactant to the dense NAPL (DNAPL), while the second serves as a means to extract the polluted liquid (Fig. 14.2).

14.3.2 Cosolvents

Cosolvent technology is similar to the surfactant enhancement technology. Instead of a surfactant, the injection well receives a solvent mixture (e.g., water plus a miscible organic solvent such as alcohol). The cosolvent mixture is injected up-gradient of the

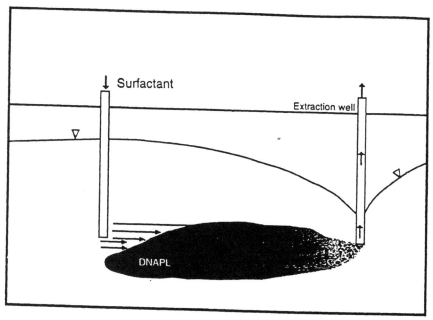

Figure 14.2. Schematic of surfactant extraction (from U.S. EPA, 1995d).

contaminated area, and the resulting dissolved contaminant is extracted down-gradient (Fig. 14.3).

Cosolvents that organisms may use as substrates could increase the microbial degradation of the pollutant if the concentration has not reached toxic levels.

14.3.3 Electrokinetics

Electrokinetics is an in situ remediation technology applicable to soil or soil-like material with low hydraulic conductivities (e.g., clay) contaminated with heavy metals, radionuclides, and selected organic pollutants. The technique has been used in the past in the oil recovery industry and to remove water from soils.

The technology involves the application of low-intensity, direct electrical current across electrode pairs that have been implanted on each side of the contaminated soil. The electrical current induces electroosmosis and ion migration between the two implanted electrodes. Depending on their charge, the contaminants accumulate on one of the electrodes and are extracted to a recovery system (Fig. 14.4). Improved performance of electrokinetics could be attained by the introduction of surfactants.

14.3.4 Hydraulic and Pneumatic Fracturing

This is a technology employed for systems with low permeability. The hydraulic fracturing involves pumping under high pressure water, sand, and a thick gel into a

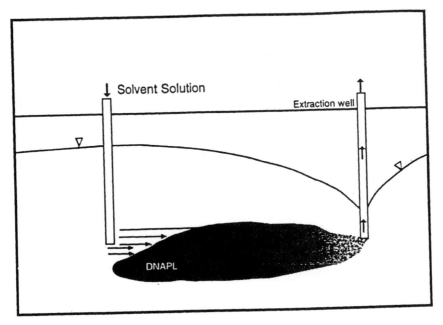

Figure 14.3. Schematic of solvent extraction (from U.S. EPA, 1995a).

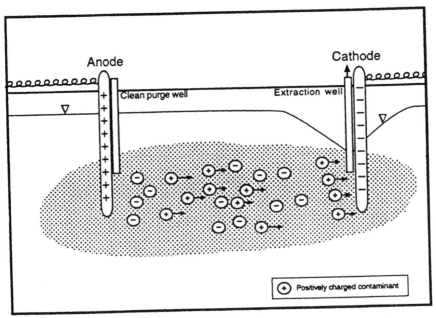

Figure 14.4. Schematic of electrokinetic extraction (from U.S. EPA, 1995b).

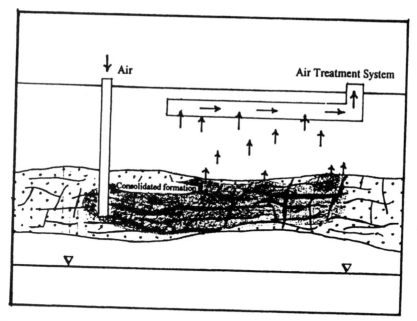

Figure 14.5. Schematic of air-induced fracturing and extraction (from U.S. EPA, 1995c).

borehole to propagate the fracture. The residual gel biodegrades and the fracture becomes highly permeable (Fig. 14.5).

The pneumatic fracturing involves injection of highly pressurized air into the contaminated consolidated sediments, thus creating secondary fissures and channels and accelerating the removal of contaminants by vapor extraction, biodegradation, and thermal treatment.

14.3.5 Treatment Walls

Treatment walls is a technology involving permanent, semipermanent, or replaceable units across the flow path of a contaminated plume. It allows the plume to move passively through while precipitating, sorbing, or degrading the contaminants (Fig. 14.6). The technology is relatively cheap because it does not require an external energy source. The effectiveness of the treatment wall depends on its makeup and the nature of the pollutants. The components making the treatment wall may include chelators, oxidants–reductants, and alkaline material.

Recent advances in this technology include the use of 2:1 clays converted to hydrophobic forms through the introduction of surfactants in the interlayer. For example, Boyd et al. (1991) introduced cationic chain surfactants into 2:1 clay minerals. Such clays were demonstrated to have high affinity for hydrophobic organic chemicals. Additionally, polyethylene oxides (PEOs) have been intercalated into aluminum-pillared montmorillonite (Montarges et al., 1995). Because PEOs have a

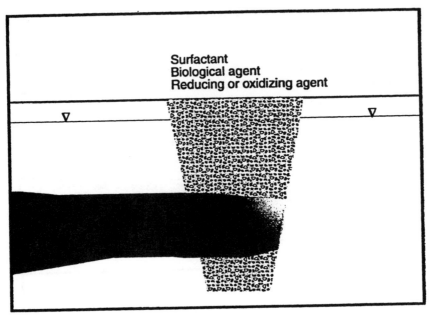

Figure 14.6. Schematic of treatment walls (from U.S. EPA, 1995f).

high affinity for cations, these PEO-treated clays have been proposed for use in scavenging heavy metals.

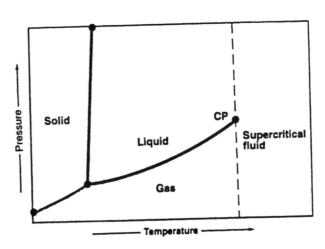

Figure 14.7. Phase diagram of a pure substance.

14.4 Supercritical Water Oxidation

Water possesses vastly different properties as a reaction medium in its supercritical state than in its standard state. The diagram in Fig. 14.7 is that of a pure substance and shows the regions of temperature and pressure where the substance exists as a solid, liquid, gas, and supercritical fluid. The supercritical point for water is met at a temperature of 400°C and above and at high pressure (about 25 MPa). At the supercritical point, water behaves as a nonpolar dense gas, and hydrocarbons exhibit generally high solubility. However, the solubility of inorganic salts is very low in such liquid. Note that the dielectric constant of water is 80 at the standard state reaches approximately 0 at the supercritical point; the K_w of 10^{-14} at the standard state reaches approximately 10^{-24} at the supercritical point.

When water, oxygen, and organics (e.g., contaminants) are brought together at temperatures and pressures above the supercritical point, the organics oxidize nearly completely, spontaneously within 1 min or less. Because of this oxidizing potential of supercritical water, some consider it as the most promising technology for destroying organic contaminants.

14.5 Public Community Water Systems

The Safe Drinking Water Act (SDWA) requires that all public water systems sample and test their water supplies for all contaminants with MCLs. The exact type and frequency of testing depends on the seriousness of any potential adverse health effects and on state and local regulations. Those concerned about their drinking water quality

TABLE 14.1. Recommended Water Tests

Test Name	MCL or SMCL
Bacteria (total coliform)[a]	None detected
Nitrate[a]	10 mg L^{-1} NO$_3^-$
Lead	0.05 mg L^{-1}
Volatile organic chemical scan[a]	If positive, retest for specific chemicals
Hardness (total)	150 mg L^{-1}
Iron	0.3 mg L^{-1}
Manganese	0.05 mg L^{-1}
Sodium	50 mg L^{-1}
pH	$6.5–8.5$
Corrosivity	Langelier index ± 1.0
Radioactivity (gross alpha)[a]	5 L^{-1} pCi
Mercury	0.002 mg L^{-1}

Source: Adapted from Shelton, 1989.
[a]Denotes an MCL based on health standard. Additional pollutants, depending on concerns, include arsenic, barium, chromium, lead, selenium, silver, and fluoride.

could obtain the complete water test results required under the SDWA directly from the local water utility. The recommended tests for such water systems are given in Tables 14.1 and 14.2.

Table 14.2. Additional Water-Testing Recommendations for Common Problems or Special Situations

Problem	Common Signs/Situations	Causes	Test Recommended
"Hard" water	Large amount of soap required to form suds. Insoluble soap curd on dishes and fabrics. Hard scaly deposit in pipes and water heaters	Calcium, magnesium, manganese, and iron (may be in the form of bicarbonates, carbonates, or chlorides)	Hardness test
Rusty colored water	Rust stains on clothing and porcelain plumbing. Metallic taste to water. Rust coating in toilet tank. Faucet water turns rust-colored after exposure to air	Iron, manganese, or iron bacteria	Iron test Manganese test
"Rotten egg" odor	Iron, steel, or copper parts of pumps, pipes, and fixtures corroded. Fine black particles in water (commonly called black water). Silverware turns black	Hydrogen sulfide gas, sulfate-reducing bacteria, or sulfur bacteria	Hydrogen sulfide test
"Acid" water	Metal parts on pump, piping, tank, and fixtures corroded. Red stains from corrosion of galvanized pipe; blue-green stains from corrosion of copper or brass	Carbon dioxide. In rare instances mineral acid– sulfuric, nitric, or hydrochloric	pH Langelier index
Cloudy turbid water	Dirty or muddy appearance	Silt, sediment, micoorganisms	Check well construction with local well driller
Chemical odor of fuel oil	Well near abandoned fuel oil tank; gas station	Leaking underground storage tank	Volatile organic chemical scan or specific fuel component
Unusual chemical odor	Well near dump, junkyard, landfill, industry or dry cleaner	Groundwater contamination, underground injection, or leaching waste site	Check with Health Dept., Organic chemical scan, heavy metals

(continued)

Table 14.2. Continued

Problem	Common Signs/Situations	Causes	Test Recommended
No obvious problem	Well located in area of intensive agricultural use	Long-term use of pesticides and fertilizers	Test for pesticides used in area, nitrate test.
Recurrent gastro-in-testinal ill-ness	Recurrent gastro-intestinal illness in guests drinking the water	Cracked well casing, cross connection with septic system	Bacteria (coliform test) nitrate test
Sodium re-stricted diet; salty, brackish, or bitter taste	Well near seawater, road salt storage site, or heavily salted roadway	Saltwater intrusion, groundwater contamination	Chloride, sodium, total dissolved solids (TDS)

Source: From Shelton, 1989.

Throughout this book, emphasis is placed on the reactions between soil and water and how such reactions control the composition and/or quality of soil and water. In this section, the focus is on water quality with respect to household uses, including drinking. A goal of the modern environmental soil chemistry discipline is to ensure that the degradation of natural water is kept to a minimum when land (soil) is used for producing food or for any other purpose (e.g., industrial development). However, even with the best intentions, water quality is affected to some degree by humans or by nature in general, and not all water on the earth's surface is suitable for household uses, including drinking. For this reason, prior to using much of the surface- or groundwater, testing and further treatments are necessary to make such water suitable for human consumption.

14.5.1   Some General Information on Water Testing

A problem that has often been reported with drinking water is the presence of lead, a heavy metal. Shelton (1989) outlined the following steps to minimize human exposure to lead:

1. Flush the plumbing to counteract the effects of "contact time." Flushing involves allowing the cold faucet to run until a change in temperature occurs (minimum of 1 min).
2. Hot water tends to aggravate lead leaching when brought in contact with lead plumbing materials.
3. Water-treatment devices for individual households include calcite filters and other devices to lessen acidity which increases lead release.
4. Use lead-free materials for repairs and installation of new plumbing.

5. Lead can be removed from tap water by installing ion-exchange filters, reverse-osmosis devices, and distillation units.

14.5.2 Microbiological Maximum Contaminant Levels

The standard bacteriological method for judging the suitability of water for domestic use is the coliform test. It detects the presence of coliform bacteria, which are found in the natural environment (soils and plants) and in the intestines of humans and other warm-blooded animals. Any food or water sample in which this group of bacteria is found is to be suspected of having come into contact with domestic sewage or animal manure. Such a water supply may contain pathogenic bacteria and viruses responsible for typhoid fever, dysentery, and hepatitis (Shelton, 1989).

The two standard methods for determining the numbers of coliform bacteria in a water sample are the *multiple-tube fermentation technique* and the *membrane filter technique*. In the multiple-tube fermentation technique, a series of fermentation tubes containing special nutrients is inoculated with appropriate quantities of water to be tested and incubated. After 24 hr, the presence or absence of gas formation in the tubes is noted. In the membrane filter technique, a quantity of water is filtered through a specially designed membrane filter which traps bacteria. The filter, with certain specified nutrients, is incubated for 24 hr. The results are usually expressed as number of coliform colonies per 100 mL of water sample. Currently, emerging waterborne pathogenic organisms of parasitic 00/cysts of *Giardia* and *Cryptosporidium* are a major concern. They are distributed ubiquitously in pristine and human impacted water. The concern about pathothegic organisms has arisen because methods of analysis are inefficient and expensive (Shelton, 1989).

When water destined for drinking does not meet federal and/or state standards, appropriate treatments are introduced. Some of the current technologies used to meet federal water drinking standards are briefly discussed below.

14.5.3 Activated Carbon Filtration

This technology is effective for certain chemicals, including pesticides, radon gas, chlorine, and trihalomethanes, as well as for odor. The carbonaceous material made from bituminous coal, lignite, peat, or wood is activated (pore formation) under high heat by steam, but in the absence of oxygen. Organic contaminants present in water are sorbed by the pore surfaces formed during activation. They are specifically effective in removing organic contaminants with low solubility (e.g., pesticides, benezene, and chlorinated hydrocarbons).

14.5.4 Air Stripping

The technology involves air-stripping columns where water flows downward by gravity while air is pumped upward. As the water flows down and passes over a packing material possessing a large air–liquid interphase, the volatile organic compounds (VOC) are transferred from the water to the air. It is effective in removing organic as well as inorganic contaminants including hydrogen sulfide. The efficiency of VOC removal by air-stripping columns depends on the type of VOCs present in the water, air:water ratio, type of packing material, height of packing material, temperature of the water and air, and concentration of VOC.

14.5.5 Disinfection

Disinfection destroys microorganisms capable of causing disease in humans. Some methods of disinfection are chlorination, chloramines, ozone, ultraviolet light, as well as chlorine dioxide, potassium permanganate, and nanofiltration. Disinfection kills disease-causing organisms and inactivates *Giardia lamblia* cysts and enteric viruses. There are two kinds of disinfection: primary, which kills or inactivates microorganisms, and secondary, which protects the finished water from regrowth of microorganisms.

The EPA Surface Water Treatment Ruse (SWTR) requires public water supplies, under the direct influence of surface water, to be disinfected. Some disinfectants produce chemical by-products; SWTR requires that their concentration remain within the MCL. Currently, one such by-product is trihalomethanes. Water disinfection is effective when combined with conventional treatment, such as coagulation, flocculation, sedimentation, and filtration. The latter is accomplished by sand or diatomaceous earth. The effectiveness of disinfection is evaluated by determining total coliform bacteria which are not pathogenic, but their presence suggests that certain pathogens may have survived. The various chemicals commonly used as disinfectants are presented below and some of their advantages and disadvantages are listed.

Chlorine. It is a toxic, yellow-green gas under normal pressures but becomes a liquid under high pressures. It is very effective as both a primary and secondary disinfectant. It is lethal, however, at concentrations as low as 0.1% by volume.

Sodium Hypochlorite Solution. It is available as a solution in concentrations of 5–15% chlorine and for this reason is easier to handle than chlorine gas. However, sodium hypochlorite is very corrosive and easily decomposes.

Solid Calcium Hypochlorite. It is a white solid in granular, powdered, or tablet form that contains 65% available chlorine, easily dissolves in water, and is very stable in the absence of moisture. However, calcium hypochlorite is corrosive with a strong odor and readily adsorbs moisture-forming chlorine gas. Reactions between calcium hypochlorite and organic material can cause a fire or explosion.

Chloramines. It is formed when water containing ammonia is chlorinated or when ammonia is added to water containing chlorine (hypochlorite or hypochlorous acid). It is an effective bactericide and produces fewer disinfection by-products. However, chloramine is a weak disinfectant (less effective against viruses or protozoa than chlorine) and is more appropriate as a secondary disinfectant to prevent bacterial regrowth. When water pH is below 5, nitrogen trichloride appears to form. It may be harmful to humans and produces undesirable taste and odor. Using the proper amount of the chemical avoids such problems.

Ozone. It is a gas composed of 3 oxygen atoms. It is a powerful oxidizing and disinfecting agent formed by passing dry air through a system of high-voltage electrodes. Ozone is widely used as a primary disinfectant in many parts of the world, partly because it does not produce halogenated organic substances unless bromide is present. However, it is very unstable and must be generated on site. Because of this instability, it does not maintain an adequate residual in water and chlorine is required as a secondary disinfectant.

Ultraviolet Light (UV). It is generated by a special lamp and disrupts cell division processes. It is highly effective against bacteria and viruses, has no residual effects, and thus does not produce any known toxic by-products. However, UV does not inactivate *Giardia lamblia* or *Cryptosporidium* cysts and it is not suitable for water with high levels of suspended solids, turbidity, color, or soluble organics. These materials absorb the UV radiation, reducing disinfection performance.

14.5.6 Distillation

In this technology, water is heated until it turns to steam, which is then condensed as distilled pure water, free of most dissolved or any solid contaminants including bacteria and viruses. Because distilled water is free of all minerals, it may not be ideal for drinking. Furthermore, certain organic contaminants with a lower boiling point than water (e.g., pesticides) evaporate with the water and end up with the processed water. Advanced distillation units could eliminate this problem.

14.5.7 Ion Exchange

This is a process by which ions that are dissolved in water are transferred to, and held by, a solid material or exchange resin. It is a process commonly used for softening water. When water containing dissolved cations associated with hardness (e.g., Ca^{2+}, Mg^{2+}, Fe^{2+}, and Mn^{2+}), as well as heavy metal or radioactive cations, contacts the resin, the cations are exchanged for the loosely held sodium ions on the resin. The process makes the water "soft."

Eventually, the resin loses all of its sodium ions; thus, no more cations associated with hardness can be removed from the incoming water. The resin at this point is "recharged" or "regenerated." This involves replacement of the cations associated with hardness by Na, having as its source NaCl. The water used during this process is discarded. Anion exchange may also contain relatively high levels of anions (e.g., nitrates and sulfates) (Shelton, 1989).

14.5.8 Mechanical Filtration

This process removes dirt, sediment, and loose scale from the incoming water by employing sand, filter paper, or compressed glass wool or other straining material. These filters will not remove any dissolved substances.

14.5.9 Reverse Osmosis

This technology successfully treats water with high salt content, cloudiness, and dissolved minerals (e.g., metal sulfate, metal chloride, metal nitrate, metal fluoride, salts, boron, and orthophosphate). It is effective with some detergents, some taste-, salt-, color-, and odor-producing chemicals, certain organic contaminants, and specific pesticides. It works by passing water under pressure through cellulosic or noncellulosic (polyamide) membranes. It removes 90–95% of most dissolved contaminants and, when membrane density is in the submicron range, may remove many types of bacteria. However, reverse osmosis is a slow and wasteful process. In some systems, for every gallon of portable water obtained, approximately 4–6 gal of water will be discarded (Shelton, 1989).

14.6 BOTTLED WATER

This solution represents drinking either natural spring water or processed water made available through any number of technologies. The Food and Drug Administration (FDA) requires that bottled water represent sources free from pollution and be "of good sanitary quality" when judged by the results of bacteriological and chemical analyses, and supplies must list the addition of salt and carbon dioxide on their labels. In the United States, bottled water is not as popular as in Europe, where bottled water

TABLE 14.3. Various Bottled Drinking Waters and Their Compositions[a]

	Metallic Water (Greece)
CATIONS	
Calcium (Ca)	65.7
Magnesium (Mg)	58.4
Sodium (Na)	179.3
Potassium (K)	31.3
Iron (Fe)	< 0.05
Aluminum (Al)	0.06
Lithium (Li)	0.35
Strontium	0.5

(continued)

TABLE 14.3. Continued

ANIONS	
Bicarbonate (HCO$_3$)	862.8
Chloride (Cl)	39.0
Sulfate (SO$_4$)	72.0
Nitrate (NO$_3$)	0.0
Phosphate (PO$_4$)	< 0.20
Fluoride (F)	0.90
Boron (B)	0.80
pH	7
Specific conductance	1.325 dS m^{-1}
Total dissolved solids	1308.6 mg L^{-1}

Table Water (Greece)

CATIONS	
Calcium (Ca)	101.6
Magnesium (Mg)	6.5
Sodium (Na)	7.5
Potassium (K)	12

ANIONS	
Bicarbonate (HCO$_3$)	286
Chloride (Cl)	10.2
Sulfate (SO$_4$)	28.3
Nitrate (NO$_3$)	28.6
pH	7.3
Specific conductance	0.541 dS m^{-1}
Total dissolved solids	470.2 mg L^{-1}
Hardness	280 mg L^{-1}

Natural Spring Water (France)

CATIONS	
Calcium (Ca)	78.0
Magnesium (Mg)	24.0
Sodium (Na)	—
Potassium (K)	—

ANIONS	
Bicarbonate (HCO$_3$)	357
Chloride (Cl)	4.0
Sulfate (SO$_4$)	10.0
Nitrate (NO$_3$)	1.0
Silica	14.0
pH	7.2
Hardness	309 mg L^{-1}

(*continued*)

TABLE 14.3. Continued

Natural Mineral Water (Bulgaria)

CATIONS

Calcium (Ca)	1.6
Magnesium (Mg)	Traces
Sodium (Na)	No value given
Potassium (K)	No value given

ANIONS

Bicarbonate (HCO_3)	73.2
Carbonate (CO_3)	12.0
Chloride (Cl)	9.4
Fluoride (F)	0.6
pH	9.33

Natural Spring Water (Canada)

CATIONS

Calcium (Ca)	38.0
Magnesium (Mg)	22.0
Sodium (Na)	6.0
Potassium (K)	2.0
Lead–Arsenic	<10 ppb
Copper–Zinc	<10 ppb

ANIONS

Bicarbonate (HCO_3)	243
Chloride (Cl)	1.0
Sulfate (SO_4)	14.0
Nitrate (NO_3)	<0.05

[a]Compositions are given in milligrams per liter or parts per million unless otherwise noted.

represents most water consumed for drinking purposes. Such waters are sold under various labels (e.g., Natural Table Water, Natural Metallic Water, or Metallic Water). Some examples of such waters and their composition are given in Table 14.3. Presently, the United States does not require water-bottling companies to label inorganic water constituents.

PROBLEMS AND QUESTIONS

1. A truck carrying H_2SO_4 overturns and spills its content on the freeway. Describe a remediation approach and identify and describe the principle on which this approach is based.

2. A water well serving a community is found to contain large amounts of $BaCl_2$. Describe a remediation approach and identify and describe the principle on which this approach is based.

3. A lake serving as a drinking water supply for a community is found to contain significant concentrations of a number of toxic organic compounds. Describe a remediation approach and identify and describe the principle on which this approach is based.

4. A farmer uses his farm land to dispose of large amounts of wheat straw. Explain the potential problems that he may run into with respect to the soil's fertility status and explain how you would remediate the situation.

5. The fuel tank of a gas station is found to leak gasoline and the surrounding area (soil) has been heavily contaminated. Describe a remediation approach and identify and describe the principle on which this approach is based.

6. The drinking water at your home is found to contain excessive amounts of lead (Pb). You decide that the immediate solution to the problem is using some form of an exchanging column. How would you answer the following questions in order to decide which column to purchase?

   a. Should there be a cationic or anionic Why?

   b. What factors would determine the column size?

   c. What factors would you consider in determining flow rate?

   d. What ion(s) would you prefer the resin to be loaded with initially? Why?

Appendix

TABLE A.1. Values for Some Physical Constants

Name	Symbol	Value
Avogadro's constant	N_A	6.022137×10^{23} mol^{-1}
Boltzmann constant	k	1.38066×10^{-23} JK^{-1}
Electron charge	e	1.602177×10^{-19} C
Faraday constant	F	96485 C mol^{-1}
Gas constant	R	8.31451 JK^{-1} mol^{-1}
Permittivity of vacuum	ε_0	8.8543×10^{-12} C^2 J^{-1} m^{-1}
Planck's constant	η	6.626075×10^{-34} Js

TABLE A.2. SI-Derived Units Expressed in Base Units

Property	Unit	Symbol	Expression in Terms of Si Base Units
Area	Square meter	m^2	
Specific surface area	Square meter per kilogram	$m^2\,kg^{-1}$	
Charge concentrations	Moles of charge per cubic meter	$mol_c\,kg^{-1}$	
Specific adsorbed charge[a]	Moles of charge per kilogram of adsorbent	$mol_c\,kg^{-1}$	
Concentration	Moles per cubic meter	$mol\,m^{-3}$	
Electrical capacitance	Faraday	F	$m^{-2}\cdot kg^{-1}\cdot s^4\cdot A^2$
Electric charge	Coulomb	C	$A\cdot s$
Electric potential difference	Volt	V	$m^2\cdot kg\cdot s^{-3}\cdot A^{-1}$
Electrical conductivity	Siemens per meter	$S\,m^{-1}$	$m^{-3}\cdot kg^{-1}\cdot s^2\cdot A^2$
Energy	Joule	J	$m^2\cdot kg\cdot s^{-2}$
Force	Newton	N	$m\cdot kg\cdot s^{-2}$
Mass density	Kilogram per cubic meter	$kg\,m^{-3}$	
Molality	Moles per kilogram of solvent	$mol\,kg^{-1}$	
Pressure	Pascal	Pa	$m^{-1}\cdot kg\cdot s^{-2}$
Viscosity	Newton-second per square meter	$N\cdot s\,m^{-2}$	
Volume	Cubic meter	m^3	
	Liter	L	$10^{-3}\cdot m^3$
Specific volume[a]	Cubic meter per kilogram	$m^3\,kg^{-1}$	

[a]Specific refers to "divided by mass."

TABLE A.3. Factors for Converting Non-SI Units to SI Units

Multiply Non-SI Unit	By	To Obtain SI Unit
Angstrom unit	10	Nanometer, nm (10^{-9} m)
Atmosphere	0.101	Megapascal, MPa (10^6 Pa)
Bar	1	Megapascal, MP (10^6 Pa)
Calorie	4.19	Joule, J
Cubic feet	0.028	Cubic meter, m^3
Cubic feet	28.3	Liter (10^{-3} m^3)
Cubic inch	1.64×10^{-5}	Cubic meter, m^3
Cubic inch	16.4	Cubic centimeter, cm^3 (10^{-6} m^3)
Dyne	10^{-5}	Newton, N
Erg	10^{-7}	Joule, J
Foot	0.305	Meter, m
Gallon	3.785	Liter, L (10^{-3} m^3)
Inch	25.4	Millimeter, mm (10^{-3} m)
Micron	1.00	Micrometer, μm (10^{-6} m)
Mile	1.61	Kilometer, km (10^3 m)
Millimhos per centimeter	1.00	Decisiemen per meter, dS m^{-1}
Ounce	28.4	Gram, g (10^{-3} kg)
Pint (liquid)	0.473	Liter, L (10^{-3} m^3)
Pound	453.6	Gram, g (10^{-3} kg)
Pound per acre	1.12	Kilogram per hectare, kg ha^{-1}
Pounds per cubic foot	16.02	Kilogram per cubic meter, kg m^{-3}
Pounds per square foot	47.88	Pascal, Pa
Pounds per square inch	6.90×10^3	Pascal, Pa
Quart (liquid)	0.946	Liter, L (10^{-3} m^3)
Square centimeter per gram	0.1	Square meter per kilogram, m^2 kg^{-1}
Square foot	9.29×10^{-2}	Square meter, m^2
Square inch	645.2	Square millimeter, mm^2 (10^{-6} m^2)
Square mile	2.59	Square kilometer, km^2
Temperature (°F) − 32	0.5555	Temperature, °C
Temperature (°C) + 273	1	Temperature, K
Ton (metric)	10^3	Kilogram, kg
Ton (2000 lb)	907	Kilogram, kg

Suggested and Cited References

Abd-Alfattah, A. and K. Wada. 1981. Adsorption of lead, copper, zinc, cobalt, and cadmium by soils that differ in cation exchange materials. *J. Soil Sci.* **32**: 271.

Abu-Sharar, T. M., F. T. Bingham, and J. D. Rhoades. 1987. Reduction in hydraulic conductivity in relation to clay dispersion and disaggregation. *Soil Sci. Soc. Am. J.* **51**:342–346.

Adams, F. 1971. Ionic concentration and activities in soil solutions. *Soil Sci. Soc. Am. Proc.* **35**:420–426.

Adams, F. 1981. Alleviating chemical toxicities: Liming acid soils. In G. F. Arkin and H. M. Taylor, Eds. *Modifying the Root Environment to Reduce Crop Stress.* ASAE Monograph 4. American Society of Agricultural Engineers, St. Joseph, MI, pp. 267–301.

Adams, F. and B. L. Moore. 1983. Chemical factors affecting root growth in subsoil horizons of Coastal Plain soils. *Soil Sci. Soc. Am. J.* **47**:99–102.

Agassi, M., I. Shainberg, and J. Morin. 1981. The effect of electrolyte concentration and soil sodicity on infiltration and crust formation. *Soil Sci. Soc. Am. J.* **45**:848–851.

Ainsworth, C. C. 1979. Pyrite forms and oxidation rates in Missouri shales. M.S. Thesis, University of Missouri,

Ainsworth, C. C. and R. W. Blanchar. 1984. Sulfate production rates in pyritic Pennsylvanian-aged shales. *J. Environ. Qual.* **13**:93–196.

Alexander, M. 1961. *Introduction to Soil Microbiology.* John Wiley and Sons, New York pp. 139–238.

Allen, P. 1993. Effects of Acute Exposure to Cadmium II Chloride and Lead II Chloride on the Haemtaological Profile of Oreochromis aurus Steindachner. *Comp. Biochem. Physiol.* **105**:213–217.

Allison, E. F. 1973. *Soil Organic Matter and its Role in Crop Production.* U. S. Department of Agriculture, Washington, D. C. Elsevier Scientific Publishing Co. Amsterdam, pp. 140–158.

Allison, J. D., D. S. Brown, and K. J. Novo-Gradac. 1994. MINTEQA2/PRODEFA2, A geochemical assessment model for environmental systems. EPA 600/3-91/021. U. S. Environmental Protection Agency, Office of Research and Development, Athens, GA.

Allison, L. E. 1965. Organic carbon. In C. A. Black (Ed.) *Methods of Soil Analysis.* Part 2. Agronomy Monograph 9. ASA, Madison, WI.

American Public Health Association et al. 1981. *Standard Methods for the Examination of Water and Wastewater*, 15th ed. American Public Health Association, Washington, DC.

Anderson, M. A. and D. T. Malotky. 1979. The adsorption of protolyzable anions on hydrous oxides at the isoelectric pH. *J. Colloid Interface Sci.* **72**:413–427.

Andriano, D. C. 1986. *Trace Elements in the Terrestrial Environment.* Springer-Verlag, New York.

Andriano, D. C., A. L. Page, A. A. Elseewi, A. C. Chang, and I. Straughan. 1980. Utilization and disposal of fly ash and other coal residues in terrestrial ecosystems: A review. *J. Environ. Qual.* **9**:333–344.

APHA. 1985. *Standard Methods for the Examination of Water and Wastewater*, 16th ed. American Public Health Association, American Water Works Association and Water Pollution Control Federation, Washington, DC.

Argersinger, W. J., Jr., A. W. Davison, and O. D. Bonner. 1950. Thermodynamics and ion exchange phenomena. *Trans. Kansas Acad. Sci.* **53**:404–410.

Armstrong, D. E., G. Chesters, and R. F. Harris. 1967. Atrazinc hydrolysis in Soil. *Soil Sci. Soc. Am. Proc.* **31**:61–66.

Arora, H. S., J. B. Dixon, and L. R. Hossner. 1978. Pyrite morphology in lignitic coal and associated strata of east Texas. *Soil Sci.* **125**:151–159.

Arora, H. S. and N. T. Coleman. 1979. The influence of electrolyte concentration on flocculation of clay suspensions. *Soil Sci.* **127**:134–139.

Arnold, P. W. 1970. The behavior of potassium in soils. *Fertil. Soc. Proc.* **115**:3–30.

Asghar, M. and Y. Kanehiron. 1981. The fate of applied iron and manganese in an oxisol and an ultisol from Hawaii. *Soil Sci.* **131**:53–55.

Aspiras, R. B., D. R. Keeney, and G. Chesters. 1972. Determination of reduced inorganic sulfur forms as suphide by zinc–hydrochloric acid distillation. *Anal. Lett.* **5**:425–432.

ASTM. 1989a. *Annual Book of ASTM Standards*, Volume 11.01, Water. American Society for Toxicity Testing and Materials, Philadelphia, PA.

ASTM. 1989b. *Standard Guide for Conducting Three-Brood, Renewal Toxicity Tests with Ceriodaphnia dubia*. E1295-89. American Society for Testing and Materials, Philadelphia, PA.

Avnimelech, Y. and M. Laher. 1977. Ammonia volatilization from soils: Equilibrium considerations. *Soil Sci. Am. J.* **41**:1080–1084.

Ayers, R. S. 1977. Quality of water for irrigation. *J. Irrig. Drain. Div., ASCE*, **103**(IR2):135–154.

Baas Becking, L. G. M., L. R. Kaplan, and D. Moore. 1960. Limits of the natural environment in terms of pH and oxidation-reduction potentials. *J. Geology* **68**:224–284.

Babcock, K. L. 1963. Theory of chemical properties of soil and colloidal systems at equilibria. *Hilgardia* **34**:417–542.

Babcock, K. L. and R. K. Schultz. 1963. Effect on anion on the sodium–calcium exchange in soils. *Soil Sci. Soc. Am. Proc.* **27**:630–632.

Baes, A. U. and P. R. Bloom. 1989. Diffuse reflectance and transmission Fourier transform infrared. *Soil Sci. Soc. Am. J.* **53**:695–700.

Baes, C. F. and R. E. Mesmer. 1976. *The Hydrolysis of Cations*. John Wiley & Sons, New York.

Bailey, L. K. and E. Peters. 1976. Decomposition of pyrite in acids by pressure leaching and anodization: The case for electrochemical mechanism. *Can. Met. Quart.* **15**:333–344.

Bailey, G. W., J. L. White, and T. Rothberg. 1968. Adsorption of organic herbicides by montmorillonite: Rolde of pH and chemical character of adsorbate. *Soil Sci. Soc. Am. J.* **32**:222–233.

Baker, B. K. 1983. The evaluation of unique acid mine drainage abatement techniques. Master's Thesis, West Virginia University, Morgantown, WV.

Baker, J. P. and C. L. Schofield. 1982. Aluminum toxicity to fish in acidic waters. *Water Air Soil Pollut.* **18**:289–309.

Banta, A. M. 1939. *Studies on the Physiology, Genetics, and Evolution of Some Cladocera.* Carnegie Institution of Washington, Washington, DC.

Barbarik, K. A. and H. J. Pirela. 1984. Agronomic and horticultural uses of zeolites: A review. In *Scientific Series by International Committee on Natural Zeolites*, Paper No. 2750, Colorado State University, Agricultural Experimental Station, Fort Collins, CO, pp. 93–103.

Barbayiannis, N., V. P. Evangelou, and V. C. Keramidas. 1996. Potassium–ammonium–calcium quantity–intensity studies in the binary and ternary modes in two soils of micaceous mineralogy of northern Greece. *Soi. Sci.* **161**:716–724.

Barfield, B. J., R. C. Warner, and C. T. Haan. 1981. *Applied Hydrology and Sedimentology for Disturbed Areas.* Oklahoma Technical Press, Stillwater, OK, pp. 603.

Barnhisel, R. I. 1977. Chlorites and hydroxy interlayer vermiculite and smectite. In J. B. Dixon et al. Eds. *Minerals in Soil Environments.* Soil Science Society of America, Madison, WI, pp. 797–808.

Barnhisel, R. I. and H. F. Massey. 1969. Chemical, mineralogical and physical properties of eastern Kentucky acid-forming coal spoil materials. *Soil Sci.* **108**:367–372.

Barrel, R. M., R. Papadopoulos, and L. V. C. Rees. 1967. Exchange of sodium in clinoptilolite by organic cations. *J. Inorg. Nucl. Chem.* **29**:2047–2063.

Barrer, R. M. and R. P. Townsend. 1984. Ion-exchange equilibria in zeolites and clay minerals. *J. Chem. Soc., Faraday Trans.* **80**:629–640.

Barriga, F. J. A. S. and W. S. Fyfe. 1988. Giant pyritic base–metal deposits: The example of Feitais Aljustrel, Portugal. *Chem. Geo.* **69**:331–343.

Bartlett, R. J. 1981. Oxidation–reduction status of aerobic soils. In *Chemistry in the Soil Environment.* M. Stelly, R. H. Dowdy, J. A. Ryan, V. V. Volk, and D. E. Baker, Eds. SSSA Special Publication No. 40, pp. 77–102. Madison: Soil Sci. Soc. Am. Press.

Barshad, I. 1954. Cation exchange in micaceous minerals. II. Replaceability of ammonium and potassium from vermiculite, biotite, and montmorillonite. *Soil Sci.* **78**:57–76.

Battelle. 1988. *Effects of Food and Water Quality on Culturing and Toxicity Testing of Ceriodaphnia dubia.* J. D. Coongy, G. M. DeGraeve, E. L. Moore, W. D. Palmer, and T. L. Pollack. Principal Investigators. EPRI EA-5820, Project 2368-2. Battelle, Columbus Division, Columbus, OH.

Baver, L. D. 1959. *Soil Physics.* John Wiley and Sons, New York.

Beckett, P. H. T. 1964. Studies on soil potassium. I. Confirmation of the ratio law: Measurement of potassium potential. *J. Soil Sci.* **15**:1–8.

Beckett, P. H. T. 1965. The cation-exchange equilibria of calcium and magnesium. *Soil Sci.* **100**:118–122.

Beckett, P. H. T. 1972. Critical cation activity ratios. *Adv. Agron.* **24**:376–412.

Behra, R. 1993. In vitro effects of cadmium, zinc, and lead on calmodulin-dependent actions in *Oncorhynchus mykiss, Mytilus sp.*, and *Chlamydomonas reinhardtii. Arch. Environ. Contam. Toxicol.* **24**:21–27.

Belander, S. E. and Cherry, D. S. 1990. Interacting effects of pH acclimation, pH and heavy metals on acute and chronic toxicity to *Ceriodaphnia dubia* Cladocera. *J. Crustacean Biol.* **102**:225–235.

Belanger, S. E., J. L. Farris, and D. S. Cherry. 1989. Effects of diet, water hardness and population source on acute and chronic copper toxicity to *Ceriodaphnia dubia. Arch. Environ. Contam. Toxicol.* **18**:601–611.

Bell, A. B. 1987. Prevention of acid generation in base metal tailings and waste rock. In *Proceedings of Acid Mine Drainage*. Seminar and Workshop. March 23–26. Halifax, Nova Scotia, pp. 391–410.

Benjamin, M. N. and J. O. Leckie. 1980. Adsorption of metals and oxide interfaces: Effect of the concentrations of adsorbate and competing metals. In R. A. Baker, Ed., *Contaminants and Sediments*, Ann Arbor Science Publishers, Ann Arbor, MI, pp. 305–322.

Bennett, J. C. and H. Tributsch. 1978. Bacteria leaching patterns on pyrite crystal surface. *J. Bacteriol.* **134**:310–317.

Bennett, R. J. and C. M. Breen. 1991. The aluminum signal: New dimensions to mechanisms of aluminum tolerance. *Plant and Soil* **134**:153–166.

Benson, S. W. 1982. *The Foundation of Chemical Kinetics*. Krieger Publishing, Malabar, FL.

Berner, R. A. 1970. Sedimentary pyrite oxidation. *Amer. J. Sci.* **268**:1–23.

Berner, R. A. 1984. Sedimentary pyrite formation: An update. *Geochim. Cosmochim. Acta* **48**:605–615.

Bernstein, L. 1975. Effect of salinity and sodicity on plant growth. *Am. Rev. of Phytopathol.* **13**:295–311.

Berrow, M. L. and J. C. Burridge. 1991. Uptake, distribution, and effects of metal compounds on plants. In E. Merian, Ed. *Metals and Their Compounds in the Environment: Occurrence, Analysis, and Biological Relevance*. VCH Publishers, New York, pp. 399–410.

Berry, L. G., and B. Mason. 1959. *Mineralogy*. W. H. Freeman, San Francisco, CA.

Bertsch, P. M. 1983. The behavior of aluminum in complex solutions and its role in the exchange equilibria of soil. Unpublished Ph.D. Dissertation. University of Kentucky, Lexington, KY.

Beveridge, A. and W. P. Pickering. 1980. Influence of humate–solute interactions on aqueous heavy metal ion levels. *Water, Air, Soil Pollut.* **14**:171.

Biggar, J. W. and D. R. Nielsen. 1967. Miscible displacement and leaching phenomena. In R. M. Hagen Ed. *Irrigation of Agricultural Lands*. Vol. 11. American Society of Agronomists, Madison, WI, pp. 254–274.

Birge, W. J. and J. A. Black. 1980. Aquatic toxicology of nickel. In J. O. Nriagu, Ed. *Nickel in the Environment*. John Wiley and Sons, New York, pp. 349–366.

Birge, W. J., J. A. Black, T. M. Short, A. G. Westerman, S. B. Taylor, and E. M. Silberhorn. 1987. *Effects of Aluminum on Freshwater Aquatic Life*. Report submitted to the Department for Natural Resources and Environmental Protection, Division of Water, Frankfort, KY.

Black, C. A. 1957. *Soil–Plant Relationships*. John Wiley and Sons, New York.

Blancher, R. W. and C. E. Marshall. 1981. Eh and pH measurement in menfro and Mexico soils. In *Chemistry in the Soil Environment*. M. Stelly, R. H. Dowdy, J. A. Ryan, V. V. Volk, and D. E. Baker, Eds. SSSA Special Publication No. 40. Soil Sci. Soc. Am. Press, Madison, WI, pp. 103–128.

Blaylock, B. G. and M. L. Frank. 1979. A comparison of the toxicity of nickel to the developing eggs and larvae of carp *Cyprinus carpio*. *Bull. Environ. Contam. Toxicol.* **21**:604–611.

Bohn, H. B., B. McNeal, and G. O'Connor. 1985. *Soil Chemistry*, 2nd ed. John Wiley & Sons, New York.

Bohn, H. L. 1970. Comparisons of measured and theoretical Mn^{2+} concentration in soil suspensions. *Soil Sci. Soc. Am. Proc.* **34**:195–197.

Bolland, M. D. A., A. M. Posner, and J. P. Quirk. 1980. pH independent and pH dependent surface charges on kaolinite. *Clays Clay Miner.* **28**:412–418.

Bolt, G. H. 1967. Cation exchange equations used in soil science. A review. *Neth. J. Ag. Sci.* **15**:81–103.

Bolt, G. H. and R. D. Miller. 1955. Swelling pressures of illite suspensions. *Soil Sci. Soc. Am. Proc.* **19**:285–288.

Bolt, G. M. and M. Peech. 1953. The application of the Gouy theory to soil–water systems. *Soil Sci. Soc. Amer. Proc.* **17**:210–213.

Bower, C. A. 1958. Cation exchange equilibria in salt affected soils. *Soil Sci.* **88**:32–35.

Bower, C. A., F. F. Reihmeier, and M. Fireman. 1952. Exchangeable cation analysis of saline and alkali soils. *Soil Sci.* **73**:251–261.

Boyd, S. A., W. F. Jaynes, and B. S. Ross. 1991. Immobilization of organic contaminants by organoclays: Application to soil restoration and hazardous waste contaminants. In R. A. Baker, Ed. *Organic Substances and Sediments in Water*, Vol. 1. Lewis Publishers, Chelsea, MI, pp. 181–200.

Boynton, D. 1941. Seasonal and soil influences on oxygen and carbon dioxide levels of New York soils. *Cornell Univ. Agr. Exp. Sta. Bul.* **763**.

Bradfield, R. 1942. Calcium in soil. *Soil Sci. Soc. Am. Proc.* **6**:8–15.

Brady, N. C. 1990. *The Nature and Properties of Soils*, 10th ed. McMillan Publishing, New York.

Bresler, E., B. L. McNeal, and D. L. Carter. 1982. *Saline and Sodic soils*. Principle–dynamics–modeling. Springer-Verlag, Berlin.

Bricker, O. 1965. Some stability relations in the system $Mn–O_2–H_2O$ at 25 °C and one atmosphere total pressure. *Am. Miner.* **50**:1296–1354.

Brierley, C. L. 1982. Microbiological Mining. *Sci. Am.* **247**:42–51.

Brine Disposal in Northeast Ohio. 1984. Report of Northeast Ohio Brine Disposal Task Force. Northeast Ohio Area-Wide Coordinating Agency.

Brock, B. R. and D. E. Kissel, Eds. 1988. *Ammonia Volatilization from Urea Fertilizers.* Tennessee Valley Authority, Muscle Shoals, AL.

Brodie, G. A., C. R. Britt, T. M. Tomaszewski, and H. N. Taylor. 1991. Use of the passive anoxic limestone drains to enhance performance of acid drainage treatment wetlands. In W. Oaks and J. Bowden, Eds. *Proceedings Reclamation 2000: Technologies for Success.* Durango, CO, pp. 211–222.

Brodie, G. A., D. A. Hammer, and D. A. Tomljanovich. 1988. Constructed wetlands for acid drainage control in the Tennessee Valley. In *Mine Drainage and Surface Mine Reclamation*, Vol. 1. U. S. Bureau of Mines Information Circular IC 9:183, 325 pp.

Bronswijk, J. J. B. and J. E. Groenenberg 1993. SMASS: A simulation model for acid sulfate soils. I. Basic principles. In D. Dent and M. E. F. van Mensvoort, Eds. *Selected Paper, Saigon Symposium on Acid Sulfate Soils, Ho Chi Minh City, Vietnam*, March 2–6 1992. ILRI Publication 52. Institute for Land Reclamation and Improvement, Wageningen, The Netherlands.

Bronswijk, J. J. B., K. Nugroho, I. B. Aribawa, J. E. Groenenberg, and C. J. Ritsema. 1993. Modeling of oxygen transport and pyrite oxidation in acid sulfate soils. *J. Environ. Qual.* **22**:544–554.

Brown, A. D. and J. J. Jurinak. 1989. Mechanism of pyrite oxidation in aqueous mixtures. *J. Environ. Qual.* **18**:545–550.

Brown, A. D. and J. J. Jurinak. 1989b. Mechanisms of nonmicrobial pyritic sulfur oxidation under alkaline conditions. *Arid Soil Res.* **3**:65–76.

Brown, K. W., B. L. Carlile, R. H. Miller, E. M. Rutledge, and E. C. A. Runge. 1985. Utilization, treatment, and disposal of waste water on land. In *Proceedings of a Workshop in Chicago, IL.* December 6–7, 1985. Soil Science Society America, Madison, WI.

Bucek, M. F. 1981. Sedimentation ponds and their impact on water quality. p. 345–354. In D. H. Graves, Ed. *Proceedings of a Symposium on Surface Mining Hydrology, Sedimentology, and Reclamation, Lexington, KY*, December 7–11, 1985. University of Kentucky, Lexington, KY.

Buffle, J. 1984. Natural organic matter and metal–organic interactions in aquatic systems. In H. Siegel, Ed. *Metal ions in biological systems.* Vol. 18. Mercer Dekker, New York, pp. 165–221.

Buffle, J. and W. Stumm. 1984. General chemistry of aquatic systems. In J. Buffle and R. R. DeVitre, Eds. *Chemical and Biological Regulation of Aquatic Systems.* CRC Press, Boca Raton, FL, pp. 1–42.

Buol, S. W., F. D. Hole, R. J. McCracken, and R. J. Southard. 1997. *Soil Genesis and Classification.* Iowa State University Press, Ames, IA.

Calvet, R. 1989. Adsorption of organic chemicals in soils, *Environ. Health Perspect.* **83**:145.

Carpenter, P. L. 1977. *Microbiology.* W. B. Saunders, *Company* Philadelphia, PA, pp. 218–220.

Caruccio, F. T. 1975. Estimating of the acid potential of the coal mine refuse. In M. J. Chadwick and G. T. Goodman, Eds. *The Ecology of Resource Degradation and Renewal.* Blackwell Scientific Publishers, London, England.

Caruccio, F. T. and G. Geidel. 1978. Geochemical factors affecting coal mine drainage quality. In F. W. Schaller and P. Sutgton, Eds. *Reclamation of Drastically Disturbed Lands.* American Society of Agronomist, Madison, WI, pp. 129–147.

Caruccio, F. T. and G. Geidel. 1980. The assessment of a stratum's capacity to produce acidic drainage. In *Proceedings of National Symposium on Surface Mine Hydrology*, Sedimentology and Reclamation. University of Kentucky, Lexington KY, pp. 437–444.

Caruccio, F. T. and G. Geidel. 1981. Estimating the minimum acid load that can be expected from a coal strip mine. In *Proceedings of National Symposium on Surface Mine Hydrology*, Sedimentology and Reclamation. University of Kentucky, Lexington, KY, pp. 117–122.

Caruccio, F. T., J. C. Ferm, J. Horne, G. Geidel, and B. Baganz. 1977. Paleoenvironment of coal and its relation to drainage quality. EPA-600/7-77-067.

Caruccio, F. T., G. Geidel, and R. Williams. 1985. Induced alkaline recharge trenches—An innovative method to bate acid mine drainage. In *Proceedings, Sixth West Virginia Surface Mine Drainage Task Force Symposium.* Morgantown, WV.

Cass, A. and M. E. Sumner. 1982. Soil pore structural stability and irrigation water quality: I. Empirical sodium stability model. *Soil Sci. Soc. Am. J.* **46**:503–506.

Chang, L. W. and L. Cockerham. 1994. Toxic metals in the environment. In L. G. Cockerham and B. S. Shane, Eds. *Basic Environmental Toxicology.* CRC Press, Boca Raton, FL, pp. 109–132.

Chao, T. T. 1972. Selective dissolution of manganese–oxides from soils and sediments with acidified hydroxylamine hydrochloride. *Soil Sci. Soc. Am. Proc.* **36**:764–768.

Chapman, H. D. 1965. Cation-exchange capacity. p. 891–900. In C. A. Black et al., Eds. *Methods of Soil Analysis.* Part. 2. ASA, Madison, WI.

Charlot, G. 1954. *Qualitative Inorganic Analysis*. John Wiley and Sons, New York.

Chelomin, V. P. and N. N. Belcheva. 1991. Alterations of microsomal lipid synthesis in gill cells of bivalve mollusc *Mizuhopecten yessoensis* in response to cadmium accumulation. *Comp. Biochem. Physiol.* **99C**:1–6.

Chelomin, V. P. and N. N. Belcheva. 1992. The effect of heavy metals on processes of lipid peroxidation in microsomal membranes from the hepatopancreas of the bivalve mollusc *Mizuhopecten yessoensis*. *Comp. Biochem. Physiol.* **103**:419–422.

Cheng. H. H., Ed. 1990. *Pesticides in the Soil Environment: Processes, Impacts, and Modeling*. Soil Science Society of America, Madison, WI.

Chien, S. H. and W. R. Clayton. 1980. Application of Elovich equation to the kinetics of phosphate release and sorption in soils. *Soil Sci. Soc. Am. J.* **44**:265–268.

Childers, N. F., Ed. *Temperate to Tropical Fruit Nutrition*. Copyright © 1966 by Horticultural Publications, Rutgers—The State University, New Brunswick, NJ. Chapter 18 is devoted to chelates.

Chiou, C. T., V. H. Freed, D. W. Schmedding, and R. L. Kohnert. 1977. Partition coefficient and bioaccumulation of selected organic chemicals. *Environ. Sci. Technol.* **11**:457.

Chu, S-Y. and G. Sposito. 1981. The thermodynamics of ternary cation exchange systems and the subregular model. *Soil Sci. Soc. Am. J.* **45**:1084–1089.

Coale, F. J., V. P. Evangelou, and J. H. Grove. 1984. Effects of saline–sodic soil chemistry on soybean mineral composition and stomatal resistance. *J. Environ. Qual.* **13**:635–639.

Coleman, N. T. and G. W. Thomas. 1964. Buffer curves of acid clays as affected by the presence of ferric iron and aluminum. *Soil Sci. Soc. Am. Proc.* **18**:187–190.

Contantz, J. 1982. Temperature dependence of unsaturated hydraulic conductivity of two soils. *Soil Sci. Soc. Am. J.* **46**:466–470.

Curtin, D. and G. W. Smillie. 1983. Soil solution composition as affected by liming and incubation. *Soil Sci. Soc. Am. J.* **47**:701–707.

Dane, J. H. and A. Klute. 1977. Salt effects on the hydraulic properties of a swelling soil. *Soil Sci. Soc. Am. J.* **4**:1043–1049.

Daniels, F. and R. A. Alberty. 1975. *Physical Chemistry*. John Wiley and Sons, New York.

Dave, G. and R. Xiu. 1991. Toxicity of mercury, copper, nickel, lead, and cobalt to embryos and larvae of zebrafish, *Brachydanio rerio*. *Arch. Environ. Contam. Toxicol.* **21**:126–134.

Dave, N. K. and A. J. Vivyurka. 1994. Water cover on acid generating uranium tailings—laboratory and field studies. In *Proceedings of International Land Reclamation and Mine Drainage Conference and Third International Conference on the Abatement of Acidic Drainage*. Pittsburgh, PA, April 24–29, 1994, pp. 1,297.

Davies, C. W. 1962. *Ion Association*. Butterworth, Washington, DC.

Deist, J. and O. Talibudeen. 1967a. Ion exchange in soils from the ion pairs K–Ca, K–Rb, and K–Na. *J. Soil Sci.* **18**(1):1225–1237.

Deist, J. and O. Talibudeen. 1967b. Thermodynamics of K–Ca exchange in soils. *J. Soil Sci.* **18**(1):1238–1248.

Derjaguin, B. V. and L. Landau. 1941. A theory of the stability of strongly charged lyophobic soils and the coalescence of strongly charged particles in electrolytic solutions. *Acta Physiocochim.* USSR. **14**:633–662.

Diamond, J. M., W. Bower, and D. Gruber. 1993. Use of man-made impoundment in mitigating acid mine drainage in the North Branch Potomac River. *Environ. Management* **17**:225–238.

Dixon, J. B. and M. L. Jackson. 1962. Properties of intergradient chlorite-expansible layer silicates of soils. *Soil Sci. Soc. Am. Proc.* **26**:358–362.

Dixon, J. D. and S. B. Weed, Eds. 1977. *Minerals in Soil Environments*, SSSA Book Series:1, 1st ed. Soil Science Society of America, Madison, WI.

Dixon, J. D. and S. B. Weed, Eds. 1989. *Minerals in Soil Environments*. SSSA Book Series:1, 2nd ed. Soil Science Society of America, Madison, WI.

Dixon, R. L. 1986. Toxic responses of the reproductive system. In C. D. Klaassen, M. O. Amdur, and J. Doull, Eds., *Casarett and Doull's Toxicology: The Basic Science of Poisons*, 3rd ed. Macmillan Publishing Co., New York, pp. 432–477.

Doefsch, R. N. and T. M. Cook. 1974. *Introduction to Bacteria and their Ecobiology*. University Park Press, Baltimore, MD.

Doepker, R. D. and P. L. Drake. 1991. Laboratory study of submerged metal mine tailings (III). Factors influencing the dissolution of metals. In *Proceedings of Second International Conference on the Abatement of Acid Mine Drainage*, Sept. 16–18, Montreal, Quebec, pp. 1 and 139.

Donahue, R. L., R. W. Miller, and J. C. Shickluna. 1983. Soils, *An Introduction to Soils and Plant Growth*, 5th ed. Prentice-Hall, Englewood Cliffs, NJ, 667 pp.

Donato, P. de, C. Mustin, J. Berthelin, and P. Marion. 1991 An infrared investigation of pellicular phases observed on pyrite by scanning electron microscopy, during its bacteria oxidation. *C. R. Acad. Sci. Paris* **321**(II):241–248.

Donato, P. de, C. Mustin, R. Benoit, and R. Erre. 1993. *Spatial distribution of iron and sulphur species on the surface of pyrite*. Appl. Surf. Sci. **68**:81–93.

Doran, J. W. and D. C. Martens. 1972. Molybdenum availability as influenced by application of fly ash to soil. *J. Environ. Qual.* **1**:186–189.

Douglas, L. A. 1977. Vermiculites. pp. 259–292. In J. B. Dixon and S. B. Weed, Eds. *Minerals in Soil Environments*. Soil Science Society of America, Madison WI.

Dowd, J. E. and D. S. Riggs. 1965. A comparison of estimates of Michaelis–Menten kinetic constants from various linear transformations. *J. Biol. Chem.* **240**:863–869.

Drever, J. I. 1982. *The Geochemistry of Natural Waters*. Prentice-Hall, Englewood, N. J.

Dudley, L. M. and B. L. McNeal. 1987. A model for electrostatic interactions among charged sites of water-soluble, organic polyions. 1. Description and sensitivity. *Soil Sci.* **143**:329.

Duffus, J. H. 1980. *Environmental Toxicology*. Halsted Press, New York.

Dugan, P. R. 1975. Bacteria ecology of strip mine areas and its relationship to production of acidic mine drainage. *Ohio J. Sci.* **75**:266–279.

Dugan, P. R. 1987. Prevention of formation of acid drainage from high-sulfur coal refuse by inhibition of iron- and sulfur-oxidizing microorganisms. II. Inhibition in "Run of Mine" refuse under simulated field conditions. *Biotech. Bioeng.* **29**:49–54.

Dugan, P. R. and W. A. Apel. 1983. Bacteria and acidic drainage from coal refuse: Inhibition by sodium lauryl sulfate and sodium benzoate. *Appl. Environ. Microbial.* **46**:279–282.

Duke, J. M. 1980b. Production and uses of nickel. In J. O. Nriagy, Ed. *Nickel in the Environment*. John Wiley and Sons, New York, pp. 51–66.

Dzombak, D. A., W. Fish, and F. M. M. Morel. 1986. Metal–humate interaction. 1. Discrete ligand and continuous distribution models. *Environ. Soil Technol.* **20**:669.

Eger, P. 1992. The use of sulfate reduction to remove metals from acid mine drainage. In *Achieving Land Use Potential Through Reclamation*. Proceedings of 9th National Meeting of the American Society for Surface Mining and Reclamation. June 14–18, 1992. Duluth, MN, pp. 563–578.

Eisler, R. 1986. *Chromium Hazards to Fish, Wildlife, and Invertebrates: A Synoptic Review.* Biological Report 85 1.6; Contaminant Hazard Reviews Report No. 6. U.S. Department of the Interior, Fish and Wildlife Service, Laurel, MD.

El Sayed, M. H., R. G. Burau, and K. L. Babcock. 1970. Thermodynamics of copper II–calcium exchange on bentonite clay. *Soil Sci. Soc. Am. Proc.* **34**:397–400.

El-Swaify, S. A. 1976. Changes in physical properties of soil clays due to precipitated Al and Fe hydroxides. II. Colloidal interaction in the absence of drying. *Soil Sci. Soc. Am. Proc.* **40**:516–520.

Elseewi, A. A., A. L. Page, and S. R. Grimm. 1980. Chemical characterization of fly ash aqueous systems. *J. Environ. Qual.* **9**:424–428.

Emerson, W. W. 1964. The slaking of soil crumbs as influenced by clay mineral composition. *Aust. J. Soil Res.* **2**:211–217.

Endell, J. 1958. Fortuna brown coal fly ash as a calcium fertilizer. *Braunkohle.* **10**:326–353; *CA* **52**:20929.

Environment Canada. 1991. *Biological Test Method: Test of Reproduction and Survival Using the Cladoceran Ceriodaphnia dubia.* Report EPS 1/RM Environment Canada, Conservation and Protection, Ottawa, Ontario.

Ephraim, J. H., H. Boren, C. Petterson, I. Arsenie, and B. Allard. 1989. A novel description of acid–base properties of an aquatic fulvic acid. *Environ. Sci. Technol.* **23**:356.

Epstein, E., and C. E. Hagen. 1952. A kinetic study of the adsorption of alkali cations by barley roots. *Plant Physiol.* **27**:457–474.

Erickson, P. M. and K. J. Ladwig. 1985. *Control of Acid Formation by Inhibition of Bacteria and by Coating Pyritic Surfaces.* Final Report to the West Virginia Department of Energy, Division of Reclamation. Charleston, WV.

Espejo, R. T. and P. Romero. 1987. Growth of *Thiobacillus ferroxidans* on elemental sulfur. *Appl. Environ. Microbial.* **53**:1907–1912.

Evangelou, V. P. 1986. The influence of anions on potassium quantity–intensity relationships. *Soil Sci. Soc. Am. J.* **50**:1182–1188.

Evangelou, V. P. 1989a. Chemical mechanisms regulating ammonia bonding with various salts in the crystal form and field implications. *Soil Sci. Soc. Am. J.* **54**:394–398.

Evangelou, V. P. 1990. Influence of water chemistry on suspended solids in coal mine sedimentation ponds. *J. Env. Qual.* **19**:428–434.

Evangelou, V. P. 1990b. Influence of water chemistry on suspended solids in coal mine sedimentation ponds. *J. Environ. Qual.* **19**:428–434.

Evangelou, V. P. 1993. Influence of sodium on soils of humid regions. In M. Passaraki, Ed. *Handbook of Plant Stress*, Marcel Dekker, New York pp. 31–62.

Evangelou, V. P. 1995a. Potential microencapsulation of pyrite by artificial inducement of $FePO_4$ coatings. *J. Environ. Qual.* **24**:535–542.

Evangelou, V. P. 1995b. *Pyrite Oxidation and its Control. Acid Mine Drainage, Surface Chemistry, Molecular Oxidation Mechanisms, Microbial Role, Kinetics, Control, Ameliorates, Limitations, Microencapsulation.* CRC/Lewis Press, Boca Raton, FL.

Evangelou, V. P. 1996a. Pyrite oxidation inhibition in coal waste by PO_4 and H_2O_2 pH-buffered pretreatment. *Int. J. Surf. Mining, Reclam. and Environ.* **10**:135–142.

Evangelou, V. P. 1996b. U. S. and Canadian Patents on "Pyrite Silica Coating," Nos. 5,494,703 and 1,431,854, respectively.

Evangelou, V. P. and A. A. Sobek. 1988. Water quality evaluation and control. L. R. Hossner, Ed. In *Reclamation of Disturbed Lands*. CRC Press, Boca Raton, FL, pp. 17–48.

Evangelou, V. P. and A. D. Karathanasis. 1986. Evaluation of potassium quantity–intensity relationships by a computer model employing the Gapon equation. *Soil Sci. Soc. Am. Proc.* **50**:58–62.

Evangelou, V. P. and A. D. Karathanasis. 1991. Influence of sodium on the kinetics of settling of suspended solids in coal mine sedimentation pond environments. *J. Env. Qual.* **20**:783–788.

Evangelou, V. P. and C. L. Garyotis. 1985. Water chemistry and its influence on dispersed solids in coal mine sediment ponds. *Reclam. Reveg. Res.* **3**:251–260.

Evangelou, V. P. and F. J. Coale. 1987. Dependence of the Gapon coefficient K_G on exchangeable sodium for mineralogically different soils. *Soil Sci. Soc. Am. J.* **51**:68–72.

Evangelou, V. P. and F. J. Coale. 1988. An investigation on the dependence of the Gapon coefficient on exchangeable sodium by three linear transformations. *Can. J. Soil Sci.* **68**:813–820.

Evangelou, V. P. and G. J. Wagner. 1987. Effects of ion activity and sugar polyalcohol osmotica on ion uptake. *J. Exp. Bot.* **38**:1637–1651.

Evangelou, V. P. and H. Huang. 1994. Peroxide Induced Oxidation Proof Phosphate Surface Coating on Iron Sulfides, U. S. Patent No. 528-522 documentation. 23 pp.

Evangelou, V. P. and J. Wang. 1993. Infrared spectra differences of atrazine between transmittance and diffuse reflectance modes and practical implications. *Spectrochim. Acta* **49**:291–295.

Evangelou, V. P. and R. E. Phillips. 1984. Ionic composition of pyritic coal spoil leachate. Interactions and effect on saturated hydraulic conductivity. *Reclam. Reveg. Res.* **3**:65–76.

Evangelou, V. P. and R. E. Phillips. 1987. Sensitivity analysis on the comparison between the Gapon and Vanselow exchange coefficients. *Soil Sci. Soc. Am. J.* **51**:1473–1479.

Evangelou, V. P. and R. E. Phillips. 1988. Comparison between the Vanselow and Gapon exchange selectivity coefficients. *Soil Sci. Soc. Am. J.* **52**:379–382.

Evangelou, V. P. and R. E. Phillips. 1989. Theoretical and experimental interrelationships of thermodynamic exchange parameters obtained by the Argersinger and Gaines and Thomas conventions. *Soil Sci.* **148**:311–321.

Evangelou, V. P. and R. L. Blevins. 1985. Soil–solution phase interactions of basic cations in long term tillage systems. *Soil Sci. Soc. Am. J.* **49**:357–362.

Evangelou, V. P. and R. L. Blevins. 1988. Effect of long-term tillage systems and nitrogen addition on quantity–intensity relationships Q/I in long-term tillage systems. *Soil Sci. Soc. Am. J.* **52**:1047–1054.

Evangelou, V. P. and X. Huang. 1992. A new technology for armoring and deactivating pyrite. In T. Younos, P. Diplas, and S. Mostaghimi, Eds. *Land Reclamation: Advances in Research and Technology*. American Society of Agricultural Engineers, Nashville, TN, pp. 291–296.

Evangelou, V. P. and X. Huang. 1994. Infrared spectroscopic evidence of an ironII–carbonate complex on the surface of pyrite. *Spectrochim. Acta* **50A**:1333–1340.

Evangelou, V. P. and Y. L. Zhang. 1995. A review: Pyrite oxidation mechanisms and acid mine drainage prevention. *Critical Rev. Environ. Sci. Technol.* **252**:141–199.

Evangelou, V. P., B. J. Barfield, and R. I. Barnhisel. 1987. Modeling release of chemical constituents in surface mine runoff and in coal mine sedimentation ponds. *Trans. Am. Soc. Agric. Eng.* **30**:82–89.

Evangelou, V. P., A. D. Karathanasis, and R. L. Blevins. 1986. Effect of soil organic matter accumulation of potassium and ammonium quantity–intensity relationships. *Soil Sci. Soc. Am. J.* **50**:378–382.

Evangelou, V. P., J. Wang, and R. E. Phillips. 1994. New developments and perspectives in characterization of soil potassium by quantity–intensity *Q/I* relationships. In D. L. Sparks, Ed. *Advances in Agronomy*, Vol. 52, Academic Press, Orlando, FL, pp. 173–227.

Evangelou, V. P., J. H. Grove, and F. D. Rawlings. 1985. Rates of iron sulfide oxidation in coal spoil suspension. *J. Environ. Qual.* **14**:91–94.

Evangelou, V. P., L. D. Whittig, and K. K. Tanji. 1984a. An automated manometric method for differentiation and quantitative determination of calcite and dolomite. *Soil Sci. Soc. Am. J.* **48**:1236–1239.

Evangelou, V. P., L. D. Whittig, and K. K. Tanji. 1984b. Dissolved mineral salts derived from mancos shale. *J. Environ. Qual.* **13**:146–150.

Evangelou, V. P., L. D. Whittig, and K. K. Tanji. 1985. Dissolution and desorption rates of Ca and Mg from mancos shale. *Soil Sci.* **139**:53–61.

Evangelou, V. P., L. W. Murdock, and F. J. Coale. 1983. Oil well salt brine contamination of soils: Their chemistry and reclamation. *Proceedings of the National Symposium on Surface Mining Hydrology, Sedimentology and Reclamation.* University of Kentucky, College of Engineering, Lexington, KY, pp. 443–446.

Evangelou, V. P., R. E. Phillips, and J. S. Shepard. 1982. Salt generation pyritic coal spoils and its effect on saturated hydraulic conductivity. *Soil Sci. Am. J.* **46**:456–460.

Evangelou, V. P., U. M. Sainju, and X. Huang. 1992. Evaluation and quantification of armoring mechanisms of calcite, dolomite and rock phosphate by manganese. In T. Younos, P. Diplas, and S. Mostaghimi, Eds. *Land Reclamation: Advances in Research and Technology*, American Society of Agricultural Engineers, Nashville, TN, pp. 304–316.

Evans, L. J. 1989. Chemistry of metal retention by soils. *Environ. Sci. Technol.* **23**:1046.

Everett, D. H. and W. I. Whitton. 1952. A general approach to hysteresis. I. *Trans. Farady Soc.* **48**:749–757.

Ferguson, J. F. and J. Gavis. 1972. A review of the arsenic cycle in natural waters. *Water Res.* **6**:1259–1274.

Ferguson, R. B., D. E. Kissel, J. K. Koelliker, and W. Basel. 1984. Ammonia volatilization from surface–applied urea: Effects of hydrogen ion buffering capacity. *Soil Sci. Soc. Am. J.* **48**:578–582.

Fletcher, I., G. Sposito, and C. S. LeVesque. 1984. Sodium–calcium–magnesium exchange reactions on a monmorillonitic soil: I. Binary exchange reactions. *Soil Sci. Soc. Am. J.* **48**:1016–1021.

Fogle, A. W., B. J. Barfield, and V. P. Evangelou. 1991. Solution sodium/calcium ratio effects on bentonite floc density. *J. Env. Sci. Health.* **A266**:1003–1012.

Follett, R. H. 1981. *Land Application of Municipal Sludge and Wastewater.* Kansas State University and the Science and Education Administration–Cooperative Extension Service, Manhattan, KS.

Foreman, J. W. 1972. *Evaluation of Mine Sealing in Butler County, Pennsylvania.* 4th Symposium on Coal Mine Drainage Resources., Louisville, KY, pp. 83–95.

Fornasiero, D., V. Eijt, and J. Ralston. 1992. An electrokinetic study of pyrite oxidation. *Colloids Surf.* **62**:63–73.

Forstner, U. and Salomons, W. 1991. Mobilization of metals from sediments. In E. Merian, Ed. *Metals and Their Compounds in the Environment: Occurrence, Analysis, and Biological Relevance*. VCH Publishers, New York, pp. 379–398.

Frenkel, H., J. O. Goertzen, and J. D. Rhoades. 1978. The effect of clay type and content, exchangeable sodium percentage, and electrolyte concentration on clay dispersion and soil hydraulic conductivity. *Soil. Sci. Soc. Am. J.* **42**:32–39.

Furrer, G. and Stumm, W. 1986. The coordination chemistry of weathering. I. Dissolution kinetics of γ-Al_2O_3 and BeO. *Geochim. Cosmochim. Acta* **50**:1847–1860.

Fyson, A., M. Kalin, and L. W. Adrian. 1994. Arsenic and nickel removal by wetland sediments. In *Proceedings of International Land Reclamation and Mine Drainage Conference and Third International Conference on the Abatement of Acidic Drainage*. Pittsburgh, PA, pp. 1 and 109.

Fytas, K., G. J. Georgopoulos, H. Soto, and V. P. Evangelou. 1996. Feasibility and cost of creating an iron phosphate coating on pyrrhotite to prevent oxidation. *Environ. Geol.* **28**:61–69

Gagen, C. J., W. E. Sharpe, and R. F. Carline. 1993. Mortality of brook trout, mottled sculpins and slimy sculpins during acadic episodes. *Trans. Am. Fish. Soc.* **1224**:616–628.

Gaines, G. L. and H. C. Thomas. 1953. Adsorption studies on clay minerals. II. A formulation of the thermodynamics of exchange adsorption. *J. Chem. Phys.* **21**:714–718.

Gamble, D. S., M. Schnitzer, and I. Hoffmen. 1970. Cu^{2+}–fulvic acid chelation equilibrium in 0.1 M KCl at 25 °C. *Can. J. Chem.* **48**:3197.

Gamble, D. S., M. Schnitzer, H. Kerndorff, and C. H. Longford. 1983. Multiple metal ion exchange equilibria with humic acid. *Geochim. Cosmochim. Acta* **47**:1311.

Gapon, E. N. 1933. On the theory of exchange adsorption in soils. *J. Gen. Chem. U.S.S.R.* 144–163; *CA* **28**:4149.

Gardner, W. R., M. S. Mayhugh, J. O. Goertzen, and C. A. Bower. 1959. Effect of electrolyte concentration and exchangeable sodium percentage on diffusivity of water in soils. *Soil Sci.* **88**:270–274.

Garrels, R. M. and C. L. Christ. 1965. Solutions minerals and equilibria. *Freeman, Cooper & Company*, San Francisco, CA.

Gast, R. G. 1972. Alkali metal cation exchange on Chambers montmorillonite. *Soil Sci. Soc. Am. Proc.* **36**:14–19.

Gast, R. G. 1977. Surface and colloid chemistry. In J. B. Dixon, and S. B. Weed (Eds.). *Minerals in Soil Environments*. Soil Science Society of America, Madison, WI, pp. 27–73.

Gast, R. G. and W. D. Klobe. 1971. Sodium–lithium exchange equilibria on vermiculite at 25° and 50 °C. *Clays Clay Miner.* **19**:311–319.

Gee, G. W. and J. W. Bauder. 1986. Particle analysis. In A. Klude, Ed. *Methods of Soil Analysis*. Part I. 2nd ed. American Society of Agronomists, Madison, WI, pp. 383–412.

Gendusa, T. C. and T. L. Beitinger. 1992. External biomarkers to assess chromium toxicity in adult *Lepomis macrochirus*. *Bull. Environ. Contam. Toxicol.* **48**:237–242.

Gillman, G. P. 1984. Using variable charge characteristics to understand the exchange cation status of toxic soils. *Aust. J. Soil Res.* **22**:71–80.

Gillman, G. P. and G. Uehara. 1980. Charge characteristics of soils with variable and permanent charge minerals. II. Experimental. *Soil Sci. Soc. Am. J.* **44**:252–255.

Girts, M. A. and R. L. P. Kleinmann. 1986. *Constructed Wetlands for Treatment of Acid Mine Drainage: A Preliminary Review.* National Symposium on Mining, Hydrology, Sedimentology, and Reclamation, University of Kentucky, Lexington, KY.

Goldberg, S. 1992. Use of surface complexation models in soil chemical systems. *Adv. Agron.* **47**:233–329.

Goldhaber M. B. 1983. Experimental study of metastable sulfur oxyanion formation during pyrite oxidation at pH 6–9 and 30 °C. *Am. J. Sci.* **283**:913–217.

Goldhaber, M. B. and I. R. Kalan. 1974. The sulfur cycle. In E. D. Goldberg Ed. *The Sea*, Vol. 5. Wiley-Interscience. New York, pp. 527–655.

Gonzalez-Erico, E., E. J. Kamprath, G. C. Naderman, and W. V. Soares. 1979. Effect of depth of lime incorporation on the growth of corn on an oxisol of central Brazil. *Soil Sci. Soc. Am. J.* **43**:1155–1158.

Goulding, K. W. T. 1983. Adsorbed ion activities and other thermodynamic parameters of ion exchange defined by mole or equivalent fractions. *J. Soil Sci.* **34**:69–74.

Goulding, K. W. T. and O. Talibudeen. 1984. Thermodynamics of K–Ca exchange in soils. II. Effects of mineralogy, residual K and pH in soils from long-term ADAS experiments. *J. Soil Sci.* **35**:409–420.

Goyer, R. A. 1986. Toxic effects of metals. In C. D. Klaassen, M. O. Amdur, and J. Doull, Eds. *Casarett and Doull's Toxicology: The Basic Science of Poisons*, 3rd ed. Macmillan, New York, pp. 582–635.

Greenland, D. J. 1971. Interaction between humic and fulvic acids and clays. *Soil Sci.* **111**:34.

Griffin, R. A. and J. J. Jurinak. 1973. Estimation of activity coefficients from the electrical conductivity of natural aquatic systems and soil extracts. *Soil Sci.* **116**:27–30.

Guy, R. D. and C. L. Chakrabarti. 1986. Studies of metal–organic interactions in model systems pertaining to natural waters. *Can. J. Chem.* **54**:2600.

Hamdy, A. A. and G. Gissel-Nielsen. 1977. Fixation of selenium by clay minerals and iron oxides. *Z. Pflanzenernaehr. Bodenkd.* **140**:63–70.

Hamilton, I. C. and R. Woods. 1981. An investigation of surface oxidation of pyrite and pyrrhotite by linear potential sweep voltammetry. *J. Electroanal. Chem.* **118**:327–343.

Hammack R. W. and G. R. Watzlaf. 1990. The effect of oxygen on pyrite oxidation. In *Proceedings of the Mining and Reclamation Conference.* Charleston, WV, April 23–26, 1990, pp. 257–264.

Hammack, R. W. and H. M. Edenborn. 1991. The removal of nickel from mine waters using bacteria sulfate reduction. In W. Oaks and J. Bowden, Eds. *Proceedings of the 1991 National Meeting of the American Society of Surface Mining and Reclamation* Vol. 1. Princeton, WV, pp. 97–107.

Handbook of Chemistry and Physics. 1960. Chemical Rubber Publishing, Cleveland, OH.

Hani, H. 1991. Heavy metals in sewage sludge and town waste compost. In E. Merian, Ed. *Metals and Their Compounds in the Environment: Occurrence, Analysis, and Biological Relevance.* VCH Publishers, New York, pp. 357–368.

Hargrove, W. L. and G. W. Thomas. 1981. Effect of organic matter on exchangeable aluminum and plant growth in acid soils. In *Chemistry in the Soil Environment.* Stelly, R. H. Dowdy,

J. A. Ryan, V. V. Volk, and D. E. Baker, Eds. SSSA Special Publication No. 10, Soil Science Society of America Press, Madison, WI, pp. 151–165.

Harris, D. C. 1982. *Quantitative Chemical Analysis*. W.H. Freeman and Company, San Francisco, CA.

Harris, J. C. 1982a. Rate of hydrolysis. In W. J. Lyman, W. F. Reehl, and D. H. Rosenblatt, Eds. *Handbook of Chemical Property Estimation Methods: Environmental Behavior of Organic Compounds*. McGraw-Hill, New York, Chapter 7, p. 48.

Harris, J. C. 1982b. Rate of aqueous photolysis. In W. J. Lyman, W. F. Reehl, and D. H. Rosenblatt, Eds. *Handbook of Chemical Property Estimation Methods: Environmental Behavior of Organic Compounds*. McGraw-Hill, New York, Chapter 8, p. 43.

Harrison, A. P. 1984. The acidophilic thiobacilli and other acidophilic bacteria that share their habitat. *Annu. Rev. Microbio.* 38:265–292.

Harsin, A. E. and V. P. Evangelou. 1989. The electrochemical properties of soil minerals. Influence on physico-chemical stability. In S. S. Augustithis, Ed. *Weathering Its Products and Deposits*. Theophrastus Publications, S.A., Athens, Greece. pp. 197–229.

Harter, R. D. 1983. Effect of soil pH on adsorption of lead, copper, zinc, and nickel. *Soil Sci. Soc. Am. J.* 47:47–51.

Harter, R. D. and J. L. Ahlrichs. 1967. Determination of clay surface acidity by infrared spectroscopy. *Soil Sci. Soc. Am. Proc.* 31:30–33.

Hatcher, P. G., M. Schnitzer, L. W. Dennis, and G. E. Maciel. 1981. Aromaticity of humic substances in soils. *Soil Sci. Soc. Am. J.* 45:1089.

Hatton, D. and W. P. Pickering. 1980. The effect of pH on the retention of Cu, Pb, Zn, and Cd by clay–humic acid mixture. *Water Air Soil Pollut.* 14:13.

Hausenbuiller, R. L. 1985. *Soil Science—Principles and Practices*, 3rd ed. Wm. C. Brown Company, Dubuque, IA.

Hayes, K. F. 1987. Equilibrium, spectroscopic, and kinetic studies of ion adsorption at the oxide/aqueous interface. Ph.D. Dissertation, Stanford University, Palo Alto, CA.

Helfferich, F. G. 1972. *Ion Exchange*. McGraw Hill, New York.

Hendershot, W. H. 1978. Measurement technique effects on the value of zero point of charge and its displacement from zero point of titration. *Can. J. Soil Sci.* 58:438–442.

Hendershot, W. H. and L. M. Lavkulich. 1983. Effect of sequioxide coatings on surface charge of standard mineral and soil samples. *Soil Sci. Soc. Am. J.* 47:1252–1260.

Hendershot, W. H. and M. Duquette. 1986. A simple barium chloride method for determining cation exchange capacity and exchangeable cations. *Soil Sci. Soc. Am. J.* 50:605–608.

Hesterberg, D. and A. L. Page. 1990. Critical coagulation concentrations of sodium and potassium illite as affected by pH. *Soil Sci. Soc. Am. J.* 54:735–739.

Hider, R. C. 1984. Siderophore mediated absorption of iron. *Struct. Bonding Berlin.* 58:26–87.

Hillel, D. 1980. *Fundamentals of Soil Physics*. Academic Press, New York.

Hiltunen, P., A. Vuorinen, P. Rehtijarvi, and O.H. Tuovinen. 1981. Release of iron and scanning electron microscopic observations. *Hydrometallurgy* 7:147–157.

Hingston, F. J., A. M. Posner, and J. P. Quirk, 1972. Anion adsorption by goethite and gibbsite. I. The role of proton in determining adsorption envelopes. *J. Soil Sci.* 23:177–192.

Hoffmann, M. R., B. C. Faust, F. A. Panda, H. H. Koo, and H. M. Tsuchiya. 1981. Kinetics of the removal of iron pyrite from coal by microbial catalysis. *Appl. Environ. Microbiol.* 42:259–271.

Hogfeldt, E. 1953. On ion exchange equilibria. II. Activities of the components in ion exchangers. *Ark. Kemi.* **5**:147–171.

Hogstrand, C., R. W. Wilson, D. Polgar, and C. M. Wood. 1994. Effects of zinc on the kinetics of branchial calcium uptake in freshwater rainbow trout during adaptation on waterborne zinc. *J. Exp. Biol.* **186**:55–73.

Hood, T. A. 1991. The kinetics of pyrite oxidation in marine systems. Ph.D. Dissertation. University of Miami, Coral Gables, FL.

Hopfer, S. M. and F. W. Sunderman, Jr. 1992. Nickel-induced derangements of thermoregulation. In E. Nieboer and J. O. Nriagu, Eds. *Nickel and Human Health: Current Perspectives.* John Wiley and Sons, New York, pp. 561–572.

Hossner, L. R. 1988. *Reclamation of Surface-Mined Land*, Vol. II. CRC Press, Boca Raton, FL.

Hourigan, W. R., R. E. Franklin, Jr., E. O. MacLean, and D. R. Bhumbla. 1961. Growth and Ca uptake by plants as affected by rate and depth of liming. *Soil Sci. Soc. Proc.* **25**:491–497.

Howarth, R. W. and J. M. Teal. 1979. Sulfur reduction in New England salt marsh. *Limnol. Oceanogr.* **24**:999–1013.

Hsu, P. H. and T. E. Bates. 1964. Formation of x-ray amorphous and crystalline aluminum hydroxides. *Mineral Mag.* **33**:749–768.

Huang, X. and V. P. Evangelou. 1994. Kinetics of pyrite oxidation and surface chemistry influences. In C. N. Alpers and D. W. Blowers, Eds., *The Environmental Geochemistry of Sulfide Oxidation.* American Chemical Society, Washington, DC, pp. 562–573.

Hunsaker, V. E. and P. F. Pratt. 1971. Calcium–magnesium exchange equilibrium in soils. *Soil Sci. Soc. Am. Proc.* **35**:151–152.

Hurlbut, C. S. Jr. and C. Klein. 1977. *Manual of Mineralogy*, 15th ed. John Wiley and Sons, New York.

Hutcheon, A. I. 1966. Thermodynamics of cation exchange on clay; Ca–K montmorillonite. *J. Soil Sci.* **17**:339–355.

Inskeep, W. P. and J. Baham. 1983. Competitive complexation of CdII and CuII by water soluble organic ligands and Na–montmorillonite. *Soil Sci. Soc. Am. J.* **47**:1109.

Ivanov, V. I. 1962. Effect of some factors on iron oxidation by cultures of *Thiobacillus ferrooxidans. Microbiol. Engl. Transl.* **31**:645–648.

Jackson, M. L. 1964. Chemical composition of soils.In F. E. Bear, Ed. *Chemistry of the Soil.* Reinhold Publishing, New York, pp. 71–141.

Jackson, M. L. 1975. Soil Chemical Analysis—Advanced Course. University of Wisconsin, Madison. Published by M. L. Jackson, Madison, WI.

Jackson, M. L., B. R. Stewart, and W. L. Daniels. 1993. Influence of flyash, topsoil, lime and rock-P on acid mine drainage from coal refuse. In *Proceedings of 1993 National Meeting of the American Society for Surface Mining and Reclamation.* Spokane, WA, May 16–19, 1993, pp. 266–276.

James, D. W., R. J. Hanks, and J. J. Jurinak. 1982. *Modern Irrigated Soils.* John Wiley & Sons, New York, 235 pp.

Jardine, P. M. and D. L. Sparks. 1984. Potassium–calcium exchange in a multireactive soil system. II. Thermodynamics. *Soil Sci. Soc. Am. J.* **48**:45–50.

Jaynes, D. B., A. S. Rogowski, and H. B. Pionke. 1984. Acid mine drainage from reclaimed coal strip mines. 1. Model description. *Water Resour. Res.* **20**:233–242.

Jenny, H. 1941. Calcium in the soil: III Pedologic relations. *Soil Sci. Soc. Am. Proc.* **6**:27–35.

Jensen, H. E. and K.L. Babcock. 1973. Cation exchange equilibria on a Yolo loam. *Hilgardia* **41**:475–487.

Jones, B. E. 1968. Effects of extending periods of osmotic stress on water relationships of pepper. *Physiol. Plant.* **21**:334–345.

Jury, W. A. and H. Fluhler. 1992. Transport of chemicals through soil: Mechanisms, models, and field applications. *Adv. Agron.* **47**:141–201.

Jury, W. A., W. R. Gardner, and W. H. Gardner. 1991. *Soil Physics.* 5-th ed. John Wiley & Sons, New York.

Just, S. R. and K. J. Stockvell. 1993. Comparison of the effectiveness of emerging in situ technologies and traditional ex-site treatment of solvent-contaminated soils. In D. W. Tedder, Ed. *Hazardous Waste Management III.* ACS Symposium Series 518, American Chemical Society, Washington, DC.

Kane, E. B. and T. J. Mullins. Thermophilic Fungi in a Municipal Waste Compost System. *Mycologia.* **65**:1087–1100.

Karathanasis, A. D. and B. F. Hajek. 1982. Revised methods for quantitative determination of minerals in soil clays. *Soil Sci. Soc. Am. J.* **46**:419–425.

Karathanasis, A. D. and V. P. Evangelou. 1986. Water sorption characteristics of Al-saturated and Ca-saturated soil clays. *Soil Sci. Soc. Am. J.* **50**:1063–1068.

Karathanasis, A. D. and V. P. Evangelou. 1987. Low temperature dehydration kinetics of Al- and Ca-saturated soil clays. *Soil Sci. Soc. Am. J.* **51**:1072–1078.

Karathanasis, A. D., H. H. Bailey, R. I. Barnhisel, and R. L. Blevins. 1986. *Descriptions and Laboratory Data for some soils in Kentucky. 2. Bluegrass Region.* University of Kentucky, Lexington, KY.

Karathanasis, A. D., V. P. Evangelou, and Y. L. Thompson. 1988. Aluminum and iron equilibrium in soil solutions and sulfate waters of acid mine watersheds. *J. Environ. Qual.* **17**:534–543.

Karathanasis, A. D., Y. L. Thompson, and V. P. Evangelou. 1991. Kinetics of aluminum and iron released from acid mine-drainage contaminated soil and spoil materials. *J. Environ. Qual.* **19**:389–395.

Kazman, Z., I. Shainberg, and M. Gal. 1983. Effect of low levels of exchangeable Na and applied phosphogypsum on the infiltration rate of various soils. *Soil Sci.* **135**:184–192.

Keay, J. and A. Wild. 1961. The kinetics of cation exchange in vermiculite. *Soil Sci.* **92**:54–60.

Keren, R. 1980. Effect of titration rate on pH and drying process on cation exchange capacity reduction and aggregate size distribution of montmorillonite hydroxy–Al complexes. *Soil Sci. Soc. Am. J.* **44**:1209–1212.

Keren, R. and I. Shainberg. 1984. Colloid properties of clay minerals in saline and sodic solution. In I. Shainberg and J. Shelhevet, Eds. *Soil Salinity Under Irrigation—Processes and Management.* Springer-Verlag, Berlin, pp. 32–47.

Keren, R. and M. J. Singer. 1988. Effect of low electrolyte concentration on hydraulic conductivity of sodium/calcium montmorillonite–sand system. *Soil Sci. Soc. Am. J.* **52**:368–373.

Keren, R. and M. J. Singer. 1989. Effect of low electrolyte concentration on hydraulic conductivity of clay–sand–hydroxy polymers systems. *Soil Sci. Soc. Am. J.* **53**:349–355.

Khan, S. U. 1969. Interaction between the humic acid fraction of soils and certain metallic cations. *Soil Sci. Soc Am. Proc.* **33**:851.

Khan, S. U. 1971. Distribution and characteristics of organic matter extracted from the black solonetzic and black chernozemic soils of Alberta: The humic acid fraction. *Soil Sci.* **112**:401.

King, L. D., Ed. 1996. *Agricultural Use of Municipal and Industrial Sludges in the Southern United States*. Southern Cooperative Series Bulletin 314. North Carolina State University Press, Raleigh, NC.

Kinniburgh, D. G. 1986. General purpose adsorption isotherms. *Environ. Sci. Technol.* **20**:895–904.

Kissel, D. E. and G. W. Thomas. 1969. Conductimetric titrations with $CaOH_2$ to estimate the neutral salt replacability and total soil acidity. *Soil Sci.* **108**:177–179.

Kissel, D. E., E. P. Gentzsch, and G. W. Thomas. 1971. Hydrolysis of non-exchangeable acidity in soils during salt extractions of exchangeable acidity. *Soil Sci.* **111**:293–297.

Kjellstroem. T. 1986. Itai–itai Disease. In L. Friberg, C. G. Elinder, T. Kjellstroem, and G. F. Nordberg, Eds. *Cadmium and Health: A Toxicological and Epidemiological Appraisal*, Vol. II. CRC Press, Boca Raton, FL, pp. 257–290.

Klassen, C. D. 1985. Heavy metals and heavy-metal antagonists. In A. G. Gilman, L. S. Goodman, T. W. Rall, and F. Murad, Eds. *Goodman and Golman's The Pharmacological Basis of Therapeutics*. Macmillan Publishing, New York, pp. 1605–1627.

Kleinmann, R.L.P. 1980. Bactericidal control of acid problems in surface mines and coal refuse. *Proceedings of National Symposium of Surface Mining Hydrology, Sedimentology and Reclamation*. University of Kentucky, Lexington, KY, pp. 31–38.

Kleinmann, R. L. P. 1981. The U. S. Bureau of Mines acid mine drainage research program. In *Proceedings, Second West Virginia Surface Mine Drainage Task Force Symposium*. Clarksburg, WV.

Kleinmann, R. L. P. 1985. Treatment of acid mine water by wetlands. In *Control of Acid Mine Drainage*. Bureau of Mines IC 9027, 61 pp.

Kleinmann, R. L. P. 1989. Acid Mine Drainage: U.S. Bureau of Mines researches and develops control methods for both coal and metal mines. *E & MJ* 16I–16M.

Kleinmann, R. L. P. and D. A. Crerar. 1979. *Thiobacillus ferrooxidans* and the formation of acidity in simulated coal mine environments. *Geomicrobiol. J.* **1**:373–388.

Kokholm, G. 1977. *Redox Measurements: Their Theory and Technique*. ST40, Radiometer A/S Emdrupvej 72 DK–2400, Copenhagen NV, Denmark.

Kononona, M. M. 1961. *Microorganisms and Organic Matter of Soils*. Academy of Science of the U.S.S.R. Dokuchaev Soil Science Institute. Israel Program for Scientific Translations. Jerusalem.

Koskinen, W. C., G. A. O'Connor, and H. H.Cheng. 1979. Characterization of hysteresis in the desorption of 2,4,5-T from soils. *Soil Sci. Soc. Am. J.* **43**:871–874.

Kuo, S. and A. S. Baker. 1980. Sorption of copper, zinc, and cadmium by some acid soils. *Soil Sci. Soc. Am. J.* **44**:969–974.

Krotz, R. M., V. P. Evangelou, and G. J. Wagner. 1989. Relationships between cadmium, zinc, Cd-peptide and organic acid in tobacco suspension cells. *Plant Physiol.* **91**:780–787.

Kruyt, H. R., Ed. 1952. *Colloid Science I: Irreversible Systems*. Elsevier, New York.

Ladwig, K. J., P. M. Erickson, and R. L. P. Kleinmann. 1985. Alkaline injection: An overview of recent work. In *Control of Acid Mine Drainage*. Bureau of Mines IC 9027. USDA, Bureau of Mines, Pittsburgh, PA, pp. 35–40.

Lagerwerff, J. V., F. S. Nakayama, and M. H. Frere. 1969. Hydraulic conductivity related to porosity and swelling of soil. *Soil Sci. Soc. Am. Proc.*, **33**:3–11.

Laudelout, H., R. van Bladel, G. H. Bolt, and A. L. Page. 1967. Thermodynamics of heterovalent cation exchange reactions in a montmorillonite clay. *Trans. Faraday Soc.* **64**:1477–1488.

Lekhakul, S. 1981. The effect of lime on the chemical composition of surface mined coal spoils, and the leachate from spoil. Ph.D. Dissertation, Agronomy Department, University of Kentucky, Lexington, KY, p.134.

LeRoux, N. W., P. W. Dacey and K. L. Temple. 1980. The microbial role in pyrite oxidation at alkaline pH in coal mine spoil. In P. A. Truidnger, M. R. Walter, and B. J. Ralph, Eds. *Biogeochemistry of Ancient and Modern Environments.* Springer–Verlag, Berlin, pp. 515–520.

Levy, R. and D. Hillel. 1968. Thermodynamic equilibrium constants of sodium–calcium exchange in some Israel soils. *Soil Sci.* **106**:393–398.

Lindsay, W. L. 1979. *Chemical Equilibria in Soil.* John Wiley & Sons, New York.

Lindsay, W. L. and W. A. Norvell. 1969. Equilibrium relationships of Zn^{+2}, Fe^{+3}, Ca^{+2}, and H^+ with EDTA and DTPA in soils. *Soil Sci. Soc. Am. Proc.* **33**:62–68.

Lizama, H. M. and I. Suzuki. 1989. Rate equation and kinetic parameters of the reactions involved in pyrite oxidation by *Thiobacillus ferrooxidans. Appl. Environ. Microbiol.* **55**:2918–2923.

Long, J. 1994. Senate's drinking water act revisions take new track on environmental issues. *Chem. Eng. News,* **7232**:21–22.

Loomis, E. C. and W. C. Hood. 1984. The effects of anaerobically digested sludge on the oxidation of pyrite and the formation of acid mine drainage. *Proceedings of National Symposium of Surface Mining Hydrology, Sedimentology and Reclamation.* University of Kentucky, College of Engineering, Lexington, KY, pp. 1–18.

Lorenz, W. C. and E. C. Tarpley. 1963. *Oxidation of Coal Mine Pyrites.* U.S. Bureau of Mines. RI 6247.

Low, P. F. 1955. The role of aluminum in the titration of bentonite. *Soil Sci. Soc. Am. Proc.* **19**:135–139.

Lowe, L. E. 1969. Distribution and properties of organic fraction in selected Alberta soils. *Can. J. Soil Sci.* **49**:129.

Lowson, R. T. 1982. Aqueous oxidation of pyrite by molecular oxygen. *Chem. Rev.* **82**:461–497.

Lumbanraja, J. and V. P. Evangelou. 1990. Binary and ternary exchange behavior of potassium and ammonium on Kentucky subsoils. *Soil Sci. Soc. Am. J.* **54**:698–705.

Lumbanraja, J. and V. P. Evangelou. 1991. Influence of acidification and liming on surface charge behavior of three Kentucky subsoils. *Soil Sci. Soc. Am. J.* **54**:26–34.

Lumbanraja, J., and V. P. Evangelou. 1992. Potassium quantity–intensity relationships in the presence and absence of NH_4 for three Kentucky soils. *Soil Sci.* **154**:366–376.

Lumbanraja, J. and V. P. Evangelou. 1994. Adsorption–desorption of potassium and ammonium at low exchange fractional loads of three Kentucky subsoils. *Soil Sci.* **157**:269–277.

Lundgren, D. G. and M. Silver. 1980. Ore leaching by bacteria. *Annu. Rev. Microbio.* **34**:263–283.

Lundgren, D. G., J. R. Vestal, and F. R. Tabita. 1972. *Water Polution Microbiology.* Wiley-Interscience, New York.

Lurtz, J. A., Jr. 1966. Ammonium and potassium fixation and release in selected soils of southeastern United States. *Soil Sci.* **102**:366–372.

Luther III, G. W. 1982. Pyrite and oxidized iron mineral phases from pyrite oxidation in salt marsh and estuarine sediments. *Geochim. Cosmochim. Acta* **46**:2665–2669.

Luther III, G. W. 1987. Pyrite oxidation and reduction: Molecular orbital theory consideration. *Geochem. Cosmochem. Acta* **51**:3193–3199.

Luther III, G. W. 1990. The frontier-molecular-orbital theory approach in geotechnical processes. In W. Stumm, Ed. *Aquatic Chemical Kinetics.* John Wiley & Sons, New York, pp. 173–198.

Luther III, G. W., J. E. Kostka, T. M. Church, B. Sulzberger, and W. Stumm. 1992. Seasonal iron cycling in the salt-marsh sedimentary environment: the importance of ligand complexes with FeII and FeIII in the dissolution of FeIII minerals and pyrite, respectively. *Marine Chem.* **40**:81–103.

Lyman, W. J., W. F. Reehl, and D. H. Rosenblatt, Eds. 1982. *Handbook of Chemical Property Estimation Methods: Environmental Behavior of Organic Compounds.* McGraw-Hill, New York.

Lynch, J. and A. Lauchli. 1988. Salinity affects intra-cellular calcium in corn root protoplasts. *Plant Physiol.* **87**:351–356.

Ma, L. and H. M. Selim. 1996. Atrazine retention and transport in soils. *Rev. Environ. Contam. Toxil.* **145**:129–173.

Maes, A., P. Peigneur, and A. Creners. 1976. Thermodynamics of transition metal ion exchange in montmorillonite. *Proc. Int. Clay Conf. Mexico City* **1975**:319–329.

Magdoff, F. R. and R. J. Bartlett. 1985. Soil pH buffering revisited. *Soil Sci. Soc. Am. J.* **49**:145–148.

Malle, K. G. 1992. Zink in der Umwelt. *Acta. Hydrochim. Hydrobiol.* **20**:196–204.

Malone, R., R. Warner, J. L. Woods, and V. P. Evangelou. 1995. Transport of benzene and trichloroethylene through a landfill soil liner mixed with coal slurry. *Waste Manag. Res.* **12**:417–428.

Manahan, S. E. 1991. *Environmental Chemistry,* 5th ed. Lewis Publishers, Chelsea, MI.

Manahan, S. E. 1992. *Toxicological Chemistry,* 2nd ed. Lewis Publishers, Chelsea, MI.

Marion G. M., D. M. Hendricks, G. R. Dutt, and W. H. Fuller. 1976. Aluminum and silica solubility in soils. *Soil Sci.* **127**:76–85.

Marion, G. M. and K. L. Babcock. 1976. Predicting specific conductance and salt concentration of dilute aqueous solution. *Soil Sci.* **122**:181–187.

Marshall, C. E. 1977. *The Physical Chemistry and Mineralogy of Soils.* John Wiley & Sons, New York.

Marshall, J. S. and D. L. Mellinger. 1980. Dynamics of cadmium stressed plankton communities. *Can. J. Fish. Aquat. Sci.* **37**:403–414.

Marsi, M. and V. P. Evangelou. 1991a. Chemical and physical behavior of two Kentucky soils: I. Sodium–calcium exchange. *J. Env. Sci. Health* **A267**:1147–1176.

Marsi, M. and V. P. Evangelou. 1991b. Chemical and physical behavior of two Kentucky soils: II. Saturated hydraulic conductivity–exchangeable sodium relationships. *J. Env. Sci. Health.* **A267**:1177–1194.

Marsi, M. and V. P. Evangelou. 1991c. Chemical and physical behavior of two Kentucky soils: III. Saturated hydraulic conductivity–Imhoff cone test relationships. *J. Env. Sci. Health.* **A267**:1195–1215.

Marsi, M. and V. P. Evangelou. 1993. Modeling brackish solution influences on the chemical and physical behavior of temperate region soils. In D. W. Tedder and F. G. Pohland, Eds. *Emerging Technologies in Hazardous Waste Management III*, American Chemical Society, Washington, DC.

Martell, A. E. and M. Calvin. 1952. *Chemistry of the Metal Chelate Compounds.* Prentice-Hall, Inc., New York.

Martens, D. C. 1971. Availability of plant nutrients in fly ash. *Compost Sci.* **12**(6):15–19.

Martens, D. C., M. G. Schnappinger, and L. W. Zelazny. 1970. The plant availability of potassium in fly ash. *Soil Sci. Soc. Am. Proc.* **34**:453–456.

Martin, J. P. and K. Haider. 1986. Influence of mineral colloids on turnover rates of soil organic carbon. p. 283–304. In P. M. Huang and M. Schnitzer, Eds. *Interactions of Soil Minerals with Natural Organics and Microbes.* Soil Science Society of America Special Publication No. 17., Madison, WI, 1986.

Martin, J. P., S. J. Richard, and P. F. Pratt. 1964. Relationship of exchangeable Na percentage at different soil pH levels to hydraulic conductivity. *Soil Sci. Soc. Am. Proc.* **28**:620–622.

Mason, C. F. 1981. *Biology of Freshwater Pollution.* Longman Group Limited, Essex.

Masterson, W. L., E. J. Slowinski, and C. L. Stanitski. 1981. *Chemical Principles*, 5th ed. Saunders College Publishing, Philadelphia, PA.

Mattigod, S. V., G. Sposito, and A.L. Page. 1981. Factors affecting the solubilities of trace metals in soils. In R. H. Dowdy, ed. *Chemistry in the Soil Environment.* ASA Special Publication No. 40. Madison, WI.

McAvoy, D. C. 1988. Seasonal trends of aluminum chemistry in a second-order Massachusetts stream. *J. Environ. Qual.* **174**:528–534.

McBride, M. B. 1989. Surface chemistry of soil minerals. In J. D. Dixon and S. B. Weed, Eds. *Minerals in Soil Environments*, 2nd ed. SSSA Book Series:1, Madison, WI.

McBride, M. B. 1994. *Environmental Chemistry of Soils.* Oxford University Press, New York.

McBride, M. B. 1994. Toxic metal accumulation from agricultural use of sewage sludge: Do USEPA regulations ensure long-term protection of soil? *Composting Frontiers*, **II**, (4) p. 27.

McDonald, L. M. and V. P. Evangelou. 1997. Choice of an optimal solid-to-solution ratio for organic chemical sorption experiments. *Soil Sci. Soc. Am. J.* November 1996.

McHardy, B. M. and J. J. George. 1990. Bioaccumulation and toxicity of zinc in the green alga, *Cladophora glomerata. Environ. Pollut.* **66**:55–66.

McKenzie, R. M. 1977. The manganese oxides and hydroxides. In J. B. Dixon and S. B. Weed, Eds. *Minerals in Soil Environments.* Soil Science Society of America, Madison, WI, pp. 181–193.

McKenzie, R. M. 1980. The adsorption of lead and other heavy metals on oxides of manganese and iron. *Aust. J. Soil Res.* **18**:61–73.

McKibben M. A. and H. L. Barnes. 1986. Oxidation of pyrite in low temperature acidic solutions: Rate laws and surface textures. *Geochim. Cosmochim. Acta* **50**:1509–1520.

McNeal, B. L. and N. T. Coleman. 1966. Effect of solution composition on soil hydraulic conductivity. *Soil Sci. Soc. Am. Proc.* **30**:308–312.

McNeal, B. L., D. A. Layfield, W. A. Norvell, and J. D. Rhoades. 1968. Factors influencing hydraulic conductivity of soils in the presence of mixed salt solutions. *Soil Sci. Soc. Am. Proc.* **32**:187–290.

McNeal, B. L., W. A. Norvell, and N. T. Coleman. 1966. Effect of solution composition on the swelling of extracted soil clays. *Soil Sci. Soc. Am. Proc.* **30**:313–315.

Means, J. C., G. S. Wood, J. J. Hassett, and W. L. Banwart. 1987. Sorption of polynuclear aromatic hydrocarbons by sediments and soils. *Environ. Sci. Technol.* **14**:1524.

Mehta, A. P. and L. E. Murr. 1983. Fundamental studies of the contribution of galvanic interaction to acid-bacterial leaching of mixed metal sulfides. *Hydrometallurgy* **9**:235–256.

Mehta, S. C., S. R. Poonia, and R. Pal. 1983. Sodium–calcium and sodium–magnesium exchange equilibria in soil for chloride- and sulfate-dominated systems. *Soil Sci.* **136** (6):339–346.

Meikle, P. G. 1975. Fly ash. In C. L. Mantell, Ed. *Solid Wastes: Origin, Collection, Processing and Disposal.* John Wiley & Sons, New York, pp. 727–749.

Meiri, A. and A. Poljakoff-Mayber. 1970. Effect of various alinity regimes on growth. Leaf expansion and transpiration rate of bean plants. *Soil Sci.* **109**:26–34.

Meites, L. 1963. *Handbook of Analytical Chemistry.* McGraw-Hill, New York, pp. 1–39.

Miller, R. H., R. K. White, T. L. Logan, D. L. Forster, and J. N. Stitzlein. 1979. *Land Application of Sewage Sludge.* The Ohio State University Cooperative Extension Service, Columbus, OH.

Millero, F. J. 1985. The effect of ionic interactions on the oxidation of metals in natural waters. *Geochim. Cosmochim. Acta* **49**:547–553.

Millero, F. J. and M. Izaguirre. 1989. Effect of ionic strength and ionic interactions on the oxidation of Fe^{2+}. *J. Solution Chem.* **18**:585–599.

Ming, D. W. and J. B. Dixon. 1987. Quantitative determination of clinoptilolite in soils by a cation-exchange capacity method. *Clays Clay Min.* **356**:463–468.

Moni, S. and S. Manohar-Dhas. 1989. Effects of water hardness on the toxicity of zinc to *Sarotherodon mossambicus* (Peters). *Uttar Pradesh J. Zool.* **9**(2):263–270.

Montarges, E., L. J. Michot, F. Lhote, T. Fabien, and F. Villieras. 1995. Intercalation of Al_{13}-polyethyleneoxide complexes into montmorillonite clay. *Clays Clay Min.* **43**:417–426.

Moore, M. V. and R. W. Winner. 1989. Relative sensitivity of *Ceriodaphnia dubia* laboratory tests and pond communities of zooplankton and benthos to chronic copper stress. *Aquat. Toxicol.* **15**:311–330.

Morel, F. M. M. and J. G. Hering. *Principles and Applications of Aquatic Chemistry.* John Wiley & Sons, New York.

Morgan, J. J. and W. Stumm. 1991. Chemical processes in the environment: Relevance of chemical speciation. In E. Merian, Ed. *Metals and Their Compounds in the Environment: Occurrence, Analysis, and Biological Relevance.* VCH Publishers, New York, pp. 67–103.

Morgun, Y. G. and Y. A. Pachepskiy. 1986. Selectivity of ion exchange sorption in $CaCl_2$–$MgCl_2$–$NaCl$–H_2O soil system. *Soviet Soil Sci.* **19**:1–10.

Morison, R. T. 1976. *Organic Chemistry,* 3rd ed., Allyn and Bacon, Boston, MA.

Mortensen, J. L. 1963. Complexing of metals by soil organic matter. *Soil Sci. Soc. Am. Proc.* **27**:179.

Mortland, M. M. 1968. *Protonation of Compounds on Clay Mineral Surfaces,* Vol. I. 9th International Congress on Soil Science, pp. 691–699.

Mortland, M. M. and K. V. Raman. 1968. Surface acidity of smectites in relation to hydration, exchangeable cation and structure. *Clays Clay Min.* **16**:393–398.

Mortland, M. M., J. J. Fripiat, J. Chaussidon, and J. Uytterhoeven. 1963. Interaction between ammonia and the expanding lattices of montmorollonite and vermiculite. *J. Phys. Chem.* **67**:248–258.

Moses, C. O. and J. S. Herman. 1991. Pyrite oxidation at circumneutral pH. *Geochim. Cosmochim. Acta* **55**:471–482.

Moses, C. O., D. K. Nordstrom, J. S. Herman, and A. L. Mills. 1987. Aqueous pyrite oxidation by dissolved oxygen and by ferric iron. *Geochim. Cosmochim. Acta* **51**:1561–1571.

Murungi, J. I. and J. W. Robinson. 1992. Uptake and accumulation of aluminum by fish: The modifying effect of added ions. *J. Environ. Sci. Health Part A: Environ. Sci. Eng.* **27**:3713–3718.

Nairn, R. W., R. S. Hedin, and G. R. Watzlaf. 1991. A preliminary review of the use of anoxic limestone drains in the passive treatment of acid mine drainage. In *Proceedings of the 12th Annual West Virginia Surface Mine Drainage Task Force Symposium*. Morgantown, WV, pp. 23–38.

Nairn, R. W., R. S. Hedin, and G. R. Watzlaf. 1992. Generation of alkalinity in an anoxic limestone drain. In *Proceedings of the 9th Annual Meeting of the American Society for Surface Mining and Reclamation*. Duluth, MN, June 14–18, 1992.

Nalewajko, C. and B. Paul. 1985. Effects of manipulations of aluminum concentrations and pH on phosphate uptake and photosynthesis of planktonic communities in two precambrian shield lakes. *Can. J. Fish Aquat. Sci.* **42**:1946–1953.

Neal, R. H., G. Sposito, K. M. Holtzclaw, and S. J. Traina. 1987. Selenite adsorption on alluvial soils: I. Soil composition and pH effects. *Soil Sci. Soc. Am. J.* **51**:1161–1165.

Nebeker, A. V., A. Stinchfield, C. Savonen, and G. A. Chapman. 1986. Effects of copper, nickel and zinc on three species of Oregon freshwater snails. *Environ. Toxicol. Chem.*, **5**:807–811.

Nelson, D. D., G. T. Fraser, and W. Klemper. 1987. Does ammonia hydrogen bond? *Science* **238**:1670–1674.

Nelson, D. W. and L. E. Sommers. 1982. Total carbon, organic carbon, and organic matter. In A. L. Page, Eds. *Methods of Soil Analysis*. Part 2. 2nd ed. ASA, Soil Science Society of America, Madison, WI, pp. 539–580.

Neumann, P. M., E. van Volkenburgh, and R. E. Cleland. 1988. Salinity stress inhibits bean leaf expansion by reducing turgor, not wall extensibility. *Plant Physiol.* **88**:233–237.

Nicholas, G. D. and E. G. Foree. 1979. Reducing iron and manganese to permissible levels in coal mine sedimentation ponds. In D. H. Graves, Ed. *Proceedings of the Symposium on Surface Mining Hydrology, Sedimentology, and Reclamation*. Lexington, KY. December 4–7, University of Kentucky, Lexington, KY, pp. 181–187.

Nicholson, R. V., R. W. Gillham, and E. J. Reardon. 1988. Pyrite oxidation in carbonate-buffered solution: 1. Experimental kinetics. *Geochim. Cosmochim. Acta.* **52**:1077–1085.

Nicholson, R. V., R. W. Gillham, and E. J. Reardon. 1990. Pyrite oxidation in carbonate-buffered solution: 2. Rate control by oxide coatings. *Geochim. Cosmochim. Acta* **54**:395–402.

Nielsen, D. R. and J. W. Biggar. 1961. Miscible displacement in soils: I. Experimental information. *Soil Sci. Soc. Am. Proc.* **25**:1–5.

Nielsen, D. R. and J. W. Biggar. 1962a. Miscible displacement: III. Theoretical considerations. *Soil Sci. Soc. Am. Proc.* **36**:216–221.

Nielsen, D. R. and J. W. Biggar. 1962b. Miscible displacement: II. Behavior of Tracers. *Soil Sci. Soc. Proc.* **26**:125–128.

Nielsen, D. R., M. Th. van Genuchten, and J. W. Biggar. 1986. Water flow and solute transport processes in the unsaturated zone. *Water Rec. Res.* **22**:89S–108S.

Nikinmaa, M. 1992. How does environmental pollution affect red cell function in fish. *Aquat. Toxicol.* **22**:227–238.

Nkedi-Kia, P., P. S. C. Rao, and J. W. Johnson. 1983. Adsorption of diuron and 2,4,5-T on soil particle-size separates. *J. Environ. Qual.* **12**:195–197.

Nordstrom, D. K. 1982a. Aqueous pyrite oxidation and the consequent formation of secondary iron minerals. In L. R. Hossner, J. A. Kittrick, and D. F. Fanning, Eds. *Acid Sulfate Weathering: Pedogeochemistry and Relationship to Manipulation of Soil Minerals.* Soil Science Society of America Press, Madison, WI, pp. 46–53.

Nordstrom, D. K. 1982b. The effect of sulfate on aluminum concentrations in natural waters: Some stability relations in the system Al_2O_3–SO_3–H_2O at 298 K. *Geochim. Cosmochim. Acta* **46**:681–692.

Norvell, W. A. 1972. Equilibria of metal chelates in soil solution. In J. J. Mortvedt, P. M. Giordano, and W. L. Lindsay, Eds. *Micronutrients in Agriculture.* Soil Science Society of America, Madison, WI, pp. 115–138.

Norvell, W. A. and W. L. Lindsay. 1969. Reactions of EDTA complexes of Fe, Zn, Mn, and Cu with Soils. *Soil Sci. Soc. Am. Proc.* **33**:86–91.

Novak, J. M. and N. E. Smeck. 1991. Comparisons of humic substances extracted from contiguous lafisols and mollisols of southwestern Ohio. *Soil Sci. Soc. Am. J.* **55**:96.

Novozamsky, I., J. Beek, and G. H. Bolt. 1976. Chemical equilibria. In G. H. Bolt and M. G. M. Bruggenwept, Eds., *Soil Chemistry*, Elsevier Scientific Publishing Company, Amsterdam, pp. 13–42.

Nye, P. H. and P. B. Tinker. 1977. Solute movement in the soil-root system. University of California Press, Berkley, CA.

Ogwada, R. A. and D. L. Sparks. 1986a. Use of mole or equivalent fractions in determining thermodynamic parameters for potassium exchange in soils. *Soil Sci.* **141**:268–273.

Ogwada, R. A. and D. L. Sparks. 1986b. A critical evaluation on the use of kinetics for determining thermodynamics of ion exchange in soils. *Soil Sci. Soc. Am. J.* **50**:300–305.

Ohio State University Research Foundation. 1971. Acid mine drainage formation and abatement. Water Pollution Control Research Series DAST-42-14210 FPR–04/71. USEPA, Washington, DC, 83 pp.

Opuwaribo, E. and C. T. I. Odu. 1974. Fixed ammonium in Nigerian soils. I. Selection of a method and amounts of native fixed ammonium. *J. Soil Sci.* **25**:256–264.

Opuwaribo, E. and C. T. I. Odu. 1978. Ammonium fixation in Nigerian Soils: 4. The effects of time, potassium and wet dry cycles on ammonium fixation. *Soil Sci.* **125**:137–145.

O'Shay, T., L. R. Hossner, and J. B. Dixon. 1990. A modified hydrogen peroxide oxidation method for determination of potential acidity in pyritic overburden. *J. Environ. Qual.* **19**:778–782.

Oster, J. D., and G. Sposito. 1980. The Gapon coefficient and the exchangeable sodium percentage–sodium adsorption ratio relation. *Soil Sci. Soc. Am. J.* **44**:258–260.

Oster, J. D., and H. Frenkel. 1980. The chemistry of the reclamation of sodic soils with gypsum and lime. *Soil Sci. Soc. Am. J.* **44**:41–45.

Oster, J. D., I. Shainberg, and J. D. Wood. 1980. Flocculation value and gel structure of sodium/calcium montmorillonite and illite suspensions. *Soil Sci. Soc. Am. J.* **44**:955–959.

Page, A. L., T. J. Logan, and J. A. Ryan. 1987. *Land Application of Sludge.* Lewis Publishers, Ann Arbor, MI.

Palencia, I. R., W. Wan, and J. D. Miller. 1991. The electrochemical behavior of a semiconducting natural pyrite in the presence of bacteria. *Metall. Trans. B* **22B**:765–773.

Parc, S., D. Nahon, Y. Tardy, and P. Vieillard. 1989. Estimated solubility products and fields of stability for cryptomelane, nsutite, birnessite, and lithiophorite based on natural lateritic weathering sequences. *Am. Miner.* **74**:466–475.

Park, C. S., and G. A. O'Connor. 1980. Salinity effect on hydraulic conductivity of soils. *Soil Sci.* **130**:167–174.

Parker, D. R., W. A. Novell, and R. L. Chaney. 1995. GEOCHEM-P: A chemical speciation program for IBM and compatible personal computers, In R. H. Leoppert, Ed., *Chemical Equilibrium and Reaction Models.* Soil Science Society of America special publication, American Society of Agronomy, Madison, WI.

Parker, J. C., L. W. Zelazny, S. Sampath, and W. G Harris. 1979. A critical evaluation of the extension of zero point of charge ZPC theory to soil system. *Soil Sci. Soc. Am. J.* **43**:668–676.

Parks, G. A. and P. L. DeBruyn. 1962. The zero point of charge of oxides. *J. Phys. Chem.* **66**:967–972.

Patrick, W. H., Jr., and F. T. Turner. 1968. Effect of redox potential on manganese transformation in waterlogged soil. *Nature* **220**:476–478.

Pavan, M. A. and F. T. Bingham. 1982. Toxicity of aluminum to coffee seedlings grown in nutrient solutions. *Soil Sci. Soc. Am. J.* **46**:993–997.

Pavan, M. A., F. T. Bingham, and P. F. Pratt. 1982. Toxicity of aluminum to coffee in Ultisols and Oxisols amended with $CaCO_3$, $MgCO_3$, and $CaSO_42H_2O$. *Soil Sci. Soc. Am. J.* **46**:1201–1207.

Patterson, J. W., H. E. Allen, and J. J. Scala. 1977. Carbonate precipitation from heavy metals pollutants. *J. Water Pollut. Control Fed.* **12**:2397–2410.

Pearson, R. G. 1963. Acids and bases. *Science* **151**:1721–1727.

Pearson, R. G. 1966. Hard and soft acids and bases. *J. Am. Chem. Soc.* **85**:3533–3539.

Pennak, R. W. 1989. *Freshwater Invertebrates of the United States.* John Wiley and Sons, New York.

Pereira, J. J., R. Mercaldo–Allen, C. Kuropat, D. Luedke, and G. Sennefelder. 1993. Effect of cadmium accumulation of serum vitellogenin levels and hepatosomatic and gonadosomatic indices of winter flounder *Pleuronectes americanus. Arch. Environ. Contam. Toxicol.* **24**:427–431.

Perrin, C. L., and R. K. Gipe. 1987. Rotation and solvation of ammonium ion. *Science* **238**:1393–1394.

Pesic, B. D., J. Oliver, and P. Wichlacz. 1989. An electrochemical method of measuring rate of ferrous to ferric iron with oxygen in the presence of *Thiobacillus ferrooxidans. Biotech. and Bioengr.* **33**:428–439.

Petering, H. G. and C. J. McClain. 1991. Silver. In E. Merian, Ed. *Metals and Their Compounds in the Environment: Occurrence, Analysis, and Biological Relevance.* VCH Publishers, New York, pp. 1191–1202.

Petersen, W. G. and. G. Chesters. 1966. Quantitative determination of calcite and dolomite in pure carbonates and limestones. *J. Soil Sci.* **17**:317–327.

Petersen, W. G., G. Chesters, and G. B. Lee. 1966. Quantitative determination of calcite and dolomite in soils. *J. Soil Sci.* **17**:329–338.

Phung, H. T., L. J. Lund, A. L. Page, and G. R. Bradford. 1979. Trace elements in fly ash and their release in water and treated soils. *J. Environ. Qual.* **8**:171–175.

Piccolo, A. and F. J. Stevenson. 1982. Infrared spectra of Cu^{2+}, Pb^{2+}, and Ca^{2+} complexes of soil humic substances. *Geoderma* **27**:195.

Pichtel, J. R. and W. A. Dick. 1991. Influence of biological inhibitors on the oxidation of pyritic mine spoil. *Soil Biol. Biochem.* **23**:109–116.

Pierzynski, G. M., J. T. Sims, and G. F. Vance. 1994. *Soils and Environmental Quality*. Lewis Publishers, Ann Arbor, MI.

Pieters, J. A. 1927. *Green Manuring*. John Wiley and Sons, Chapman and Hall, Limited, London.

Pionke, H. B., A. S. Rogowski, and R. J. Deangelis. 1980. Controlling the rate of acid loss from strip mine spoil. *J. Environ. Qual.* **9**:694–699.

Pleysier, J. L., A. S. R. Juo, and A. J. Herbillon. 1979. Ion exchange equilibria involving aluminum in a kaolinitic ultisol. *Soil Sci. Soc. Am. J.* **43**:875–880.

Ponnamperuma, F. N., T. A. Loy, and E. M. Tiaco. 1969. Redox equilibria in flooded soils. II. The MnO_2 systems. *Soil Sci.* **108**:48–57.

Porath, E. and A. Poljakoff-Mayber. 1964. Effect of salinity on metabolic pathways in pea root tips. *Isr. J. Bot.* **13**:115–121.

Posner, A. M. 1966. The humic acids extracted by various reagents from soil. *J. Soil Sci.* **17**:65–78.

Potter, R. M. and G. R. Rossman. 1979a. The tetravalent manganese oxides: Identification, hydration, and structural relationship by infrared spectroscopy. *Am. Miner.* **64**:1199–1218.

Potter, R. M. and G. R. Rossman. 1979b. Mineralogy of manganese dendrites and coatings. *Am. Miner.* **64**:1219–1226.

Powlesland, C. and J. George. 1986. Acute and chronic toxicity of nickel to larvae of *Chironomus riparius (Meigen)*. *Environ. Pollut.* **42A**:47–64.

Pratt, P. F., L. D. Whittig, and B. L. Grover. 1962. Effect of pH on the sodium–calcium exchange equilibria in soils. *Soil Sci. Soc. Am. Proc.* **26**:227–230.

Quirk, J. P. and J. H. Chute. 1968. Potassium release from mica-like clay minerals. *Trans. Inst. Congr. Soil Sci. 9th.* **3**:671–681.

Quirk, J. P. and R. K. Schofield. 1955. The effect of electrolyte concentration on soil permeability. *J. Soil Sci.* **6**:163–178.

Radcliffe, D. E., R. L. Clark, and M. E. Sumner. 1986. Effect of gypsum and deep-rooting perennials on subsoil mechanical impedance. *Soil Sci. Soc. Am. J.* **50**:1566.

Ramirez, P., G. Barrera, and C. Rosas. 1989. Effect of chromium and cadmium upon respiration and survival of *Callinectes similis. Bull. Environ. Contam. Toxicol.* **43**:850–857.

Rao, T. S., A. L. Page, and N. T. Colemam. 1968. The influence of ionic strength and ion-par formation between alkaline–earth metals and sulfate on Na–divalent cation-exchange equilibria. *Soil Sci. Soc. Proc.* **32**:543–639.

Raspor, B. 1991. Metals and metal compounds in waters. In E. Merian, Ed. *Metals and Their Compounds in the Environment: Occurrence, Analysis, and Biological Relevance*. VCH Publishers, New York, pp. 233–256.

Rastogi, V., R. Krecic, and A. Sobek. 1986. ProMac systems for reclamation and control of acid production in toxic mine waste. In *Proceedings, Seventh West Virginia Surface Mine Drainage Task Force Symposium*, Morgantown, WV.

Ray, S. S. and F. G. Parker. 1977. Characterization of ash from coal-fired power plants. TVA Office of Power Rept. No. PRS-18. (EPA-600/7-77-010).

Reader, J. P., N. C. Everall, M. D. J. Sayer, and R. Morris. 1989. The effects of eight trace metals in acid soft water on survival, mineral uptake and skeletal calcium deposition in yolk-sac fry of brown trout, *Salmo trutta L. J. Fish. Biol.* **35**:187–198.

Rechcigl, J. E., Ed. 1995. *Soil Amendments and Environmental Quality*. Lewis Publishers, Ann Arbor, MI.

Rees, W. J. and G. H. Sidrak. 1956. Plant nutrition on fly ash. *Plant Soil* **8**:141–159.

Reeve, R. C. and C. A. Bower. 1960. Use of high-salt water as a flocculant and source of divalent cations for reclaiming sodic soils. *Soil Sci.* **90**:139–144.

Reilley, C. N. and R. W. Schmid. 1958. Chelometric titrations with potentiometric end point detection, mercury as pM indicator electrode. *Anal. Chem.* **30**:947–953.

Reuss, J. O. and D. W. Johnson. 1985. Effect of soil processes on the acidification of water by acid deposition. *J. Environ. Qual.* **14**:26–31.

Rhoades, J. D. 1982. Cation exchange capacity, In A. L. Page, Ed. *Methods of Soil Analysis*, Part 2, 2nd ed. Soil Science Society of America, Madison, WI, pp. 149–157.

Rhoades, J. D., D. B. Krueger, and M. J. Reed. 1968. The effect of soil–mineral weathering on the sodium hazard of irrigation waters. *Soil Sci. Soc. Am. Proc.* **32**:643–647.

Rich, C. I. 1968. Mineralogy of soil potassium. In V. J. Kilmer, S. E. Younts, and N. C. Brady, Eds. *The Role of Soil Potassium in Agriculture*, American Society of Agronomy, Madison, WI, pp. 79–96.

Rich, C. I. 1970. Conductimetric and potentiometric titration of exchangeable aluminum. *Soil Sci. Soc. Am. J.* **34**:31–38.

Rich, C. I. and W. R. Black. 1964. Potassium exchange as affected by cation size, pH, and mineral structure. *J. Soil Sci.* **97**:384–390.

Richard, D. T. 1975. Kinetics and mechanism of pyrite formation at low temperatures. *Am. J. Sci.* **275**:636–652.

Ritcey, G. M. 1991. Deep water disposal of pyritic tailings. In *Proceedings of the Second International Conference on the Batement of Acidic Drainage.* Sept. 16–18. Montreal, Quebec, pp. 421–442.

Ritsema, C. J. and J. E. Groennenberg. 1993. Pyrite oxidation, carbonate weathering, and gypsum formation in a drained potential acid sulfate soil. *Soil Sci. Am. J.* **57**:968–976.

Roberts, K., V. P. Evangelou, and W. Szekeres. 1984. A rapid kinetic technique for quantifying polycarbonate species in coal spoils. *Min. Environ.* **6**:72–76.

Rowley, M. V., D. D. Warkentin, V. T. Yan, and B. M. Piroshco. 1994. The biosulfide process: Integrated biological/ chemical acid mine drainage treatment—Result of laboratory piloting. In *Proceedings of International Land Reclamation and Mine Drainage Conference and Third International Conference on the Abatement of Acidic Drainage*, Vol 1. 205 pp. Pittsburgh, PA.

Roy, W. R. and R. A. Griffin. 1982. A proposed classification system for coal fly ash in multidisciplinary research. *J. Environ. Qual.* **11**:563–568.

Russo, D. and E. Bresler. 1977. Effect of mixed Na/Ca solution on the hydraulic properties of unsaturated soils. *Soil Sci. Soc. Am. J.* **41**:713–717.

Saar, R. A. and J. H. Weber. 1982. Fulvic acid: Modifier of metal–ion chemistry. *Environ. Sci. Technol.* **16**:510A.

Sadusky, M. C. and D. L. Sparks. 1991. Anionic effects on potassium reactions in variable-charge Atlantic Coastal Plain soils. *Soil Sci. Soc. Am. J.* **55**:371–375.

Sajwan, K. S., and V. P. Evangelou. 1991. Apparent activation energies of acid dissolution of carbonates by an isothermal automanometric apparatus. *Soil Sci.* **152**:243–249.

Sajwan, K. S., V. P. Evangelou, and J. Lumbanraja. 1991. A new rapid approach for evaluating limestone quality by automanometric isothermal apparatus. *Soil Sci.* **151**:444–451.

Sajwan, K. S., V. P. Evangelou, and J. Lumbanraja. 1994. A gasometric technique for evaluating kinetic stability and apparent activation energies of manganese oxide minerals. *Soil Sci.* **157**:19–25.

Salmon, R. C. 1964. Cation exchange reactions. *J. Soil Sci.* **15**:273–283.

Satawathananont, S., W. H. Patrick, Jr., and P. A. Moore, Jr. 1991. Effect of controlled redox conditions on metal solubility in acid sulfate soils. *Plant Soil* **133**:281–290.

Sayer, M. D. J., J. P. Reader, and R. Morris. 1991. Embryonic and larval development of brown trout, *Salmo trutta L.*: Exposure to trace metal mixtures in soft water. *J. Fish. Biol.* **385**:773–788.

Scharer, J. M., V. Garga, R. Smith, and B. E. Halbert. 1991. Use of steady state models for assessing acid generation in pyritic mine tailings. In *The Second National Conference on the Abatement of Acidic Drainage.* Vol. 2 Sept. 16, 17, 18, 1991. Montreal, Canada, pp. 211–229.

Scheinberg, I. H. 1991. Copper. In E. Merian, Ed. *Metals and Their Compounds in the Environment: Occurrence, Analysis, and Biological Relevance.* VCH Publishers, New York, pp. 893–908.

Schindler, P. W., B. Furst, B. Dick, and P. U. Wolf. 1976. Ligant properties of surface silanol groups. I. Surface complex formation with Fe^{3+}, Cu^{2+}, and Pb^{2+}. *J. Colloid Interface Sci.* **55**:469–475.

Schnitzer, M. 1969. Reaction between fulvic acid, a soil humic compound and inorganic soil constituents. *Soil Sci. Soc. Am. Proc.* **33**:75.

Schnitzer, M. 1986. Binding of humic substances by soil mineral colloids. In M. Huang and M. Schnitzer, Ed. *Interactions of Soil Minerals with Natural Organics and Microbes.* Special Publication No. 17. Soil Science Society of America Madison, WI, pp. 77–101.

Schnitzer, M. 1991. Soil organic matter–The next 75 years. *Soil Sci.* **151**:41.

Schnitzer, M. and H. Kodama. 1977. Reaction of minerals with humic substances. In J. B. Dixon et al. Eds. *Minerals in Soil Environments.* Soil Science Society of America, Madison, WI, pp. 741–770.

Schnitzer, M. and S. I. M. Skinner. 1965. Organo-metallic interactions in soils: 4. Carboxyl and hydroxyl groups in organic matter and metal ion retention. *Soil Sci.* **99**:278.

Schnitzer, M. and S. U. Khan. 1972. *Humic Substances in the Environment.* Dekker, New York, 327 pp.

Schofield, R. K. and H. R. Samson. 1953. Deflocculation of kaolinite clay. *Min. Bull.* **2**:45–52.

Schofield, R. K., and H. R. Samson. 1954. Flocculation of kaolinite due to the attraction of oppositely charged crystal faces. *Discuss. Faraday Soc.* **18**:135–145.

Schorr, P. 1988. *Point of Use Devices—Air Stripping, Activated Alumina, Reverse Osmosis, Activated Carbon Distillation, Ion Exchange.* NJDEP–Division of Water Resources, CN 029, Trenton, NJ 08625.

Schubauer-Berigan, M. K., J. R. Dierkes, P. D. Monson, and G. T. Ankley. 1993. pH-Dependent toxicity of Cd, Cu, Ni, Pb, and Zn to *Ceriodaphnia dubia, Pimephales promelas, Hyalella azteca* and *Lumbriculus variegatus. Environ. Toxicol. Chem.* **12**:1261–1266.

Schulthess, C. P., and D. L. Sparks. 1989. Back titration technique for proton isotherm modeling of oxides surface. *Soil Sci. Soc. Am. J.* **50**:1406–1411.

Shulze, D. G. 1989. An Introduction to Soil Mineralogy. In J. B. Dixon and S. B. Weed, Eds. *Minerals in Soil Environments.* SSSA Book Ser. No. 1. Soil Sci. Soc. Am. Press, Madison, WI, pp. 1–34.

Segel, I. H. 1976. *Biochemical Calculations*, 2nd ed. John Wiley & Sons, New York.

Shainberg, I. and A. Caiserman. 1971. Studies on Na/Ca montmorilonite systems. 2. The hydraulic conductivity. *Soil Sci.* **111**:276–281.

Shainberg, I. and J. Letey. 1984. Response of soils to sodic and saline conditions. *Hilgardia* **52**:1–57.

Shainberg, I. and J. E. Dawson. 1967. Titration of H-clay suspensions with salt solutions. *Soil Sci. Soc. Am. Proc.* **31**:619–626.

Shainberg, I. and W. D. Kemper. 1966. Hydration status of adsorbed cations. *Soil Sci. Soc. Am. Proc.* **30**:707–713.

Shainberg, I., E. Bresler, and Y. Klausner. 1971. Studies on Na/Ca montmorillonite systems. I. The swelling pressure. *Soil Sci.***111**:214–219.

Shainberg, I., J. Oster, and J.D. Wood. 1980. Sodium/calcium exchange in montmorillonite and illite suspensions. *Soil Sci. Soc. Am. J.* **44**:960–964.

Shalhevet, J. and B. Yaron. 1973. Effect of soil and water salinity on tomato growth. *Plant and Soil.* **39**:285–292.

Shalhevet, J., P. Reineger, and D. Shimshi. 1969. Peanut response to uniform and non-uniform soil salinity. *Agron. J.* **61**:384–387.

Shellhorn, M. and V. Rastogi. 1985. Practical control of acid mine drainage using bactericides. In *Proceedings, Sixth West Virginia Surface Mine Drainage Task Force Symposium*, Morgantown, WV.

Shelton, T. H. 1989. *Interpreting Drinking Water Quality Analysis. What Do the Numbers Mean?* Rutgers Cooperative Extension, New Brunswick, NJ.

Shoemaker, H. E., E. O. McLean, and P. F. Pratt. 1961. Buffer methods for determining lime requirement of soils with appreciable amounts of extractable aluminum. *Soil Sci. Soc. Am. Proc.* **25**:274–277.

Short, T. M., J. A. Black, and W. J. Birge. 1990. Effects of acid-mine drainage of the chemical and biological character of an alkaline headwater stream. *Arch. Environ. Contam. Toxicol.* **19**:241–248.

Sillen, L. G. and A. E. Martell. 1974. Stability constants. Special Publication No. 25, The Chemical Society, London.

Silverman, M. P. 1967. Mechanism of bacterial pyrite oxidation. *J. Bacteriol.* **99**:1046–1051.

Simmons, C. F. 1939. The effect of CO_2 pressure upon the equilibrium of the system hydrology colloidal clay–H_2O–$CaCO_3$. *J. Am. Soc. Agron.* **31**:638–648.

Singer, P. C. and W. Stumm. 1970. Acid mine drainage: Rate-determining step. *Science.* **167**:1121–1123.

Singh, U. and G. Uehara. 1986. Electrochemistry of the double–layer: Principles and applications to soils. In D. L. Sparks, Ed. *Soil Physical Chemistry*. CRC Press, Boca Raton, FL, pp. 1–38.

Sittig, M. 1979. *Hazardous and Toxic Effects of Industrial Chemicals.* Noyes Data Corporation, Park Ridge, NJ.

Siu, R. G. 1951. *Microbial Decomposition of Cellulose.* Reinhold Publishing Corporation, New York.

Skinner, S. I. M. and R. L. Halstead. 1958. Note on rapid method determination of carbonates in soils. *Can. J. Soil Sci.* **38**:187–188.

Skinner, S. I. M., R. L. Halstead, and J. E. Boydon. 1959. Quantitative manometric determination of calcite and dolomite in soils and limestones. *Can. J. Soil Sci.* **39**:197–204.

Skjemstad, J. O., R. L. Frost, and P. F. Barron. 1983. Structural units in humic acids from Southeastern Queensland soils as determined by ^{13}C NMR spectroscopy. *Aust. J. Soil Res.* **21**:539.

Skoog, D. A. and D. M. West. 1976. *Fundamentals of Analytical Chemistry*, 3rd ed. Holt, Rinehart and Winston, New York.

Skousen, J. G., J. C. Sencindiver, and R. M. Smith. 1987. *Procedures for Mining and Reclamation in Areas with Acid-Producing Materials*. The Surface Mine Drainage Task Force and the West Virginia University Energy Research Center, p. 43. Morgantown, WV.

Smith, J. M. and H. C. Van Ness. 1987. *Introduction to Chemical Engineering Thermodynamics*, 4th ed. McGraw-Hill Book Company, New York.

Smith, J. V. 1964. *Index to the Powder Diffraction File.* ASTM Special Technical Publication 48–N2. Philadelphia, PA.

Smith, M. A., Ed. 1985. *Contaminated Land: Reclamation and Treatment*, Plenum Press, NY, 433 pp.

Smith, R. M. and A. A. Sobek. 1978. Physical and chemical properties of overburdens, spoils, wastes, and new soils. In F. W. Schaller and P. Sutton, Eds. *Reclamation of Drastically Disturbed Lands*. American Society of Agronomy, Madison, WI. pp.149–169.

Sobek, A. A., L. R. Hosser, D. L. Sorensen, P. J. Sullivan, and D. F. Fransway. 1987. Acid–base potential and sulfur forms. In R. D. Williams and G. E. Schuman, Eds., *Reclaiming Mine Soils and Overburden in the Western United States, Analytic Parameters and Procedures*. Soil Conservation Society of America, Iowa. 233 pp.

Sobek, A. A., W. A. Shuller, J. R. Freeman, and R. M. Smith. 1978. *Field and Laboratory Methods Applicable to Overburdens and Mine Soils*. EPA-600/2-78–054. U.S. EPA, Cincinnati, OH.

Sparks, D. L. 1989. *Kinetics of Soil Chemical Processes*, Academic Press, Boca Raton, FL.

Sparks, D. L. 1995. *Environmental Soil Chemistry.* Academic Press, San Diego, CA.

Sparks, D. L. and P. M. Huang. 1985. Physical chemistry of soil potassium. In *Potassium in Agriculture*. R. D. Munson, Ed. American Society of Agronomy, Madison, WI.

Spehar, R. L. and J. T. Fiandt. 1986. Acute and chronic effects of water quality criteria-based metal mixtures on three aquatic species. *Environ. Toxicol. Chem.* **5**:917–931.

Spicer, J. I. and R. E. Wever. 1991. Respiratory impairment in crustaceans and molluscs due to exposure to heavy metals. *Comp. Biochem. Physiol.* **100C**:339–342.

Sposito, G. 1977. The Gapon and the Vanselow selectivity coefficients. *Soil Sci. Soc. Am. J.* **41**:1205–1206.

Sposito, G. 1981a. The operational definition of zero point of charge in soils. *Soil Sci. Soc. Am. J.* **45**:292–297.

Sposito, G. 1981b. Cation exchange in soils: A historical and theoretical perspective. In R.H. Dowdy, Ed. *Chemistry in the Soil Environment.* Soil Science Society of America, Madison, WI, pp. 13–30.

Sposito, G. 1981c. *The Thermodynamics of Soil Solutions.* Clarendon Press, Oxford, London.

Sposito, G. 1984a. *The Surface Chemistry of Soils.* Oxford University Press, London.

Sposito, G. 1984b. The future of an illusion: Ion activities in soil solutions. *Soil Sci. Soc. Am. J.* **48**:451–531.

Sposito, G. 1985. Sorption of trace metals by humic materials in soils and natural waters. *CRC Crit. Rev. Environ. Control* **15**:1.

Sposito, G. and C. S. LeVesque. 1985. Sodium–calcium–magnesium exchange on Silver Hill illite. *Soil Sci. Soc. Am. S.* **49**:1153–1159.

Sposito, G. and J. Coves. 1988. *SOIL CHEM: A Computer Program for the Calculation of Chemical Equilibria in Soil Solutions and Other Natural Water Systems,* The Kearney Foundation of Soil Science, University of California, Riverside, CA.

Sposito, G, and S. V. Mattigod. 1979a. *GEOCHEM: A Computer Program for Calculation of Chemical Equilibria in Soil Solutions and Other Natural Water Systems.* The Kearney Foundation of Soil Science, University of California, Riverside, CA.

Sposito, G. and S. V. Mattigod. 1979b. Ideal behavior in Na–trace metal cation exchange on Camp Berteau montmorillonite. *Clays Clay Mineral* **27**:125–128.

Sposito, G., C. S. LeVesque, and D. Hesterberg. 1986. Calcium–magnesium exchange on illite in the presence of adsorbed sodium. *Soil Sci. Soc. Am. J.* **50**:905–909.

Sposito, G., K. M. Holtzclaw, and C. S. LeVesque-Madore. 1978. Calcium ion complexation by fulvic acid extracted from sewage sludge–soil mixtures. *Soil Sci. Soc. Am. J.* **42**:600.

Sposito, G., K. M. Holtzclaw, and C. S. LeVesque-Madore. 1979. Cupric ion complexation by fulvic acid extracted from sewage sludge-soil mixtures. *Soil Sci. Soc. Am. J.* **43**:1148.

Sposito, G., K. M. Holtzclaw, and J. Baham. 1976. Analytical properties of the soluble, metal complexing fractions in sludge-soil mixtures. II. Comparative structural chemistry of the fulvic acid. *Soil Sci. Soc. Am. J.* **40**:691.

Sposito, G., K. M. Holtzclaw, C. T. Johnston, and C. S. LeVesque-Madore. 1981. Thermodynamics of sodium–copper exchange on Wyoming bentonite at 298° K. *Soil Sci. Soc. Am. J.* **45**:1079–1084.

Sposito, G., K. M. Holtzclaw, L. Charlet, C. Jouany, and A. L. Page. 1983. Sodium–calcium and sodium–magnesium exchange on Wyoming bentonite in perchlorite and chloride background ionic media. *Soil Sci. Soc. Am. J.* **47**:51–56.

Spotts, E. and D. J. Dollhopf. 1992. Evaluation of phosphate materials for control of acid production in pyritic mine overburden. *J. Environ. Qual.* **21**:627–634.

Staub, M. W. and R. R. H. Cohen. 1992. A passive mine drainage treatment system as a bioreactor: Treatment efficiency, pH increase, and sulfate reduction in two parallel reactors. In *Achieving Land Use Potential Through the Reclamation.* Proceedings of the 9th National Meeting of the American Society of Surface Mining and Reclamation. June 14–18, 1992, Duluth, MN, pp. 550–562.

Stevenson, F. J. 1976. Stability constants of Cu^{2+}, Pb^{2+}, and Cd^{2+} complexes with humic acids. *Soil Sci. Soc. Am. J.* **40**:665.

Stevenson, F. J. 1982. *Humus Chemistry: Genesis, Composition, and Reactions.* John Wiley & Sons, New York, 443 pp.

Stevenson, F. J. 1985. Geochemistry of soil humic substances. In G. R. Aiken et al. Eds. *Humic Substances in Soil, Sediment, and Water: Geochemistry, Isolation, and Characterization.* John Wiley & Sons, New York, pp. 13–52.

Stiller, A. H., J. J. Renton, and T. E. Rymer. 1986. The use of phosphates for ameliorization. In *Proceedings, Seventh West Virginia Surface Mine Drainage Task Force Symposium.* Morgantown, WV.

Stiller, A. H., J. J. Renton, T. E. Rymer, and B. G. McConaghy. 1984. The effect of limestone treatment on the production of acid from toxic mine waste in barrel scale weathering experiments. In *Proceedings, Fifth West Virginia Surface Mine Drain Task Force Symposium.* Morgantown, WV. p. 9.0.

Stone, A. T. and J. J. Morgan. 1987. Reductive dissolution of metal oxides. In W. Stumm, Ed. *Aquatic Surface Chemistry.* John Wiley & Sons, Inc., New York, pp. 230–237.

Stone, A. T. and J. J. Morgan. 1990. Kinetics of chemical transformations in the environment. In W. Stumm, Ed. *Aquatic Chemical Kinetics,* John Wiley & Sons, New York, pp. 13–15.

Stumm, W. and C. O. O'Melia. 1968. Stoichiometry of coagulation. *J. Am. Water Works Assoc.* **60**:439–514.

Stumm, W. and H. Bilinski. 1973. Trace metals in natural waters: Difficulties of interpretation arising from our ignorance on their speciation. *Adv. Water Pollut. Res.* **6**:39–49.

Stumm, W. and J. J. Morgan. 1970. *Aquatic Chemistry.* John Wiley and Sons, New York.

Stumm, W. and J. J. Morgan. 1981. Aquatic Chemistry 2nd Ed. John Wiley & Sons, New York.

Stumm, W. and R. Wollast. 1990. Coordination chemistry of weathering, kinetics of the surface-controlled dissolution of oxide minerals. *Rev. Geophys.* **28**:53–96.

Sturey, C. S., J. R. Freeman, T. A. Keeney, and J. W. Sturm. 1982. Overburden analysis by acid–base accounting and simulated weathering studies as a means of determining the probable hydrological consequences of mining and reclamation. In *Proceedings, Symposium on Surface Mining, Hydrology, Sedimentology, and Reclamation.* University of Kentucky, Lexington, KY, pp. 163–181.

Suarez, D. L., J. D. Rhoades, R. Lavado, and C. M. Grieve. 1984. Effect of pH on saturated hydraulic conductivity and soil dispersion. *Soil Sci. Soc. Am. J.* **48**:50–55.

Subba, Rao, H. C. and M. M. David. 1957. Equilibrium in the system Cu^{++}–Na^{++}–Dowex-50. *A. I. Ch. E. J.* **3**:187.

Sue, R. G. and E. T. Reese. 1953. Decomposition of cellulose by microorganisms. *Bot. Rev.* **19**:377–416.

Sullivan, P. J. 1977. The principle of hard and soft acids and bases as applied to exchangeable cation selectivity in soils. *Soil Sci.* **124**:117–121.

Suzuki, I., H. M. Lizama, and P. D. Tackaberry. 1989. Competitive inhibition of ferrous iron oxidation by *Thiobacillus ferrooxidans* by increasing concentrations of cells. *Appl. Environ. Microbiol.* **55**:1117–1121.

Swenson, H. S. and H. L. Baldwin. 1965. *A Primer on Water Quality.* Superintendent of Documents, U. S. Government Printing Office. U. S. Department of the Interior, Washington, DC.

Symposium on the Use of Metal Chelates in Plant Nutrition. A. Wallace, Ed. A report based on a series of papers presented before the 1956 meeting of the Western Society of Soil Science. Available from the editor at $1.50 per copy.

Tabak, L. M. and K. E. Gibbs. 1991. Effects of aluminum, calcium and low pH on egg hatching and nymphal survival of *Cloeon triangulifer McDunnough Ephemeroptera: Baetidae. Hydrobiologia* **2182**:157–166.

Talibudeen, O. 1981. Cation exchange in soils. In D. J. Greenland and M. A. B. Hayes, Eds. *The Chemistry of Soil and Processes*. John Wiley & Sons, New York, pp. 115–177.

Tanji, K. K. 1969a. Predicting specific conductance from electrolytic properties and ion association in some aqueous solutions. *Soil Sci. Soc. Am. J.* **33**:887–890.

Tanji, K. K. 1969b. Solubility of gypsum in aqueous electrolytes as affected by ion association and ionic strengths up to 0.15 M at 25 C. *Environ. Sci. Technol.* **3**:656–661.

Taylor, B. E., M. C. Wheeler, and D. K. Nordstrom. 1984a. Stable isotope geochemistry of acid mine drainage: Experimental oxidation of pyrite. *Geochim. Cosmochim. Acta.* **48**:2669–2678.

Taylor, B. E., M. C. Wheeler, and D. K. Nordstrom. 1984b. Oxygen and sulfur compositions of sulfate in acid mine drainage: Evidence for oxidation mechanism. *Nature* **308**:538–541.

Taylor, R. M., R. M. McKenzie, and K. Norrish. 1964. The mineralogy and chemistry of manganese in some Australia soils. *Aust. J. Soil Res.* **2**:235–248.

Taylor, S. A. and G. L. Ashcroft. 1972. *Physical Edaphology*. W. H. Freeman and Company, San Francisco, CA.

Terman, G. L. 1978. *Solid Wastes from Coal-Fired Power Plants—Use of Disposal on Agricultural Lands*. National Fertilization Development Center, Muscle Shoals, AL.

Tester, J. W., H. R. Holgate, F. J. Armellini, P. A. Webley, W. R. Killilea, G. T. Hong, and H. E. Barner. 1993. Supercritical water oxidation technology: Process development and fundamental research. In D. W. Tedder, Ed. *Hazardous Waste Management III*. Symposium Series 518. American Chemical Society, Washington, DC.

Thom, W. O. 1990. *AGR-120. Land Application of Wastewater Treatment Sludge*. University of Kentucky, College of Agriculture, Cooperative Extension Service, Lexington, KY.

Thomas, C. L., Ed. 1978. *Taber's Cyclopedic Medical Dictionary*. F. A. Davis Co., Philadelphia, PA.

Thomas, G. W. and W. L. Hargrove. 1984. The chemistry of soil acidity. In F. Adams, Ed. *Soil Acidity and Liming*. 2nd ed. American Society of Agronomy, Madison, WI.

Thomas, R. G. 1982. Volatilization from soil. In W. J. Lyman, W. F. Reehl, and D. H. Rosenblatt, Eds. *Handbook of Chemical Property Estimation Methods: Environmental Behavior of Organic Compounds*. McGraw-Hill, New York, p. 50.

Tiller, K. G., J. Gerth, and G. Brummer. 1984. The relative affinities of Cd, Ni, and Zn for different soil clay fractions and goethite. *Geoderma* **34**:17.

Tokashiki, Y., J. B. Dixon, and D. C. Golden. 1986. Manganese oxide analysis in soils by combined x-ray diffraction and selective dissolution methods. *Soil Sci. Soc. Am. J.* **50**:1079–1084.

Torma, A. E. 1988. Leaching of metals. In H. J. Rehm and G. Reed, Eds., *Biotechnology*, Vol. 6B. VCH Verlagsgesellschaft, Weinheim, Germany, pp. 367–399.

Torma, A. E., A. S. Osoka, and M. Valayapetre. 1979. Electrochimcal method in recovery of metal from sulfide minerals. *Res. Assoc. Miner. Sarda* **84**:5–24.

Tort, L. and L. H. Madsen. 1991. The effects of the heavy metals cadmium and zinc on the contraction of ventricular fibres in fish. *Comp. Biochem. Physiol.* **99C**:353–356.

Turner, D. and D. McCoy. 1990. Anoxic alkaline drain treatment system, a low cost acid mine drainage treatment alternative. In D. H. Graves and R. W. De Vore, Eds. *Proceedings of the 1990 National Symposium on Mining.* Lexington, KY, pp. 73–75.

Turner, R. 1966. Kinetic studies of acid dissolution of montmorillonite and kaolinite. Ph.D. Dissertation, University of California, Davis, CA.

Turner, R. C. 1958. A Theoretical Treatment of the pH of Calcareous Soils. *Soil Sci.* **86**:32–34.

Turner, R. C. 1959. An investigation of the intercept method for determining the proportion of dolomite and calcite in mixtures of the two. I. Theoretical aspects of the rate of solution of dolomite when a number of crystals are present. *Can. J. Soil Sci.* **40**:219–231.

Turner, R. C. and S. I. M. Skinner. 1959. An investigation of the intercept method for determining the proportion of dolomite and calcite in mixtures of the two. II. Experimental rate of solution of dolomite and calcite in samples consisting of a number of crystals. *Can. J. Soil. Sci.* **40**:232–241.

U.S. EPA. 1980a. *Ambient Water Quality Criteria for Chromium.* EPA 440/5-80-035. U.S. Environmental Protection Agency, Office of Water Regulation and Standards, Washington, DC.

U.S. EPA. 1980b. *Ambient Water Quality Criteria for Copper.* EPA 440/5-80-036. U.S. Environmental Protection Agency, Office of Water Regulation and Standards, Washington, DC.

U.S. EPA. 1980c. *Ambient Water Quality Criteria for Nickel.* EPA 440/5-80-060. U.S. Environmental Protection Agency, Office of Water Regulation and Standards, Washington, DC.

U.S. EPA. 1980d. *Ambient Water Quality Criteria for Silver.* EPA 440/5-80-071. U.S. Environmental Protection Agency, Office of Water Regulation and Standards, Washington, DC.

U.S. EPA. 1980e. *Ambient Water Quality Criteria for Zinc.* EPA 440/5-80-079. U.S. Environmental Protection Agency, Office of Water Regulation and Standards, Washington, DC.

U.S. EPA. 1983. *Design Manual: Neutralization of Acid Mine Drainage.* EPA 600/2-83-001. U.S. Environmental Protection Agency, Washington, DC.

U.S. EPA. 1985a. *Ambient Water Quality Criteria for Copper–1984.* U.S. Environmental Protection Agency, Office of Water Regulations and Standards, Washington, DC.

U.S. EPA. 1985b. *Ambient Water Quality Criteria for Lead–1984.* EPA 440/5-84-027. U.S. Environmental Protection Agency, Office of Water Regulations and Standards, Washington, DC.

U.S. EPA. 1986. *Ambient Water Quality Criteria for Nickel–1986.* EPA 440/5-86-004. U.S. Environmental Protection Agency, Office of Water Regulation and Standards, Washington, DC.

U.S. EPA. 1991a. *Methods for Aquatic Toxicity Identification Evaluations: Phase I Toxicity Characterization Procedures,* 2nd ed. EPA 600/6-91-003. U.S. Environmental Protection Agency, Environmental Research Laboratory, Duluth, MN.

U.S. EPA. 1995a. In situ remediation technology status report: Cosolvents. EPA542-K-94-006.

U.S. EPA. 1995b. In situ remediation technology status report: Electrokinetics. EPA542-K-94-007.

U.S. EPA. 1995c. In situ remediation technology status report: Hydraulic and pneumatic fracturing. EPA542-K-94-005.

U.S. EPA. 1995d. In situ remediation technology status report: Surfactant enhancements. EPA542-K-94-003.

U.S. EPA. 1995e. In situ remediation technology status report: Thermal enhancements.

U.S. Government Publication. 1969. *Oxygenation of Ferrus Iron*. Water Pollution Control Research Series 14010–06/69. U.S Dept. of Interior, Federal Water Quality Administration.

U.S. Salinity Laboratory Staff. 1954. Diagnosis and improvement of saline and alkali soils. USDA-Agricultural Handbook No. 60. U. S. Government Printing Office, Washington, DC.

Udo, E. J. 1978. Thermodynamics of potassium–calcium and magnesium–calcium exchange reactions on a kaolinitic soil clay. *Soil Sci. Soc. Am. J.* **42**:556–560.

Uehara, G. and G. P. Gillman. 1980. Charge characteristics of soils with variable and permanent charge minerals: I. Theory. *Soil Sci. Soc. Am. J.* **44**:250–252.

Uehara, G. and G.P. Gillman. 1981. *The Mineralogy, Chemistry, and Physics of Tropical Soils with Variable Charge Clays*. Westview Press, Colorado.

USDA. 1976. Soil survey of Carrol, Gallitin, and Owen Counties, Kentucky. USDA, Soil Conservation Service, in cooperation with the Kentucky Agricultural Experiment Station.

Van Bladel, R. and H. R. Gheyi. 1980. Thermodynamic study of calcium-sodium and calcium-magnesium exchange in clacareous soils. *Soil Sci. Soc. Am. J.* **44**:938–942.

Van Bladel, R., G. Gavira, and H. Laudelout. 1972. A comparison of the thermodynamic, double-layer theory and empirical studies of the Na–Ca exchange equilibria in clay water systems. *Proc. Int. Clay Conf.* 385–398.

Van Dijk, H. 1971. Cation binding of humic acids. *Geoderma* **5**:53.

Van Olphen, H. 1977. *An Introduction to Clay Colloid Chemistry*. 2nd ed. John Wiley & Sons, New York.

Van Raij, B. and M. Peech. 1972. Electrochemical properties of some oxisols and alfisols of the tropics. *Soil Sci. Soc. Am. Proc.* **36**:587–593.

Vanselow, A. P. 1932. Equilibria of the base-exchange reactions of bentonites, permutites, soil colloids, and zeolites. *Soil Sci.* **33**:95–113.

Varadachari, C., A. H. Mondal, and K. Ghosh. Some aspects of clay–humus complexation: Effect of exchangeable cations and lattice charge. *Soil Sci.* **151**:220.

Venugopal, B. and T. D. Luckey. 1978. *Metal Toxicity In Mammals*, Vol. 2. Plenum Press, New York.

Verma, S. K., R. K. Singh, and S. P. Singh. 1993. Copper toxicity and phosphate utilization in the cyanobacterium *Nostoc calcicola*. *Bull. Environ. Contam. Toxicol.* **50**:192–198.

Verwey, E. J. W. and J. Th. G. Overbeek. 1948. *Theory of the Stability of Lyophobic Colloids*. Elsevier, Amsterdam.

Wakao, N., M. Mishina, Y. Sakurai, and H. Shiota. 1982. Bacteria pyrite oxidation. I. The effect of the pure and mixed cultures of *Thiobacillus ferrooxidans* and *Thiobacillus thiooxidans* on release of iron. *J. Gen. Appl. Microbiol.* **28**:331–343.

Wakao, N., M. Mishina, Y. Sakurai, and H. Shiota. 1983. Bacteria pyrite oxidation. II. The effect of various organic substances on release of iron from pyrite by *Thiobacillus ferrooxidans*. *J. Gen. Appl. Microbiol.* **29**:177–185.

Wakao, N., M. Mishina, Y. Sakurai, and H. Shiota. 1984. Bacteria pyrite oxidation. III. Adsorption of *Thiobacillus ferrooxidans* on solid surfaces and its effect on iron release from pyrite. *J. Gen. Appl. Microbiol.* **30**:63–74.

Wallace, A., Ed. *A Decade of Synthetic Chelating Agents in Inorganic Plant Nutrition*. Copyright © 1962 by Arthur Wallace.

Walton, H. F. 1949. Ion exchange equilibria. In F. C. Nackod, Ed. *Ion Exchange Theory and Practice*. Academic, New York.

Wang, J. and V. P. Evangelou. 1993. Plant metal tolerance aspects of cell wall and vacuole. In M. Passarakli, Ed. *Handbook of Plant and Crop Physiology*. Marcel Dekker, New York.

Wang, J., M. T. Nielsen, and V. P. Evangelou. 1994. A solution culture study of Mn-tolerant and sensitive tobacco genotypes. *J. Plant Nutr.* **17**:1079–1093.

Wang, J., V. P. Evangelou and B. Creech. 1993. Characteristics of Mn–Ca exchange behavior on kaolinite and illite and pH influence. *J. Environ. Sci. Health*, **A286**:1381–1391.

Wang, J., V. P. Evangelou, and M. T. Nielsen. 1992. Surface chemical properties of root cell walls from two tobacco genotypes with different tolerance to MnII toxicity. *Plant Phys.* **100**:496–501.

Wang, J., V. P. Evangelou, M. T. Nielsen, and G. J. Wagner. 1991. Computer–simulated evaluation of possible mechanisms for quenching heavy metal ion activity in plant vacuoles. I. Cadmium. *Plant Phys.* **97**:1154–1160.

Wang, J., V. P. Evangelou, M. T. Nielsen, and G. J. Wagner. 1992. Computer-simulated evaluation of possible mechanisms for sequestering metal ion activity in plant vacuoles. II. Zinc. *Plant Phys.* **99**:621–626.

Wang, W. 1987. Toxicity of nickel to common duckweed lemna minor. *Environ. Toxicol. Chem.* **6**:961–967.

Wann, S. S. and G. Uehara. 1978. Surface charge manipulation of constant surface potential soil colloids: I. Relation to sorbed phosphorus. *Soil Sci. Soc. Am. J.* **42**:565–570.

Warden, B. T. and H. M. Reisenauer. 1991. Fractionation of soil manganese forms important to plant availability. *Soil Sci. Soc. Am. J.* **55**:345–349.

Watzlaf, G. R. 1992. Pyrite oxidation in saturated and unsaturated coal waste. In *Proceedings of 9th National Meeting of the American Society for Surface Mining and Reclamation*. June 14–18, 1992. Duluth, MN, pp. 191–205.

Watzlaf, G. R. and R. S. Hedin. 1993. A method for predicting alkalinity generated by anoxic limestone drains. In *Proceedings of the 1993 West Virginia Surface Mine Drainage Task Force Symposium*, April 27–28, Morgantown, WV.

Weber, J. B. 1966. Molecular structure and pH effects on the adsorption of 13 s-triazine compounds on montmorillonite clay. *The Am. Mineralogist.* **51**:1657–1670.

Weber, M. A., K. A. Barbaric, and D. G. Westfall. 1983a. Ammonium adsorption by a zeolite in a static and dynamic system. *J. Environ. Qual.* **12**:549–552.

Weber, M. A., K. A. Barbaric, and D. G. Westfall. 1983b. Application of clinoptilolite to soil amended with municipal sewage sludge, pp. 263–271. In *Scientific Series by International Committee on Natural Zeolites*. Paper No. 2767, Colorado State University, Agricultural Experimental Station, Fort Collins, CO, pp. 263–271.

Weber, W. J., Jr., P. M. McGinley, and L. E. Katz. 1992. A distributed reactivity model for sorption by soils and sediments. 1. Conceptual basis and equilibrium assessments. *Environ. Sci. Technol.* **26**:1955–1962.

Welch, L. F. and A. D. Scott. 1960. Nitrification of fixed ammonium in clay minerals as affected by added potassium. *Soil Sci.* **90**:79–85.

Welcher, F. J. 1958. *The Analytical Uses of Ethylenediamine–Tetraacetic Acid*. D. Van Nostrand, Princeton, NJ.

Wells, L. G., A. D. Ward, and R. E. Phillips. 1982. *Proceedings of the National Symposium on Surface Mining Hydrology, Sedimentation, and Reclamation.* University of Kentucky, Lexington, KY, pp. 445–456.

Wentsel, R., A. McIntosh, and W. P. McCafferty. 1978. Emergence of the midge *Chironomus tentans* when exposed to heavy metal contaminated sediment. *Hydrobiologia* **57**:195–196.

Wentzler, T. H. and F. F. Aphan. 1972. Kinetics of limestone dissolution by acid waste waters. In C. Rampacek, Ed. *Environmental Control.* San Francisco, CA, pp. 513–523.

Westall, J. C. 1980. Chemical equilibrium including adsorption on charged surfaces. *Adv. Chem. Ser.* **189**:33–44.

White, G. N. and L. W. Zelazny. 1986. Charge properties of soil colloids. In D. L. Sparks, Ed. *Soil Physical Chemistry.* CRC Press, Boca Raton, FL, pp. 39–81.

Whitney, R. S. and R. Gardner. 1943. The effect of carbon dioxide on soil reduction. *Soil Sci.* **55**:127–141.

Wiersma, C. L. and J. D. Rimstidt. 1984. Rate of reaction of pyrite and marcasite with ferric iron at pH 2. *Geochim. Cosmochim. Acta* **48**:85–92.

Wild, A. and J. Keay. 1964. Cation-exchange equilibria with vermiculite. *J. Soil Sci.* **15**(2):135–144.

Wildeman, T. R. 1991. Drainage from coal mine: Chemistry and environmental problems. In D. C. Peters, Ed. *Geology in Coal Resource Utilization.* Techbooks, Fairfax, VA, pp. 499–512.

Wilmoth, R. C., S. J. Hubbard, J. O. Burckle, and J. F. Martin. 1991. Production and processing of metals: Their disposal and future risks. In E. Merian, Ed. *Metals and Their Compounds in the Environment: Occurrence, Analysis, and Biological Relevance.* VCH Publishers, New York, pp. 19–65.

Wolfe, N. L., U. Mingelrin, and G. C. Miller. 1990. Abiotic transformations in water, sediments, and soil. In H. H. Cheng, Ed. *Pesticides in the Soil Environment: Processes, Impacts, and Modeling.* Soil Science Society of America, Madison, WI, 103.

Woodruff, C. M. 1955. The energies of replacement of calcium by potassium in soils. *Soil Sci. Soc. Am. Proc.* **19**:167–171.

Yaalon, D. H. 1954. *Physico-Chemical Relationships of $CaCO_3$, pH, and CO_2 in Calcareous Soils.* International Congress of Soil Science, 5th Congress, pp. 356–362.

Yadav, J. S. P. and I. K. Girdhar. 1980. The effects of different magnesium:calcium ratios and sodium adsorption ratio values of leaching water on the properties of calcareous versus noncalcareous soils. *Soil Sci.* **131**:194–198.

Yousaf, M., O. M. Ali, and J. D. Rhoades. 1987. Clay dispersion and hydraulic conductivity of some salt affected arid land soils. *Soil Sci. Soc. Am. J.* **51**:905–907.

Zelazny, L. W., D. A. Leitzke, and H. L. Barwood. 1980. Septic tank drainfield failure resulting from mineralogical changes. *Va. Water Resour. Res. Center, Bull.* **129**:118.

Zhang, Y. L. and V. P. Evangelou. 1996. Influence of iron oxide forming conditions on pyrite oxidation. *Soil Sci.* **161**:852–864.

Zhang, Y. L., R. W. Blanchar, and R. D. Hammer. 1993. Composition and pyrite morphology of materials separated from coal. In *Proceeding of 10th National Meeting of American Society of Surface Mining and Reclamation*, Vol. 2. Spokane, WA, May 16–19, 1993. pp. 284–297.

Zuiderveen, J. A. 1994. Identification of critical environmental toxicants using metal-binding chelators. Ph.D Dissertation, University of Kentucky, Lexington, KY.

INDEX

A